Pressure-Sensitive Adhesives Technology

Pressure-Sensitive Adhesives Technology

Istvan Benedek
Industrial Consultant
Wuppertal, Germany

Luc J. Heymans
MACtac
Grimbergen, Belgium

Marcel Dekker, Inc.   New York•Basel•Hong Kong

Library of Congress Cataloging-in-Publication Data

Benedek, Istvan.
   Pressure–sensitive adhesives technology / Istvan Benedek, Luc J. Heymans.
      p. cm.
   Includes bibliographical references and index.
   ISBN 0-8247-9765-5 (hardcover : alk. paper)
   1. Adhesives. I. Heymans, Luc J. II. Title.
TP968.B424 1996
668'.3—dc20   96–31582
                                                                                                                   CIP

The publisher offers discounts on this book when ordered in bulk quantities. For more information, write to Special Sales/Professional Marketing at the address below.

This book is printed on acid-free paper.

Copyright © 1997 by MARCEL DEKKER, INC. All Rights Reserved.

Neither this book nor any part may be reproduced or transmitted in any form or by any means, electronic or mechanical, including photocopying, microfilming, and recording, or by any information storage and retrieval system, without permission in writing from the publisher.

MARCEL DEKKER, INC.
270 Madison Avenue, New York, New York 10016

Current printing (last digit):
10 9 8 7 6 5 4 3 2 1

PRINTED IN THE UNITED STATES OF AMERICA

Preface

Since their introduction half a century ago, pressure-sensitive adhesives have been successfully applied in many fields. They are used in self-adhesive labels, and tapes and protective films, as well as in dermal dosage systems for pharmaceutical applications, the assembly of automotive parts, toys, and electronic circuits and keyboards. They have experienced an astonishing growth rate, and the installed manufacturing and converting capacity has also sharply increased. A specific engineering technology for pressure-sensitive adhesives, surprisingly a special science, appears to be lacking. Very few books deal with the intrinsic features of pressure-sensitive adhesives.

The application of pressure-sensitive adhesives requires a thorough knowledge of basic rheological and viscoelastic phenomena. Adhesive and polymer scientists, however, are not very often employed as industrial managers or machine operators. Therefore the need arises to investigate and summarize the most important features of pressure-sensitive adhesive technology and to explain the phenomena scientifically. This book covers all the fields of manufacturing, conversion, and application and end-uses of pressure-sensitive adhesives.

The classical approach would be to compile a treatise based on the work of various experts, theoreticians, chemists, and engineers, thereby coming up with a book consisting of a series of papers with a common title only. We have, however, chosen a different approach. Based on our experience as engineers (in both scientific activity and industrial areas), and using the available technical literature, we have addressed all aspects of pressure-sensitive adhesives. We have included the scientific basis of suitability for specific applications (i.e., chemical and physical, rheology), the raw materials, the manufacture (formulation) of the adhesive and of the labelstock (converting the adhesive). We have selected self-adhesive labels as the most complex self-adhesive laminate; we mainly discuss labels, but, whenever possible, a comparison with and extension to other applications is included. In order to illustrate the different topics and issues discussed, we have referred to a number of comercially available products. It should be kept in mind that these products are only mentioned in order to clarify the discussion and in no way does it constitute any judgment about inherent performance characteristics or their suitability for specific applications or end-uses.

It is not the aim of this book to establish or complete the science of pressure-sensitive adhesives, nor does it constitute a series of recipes. Rather, it serves as a practical aid to convertors and those involved in the design and use of pressure-sensitive adhesives.

Istvan Benedek
Luc J. Heymans

Contents

Preface		*iii*
1.	**Introduction**	1
	References	3
2.	**Rheology of Pressure-Sensitive Adhesives**	5
	2.1 Rheology of Uncoated PSAs	5
	2.1.1 Properties of PSAs	6
	2.1.2 Influence of Viscoelastic Properties on the Adhesive Properties of PSAs	11
	2.1.3 Influence of Viscoelastic Properties on the Converting Properties of PSAs and PSA Laminates	21
	2.1.4 Influence of Viscoelastic Properties on End-Use Properties of PSAs	22
	2.1.5 Factors Influencing Viscoelastic Properties of PSAs	25
	2.2 Rheology of PSA Solutions and Dispersions	34
	2.2.1 Rheology of PSA Solutions	35
	2.2.2 Rheology of PSA Dispersions	38
	2.3 Rheology of the Pressure-Sensitive Laminate	44
	2.3.1 Influence of Liquid Components of the Laminate	46
	2.3.2 Influence of the Solid Components of the Laminate	48
	2.3.3 Influence of the Composite Structure	70
	References	70

3.	**Physical Basis for the Viscoelastic Behavior of Pressure-Sensitive Adhesives**		**75**
	3.1	The Role of the T_g in Characterizing PSAs	76
		3.1.1 Values of T_g for Adhesives	77
		3.1.2 Factors Influencing T_g	80
		3.1.3 Adjustment of T_g	86
		3.1.4 Correlation Between the Main End-Use/Converting Properties of PSAs and T_g	88
	3.2	Role of the Modulus in Characterizing PSAs	91
		3.2.1 Factors Influencing the Modulus	92
		3.2.2 Adjustment of the Modulus	99
		3.2.3 Modulus Values	101
	References		103
4.	**Comparison of PSAs**		**107**
	4.1	Comparison of PSAs with Thermoplastics and Rubber	108
		4.1.1 Cold Flow	108
		4.1.2 Relaxation Phenomena	113
		4.1.3 Mechanical Resistance	114
	4.2	Comparison Between PSAs and Other Adhesives	115
	References		116
5.	**Chemical Composition of PSAs**		**119**
	5.1	Raw Materials	120
		5.1.1 Elastomers	120
		5.1.2 Viscous Components, Plasticizers, and Tackifiers	142
		5.1.3 Other Components	143
	5.2	Factors Influencing the Chemical Composition	145
		5.2.1 Synthesis	145
		5.2.2 Formulation	147
		5.2.3 Physical State of PSAs	147
		5.2.4 End-Use	151
		5.2.5 Coating Method	155
		5.2.6 Solid State Components of the Laminate	156
	References		157
6.	**Adhesive Performance Characteristics**		**163**
	6.1	Adhesion-Cohesion Balance	163
		6.1.1 Tack	163
		6.1.2 Peel Adhesion	186
		6.1.3 Shear Resistance (Cohesion)	213

Contents   vii

    6.2   Influence of Adhesive Properties on Other
           Characteristics of PSAs   223
           6.2.1 Influence of Adhesive Properties on the
                  Converting Properties   223
           6.2.2 Influence of Adhesive Properties on End-Use
                  Properties   225
           6.2.3 Influence of Peel Adhesion   225
           6.2.4 Influence of Shear   225
    6.3   Comparison of PSAs on Different Chemical Bases   226
           6.3.1 Rubber-Based Versus Acrylic-Based PSAs   226
           6.3.2 Acrylics and Other Synthetic Polymer-Based
                  Elastomers   227
    References   248

7. Converting Properties of PSAs   **253**
    7.1   Convertability of the Adhesive   253
           7.1.1 Convertability of Adhesive as a Function of the
                  Physical State   254
           7.1.2 Convertability of Adhesive as a Function of
                  Adhesive Properties   260
           7.1.3 Convertability of Adhesive as a Function of the
                  Solid State Components of the Laminate   261
           7.1.4 Convertability of Adhesive as a Function of
                  Coating Technology   261
           7.1.5 Convertability of Adhesive as a Function of
                  End-Use Properties   261
    7.2   Converting Properties of the Laminate   261
           7.2.1 Definition and Construction of the Pressure-
                  Sensitive Laminate   262
           7.2.2 Printability of the Laminate   289
    References   318

8. Manufacture of Pressure-Sensitive Adhesives   **323**
    8.1   Manufacture of PSA Raw Materials   323
           8.1.1 Natural Raw Materials   323
           8.1.2 Synthetic Raw Materials   325
    8.2   Formulating PSAs   327
           8.2.1 Adhesive Properties   329
           8.2.2 Formulating Opportunities   331
           8.2.3 Tackification   331

	8.2.4	Rosin-Based Tackifiers	357
	8.2.5	Hydrocarbon-Based Tackifiers	365
	8.2.6	Coating Properties	374
	8.2.7	Converting Properties	377
	8.2.8	End-Use Properties	381
	8.2.9	Influence of Adhesive Technology	397
	8.2.10	Technological Considerations	433
	8.2.11	Comparison Between Solvent-Based, Water-Based, and Hot-Melt PSAs	435
References			442

9. Manufacture of Pressure-Sensitive Labels — 449

- 9.1 Coating Technology — 450
- 9.2 Coating Machines — 453
  - 9.2.1 Adhesive Coating Machines — 454
  - 9.2.2 Coating Devices/Coating Systems — 455
  - 9.2.3 Choice of Coating Geometry — 476
  - 9.2.4 Other Coating Devices — 484
- 9.3 Coating of Hot-Melt PSAs — 486
  - 9.3.1 Roll Coaters for Hot-Melt PSAs — 486
  - 9.3.2 Slot-Die Coating for Hot-Melt PSAs — 487
- 9.4 Drying of the Coating — 489
  - 9.4.1 Adhesive Drying Tunnel — 489
- 9.5 Simultaneous Manufacture of PSAs and PSA Laminates — 494
  - 9.5.1 Radiation Curing of PSAs — 494
  - 9.5.2 Siliconizing Through Radiation — 495
- 9.6 Manufacture of the Release Liner — 498
  - 9.6.1 Nature of the Release Liners — 498
  - 9.6.2 Coating Machines for Silicones — 500
  - 9.6.3 Technology for Solvent-Based Systems — 500
  - 9.6.4 Technology for Solventless Siliconizing — 501
- 9.7 Rehumidification/Conditioning — 502
- References — 502

10. Test Methods — 509

- 10.1 Evaluation of the Liquid Adhesive — 509
  - 10.1.1 Hot-Melt PSAs — 510
  - 10.1.2 Solvent-Based PSAs — 510
  - 10.1.3 Water-Based PSAs — 511
- 10.2 Evaluation of the Solid Adhesive — 518
  - 10.2.1 Test of Coating Weight — 518
  - 10.2.2 Other Properties — 519

10.3	Laminate Properties	522
	10.3.1 General Laminate Properties	522
	10.3.2 Special Laminate Properties	552
References		564

Abbreviations and Acronyms   *569*

Index   *573*

Pressure-Sensitive Adhesives Technology

1
Introduction

Adhesives are nonmetallic materials [1] used to bond other materials, mainly on their surfaces through adhesion and cohesion. Adhesion and cohesion are phenomena which may be described thermodynamically, but actually they cannot be measured precisely. It was shown [2] that the most important bonding processes are bonding by adhesion and bonding with pressure sensitive adhesives (PSAs). For adhesives working through adhesion phenomena the adhesive fluid is transformed after bonding (i.e., the build-up of the joint) into a solid. In the case of PSAs, the adhesive conserves its fluid state after the bond building too. Thus its resistance to debonding is moderate and the joint may be delaminated without destroying the laminate components in most cases.

Pressure-sensitive adhesives were in wide use since the late 19th century, starting with medical tapes and dressings. The first U.S. patent describing the use of a PSA—for a soft, adhering bandage—was issued in 1846 [3]. Ninety years later Stanton Avery developed and introduced the self-adhesive label [4]. Two major industries resulted from these innovations: pressure-sensitive tapes and labels. Industrial tapes were introduced in the 1920s and 1930s followed by self-adhesive labels in 1935. The history of PSAs was described by Villa [5]. First, solvent-based PSAs using natural rubber were developed (19th century). In the 1940s hot-melt adhesives were introduced. Pressure-sensitive adhesives are adhesives that form films exhibiting permanent tack, and display an adhesion which does not strongly depend on the substrate [6]. The term PSAs has a very precise technical definition and was dealt with extensively in the chemical literature [7,8].

The function of PSAs is to ensure instantaneous adhesion upon application of a light pressure. Most applications further require that they can be easily removed from the surface to which they were applied through a light pulling force. Thus PSAs are characterized by a built-in capacity to achieve this instantaneous adhesion to a surface without activation, such as a treatment with solvents or heat, and also by having sufficient internal strength so that the adhesive material will not break up before the bond between the adhesive material and the surface ruptures. The bonding and the debonding of PSAs are energy driven phenomena. Pressure-sensitive adhesives must possess viscous properties in order to flow and to be able to dissipate energy during the adhesive bonding process. However, the adhesive must also be elastic (i.e., it must resist the tendency to flow) and, in addition, store bond rupture energy in order to provide good peel and tack performance. Pressure-sensitive adhesives should possess typical viscoelastic properties that allow them to respond properly to both a bonding and a debonding step. For satisfactory performance in each of these steps the material must respond to a deforming force in a prescribed manner.

Polymers employed as PSAs have to fulfill partially contradictory requirements; they need to adhere to substrates, to display high shear strength and peel adhesion, and not leave any residue on the substrate upon debonding. In order to meet all these requirements, a compromise is needed. When using PSAs there appears another difference with wet adhesives, namely the adhesive does not change its physical state because film forming is inherent to PSAs. Thus, PSAs used in self-adhesive laminates are adhesives which, through their viscoelastic fluid state, can build up the joint without the need to change this flow state during or after application. On the other hand, their fluid state allows controlled debonding giving a temporary character to the bond. Because of the fluid character of the bonded adhesive, the amount of adhesive (i.e., the dimensions of the adhesive layer) is limited; the joint works as a thin-layer laminate or composite. Because of this special, thin-layer structure of the composite, the solid state components of the laminate exert a strong influence on the properties of the adhesive in the composite. Therefore, there exists a difference between the measured properties of the pristine adhesive and of the adhesive enclosed within the laminate.

Adhesives, in general, and PSAs, in particular, have to build up a continuous, soft (fluid), and tacky (rubbery) layer. The latter will adhere to the substrate. On the other hand, the liquid adhesive layer of the PSAs working in the bond has to offer a controlled bond resistance. This special behavior requires materials exhibiting a viscoelastic character. The properties which are essential in characterizing the nature of PSAs comprise: tack, peel adhesion, and shear. The first measures the adhesive's ability to adhere quickly,

Introduction 3

the second its ability to resist removal through peeling, and the third its ability to hold in position when shear forces are applied [9].

These properties will be discussed in more detail in Chapter 6, which describes the adhesive properties of PSAs. In order to understand the importance of these properties, it is absolutely necessary to answer the following questions:

- What does the viscoelastic character of a PSA comprise?
- What is the material basis (main criteria) for the viscoelastic behavior of a PSA?

REFERENCES

1. DIN 16921.
2. R. Köhler, *Adhäsion*, (3) 90 (1970).
3. J. A. Fries,"New Developments in PSA", in *TECH 12, Advances in Pressure Sensitive Tape Technology, Technical Seminar Proceedings*, Itasca, IL, May 1989.
4. *Der Siebdruck*, (3) 69 (1986).
5. G. J. Villa, *Adhäsion*, (10) 284 (1977).
6. *Vinnapas, Eigenschaften und Anwendung*, 7.1. Teil, Anwendung, Wacker, München, 1976.
7. R. Houwink and G. Salomon, *Adhesion and Adhesives*, Vol. 2., Chapter 17, Elsevier Co., New York, 1982.
8. D. Satas, *Handbook of Pressure Sensitive Technology*, Van Nostrand-Rheinhold Co., New York, 1982.
9. J. P. Keally and R. E. Zenk, (Minnesota Mining and Manuf. Co., USA), Canad. Pat., 1224.678/19.07.92 (USP. 399350).

2
Rheology of Pressure-Sensitive Adhesives

Pressure-sensitive adhesives are viscoelastic materials with flow properties playing a key role in the bond forming; their elasticity plays a key role in the storage of energy (i.e., the debonding process). The balance of these properties governs their time dependent repositionability and bonding strength (i.e., their removability). Their flow properties are useful in the coating technology and at the same time detrimental to the converting technology of the labels.

Generally PSAs are used as thin layers, therefore their flow is limited by the physico-mechanical interactions with the solid components of the laminate (liner and face) materials. On the other hand the solid components of the laminate are generally thin, soft, viscous, and/or elastic layers, allowing a relatively broad and uniform distribution of the applied stresses. Thus the properties of the bonded adhesive (i.e., its flow characteristics) may differ from those of the pure (unbonded) adhesive. Therefore in this chapter the rheology of pure and coated PSAs will be dealt with separately.

1 RHEOLOGY OF UNCOATED PSAs

It remains difficult to examine the properties of pure (i.e., uncoated or unbonded) PSAs, and to obtain generally valid information. Pressure-sensitive adhesives are seldom used as thick layers between motionless rigid surfaces (i.e., as fluids). On the other hand, as known from industrial experience, the nature of the face stock material or of the substrate used, and their characteristics and dimensions may significantly influence the properties of the PSA laminate. Practically this disadvantage is eliminated by the use of normalized or standard solid state components. However, a theoretical approach may be used for the investigation of pristine PSAs.

1.1 Properties of PSAs

The adhesive and end-use properties of PSAs require a viscoelastic, non-Newtonian flow behavior which is based on the macromolecular nature of the adhesive. In order to understand the needs and means of viscoelastic behavior one needs to summarize the most important material properties specifically related to PSAs. Generally, adhesives in a bond behave like a fluid or a solid. Fluids are characterized by their viscosity which influences their mobility, whereas solids are characterized by their modulus which determines their deformability. In an ideal case, for Newtonian fluids (or for solids obeying Hooke's law) the applied force (load) will be balanced by the materials' own mechanical characteristics, that is, the viscosity η or the Young's modulus E:

$$\eta = \frac{\tau}{\dot{\gamma}} \qquad (2.1)$$

$$E = \frac{\sigma}{\gamma} \qquad (2.2)$$

where τ and σ are the applied stresses, and γ and $\dot{\gamma}$ are the strain and shear rate, respectively.

As indicated earlier, PSAs originate from a film-forming, elastomeric material, which combines a high degree of tack with an ability to quickly wet the surface to which it is applied, to provide instant bonding at low to moderate pressure as a result of its flow characteristics. On the other hand, PSAs possess sufficient cohesion and elasticity, so that despite their agressive tackiness they can be handled with the fingers and removed from smooth surfaces without leaving any residue. Moreover, in order to achieve bond strength they have to store energy (i.e., they must be elastic). Fundamentally PSAs require a delicate balance between their viscous and elastic properties. One should note that PSAs have to satisfy these contradictory requirements under different stress rate conditions, that is, at low shear rates they must flow (bonding) and at high peeling rates they have to respond elastically (debonding).

Consequently, according to their adhesive and end-use properties, PSAs cannot be Newtonian systems: they do not obey Newton's law (i.e., there is no linear dependence between their viscosity and the shear rate). Their viscosity is not a material constant, but depends on the stress value or shear rate:

$$\eta(\tau) = \frac{\tau}{\dot{\gamma}} \qquad (2.3)$$

Rheology of Pressure-Sensitive Adhesives

That is:

$$\tau = \eta_a \cdot \gamma^n \tag{2.4}$$

where η_a is the apparent viscosity, and n denotes the flow index [1]. For Newtonian systems the exponent n is one, which implies that the viscosity does not depend on the shear rate. As pointed out, the viscosity of PSAs depends on the shear rate. This is possibly due to their macromolecular character. Pressure-sensitive adhesives are polymers containing long-chain entangled molecules with intra- and intermolecular mobility. At low strain rates, the viscous components of the polymer dissipate energy, and as a result resistance to debonding forces is low. At higher strain rates, the molecules have less time to disentangle, and slide past one another; in this case viscous flow is reduced, but the elastic modulus or stiffness of the polymer increases [2]. This behavior results in additional stored energy, and the debonding resistance intensifies accordingly.

Practically, the dependence of the adhesive performance characteristics on the stress rate may be observed by peeling off removable PSAs at different peel rates: at higher rates paper tear may occur. The stress rate-dependent stiffening is an increase in the elastic contribution to the rheology of the polymer. When the elastic components are predominant more of the bond rupture energy is stored, resulting in higher peel and tack properties.

The end-use properties of PSAs result from the nonlinear viscoelastic behavior of the adhesive material, and the elastomeric polymer basis of PSAs imparts them such a viscoelastic behavior. It is evident that the same stiffening effect is apparent when the polymer temperature decreases. In this case the polymer molecules are again restricted in their ability to flow, and the modulus increases. Consequently the adhesive properties of PSAs are also temperature dependent. Thus one always has to take into account that the viscoelastic properties of PSAs are strain-rate and temperature dependent.

Zosel [3] demonstrated that the separation or debonding energy of the adhesive joint is a function of the thermodynamical work of adhesion and of a temperature and rate dependent function (depending on the viscoelastic properties). Accordingly, PSAs would absorb less or more energy depending on the rate (frequency ω) of the applied stress. Practically, end-use situations with different stress rates may be simulated experimentally by applying a strain to a thin sample of the material and measuring the output stress. If the material is an ideal solid, its response is completely in phase with the applied strain. A viscoelastic fluid, such as a PSA, displays a mixture of solid-like and liquid-like responses. Therefore the output stress curve is deconvoluted into an in-phase part (related to energy storage) and an out-of-phase part (related to energy loss). The coefficient of the in-phase and out-of-phase part are called the energy storage modulus and the energy loss modulus [4].

According to the theory of Lodge [5] the rheological state of a viscous liquid subject to a sinusoidal deformation will be described by the following equation:

$$\tau_{12} = \eta(\omega) \cdot \gamma_o \cdot \omega \cos \omega t \tag{2.5}$$

The response of the elastical solid may be described as follows:

$$\tau_{12} = G(\omega) \cdot \gamma_o \cdot \sin \omega t \tag{2.6}$$

where G is the shear modulus. For the reversible and irreversible work of deformation one can write:

$$\tau_{12} = G \gamma_{o\,rev} \sin \omega t + \eta_{rev} \gamma_{o\,rev} \cdot \omega \cos \omega t \tag{2.7}$$

$$\tau_{12} = \eta_{irr} \cdot \gamma_{o\,irr} \cdot \omega \cos \omega t \tag{2.8}$$

where τ denotes the stress, ω the angular speed, η the viscosity, and γ_o the amplitude of the deformation.

The storage modulus G' increases with the frequency:

$$G' = (\eta_{rev} \cdot \omega + G) \cos \delta \tag{2.9}$$

whereas the viscosity decreases with the frequency; δ is the loss angle.

Pressure-sensitive adhesives must display irreversible work of deformation during bonding and reversible deformation work upon debonding. The ratio of both kinds of deformation work (i.e., of stored and dissipated energy) characterizes the behavior of PSAs. In general the energy state of the viscoelastic polymer may be described as follows:

$$\sigma(t) \sim [G'(\omega) \cos \omega t + G''(\omega) \sin \omega t] \tag{2.10}$$

where G' the storage modulus and G'' the loss modulus, and

$$\text{loss tan } \delta = \text{loss modulus/storage modulus} = \frac{G''}{G'} \tag{2.11}$$

One can suppose that the term "loss tan δ", as a index of the amount of stored or lost energy (i.e., of the contribution of the elastic and viscous part of the polymer) may also characterize the adhesive properties. It was shown that loop, peel, and quick stick show a good correlation with loss tan δ [4]. It was demonstrated that PSAs intended for similar applications also exhibit similar rheological properties. The correlation between adhesive properties and the dynamic shear storage modulus appears quite good [6]: hence the concept of "window of performance" as a function of the storage modulus of

Rheology of Pressure-Sensitive Adhesives

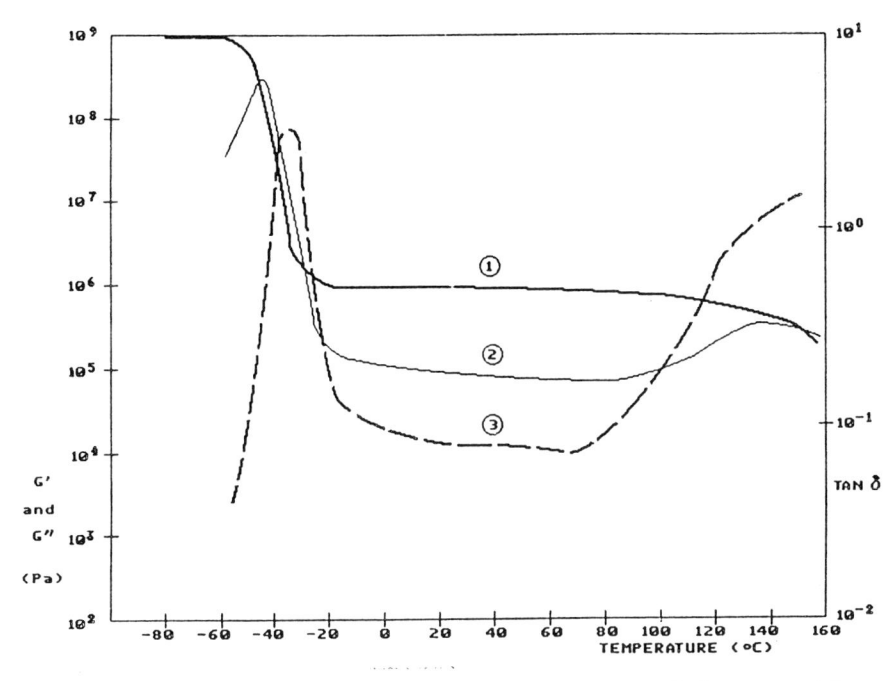

Figure 2.1 Dependence of the modulus on the temerature. ① Storage modulus; ② Loss modulus; ③ Tan δ.

the adhesive was developed [7]. These moduli, the storage and the loss moduli, can be displayed as a function of temperature (Figure 2.1).

The storage modulus starts high at low temperatures where all motion within the polymer is frozen and the material behaves like a glass. At higher temperatures it drops off and exhibits a plateau region which represents the elastomeric response generally encountered at normal end-use temperatures; the storage modulus then decreases further when softening begins. The temperature region through which the polymer changes from a glassy (hard) state into a liquid (rubber-like) state, this second-order transition point (with a continuous differential of the free enthalpy, but discontinuous, second order differential of the Gibb's free energy) is called the *glass transition temperature* (T_g), and has a special significance in the characterization of PSAs. Above the T_g the time temperature superposition principle can be applied. Differences in viscoelastic parameters around the glass transition can be directly related to the side chain size and mobility of the polymer. At T_g there occurs a change in the thermodynamic state which can be related to a mechanical energy loss function such as the loss modulus. The loss tan δ peak does not occur at the glass transition but in the transition zone between the glassy and rubbery regions. Ferry [8] pointed out that in this region of transition there is no change in the thermodynamic state. The loss tan δ peak is the midpoint of this transition zone where the ratio of loss modulus and storage modulus reaches a maximum. The energy loss maximum at this point has considerable influence on the tack of the system. The storage modulus definition can be simplified as a hardness parameter [9]. Optimum PSA performance can be quantified using the storage modulus; ideally the value of the storage modulus should vary between 20 and 80 kPa.

Chu [7] correlated PSA performance and dynamic mechanical performance (DMA) properties, whereby PSA performance, especially for tapes, was related to the storage modulus G' at room temperature and the loss tan δ peak temperature of the system. He also showed that the PSAs application window for a high cohesive strength tape adhesive require G' values at room temperature between 50 kPa and 200 kPa with loss tan δ peak temperature limits between −10 and +10°C. Optimum G' values for permanent PSAs labels were determined to be around 20 kPa at room temperature.

Pressure-sensitive adhesives were defined using viscoelastic application windows relating the storage modulus G' at room temperature to the loss tan δ peak of the adhesive. In water-based adhesives the viscoelastic relationship is not as simple, that is, it was determined that for the most commonly used polymers in water-based dispersions these relationships may not apply. In hot-melt and solvent-based PSAs a close and predictable relationship exists between the loss tan δ peak, defined as the dynamic T_g, and the T_g as measured by differential scanning calorimetry (DSC). For water-based adhesives the relationship varies depending upon the polymer type used. The loss tan δ peak temperature of an acrylic can differ from the T_g (DSC) by as much as 30°C. The phenomenon is much less pronounced for styrene-butadiene rubber (SBR)-type polymer dispersions. This variation is also valid for an adhesive dispersion containing a tackifier. The consequence of this is that the loss tan δ peak temperature cannot be used to predict and define PSAs performance in a viscoelastic application window. According to Bamborough [9] and applying the rule proposed by Chu, it may appear that an SBR would require considerably more compatible resin than an acrylic polymer; a soft acrylic (AC) PSA (like Acronal V 205) would require a higher amount of compatible resin than a hard one (like Acronal 80 D). Adhesive formulators know this not to be the case. Experienced adhesive formulators know that for Acronal V 205 an optimum concentration of tackifier would be 30 parts per 100 parts polymer whereas for Acronal 80 D one needs 80 parts (see Chapter 6.) The loss tan δ peak temperature for water-based adhesive systems is not a reliable predictor of PSA performance characteristics.

It is proposed that the loss modulus peak temperature of the adhesive should be 50°C below the operating temperature of the adhesive. Bamborough [9] proposed using the loss modulus peak temperature in the glass transition region instead of the loss tan δ peak temperature as a means of predicting PSA performance for water-based adhesives. Despite the above illustrated discrepancies concerning DMA for adhesive characterization, the use of dynamic mechanical spectroscopy to measure modulus changes, and differential scanning calorimetry to measure shifts in the T_g of the adhesive are now common methods for rheological studies of adhesives [10]. In order to understand the practical benefits of such investigations the rheology of PSAs needs to be studied.

1.2 Influence of Viscoelastic Properties on the Adhesive Properties of PSAs

The essential performance characteristics when characterizing the nature of PSAs are tack, peel adhesion, and resistance to shear. The first property represents the adhesive's ability to adhere quickly (initial grab), the second measures its ability to resist removal by peeling, and the third characterizes the adhesive's ability to resist flow when shearing forces are exerted. Generally speaking the first two characteristics are directly related to each other, but inversely related to the third one [11].

Tack and peel tests imply a high loading rate, whereas shear will be mostly measured in a static manner. The first phase in the use of PSAs (bonding) generally occurs slowly, whereas the second step (debonding during converting or end-use) imposes higher stress rates. The balance of the adhesive properties, and of the adhesive/converting/end-use properties reflects at the same time the need for the balance of the viscoelastic characteristics. In this chapter the influence of the viscoelastic properties on the adhesive properties will be briefly examined.

Influence of Viscoelastic Properties on the Tack of PSAs

According to Rivlin [12] the separation energy after a short contact time and low pressure is a measure of the tack. A short contact time and low pressure during application of PSAs imply a high wetting ability. For bonding to occur there is an a priori need for wetting of the substrate. As confirmed by Sheriff [13] and by Counsell [14] tack is a function of wetting. Good wetting supposes sufficient fluidity of the adhesive, and fluidity is characterized by viscosity.

Tack Dependence on the Viscosity. According to Zosel [15] tack is measured in two steps, namely the contact step and the separation step. During the first step, contact is made in the geometrical surface points, which increase to a larger area through wetting out, viscous flow, and elastic deformation. Wetting out implies high fluidity, as characterized by an adequate viscosity of the adhesive. Wetting out (i.e., covering the surface by the fluid adhesive) is followed by bonding due to the viscoelastic deformation of PSAs. On the other hand, debonding assumes the deformation of the laminate, the creation of two new surfaces, and deformation of the new surfaces. Thus, it may be concluded that for bond forming a high deformation with a medium elasticity is required, whereas for debonding a medium deformation with a high elasticity are required.

The tack may also be characterized as separation energy. During debonding high tack means that the adhesive absorbs a high amount of deformation energy, which dissipates on the break of the bond [16]: a high ability to store energy implies elasticity, a high energy at break means high cohesion. Thus, tack depends on elasticity and cohesion. Therefore, for high tack, a low bonding viscosity, a high debonding viscosity, and high elasticity are required. Factors influencing the viscosity and the elasticity of the polymer will also influence the tack. The polymer's own characteristics and the environmental conditions (experimental parameters) influence its viscous and elastic behavior. Studying the performance of carboxylated styrene-butadiene rubber (CSBR) latexes, Midgley shows that tack depends on the Mooney viscosity (i.e., on the molecular weight, MW) [17].

Tack Dependence on the Modulus of Elasticity. Loss moduli correlate with PSAs debonding tests; Mc Elrath [18] studied the debonding frequency of PSAs tests and their location on the loss modulus master curve. A good agreement between adhesive properties and loss modulus was demonstrated. Although absolute correlations have not been established, Class [19] suggested that the location and the shape of the modulus curve in the transition zone is important for PSA performance. In accordance with Dahlquist's criterion for a minimum value of compressive creep compliance to achieve tack, Class said his data indicated a maximum modulus value. Dahlquist [20,21] related tack to modulus, showing that the compression modulus should not be much higher than 10^5 Pa. Very high modulus adhesives do not possess sufficient conformability to exhibit pressure-sensitive tack. The optimum tack properties of PSAs are obtained when the room temperature modulus falls within the range of $\sim 5 \times 10^5$ to 1×10^5 Pa, and the T_g lies between -10 to $+10°C$ [22].

Hamed and Hsieh [23] showed that for a given total bonded area, test specimens containing noncontact regions of sufficiently small scale can exhibit a higher peeling force than those in which the noncontact ones are large. For an elastomer there is a critical flaw size below which adhesive strength will remain unchanged. Rubbery materials with higher elasticity have a smaller critical flaw size. Thus the modulus influences the tack. Rubbery materials with a more elastic response exhibit tack properties that are more sensitive to interfacial flaws, compared to those that respond more by viscous flow. When failure occurs by viscous flow, the tack is relatively insensitive to interfacial flaws, but when the strain rate is sufficiently high, so that the elastomer responds elastically, stresses may be concentrated at the edges of interfacial flaws causing a reduction of the strength.

Tack Dependence on Experimental Parameters. Viscosity and elastic moduli of PSAs are not intrinsic material characteristics (i.e., they depend on the experimental parameters used, such as the temperature and time, and the strain rate). Thus, a similar dependence of the tack on time, temperature, and the strain rate has to be taken into account. This dependence is illustrated by the quite different values of the tack obtained using different experimental techniques (rolling ball, quick stick, or loop tack) which are characterized by different time and strain rates, and by the sensitivity of the adhesive properties to environmental conditions (see Chapter 10).

Tack Dependence on Temperature. In reality it is very difficult to apply PSA labels at low temperature conditions. Deep freeze label (or tape) adhesives must be specially formulated, with a low viscosity at low temperatures. This behavior is due to the limited flow of the adhesive at lower temperatures, a phenomenon governed by the strong temperature dependence of the viscosity and of the elastical modulus. As shown by Hamed and Hsieh [23] tack is a function of test temperature and strain rate, and the experimental data may be shifted from a master curve. However the tack behavior as a function of the temperature and strain rate is more complex than that of the cohesion.

Tack Dependence on Strain Rate. It was shown earlier that tack depends on the bonding and debonding process, on debonding (separation) work, and thus tack also depends on the strain rate. This phenomenon may be observed by testing the tack using methods measuring the debonding resistance (force), like loop tack or Polyken tack. Tack varies as the separation speed of the Polyken test is changed [4]. The dependence of the tack value on the increasing speed was confirmed by Hamed and Hsieh [23] too. They observed a nonlinear, discontinuous increase of the tack with the measurement speed. Tack values rise to a first maximum and then, after a period of slight decrease, rise continuously with the measurement speed. Data obtained from Polyken tack measurements show a correlation between tack and loss tan δ peak values [4]. Tack dependence on the debonding rate was also confirmed by Mc Elrath [18]. He demonstrated that loss modulus values depend on the applied frequency, and different tack test methods (loop, quick stick, and probe tack) exhibit maxima at different frequencies.

Influence of Viscoelastic Properties on the Peel of PSAs

Peel and peel strength are measured by separating an adhesive applied to a substrate at some angle with respect to the substrate, usually at an angle of 90° or 180°C [24]. Similar to tack, the measurement of the peel adhesion involves a bonding step before the debonding or peeling step.

Tack measures the resistance to separation of the adhesive after a short contact time, or by light pressure. Peel is measured after a relatively long or very long contact time (at least 10^2 longer contact time than in the tack test) after application of a light or medium pressure. The time available for bond forming (wetting and penetration of the surface) during the first contact step is longer for peel measurements than for tack. It follows that flow properties of the adhesive during the bonding step are less critical than for the tack measurement. On the other hand, the debonding resistance will depend on the viscous nature/elasticity balance requiring its precise adjustment in order to achieve peelability (removability or repositionability) and on the strain rate, which influences the separation resistance in a more pronounced manner than during the measurement of the tack.

Peel Dependence on Viscosity. Like tack, peel implies a bonding and debonding step, with the time for the latter lasting longer. Supposedly the influence of the viscosity on the bonding step during a peel measurement is less important than for the tack. On the other hand, the debonding resistance of the joint is increasingly proportional to the viscous flow of the adhesive (i.e., a high peel needs a solid-like adhesive). The value of the peel is also a criterion for the distinction between removable and permanent labels. The peel dependence on the viscosity (modulus) and its theoretical basis will be discussed in more detail in Section 1.4.

Peel Dependence on the Elastic Modulus. Special PSAs intended for medical and surgical applications should display a low value of the modulus (and a high value of the creep compliance) in order to allow removability [25]: the higher the creep compliance, the greater the adhesive residue left on the substrate. Creep compliance values greater than 2.3×10^{-5} Pa^{-1} are not preferred.

The peel force is related to the storage modulus G' of the adhesive. High tack, removable adhesives should have a low G', and the storage modulus should not vary much with the peel rate. On the other hand, the debonding takes place in a much higher frequency range than the bonding process. In order to maximize the peel force, the highest possible G' value in the high frequency range is needed. Satas [6] showed, that solution polymers with a higher G' slope at higher frequencies than emulsion polymers, exhibit higher peel adhesion, as is generally the case with acrylic solution polymers. Chang-

Table 2.1 Peel Adhesion as a Function of Increasing Dwell Times

Dwell time (week)	Peel adhesion (N/25 mm)		
	Sample 1	Sample 2	Sample 3
0	0.32	0.12	0.09
1	0.44	0.24	0.21
2	0.47	0.30	0.30
3	0.55	0.35	0.35
4	0.60	0.40	0.47

ing the chemical composition (e.g., sequence distribution) may lead to a change (increase) in modulus and eventually to a decrease of the peel force [26].

Peel Dependence on Dwell Time and Strain Rate. Contact forming or bonding of the adhesive assumes its viscoelastic deformation. Viscous, slow flow is time dependent and thus bond forming will also be time dependent. The time dependence of the bonding step is illustrated by the influence of the dwell time on the peel values.

Like the pressure used during application, the contact time (i.e., the interval between bonding and debonding) takes into account the relatively low mobility (flow rate) of PSAs. A higher pressure or a longer dwell time should help adhesive flow and increase the peel resistance. Generally, until equilibrium is reached, the peel resistance increases with increasing contact pressure and time [23, 27] (Table 2.1; Figure 2.2).

As can be seen from Figure 2.2, the peel force increases with the dwell time of the adhesive, that is, the viscous and elastic deformation of the adhesive need a period of time (depending on the viscosity and experimental conditions). As will be discussed later, the time dependence of the bonding imposes the use of normalized dwell times for peel measurement purposes. Generally the peel/dwell time dependence is not linear, but partially cross-linked silicone rubber exhibits peel values which increase linearly with the dwell time [28]. The dependence of the separation energy on the contact time was also demonstrated by Zosel [3]. The build-up of the peel force in time has a special importance in the design of removable adhesives. On the other hand the time dependence of the flow/deformation of the adhesive and its influence on the peel may also be observed during the debonding process. The "memory effect" in the adhesion of rubber to rigid substrates is well known [29]. After re-adhering the adhesive, the peel force is lower. After

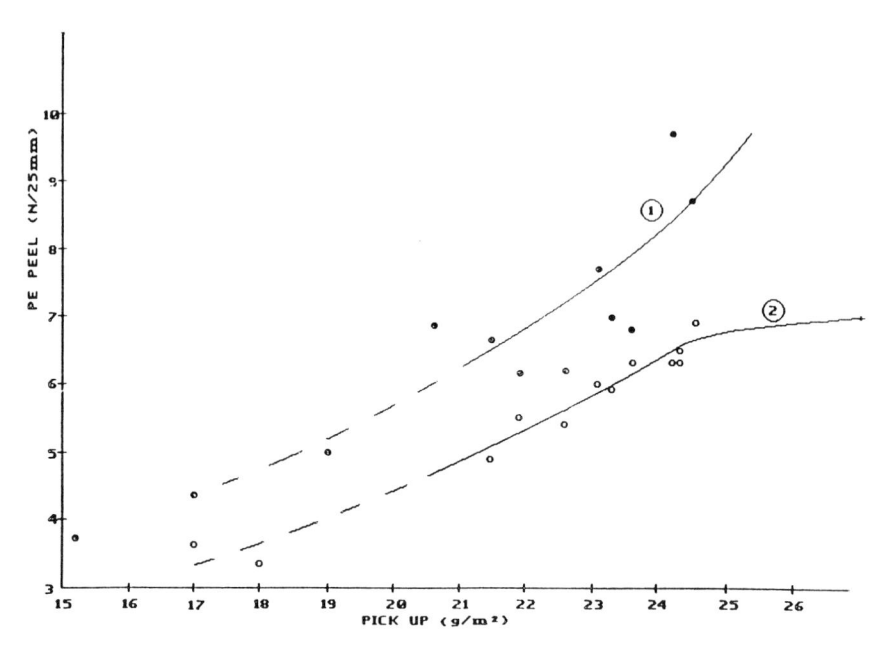

Figure 2.2 Dependence of the peel values on the dwell time. Peel from polyethylene as a function of the coating weight for different dwell time values: Dwell time of 1) 20 min; 2) 0 min.

initial peeling off the adhesive, it regains its original peeling resistance only after a period of time (or much faster with the help of some solvent). The memory as a function of the dwell time is associated with some rearrangements of the molecular structure of the rubber at the interface. Another influence of the dwell time may be observed, when examining the influence of the storage and aging on the peel adhesion value. In general there is a build-up of the peel resistance with time.

Influence of Peeling Rate. At low peel rates the viscous flow, and the deformation of the adhesive layer are dominant for the peel resistance. Therefore the peel resistance increases with increasing peeling rate. At high peeling rates the elastical character of the adhesive dominates, thus in this region the peel resistance does no longer depend on the peeling rate. Of practical importance is the very pronounced dependence of the peel force on the peeling rate (Figure 2.3); the peel force increases with the peel rate [30]. This behavior requires the use of normalized peeling rates for the measurement of the peel adhesion force.

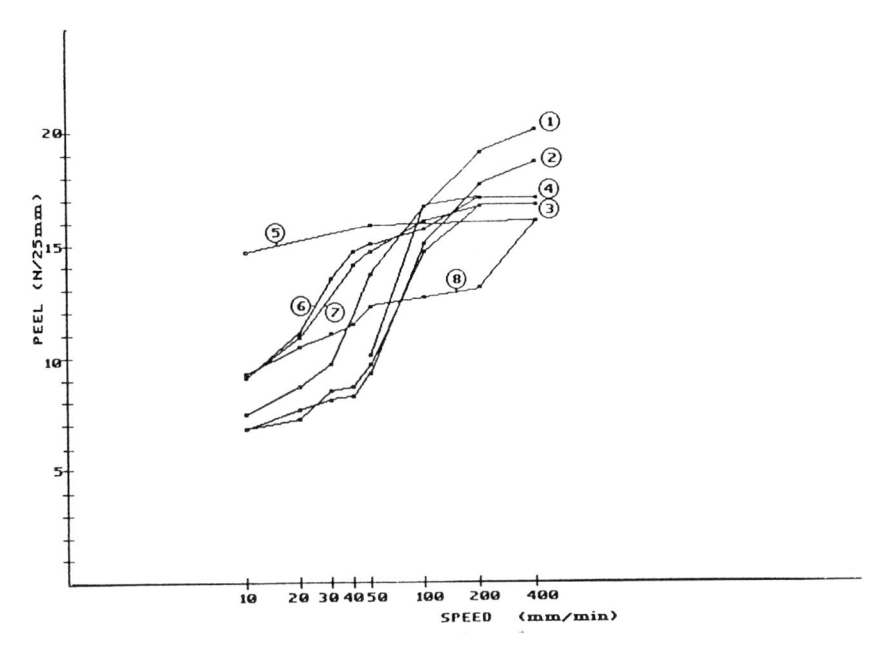

Figure 2.3 Dependence of the peel force on the peeling rate. 1 through 8 are different tackifier water-based acrylic formulations; the peel adhesion from glass was measured.

The dependence of the peel on the peeling rate may also be observed for very low peel forces (peeling from release liner) [31]. McElrath [18] showed that (as theoretically supposed) the loss modulus depends on the frequency, as does peel adhesion, and shows a maximum for a given frequency value. Removable adhesives displaying a plateau region in the storage modulus/ frequency plot, exhibit a peel force independent of the peel rate [6]. Kendall and Chu [22] studied the dependence of the peel on the T_g; they also stated that the peel maxima change with the testing rate. Higher test rates or lower ambient temperatures produce maxima at lower resin levels, lower resin softening points or lower elastomer T_g. Sehgal [32] measured the peel of different water-based PSAs after storage at 100°C, and found large differences between the obtained peel values. This indicates the improvement of the wetout through flow, underlining the importance of rheology. Sehgal also showed, that adhesive break (in the adhesive layer) appears above a certain peeling rate (2.5 mm/min), or below a certain temperature (+25°C) only.

It may be possible to remove a label from a paper substrate if it is peeled off very slowly, but the same label will certainly tear if it is peeled off quickly.

Since the peel force is related to the storage modulus of the adhesive, high-tack removable adhesives should possess a low storage modulus which does not vary much with frequency (rate of peel). At any given temperature peel adhesion is observed to increase as the peel rate is increased [2]. At low strain rates the peel forces are much lower. Under these conditions, the viscous components of the polymer dissipate significant amounts of energy and, as a result, resistance to peel forces is low. At higher strain rates the molecules have less time to disentangle and slide past one another. This behavior results in more stored energy and peel forces intensify accordingly. One can conclude that at least theoretically, each adhesive may be considered as a low peel adhesion adhesive (see Chapter 6) provided a very low peel rate is applied.

Temperature Dependence of the Peel. With the increase of the temperature the viscosity of PSAs decreases. Therefore the increase of the temperatures improves the tack and instantenous peel, and exerts a negative influence on the cohesion. It is well known in packaging applications that the winding resistance for tapes depends on the temperature and the winding rate [33] (i.e., the peel is temperature dependent). It was shown for ethylene acrylic acid copolymers that peel adhesion depends on the temperature [24]. As known, a maximum peel strength implies a certain modulus value and viscosity. Both modulus and viscosity depend on the temperature.

Influence of Viscoelastic Properties on the Shear of PSAs

Cohesive strength is measured as shear or shear strength, which is the resistance of adhesive joints to shear stress, and is measured as a force per unit area at failure. The shear force is applied in a plane parallel to the adhesive joint.

The shear resistance depends on the adhesive's viscosity and elasticity. Different authors have formulated a definition of the cohesive strength. If deformation by shearing is considered like a flow, the flow limit (F_L) is given by the following correlation [35]:

$$F_L = \frac{E \cdot W_e \cdot b^2}{4h^3} \tag{2.12}$$

where W_e denotes the maximum elastical deformation of the sample, b the width of the adhesive surface in the stress direction, and h the thickness of the adhesive layer. It can be seen from the above relation that the flow limit (i.e., the cohesion) is a function of the elastic modulus of the adhesive. Considering that the strength of a PSA joint depends on the viscosity η, the adhesive layer thickness h, and time t, the interdependence of these factors may be formulated as follows [36]:

$$F = \frac{\eta}{h^2 \cdot t} \tag{2.13}$$

The influence of the viscoelastic properties on the adhesive characteristics of PSAs depends on the nature of the stresses applied during tack, peel, or cohesion measurements. Köhler [2] demonstrated that tensile or shear stresses produce different levels of deformation in PSAs. When applying a tensile force on a circular, laminated PSA layer, with a viscosity of 10^3 Pa, according to the relation:

$$P = \frac{3\pi \cdot \eta\, R^4}{4 t h^2} \tag{2.14}$$

where P is the applied force, R is the sample radius, t is the time, and h is the thickness of the adhesive layer, one needs a force of 100 kg/cm^2 during 1 sec in order to achieve a certain deformation; but applying a shear stress, according to the Newtonian relation:

$$\tau = \eta\,\dot{\gamma} \tag{2.15}$$

where $\dot{\gamma} = 1$ sec^{-1}, only a 2.5 kg/cm^2 force is needed. Thus it may be concluded that shear-stressed pressure-sensitive joints need lower force levels in order to undergo a deformation (i.e., pressure-sensitive joints are weaker than classical ones).

Shear Dependence on Viscosity. Increasing the molecular weight of CSBRs increases the Mooney viscosity [17]. At the same time an increase in the molecular weight will improve the cohesive strength (shear) of carboxylated SBR. If the adhesive is considered a Newtonian fluid, the shear resistance δ_{SH} is given as a function of the viscosity η, the adhesive thickness h and the tensile rate v:

$$\delta_{SH} = \frac{v}{h} \cdot \eta \tag{2.16}$$

If the solid obeys Hooke's law the shear resistance depends on the shear modulus G and the tensile rate v:

$$\delta_{SH} = \frac{v}{h} \cdot G \tag{2.17}$$

Shear Dependence on the Modulus. Milled rubber possesses a lower elastic modulus and shorter modulus plateau than rubber from dried latex [22]. The thermo-mechanical degradation of the rubber through milling does not change the tan δ peak temperature (T_g). It does reduce the modulus at high temperatures. This modulus reduction relates to the lower shear performance of solvent-based systems.

Correlations between the viscoelastic and PSAs performance properties can be made by looking at those temperature regions which correlate (via the time-temperature superposition principle) to the time scales of the adhesive performance parameters. The magnitude of G' in the upper temperature range is indicative of internal strength or cohesion. Initial peel, which is largely dependent upon the wetout characteristics of the polymer, is governed by G' and tan δ in this same temperature range.

Similarly, loop tack and quick stick properties are wetout dependent, but are associated with faster relaxation times and thus correlate with room temperature viscoelastic properties [38]. The possibility to characteristize application field-related properties of PSAs and to correlate them with the rheology of the adhesive is illustrated by formulating adhesives for medical tapes [39]. The adhesive used for medical tapes is characterized by a dynamic shear modulus of about $1-2 \times 10^4$ Pa, a dynamic loss modulus of $0.6-0.9 \times 10^4$ Pa, and a modulus ratio or tan δ of about 0.4–0.6 as determined at an oscillation frequency sweep of 1.0 rad/sec at 25% strain and 36°C. Adhesives with moduli higher than the acceptable range display poor adhesive strength, while adhesives with moduli below the acceptable range exhibit poor cohesive strength and transfer large amounts of adhesive to the skin upon removal [39].

Shear Dependence on Time and Strain Rate. The shear resistance of the adhesive depends on its internal cohesion. The cohesion is a function of the inherent viscosity or modulus and thus it depends on the parameters influencing the viscosity or modulus. Both viscosity and modulus are not intrinsic material characteristics, they depend on the temperature and time (i.e., on the nature and time history of the applied forces). Therefore, the shear resistance will depend on the temperature and the time: in the labeling practice this behavior is illustrated by the low temperature resistance of the adhesive joints and by the differences between statical and dynamical shear.

The cohesive strength (shear) increases with increasing test rate or with decreasing temperature [23]. When non-Newtonian liquids are subjected to variable shear rates, the plot of the shear stress/shear rates no longer shows a linear relationship [40]. In a diagram with double logarithmic scaling this curve becomes a straight line. The mathematical equivalent of these two curves is given by the following equations:

$$\tau = K \cdot D^n \tag{2.18}$$

$$\ln \tau = \ln K + n \cdot \ln D \tag{2.19}$$

where D denotes the shear rate, K is a viscosity related coefficient, and n is an exponent of the "power law equation" defining the shear rate dependence on the viscosity. This exponent is known to vary for polymers between 0.3–1.0. Practically, it was demonstrated that the increase of the test rate produces an increase of the cohesion and of the peel up to a critical value [23]. Bonding is a diffusion and time-dependent process. This is illustrated by the pressure dependence of the lap shear adhesion for adhesives; the higher the pressure the greater the bond strength [41]. As shown for cohesive strength measurements as a function of temperature and shear rate [23], the principle of strain rate-temperature equivalence can be applied.

Shear Dependence on Temperature. As discussed earlier, the cohesive strength increases with increasing test rate or with decreasing temperature. Shear data obtained as a function of the test temperature and shearing rate can be shifted horizontally to form a single master curve, illustrating the principle of time-temperature equivalence [23]. Hamed and Hsieh [23] found a good agreement between experimental and calculated values, using the universal Williams-Landel-Ferry relationship for an amorphous rubber. The finding that data can be shifted to form a master curve is evidence of the importance of chain segmental mobility in controlling the shear strength [23]. If the chain segmental mobility (i.e., the ability to relax) is high (high temperature or low test rate) then the fracture stress is small. On the other hand, a large stress is required to rapidly tear apart a PSA sample, since the chains have little time to rearrange their microstructure in order to accomodate the applied stress. This statement appears valid for peel adhesion too. One should recall that the removability of a PSA label strongly depends on the peeling rate.

1.3 Influence of Viscoelastic Properties on the Converting Properties of PSAs and PSA Laminates

The converting properties of the adhesive and of the laminate depend on the rheology of the adhesive. It has to be pointed out that, except for hot-melt PSAs, the rheology of the uncoated adhesive (with an inherent fluidity required for processing purposes) is different from that of the converted material.

Converting Properties of the Adhesive

The liquid adhesive must be coated onto a release liner or face material. Good coatability implies adequate machinability or processing properties on the coater (metering roll, drying tunnel). During manufacturing, transport, and coating, the adhesive fluid is subjected to shear forces, going through a more or less pronounced change of the viscosity. The coated shear-thinned adhesive has to wet the web, and the wetout depends on its viscosity. Except for hot-melt PSAs, the coated liquid adhesive layer has to allow the elimination of the carrier liquid (drying) in order to form a solid adhesive layer. Evaporation of the carrier depends on its diffusion through the adhesive layer (i.e., on the viscosity of the adhesive). It may be concluded that the viscosity of the adhesive, the time (shear rate)/temperature dependence of the viscosity influences the coatability and convertability of PSAs. It was mentioned earlier that, except for hot-melt PSAs, the other PSAs are dispersed or diluted systems, their rheology being different from that of the converted material. The coating-related rheology will be covered in Section 3.

Convertability of Laminate

Convertability of the PSA laminate means its ability to be processed into finished products (labels, decals) mostly by cutting (or die cutting) and printing. The flow properties of the adhesive influence the migration or penetration, oozing, and cold flow, thereby limiting the convertability of the laminate. Thus the viscosity of the adhesive and its time/temperature dependence (i.e., a nonlinear character) determine the converting properties of the PSA laminate. It should be mentioned that the converting properties of the laminate really depend on the interaction of the PSA–laminate components. These properties will be discussed in more detail in Section 3.2.

1.4 Influence of Viscoelastic Properties on End-Use Properties of PSAs

The most important end-use properties of PSAs are the propensity to labeling and bonding behavior. Labeling is influenced by the adhesive properties (peel and tack) and by the dispensing properties of the label. Removability or peeling off is influenced by the adhesive properties. Thus the parameters influencing the adhesive properties will also characterize the end-use properties.

Label Application Technology

Label application technology refers to labeling. Labels are either applied by hand or with mechanical processes. Generally the label dimensions are decisive for the choice of the application technology. Large labels will only be applied by hand. On the other hand, according to the application technology, reel and sheet laminates are manufactured. For PSAs reels and sheets, a quite different adhesive/cohesive balance is required (i.e., quite different tack/peel/shear values or flow properties). High speed labeling guns need high tack PSAs, whereas for high speed cutting high modulus and high cohesion PSAs are required.

In the label industry a basic difference exists between roll and sheet supplied laminates (face material/adhesive/release liner). Pressure-sensitive adhesives for sheet applications must resist flying knife and guillotine cutting. A poor selection of the adhesive results in part in oozing of the adhesive (gum deposits) on the cut surfaces and smearing of the edges of the label stock with subsequent poor feed to the printing press. The requirements are generally less critical for roll applications. Converting in the paper label industry involves processes such as die cutting of roll stock, and guillotining of sheet stock. Formulating approaches that improve the high frequency modulus of the adhesive will enhance converting [2] since die cutting and guillotining are such high frequency processes. The less viscous and the more rigid the response of the polymer during the processes, the cleaner the process tends to be. If viscous flow within the polymer is significant during the converting operation, poor die cutting or poor guillotining (knife fouling) can result.

Removability of PSAs

Conventional PSAs can be classified as either nonpermanent (2.7–9 N/25 mm for 180° peel adhesion) or permanent (above 9 N/25 mm for 180° peel adhesion). Adhesives included in the former category are used in the manufacture of removable tapes and labels, protective laminates, and other less durable products [42].

For nonpermanent, so called removable adhesives, the flow properties and cohesion of the adhesive as well as the anchorage of the adhesive to the face stock are critical. In an ideal case, if the bond to the substrate is nonpermanent, then a clean release from that substrate is encountered and the adhesive remains on the face material. Another requirement for good removable adhesives is the low peel level with a permanent character (i.e., no build-up of the peel in time is acceptable). A clean release from the substrate and no build-up of the peel with time are the minimal requirements for removable PSAs. On the basis of the adhesive characteristics it is possible to formulate the rheological criteria necessary for removable and permanent adhesives.

Criteria for Removability. Generally a special balance between tack, peel adhesion, cohesion, and anchorage is required in order to ensure removability [43]. Some low tack, crosslinkable dispersions are good removable PSAs. Their good elasticity yields low cold flow and low build-up of the peel during storage [44]. On the other hand, a certain degree of cold flow is needed in order to obtain removability. Adding 2–20% plasticizer during the polymerization of acrylics leads to a removable adhesive but the plasticizer may migrate out after storage [45]. Furthermore cohesive strength is necessary to avoid stringing. It must be stated that the adhesion depends both on the rheological features and chemical affinity. It was shown [46] for certain PSAs that the apparent separation energy was approximately the same when either pulled away from a glass or a Teflon surface. The adhesion of these adhesives must be attributed largely to their rheological features rather than to selective wettability.

How to Achieve Removability? A special adhesion/cohesion and/or plasticity/elasticity balance is required in order to achieve removability. Flow properties allow rearrangement, relaxation, and transformation of the debonding energy into viscous flow, therefore softening of the adhesive is a prerequisite for removability. It as well known from the field of caseine-based adhesives [47] that polyethylene glycols added to caseine and dextrine glues control the humidity absorption, and thus the tear of gummed papers at low environmental humidity, and avoid blocking at high humidity, thereby acting as plasticizers (at a loading rate of 1–2%).

Flow properties are correlated to creep which, in turn, is a function of molecular weight and plasticizer loading. Higher molecular weights and low plasticizer levels reduce the tendency to creep. For removable PSAs the molecular weight should be limited. Permanent adhesives need no fluidity during debonding, as they must display debonding resistance when subject to high stress rates and large forces. They need to display a modulus and viscosity increase as a function of the stress rate. On the other hand, removable adhesives need easy debonding and viscous flow. The applied stress must be able to cause movement within the bulk of the material. Theoretically the applied stress for debonding can be minimized by using stress resistant polymers, fillers, and primers; stress-resistant polymers are those which develop a controlled crosslink density or are soft internally. Primers are somewhat flexible and generally promote good anchorage to the face stock. Thus, they can absorb expansion or impact stresses without adhesive failure.

Influence of the Viscosity/Modulus on Removability. According to Satas [6] there are two basic types of removable adhesives, namely PSAs which wet the surface poorly and PSAs which adhere easily but are easily removed. The first type of PSAs is highly elastic and has low tack. Its removability depends on the adhesive not establishing a good contact with the surface. Sometimes the poor contact is built-in mechanically. These PSAs might have inclusions that prevent the adhesive from achieving full contact with the surface, as the adhesive contains physical particles that do not deform completely and therefore limit the contact area between adhesive and substrate. These PSAs possess a high modulus. Thus, hardening of the polymer (modulus increase) is a possibility to obtain removable adhesives. The second type displays high tack and is characterized by a low modulus.

As shown earlier, hardening of the adhesive increases its modulus (i.e., decreases its ability to creep) and its ability to wet the surface, and thus decrease the tack (i.e., peel adhesion after very short dwell times) as well as the peel resistance. It can therefore be assumed that hard adhesives display low peel adhesion (i.e., are removable).

In order to soften the adhesive, low molecular-weight compounds such as plasticizers are often added. The presence of 8% plasticizer may improve the elongation by 3000% [44]. The softening effect is illustrated by the lowering of the minimum film-forming temperature (MFT) (e.g., 6% plasticizer decreases the MFT from 23°C to 0°C), or the decrease of the tensile yield strength (e.g., 6% plasticizer decreases the tensile yield strength from 13.0 to 1.0 N/mm^2). The elongation of acetate PVAc PSAs (normally 6–10% at break) was significantly increased by adding plasticizers to up to 100–1000% [48]. Differences in plasticity may be due to the molecular weight of the plasticizer or different interactions between chain segments and plasticizers. Long chain plasticizers exert a strong plasticizing effect. Generally plasticizers decrease the modulus value and influence the frequency dependence of the loss tan δ [4]. For nonplasticized adhesives loss tan δ is low at low frequencies. For plasticized adhesives loss tan δ shows a maximum at lower frequencies, and peel and quick stick also exhibit maxima at lower frequencies.

Plasticizers may be incorporated in peelable PSAs to soften the adhesive and thus improve peelability. However, some plasticizers can exert a tackifying effect on adhesive polymers and this may limit the amount used. Typical plasticizers include phthalate esters and polyalkylene ether derivatives or phenols, the principal requirement being compatibility with the main adhesive polymer and the tackifier as to avoid or minimize migration of the plasticizer. It is possible to add quite large quantities of plasticizer, even up to 50% by weight on the solid adhesive; however, usually 10–20% by weight are added [6]. Other plasticizers which can be employed, include the well-known extruder oils (aromatic, paraffinic or naphtenic) as well as a wide variety of liquid polymers. Satas [6] showed that the addition of a plasticizing oil which is compatible with the mid-block of a styrene-isoprene-styrene (S-I-S) block copolymer results in an adhesive with lower peel force.

Influence of Viscoelasticity on Applied Labels

After some time labels show a higher peel resistance then freshly applied ones (i.e., the peel resistance increases with the dwell time.) This build-up of the peel resistance is due to the flow and contact build-up of the adhesive. Therefore, one should pay attention when evaluating the peel force, that is, peel should always be measured after a well-defined (normalized) dwell time.

1.5 Factors Influencing Viscoelastic Properties of PSAs

The material characteristics, chemical composition/structure, and environmental and experimental conditions influence the viscoelastic properties of PSAs. The influence of the material characteristics on the viscoelastic properties of the PSAs will be discussed first. Numerous empirical correlations between the molecular structure, molecular weight, molecular weight distribution (MWD), branching, and the rheology are generally valid for chemically quite different polymers [49].

Influence of Material Characteristics on Viscoelastic Properties of PSAs

The molecular character of the base polymer, its chemical composition and structure influence its viscoelastic properties. The influence of the molecular weight on the viscoelastic properties of the PSAs will be examined first. Viscosity and the elastic modulus are the most important parameters characterizing the viscoelastic behavior of PSAs. Both parameters depend on the molecular weight of the base polymer.

Dependence of the Viscosity of PSAs on the Molecular Weight. The viscosity of macromolecular compounds is a function of the molecular weight. Their viscosity depends on the molecular weight according to a correlation of the form:

$$\eta = f(MW)^\alpha \tag{2.20}$$

Zosel [15] demonstrated the strong influence of the molecular weight of acrylic PSAs on their viscosity and on their tack. The increase of the molecular weight in the range of 10^3–10^7 produces an increase of the viscosity from 10^2 to 10^{10} Pa·s. He also showed the existence of a maximum in the tack/molecular weight graph for polyisobutylene. The pronounced increase of the viscosity with the molecular weight limits the usable molecular weight range for hot-melt PSAs, limiting (unfortunately) their cohesion as well. It is well known to skilled formulators that block copolymers have lower viscosities than natural rubbers. The solution viscosity of the high polymers also increases with their molecular weight. Thus a viscosity imposed balance between processing and end-use properties, that is, a solid content/molecular weight balance for solvent-based PSAs is always a limiting factor, keeping in mind that the solid content of common solvent-based PSAs does not exceed 30–40%. On the other hand the additives in an adhesive formulation, their nature and concentration also influence the viscosity of the blend. Practical examples are given by plasticizers and tackifiers. Micromolecular tackifiers (plasticizers) may be considered as diluting agents. Thus the viscosity of the polymer "solution" obeys an exponential law, depending on the polymer concentration C, namely

$$\eta = f(C)^\beta \tag{2.21}$$

Rheology of Pressure-Sensitive Adhesives

In a similar manner the level of the tackifier resin influences the resulting viscosity. As an example, Zosel [15] demonstrated that the "zero viscosity" of tackified PSAs decreases up to 80% resin loading level, resulting in better flow properties. The diluting effect of the tackifier resin depends on its own viscosity too. Generally, soft resins impart agressive grab and quick stick, while hard resins help retain good cohesive strength and creep resistance. Thus resins with a softening point below about 50°C impart tack, but poor cohesive strength, while those above 70°C lead to poor tack properties [50].

It was shown earlier that the peel adhesion depends on the viscosity. Peel is a function of molecular contacts which depend on the diffusion. The rate of the adhesive diffusion into the face stock is inversely proportional to $(MW)^2$. Thus the number of contact points decreases with $(MW)^2$, and thus tack and peel decrease with the molecular weight [23]. On the other hand the chain elasticity increases with the molecular weight as well as the so called "critical flaw size," resulting in a lower peel at a higher molecular weight.

Dependence of the Elastic Modulus on the Molecular Weight. Raw materials for PSAs are characterized by two special features: the T_g and the modulus of elasticity. Both of these properties and their dependence on the base polymer will be examined in Chapter 3. The dependence of the elastic modulus on the molecular weight will be studied in this chapter. It should be mentioned that toughness properties of polymers generally increase with increasing molecular weight up to a point, then level off. Milled rubber used for solvent-based PSAs has a lower modulus than rubber obtained from dried latex [22]. An extension towards higher temperatures of the plateau region of the dynamic modulus master curve can be achieved by the increase of the molecular weight of CSBR [17].

Influence of Chemical Composition/Structure on Viscoelastic Properties of PSAs

As discussed earlier and detailed in the next chapter, the rheological properties of PSAs are characterized by viscosity and modulus, and the T_g of the base polymers. All these parameters depend on the chemical composition/structure of the PSAs. The dependence of the viscosity on the chemical composition/structure is discussed next.

Viscosity of the Base Elastomer. As known from polymer chemistry the inherent viscosity of polymers depends on the intermolecular mobility of the polymer molecules. This mobility is a function of the chain length (molecular weight) and structure (e.g., branching) and reciprocal interaction of the polymer molecules (polar forces). It must be noted that both the viscosity and its dependence on the temperature are a function of the chemical composition/structure of the polymer. The flow activation energy (e.g., Arrhenius) depends on branching [49]. Styrenic block copolymers have a relatively low solution or melt viscosity, since in solution or at temperature above the T_g of polystyrene, the systematic polystyrene domain within the block copolymer no longer exists.

Influence of Tackifier Resin. Exhibiting a surface free energy close to that of the elastomer the resin (filler) is perfectly wetted by the organic matrix, and as a consequence, good dispersions (i.e., lower viscosities) are obtained [51]. Addition of resins to rubber-based adhesives decreases the viscosity [52]. The chemical composition of the base elastomer also influences the viscosity of the PSAs. Acrylic hot-melt PSAs with non-Newtonian behavior differ from compounded hot-melt PSAs and from conventional solvent-based rubber/resin PSAs, and allow for the development of high shear adhesives with a low processing viscosity [53].

Influence of Environmental and Experimental Conditions on the Viscoelastic Properties of PSAs

The coating of PSAs, the converting of the laminate, and the end-use of the pressure-sensitive labels occur under very different temperature and stress rate conditions. Due to their macromolecular chemical basis, the properties of the PSAs are time/temperature dependent. Thus, for adequate manufacturing and use their time/temperature sensitivity must be studied. For a better understanding of this dependence the temperature and time dependence of the viscoelastic properties will be examined separately. First, the dependence of the viscosity and of the modulus as a function of the temperature will be reviewed.

The physical and mechanical characteristics of a polymer depend on its chemical composition/structure. Chemical and macromolecular characteristics determine the internal mobility of the polymer chains. On the other hand, the ambient temperature influences the rate of chemical and physical interactions (i.e., the chain mobility) and thus the viscoelasticity of the polymer. Liquid characteristics (viscosity) or solid behavior (modulus) of the macromolecular compound are both functions of the temperature. Adhesion properties require a special balance of the viscous liquid/elastic solid state. Therefore the temperature also influences the adhesive properties.

Dependence of Viscosity on Temperature. In general the viscosity is a function of temperature. For low molecular weight, nonassociated liquids (Newtonian systems) the viscosity depends on the temperature according to the equation of Arrhenius:

$$\eta = A \exp(E_o/RT) \tag{2.22}$$

where E_o represents the activation energy, R is the universal gas constant, and T the absolute temperature. For highly viscous liquids [54]:

$$\ln \eta = A \exp [B/(T + C)] + D \tag{2.23}$$

where A, B, and D are experimental parameters. For non-Newtonian systems the Williams-Landel-Ferry equation is valid, that is:

$$\eta = A \exp [B/(T - T_o)] + \frac{E_o}{RT} \tag{2.24}$$

where A and B are experimental parameters, and T_o is the temperature at which the free volume is zero. This equation remains valid for $T < T_g + 100°K$. Generally the temperature-dependent shift factor a_T for the viscosity is given by [55]:

$$a_T = \frac{\eta_o(T)}{\eta_o(T_o)} \tag{2.25}$$

where η_o is the viscosity at zero shear rate. For amorphous polymers (e.g., PSAs):

$$\ln a_T = -\frac{C_1(T - T_o)}{C_2 + (T - T_o)} \tag{2.26}$$

where C_1 and C_2 are material-dependent parameters. For partially crystalline polymers (e.g., resins):

$$\ln a_T = \frac{E_o}{R}\left(\frac{1}{T} - \frac{1}{T_o}\right) \tag{2.27}$$

For the temperature dependence of the dynamic viscosity, two models generally are considered. The model of Eyring et al. [56] indicates the existence of an energy barrier between holes in a liquid and this barrier is the rate determining step of viscous flow; the model applies transition state theory to explain the temperature dependence of viscosity, namely:

$$\eta = \beta' \exp (E/RT) \tag{2.28}$$

The constant β' includes both the molecular volume of the liquid and the activation entropy of flow and is generally assumed constant. Another model [57] supposes the formation of holes (i.e., the free volume as the rate determining step), introducing the expression:

$$\eta = A' \exp (B \cdot V_o/V_f) \tag{2.29}$$

A' and B are two constants, and V_o and V_f are the occupied and the free specific volume of the liquid, respectively. The temperature dependence of the viscosity is explained by that of the volumes; by rearranging the well-known Williams-Landel-Ferry equation [58]:

$$\log\left(\frac{\eta}{\eta_o}\right) \cong \log a_T = -\frac{C_1(T-T_g)}{C_2(T-T_g)} \tag{2.30}$$

where the constants C_1 and C_2 depend on the fractional free volume.

Influence of Time on the Viscoelastic Properties of PSAs. The intermolecular mobility of the macromolecules is a time dependent, relatively slow process. This means that molecular rearrangements require a certain time. Practically this time may be longer, equal to, or (rarely) shorter than the duration of the external stress. Therefore, internal (molecular) motion does not always occur synchronously to the external one. Macroscopic flow behaviour and elastic recovery are characterized by viscosity and modulus. Both of these properties are the result of internal microscopic interactions (i.e., of the time needed for these ones). Therefore, the viscoelasticity of PSAs depends on the time and on the rate of the applied external stress.

Williams, Landel and Ferry [58] defined a shift factor allowing the correspondence between time and temperature influences [58]. The shift factor depends on the T_g and on the reference temperature [59]. At relatively low strain rates, such as those acting during initial bonding or wetout of the surface, the polymer molecules have time to slide past one another and undergo viscous flow [2]. This effect, as would be expected, is more pronounced at higher temperatures, where molecular mobility is greater. Storage and end-use of PSAs are both low strain processes. Alternatively, strain rates are significantly higher during peel measurements. Here the molecules are not able to flow past one another sufficiently fast, and the polymer stiffens. This stiffening constitutes an increase in the elastic contribution to the rheology of the polymer. When the elastic components predominate, more of the bond rupture energy is stored resulting in a higher peel and tack. High speed converting and end-use operations (e.g., die cutting, removal of labels) are high strain processes [2].

The time and temperature dependence of the viscoelastic behavior of polymers may be interpreted by the corresponding dependence of the relaxation spectra and by conformational changes within the macromolecular chains as shown by Schneider and Cantow [60]. The corresponding contributions to the viscoelastic functions are expressed by the frequency shift factor a_T which allows the superposition of viscoelastic isotherms into composite curves at T_o, the reference temperature [60]:

$$a_T = \frac{[a^2 \cdot \zeta_o]_T \cdot T_o}{[a^2 \cdot \zeta_o]_{T_o} \cdot T} \quad (2.31)$$

From Equation 2.31 it follows that the temperature dependence is connected mainly to the translational friction coefficient per monomeric unit ζ_o, although the mean square end-to-end distance per monomeric unit a^2 may vary significantly with temperature for most polymers. The shift factor is related to the dynamic viscosity η' according to:

$$a_T = \frac{\eta' T_o \rho^o}{\eta_o' T \rho} \quad (2.32)$$

where ρ is the density of the polymer. Intermolecular polymeric motion influences the polymer flow, and this motion is a function of time and temperature. From the mechanics of fluids or solids, it is known that mechanical properties decrease as temperature increases (i.e., the molecular motion is enhanced). In a similar manner there is an opportunity for molecular motion if the rate of the applied force is lowered. Thus high temperature and low forces act similarily on the molecular motion and on phenomena related to it (e.g., viscosity).

Dependence of the Viscosity on the Shear Rate. Pressure-sensitive adhesives are viscoelastic materials, that is, their mechanical properties are temperature and time dependent. The material characterization requires the measurement of a material function like viscosity or modulus as a function of temperature and time. As can be seen from Figure 2.4, the viscosity is a function of the shear rate.

At low shear rates there is a short interval, where the viscosity does not depend on the shear rate (this is the so-called zero-viscosity). At higher shear rates, there is a linear or nonlinear decrease of the viscosity as a function of the shear rate (Newtonian or non-Newtonian systems.) The decrease of the viscosity with the increase of the shear rate is given by the elastical component of the adhesive [61]. The existence of a shear rate-dependent viscosity domain implies that static methods for testing the cohesive strength of PSAs cannot yield real information about their behavior at high stress rates (peel, cutting). This statement can be confirmed through the definition of the adhesion failure energy (separation energy W), upon debonding through peel or loop tack [62]:

$$W = \int F \cdot v \cdot dt \quad (2.33)$$

where v is the debonding (separation) rate and F the applied force. The exponent value for a Newtonian liquid that has a shear rate-independent viscosity is $n = 1$ (see Equation 2.4). Changes in the polymer structure may

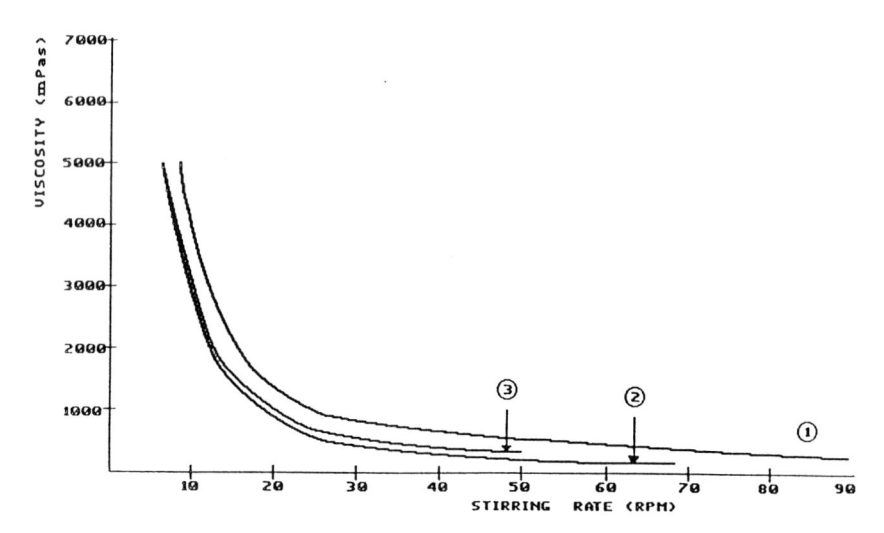

Figure 2.4 Dependence of the viscosity on the shear rate. Decrease of the Brookfield viscosity for different (1 through 3) water-based acrylic PSAs as a function of the stirring rate.

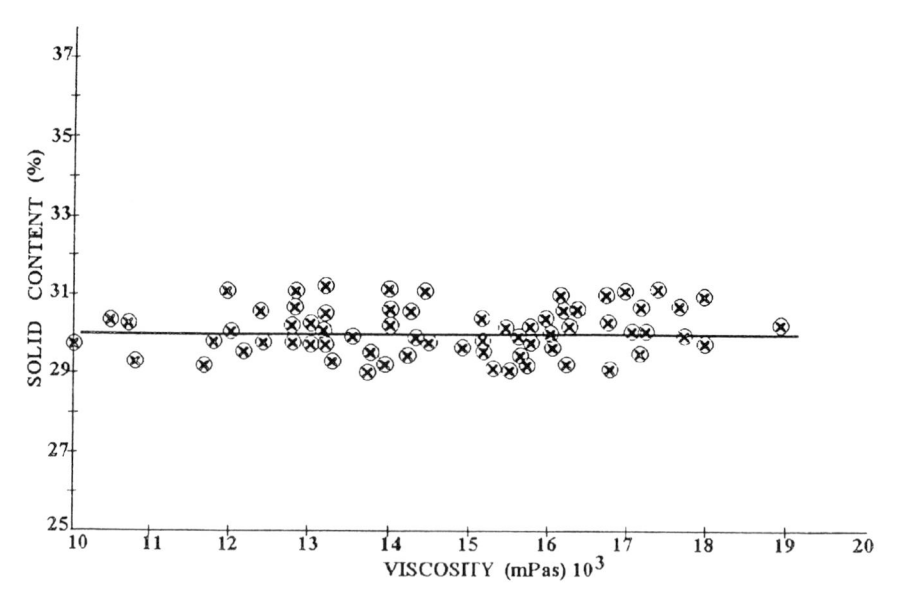

Figure 2.5 Range of solution viscosity values. Tackified, unmilled rubber/resin PSA: solution Brookfield viscosity as a function of the solid content.

be associated with changes in the viscosity-shear rate dependence. Detailed studies of elastomers showed mastication increased n from 0.25 to 0.5 [63]. Mastication in a mixer yields less change of the polymer elasticity than mastication on a roll mill. Consequently rubber-based PSA formulations manufactured without a priori calendering or roll mixing exhibit a higher viscosity and a very broad range of viscosity values (see Figure 2.5) as a function of solids content.

The viscosity of a PSA influences the contact build-up of the liquid adhesive. For easy bonding low viscosity values are required. In order to obtain a high debonding resistance (high peel and shear) high viscosity values are needed. Thus one can wonder what low and high viscosity values mean? What should be chosen, the low or the high viscosity version of the adhesive? Generally, with common formulations the elastic component of the polymer predominates, that is, high peel and high shear as exhibited by high viscosity polymers are used, and bonding (flow) of the adhesive is obtained by the use of longer dwell times. For those highly viscous polymers the value of the viscosity is given indirectly as cohesion (shear). On the other hand, removable PSAs are more plastic and possess a lower viscosity, causing stringing in many cases. Viscosities higher than 10^5 Pa·s are required in order to avoid this phenomenon [43]. Generally, viscosities of PSAs vary between 10^4 and 10^5 Pa [37].

Dependence of the Modulus on the Stress Rate. Like the viscosity the modulus of the PSAs also depends on the stress rate. As an example, Zosel [15] measured the modulus of tackified PSAs as a function of the resin concentration, at a rate above 10^{-2} sec, and demonstrated the existence of a maximum at a resin concentration of 50%. On the other hand Dahlquist working at faster rates ($t < 10^{-2}$ sec) showed the existence of a maximum at 60% resin concentration [20].

Influence of the Elasticity/Plasticity Balance on the Properties of PSAs.
If the elastic components are predominant in the rheology of the polymer, more of the bond rupture energy is stored, resulting in higher peel and tack properties [2]. Some of the energy supplied by the application of stress to a material is stored energy, in contrast to that which is dissipated in overcoming viscous drag. The two recognized means of storing mechanical energy are inertia (kinetic) and elasticity (potential) of which the latter commands the most attention as the determining factor which allows to distinguish permanent adhesives from removable ones. If the time scale of the measurement is short enough, any liquid will respond elastically to prevent relaxation of the applied stress (e.g., high speed peel).

2 RHEOLOGY OF PSA SOLUTIONS AND DISPERSIONS

In order to be coated onto the face stock material the adhesive must be in the liquid state. This state may be achieved by melting (hot-melt PSAs), dissolving (solvent-based adhesives), or dispersing (adhesive dispersions) the PSAs. Dissolving and dispersing the adhesive produce diluted adhesive systems, where flow properties are due to the carrier liquid. The rheology of diluted PSAs also depends on the carrier liquid. Generally, there are aqueous and solvent-based adhesive solutions, and aqueous or solvent-based adhesive dispersions. The most important are the organic solvent solutions of adhesives (solvent-based adhesives) and the water-based dispersions of adhesives (water-based adhesives). In fact most solvent-based adhesives like rubber resin adhesives, contain some dispersed material, and most of the dispersion-based aqueous adhesives contain some dissolved substances (e.g., thickeners, protective colloids, tackifiers, ...).

The rheology of PSAs in the liquid state (molten or diluted) is very important for the coating process. In this case one needs to be concerned with coating rheology. Wetting out, coating rheology, coating speed, and versability are the most important flow parameters influencing the convertability (coatability) of the adhesive. All of these parameters depend on the physical state of the adhesive, either as a diluted liquid system, yielding an adhesive layer by evaporation of the carrier liquid (water or solvent), or as a 100% solid system, giving a solid adhesive layer by a physical (hot-melt) or chemical [ultraviolet (UV)- or electron beam (EB)-cured] process. Therefore the coatability and the coating rheology of diluted and undiluted systems will be discussed separately.

The molten state of hot-melt PSAs differs from the coated, solid state by the value of their viscosity only. Therefore, it may be assumed that the rheological properties of hot-melt PSAs during converting do not differ from those of solid PSAs. Hot-melt flow is not an isothermal process; it is more sensitive to thermal transfer and temperature than the room temperature cold flow of the solid, coated adhesive layer. However, it is designed to possess an adequate fluidity at a given temperature (i.e., an optimized viscosity). Because of the relatively high viscosity of the fluid hot-melt PSA and the short period of the liquid state, phenomena encountered by diluted PSA systems (e.g., wetting out, migration, mechanical stability and volatility of low molecular weight components) appear less important. Diffusion problems for hot-melt PSAs are due not to the carrier liquid (and do not occur during the coating) but mainly to the low molecular weight polymers. Hence there is no need to discuss the coating rheology of hot-melt PSAs separately. Unfortunately, the nature and concentration of the carrier liquid in a diluted PSA play a decisive role in their rheology. Thus it appears necessary to discuss the rheological features of those systems. Because of the basic differences between solution adhesives and polymer dispersions, considerable differences in rheology exist and it is necessary to discuss the coating rheology of both these diluted PSA systems separately.

2.1 Rheology of PSA Solutions

The rheological behavior of polymer solutions is a result of solute/solvent concentration; it depends on the properties of the dissolved PSAs and of the solvent system. The influence of the dissolved polymers and their chemical composition and macromolecular characteristics will determine their interaction with the solvent. On the other hand, like the rheology of the solid polymer, the flow characteristics of the polymer solution also depend on time and temperature.

Coating rheology includes the flow properties of the wet and dried adhesive, leading to a wettable, stable running time-dependent liquid layer, with a smooth surface after drying under the applied shear forces. Coating rheology of solvent-based PSAs is the rheology of a true solution, thus mainly viscosity driven and more simple than that of water-based systems. The first step of the coating process is the wetout of the solid surface by the flowing adhesive. Generally, wetting out is a function of surface tension, viscosity, and density of the adhesive layer. It should be kept in mind that solvent-based adhesives are usually highly viscous systems (more than 10–15,000 mPa·s) with low surface tension solvents as carriers, and being (at least partially) directly coated on high surface energy webs. Thus, the wetout of solvent-based PSAs is less difficult and less critical than that of water-based ones. Surface tension depends on polymer and solvent nature, viscosity and density.

Synthetic raw materials for solvent-based adhesives are homogeneous, whereas natural raw materials are not. In general rubber/resin adhesives display a broader variation of the solution properties than acrylic-based ones. On the other hand solution polymerization allows better chemical control, thus some macromolecular properties of solvent-based polymers are superior to those of water-based ones (e.g., MWD, sequence distribution, stereoregularity, gel content, etc.). The use of special solvents or solvent mixtures constitutes an additional possibility for surface tension control of solvent-based adhesives. Generally solvents used for solvent-based PSAs have a low surface tension (25–30 dyn/cm), much lower than that of water (72 dyn/cm) or even of aqueous dispersions (usually 35–42 dyn/cm). The surface tension of solvent-based PSAs causes less concern than that of aqueous dispersions. Viscosity control of solvent-based PSAs can also be realized by the use of fillers.

Special Features of the Rheology of PSA Solutions

Polymer solutions are multicomponent systems, where the rheological properties of the components and their mutual interactions may play an important role. The rheology of the basic polymer and of the solvent system interface as a function of the nature and concentration of the components. In some cases the procedure for dissolving the basic polymers, respectively for eliminating the carrier solvent, also influence the final adhesive properties. These aspects will be examined in more detail in Chapter 9. In this section the most important features of the rheology of polymer solutions are summarized. The flow properties of PSA solutions are important for their coating behavior.

The coating rheology of the polymer solutions may be different from that of the polymer without the solvent carrier, due to the multicomponent (system) character, and to the quite different manufacturing and end-use conditions (stress rate and temperature) for PSA solutions.

Polymer-Dependent Characteristics. Polymer solutions are viscoelastic, with a non-Newtonian character, their behavior being determined by the nature and concentration of the components. Theoretically the most important rheological characteristics of PSA solutions will be described by a correlation depending on the macromolecular and chemical characteristics of the polymer (solute) and the chemical nature and concentration of the solvents used. The time/temperature dependence of the viscosity is also pronounced.

Solvent-based adhesives are mostly solutions of rubber/resin blends or acrylic copolymers. Their viscosity depends on the chemical composition and molecular weight of the base polymers. For rubber/resin-based PSAs viscosity control is limited to the change of the MW/gel content ratio. Because of the high dispersion of these values, the use of an additional 10–15% of solvent in these formulations is common practice. The molecular weight of solvent-based adhesives is lower than that of water-based dispersions. However these relatively low molecular weights are high enough to yield high viscosities. As shown earlier viscosity directly influences the wetting out. The wetout is a function of surface tension, viscosity, and density. Surface tension depends on the solvents used, and on the density of both polymer and solvent.

Solvent-Dependent Characteristics. Generally polymer solubility depends on solute/solvent interactions, thus the viscosity of PSA solutions depends on the nature and concentration of the solvents used. The viscosity of solvent-based PSAs depends on the molecular weight, chemical nature and solids content of the formulation. With suitable molecular weight values and a solids content of 20–25% solvent-based PSAs display viscosities of more than 10,000 mPa·s. It should be kept in mind that solvent-based PSAs differ from the water-based PSAs not only by their higher viscosity but also through the easier viscosity control. As known from the practice, it is possible to formulate solvent-based PSAs according to their viscosity (for doctoring ease) or to their solids content (for high drying speeds).

Rubber/resin PSAs mainly use hydrocarbon-based solvents, thus the choice of the solvent as an aid cannot play an important role in the control of their viscosity. Currently binary or ternary solvent mixtures allow better viscosity control. On the other hand, the rheology of the solvent-based PSAs is sensitive towards drying speed (i.e., the rate of the diffusion of the solvents).

Technology-Dependent Characteristics. The rheology of the polymer solutions may depend on the manufacturing and coating technology. Solvent-based PSAs are made by dissolving the basic polymer. Because of the sensitivity of the macromolecular compounds towards mechano-chemical destruction, the dissolving technology will influence the spread of the viscosity values. On the other hand the same PSA solution may give coatings of different quality, according to the geometry (shear) of the metering device. Another feature is given by the sensitivity of the coated adhesive layer towards drying speed (rate of diffusion of the solvents used). Mainly for rubber/resin adhesives based on hydrocarbon solvents, superficially stripped, textured surfaces may be obtained if the solvent "boils" during drying and the fluidity of the dry layer is lost.

Adherent-Dependent Characteristics. Liquid components of solvent-based PSAs may interact (physically or chemically) with the contact surface of the adherent, resulting in better adhesion.

Time/Temperature Dependence of the Viscosity of PSA Solutions

A shear-induced viscosity decrease of polymer solutions is governed by molecular orientation and disentangling. For Newtonian liquids at low shear rates it is possible to restore the polymer network in the solution through relaxation. High shear implies the decrease of the network points. Therefore, the shear modulus G is given, as a function of the network points v_c,

$$G = v_c \cdot RT \tag{2.34}$$

The dependence of the viscosity on the shear is a general phenomenon also valid for polymer (adhesive) dispersions. It should be mentioned, however, that in a manner different from adhesive dispersions, the mechanical stability of the system for polymer solutions is viscosity (or solids content) independent.

2.2 Rheology of PSA Dispersions

The coatability of liquid adhesives is a function of wetting out, coating versatility, coating rheology, and coating speed. On the other hand the coating rheology influences the wetout and the coating versatility, and depends on the coating speed. The wetting out of water-based systems constitutes a difficult problem because of the high surface tension of the carrier liquid (water). Solvent/solute interactions in "real" solutions give rise to the increase in viscosity. Water-based dispersions generally possess a lower viscosity while ready-to-use PSA dispersions possess water-like low viscosities. Their low density and the transfer coating technique used also contribute to the coating difficulties. In a different manner from solvent-based PSAs, improvement of the wetout by the use of surface active agents causes irreversible changes in the adhesive properties. Thus an adequate coating rheology appears important.

Water-based dispersions have a limited shelf life and thus the coating rheology also depends on the age of the dispersion. This one is a function of the dispersed polymer particles (macromolecular and dispersion characteristics) and of the carrier liquid as well. Only small changes in the composition of the carrier liquid are allowed. It may be assumed that the two main parameters influencing the coating rheology of the polymer solutions (polymer and solvent nature) have a reduced importance for water-based PSA dispersions. More important seem the dispersion nature and the inhomogeneous character of the liquid, dispersed adhesive.

The flow of a dispersion depends on the concentration and characteristics of the dispersed solid particles (particle size and particle size distribution). On the other hand, the inhomogeneous, anisotropic character of the adhesive layer persists after the coating (wetting out and drying) of the liquid layer. The final water content of the dried, coated adhesive (residual humidity) depends on the humidity of the environment (i.e., it reaches an equilibrium value). The dried dispersion-based PSAs will never have the rheology of the bulky adhesive (the adhesive from the polymer particles) but remain in an intermediate state between the rheology of the dispersed system and of the base polymers.

Furthermore the wetting additives left in the adhesive layer also contribute to the composite character of the PSA layer. Therefore it can be assumed that the rheology of the water-based PSA layer is that of a composite material (i.e., polymer particles surrounded by a layer of additives). From the experimental practice it is known that the rheology of water-based PSAs is shear sensitive (i.e., dynamic wetting needs a certain level of coating speed and a certain level of shear on the metering/coating device). The most important rheological parameter of a water-based PSA dispersion is its viscosity. The viscosity influences the coatability, the film-forming characteristics, and the properties of the adhesive layer thus formed.

Viscosity of PSA Dispersions

In order to be transfer coated onto the face stock (or release liner) the adhesive dispersion must wetout the solid surface. Furthermore, the coated adhesive has to permit the withdrawal of the carrier liquid (water), and the coalescence of the solid, dispersed particles (film forming). The wetout of the adhesive and the film building process are strongly influenced by the viscosity of water-based PSAs. On the other hand, the coating technology (coating machines and coating speed) is viscosity dependent too. Coating devices are very sensitive towards viscosity, thus the coating versatility (i.e., the ability of an adhesive to be used on different coating machines) depends on the viscosity. The coating speed contributes to the wetout process. Excessive speed may produce foam or adhesive build-up on the machine. Both phenomena are function of the viscosity of the adhesive. The adhesive's flow on the metering device (i.e., the viscosity) also influences the coating weight of the PSAs.

Wetout of Water-Based Adhesives

As will be shown later, the wetout is a function of the quality of the face stock and/or release liner. Water-based PSAs are coated mainly by transfer, that is, coated onto the release liner and subsequently transferred onto the face stock. The surface quality of the release liner is a function of the siliconizing technology (condensation/addition; solventless/solvent-based/water-based silicones). The fast curing rate at low temperatures (ca. 5 sec at 110°C) of the addition-based systems allows mild coating conditions for the paper (better paper/moisture equilibrium) [64]. More recently, solventless systems were developed, permitting better release properties and economical advantages. Today, they represent about 40% of the world usage in self-adhesive applications [37]. These systems are based on very low molecular weight polymers, which can only be applied to very high grade papers in terms of pore sealing and smoothness of the surface, in addition to which special (better) wetting properties of the water-based adhesives are required. Generally, suitable water-based PSAs have to display good wetting on the worst liners (i.e., on siliconized liners) and especially on solventless silicone-coated release liners.

Theoretical Basis of Wetting Behavior. When in its liquid state, the adhesive must be able to uniformly wet the web. Equilibrium spreading occurs when the surface energy of the substrate γ_{SV} is higher than the surface tension of the latex γ_{LV} and the internal surface tension between latex and substrate surface γ_{SL}, as summarized by the following formula and illustrated in Figure 2.6:

$$\gamma_{SV} = \gamma_{LV} \cos \theta + \gamma_{SL} \tag{2.35}$$

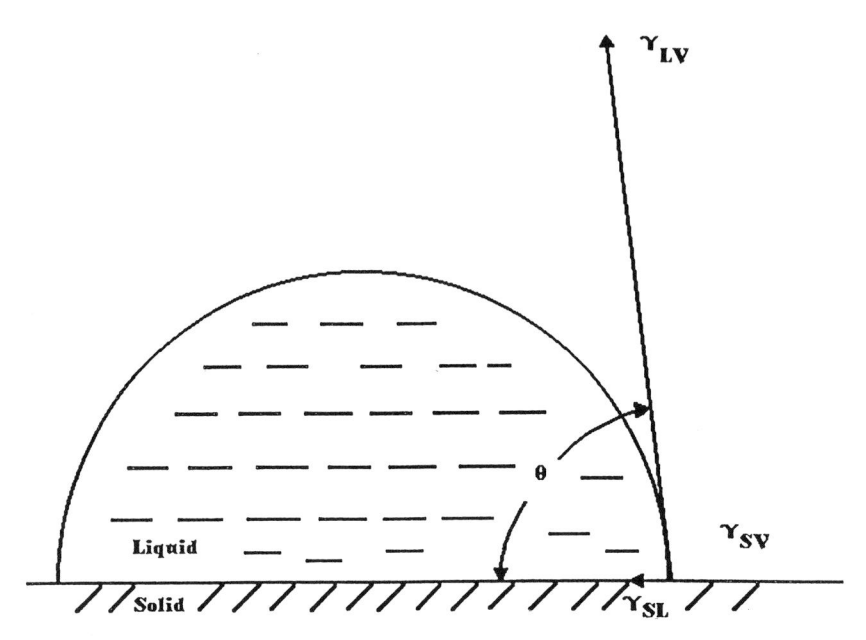

Figure 2.6 The contact angle of a high surface tension liquid on a nonpolar surface.

Thus, for a given substrate, wetting is facilitated if the surface tension of the latex or/and the interfacial surface tension is minimized. Industrial coating requires not only wetting but also the stability of the coated liquid layer. Unfortunately, this liquid layer formed by wetting the surface is not stable, it suffers dewetting. Two types of dewetting phenomena are possible, namely retraction of the latex coating from the edge and surface cracking.

Static/Dynamic Wetout. In the industrial practice the wetting failure of greatest concern is the dewetting of the adhesive from the primary web, between the gravure coating station and the drying tunnel. In the process of latex coating at high speed the dynamic aspect of the wetting/dewetting is also critical.

The real wetting/dewetting behavior of the latex cannot be approximated in a laboratory because of shearing of the dispersion on the coating (metering) roll, which improves wetting, and the competition between dewetting and drying imposed coalescence. Therefore dewetting on the coating machine is diminished by shearing of the dispersion layer and coalescence of the adhesive. Industrial wetting is always better than on a laboratory scale.

Surface tension of the latex and its contact angle influence the wetting/dewetting process [65,66]. Dewetting problems can be overcome by reducing both the surface tension and the contact angle of the latex. In order to evaluate the coatability, both parameters (surface tension and coating angle) have to be measured. Generally, the contact angle of the latex is plotted against the surface tension of the latex for different surfactant systems. The best wetout is obtained for the lowest contact angle and surface tension. Other factors affecting wetting/dewetting are the viscosity and the density of the liquid adhesive.

Surface Tension (Static/Dynamic Values). The influence of the lower surface tension values is especially noticeable, when trying to coat onto low surface energy substrates. If the surface energy of the web is too high, and the formula is not thickened, defects show up in the film (e.g., cratering or pinholes). Reducing the surface tension of the latex helps to get good wetting. On the other hand, the coated liquid layer tends to retract. In order to avoid film retraction surfactants have to migrate rapidly to the interface between the adhesive and the release liner so as to avoid dewetting. Thus for PSAs applications where the coating process constantly creates new interfaces, a low static surface tension and a low dynamic surface tension are needed. Generally the surface tension is a function of the following parameters, namely, the age of the latex [67], the polymerization yield [68], the solid content of the latex [69], the temperature [69], the viscosity [69], the particle size, and particle size distribution [77]. The influence of the viscosity on the surface tension can be determined by means of the surface amplitude decay through the Orchard [71] equation for levelling:

$$\ln \frac{A_o}{A_1} = \frac{16\pi^4}{3\lambda^4} \gamma' \cdot h_1^3 \int \frac{dt}{\eta}$$

(2.36)

where A_x denotes the surface amplitude at any given time, γ denotes effective surface tension, h is the mean film thickness, λ is the wavelength, and η is the dynamic viscosity. From the above relationship it may be concluded that the surface tension γ_s is correlated to the viscosity,

$$\gamma_s = f(\eta)$$

(2.37)

Contact Angle. A low contact angle enhances the wetting out. For the evaluation of the coatability the contact angle needs to be measured. Unfortunately measuring the contact angle faces some difficulties as the contact angle depends on the surface tension and on the smoothness of the surface; furthermore the contact angle is time dependent (i.e., there is a wetting and a dewetting angle) and the contact angle also depends on the dimension of the adhesive drop.

Influence of Viscosity on Wetting

Dewetting may be reduced by increasing the standing ability of the liquid adhesive layer. The mobility of the liquid layer is inversely proportional to its viscosity and density. This is confirmed by thickening and incorporating fillers into the water-based formulations. Only slight density differences may be obtained, hence wetting out will depend mainly on the viscosity. Highly viscous aqueous dispersions are characterized by good wetting properties, and generally the increase of the viscosity is a method to improve the wetout. Thus, one can write:

$$\text{Wetting out} = f(\text{viscosity}) \tag{2.38}$$

For aqueous solutions of surface active agents Kurzendörfer [72] formulated a special dependence for the amount of rest liquid G (i.e., liquid which does not dewet) and dynamic viscosity η:

$$G = \frac{8}{3} r \pi h^{3/2} \cdot \frac{\eta^{0.5} \delta^{0.5}}{g^{0.5}} \cdot (t_w - t_d)^{-0.5} \tag{2.39}$$

where r is the average cylinder radius (for the test equipment used for vertical immersion-wetting out tests), h is the pull-out length of the cylinder, δ is the specific weight of the liquid, t_w is the pull-out time of the cylinder, t_d is the dewetting time of the liquid, and g is the gravity constant. Thus,

$$G = C_1 \cdot \eta^{0.5} \cdot \delta^{0.5} (t_w - t_d)^{-0.5} \tag{2.40}$$

This correlation confirms the dependence of the wetout on the viscosity and density of the latex. As will be shown in Section 1.4 of Chapter 3, aqueous dispersions are characterized by a minimum film-forming temperature. This temperature depends on the viscosity of the bulk polymer in the interior of the emulsion particles.

Influence of Viscosity on Coating Versatility. Coating versatility refers to the ability of water-based PSAs to be coated by different coating techniques. In Europe two methods dominate the coating technology, namely the metering bar system and the reverse gravure. The metering bar system (Meyer bar) is used mainly for tapes. Reverse gravure is recommended for labels. For specific viscosity ranges different coating devices are used. Reverse roll coating by rotogravure needs low viscosities of 150–400 mPa·s (17–30 sec Ford Cup 4); thickening makes this interval broader. The base viscosity of the dispersion is a function of the polymerization recipe. Adjusting the viscosity of an aqueous dispersion requires knowledge of the thickening/diluting response of the dispersion. On the other hand, dispersions with different viscosities may be coated directly or by transfer. The coating technology (transfer or direct) influences the anchorage and migration of the adhesive and therefore the

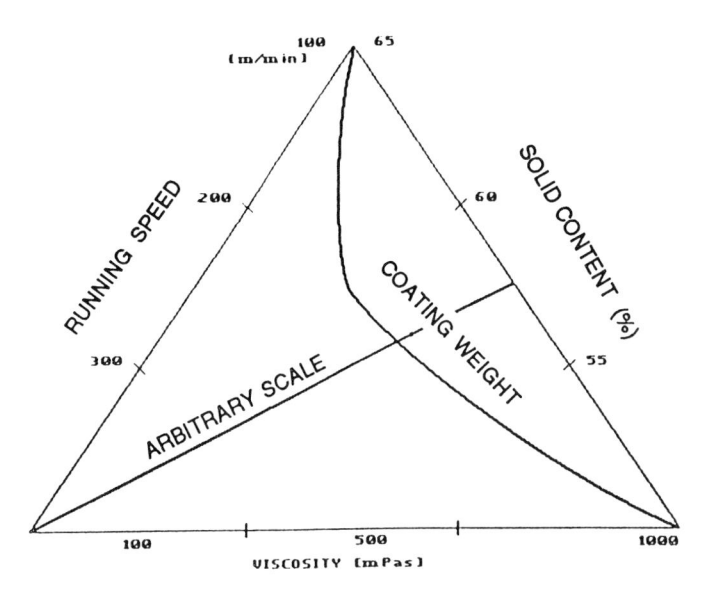

Figure 2.7 The interdependence of the coating weight, the solid content, the viscosity, and the coating speed of PSA.

end-use properties. Therefore an indirect influence of the viscosity on these properties must be taken into account.

Influence of Viscosity on Coating Weight. To keep the coating-weight tolerance within a maximum of ±5% the viscosity of the emulsion must be kept constant. For an emulsion PSA on a reverse gravure coater it is ideal to have a high solid content with a low viscosity. A low viscosity implies a high coating weight. The minimum viscosity depends on the surface tension of the adhesive too. Common viscosities for reverse gravure are situated in the range of 17–22 sec as determined with a Ford Cup 4. In practice the coating weight depends on several parameters such as the viscosity, running speed, and solids content. Viscosity is a function of the solids content; on the other hand it influences the coating speed. The interdependence of these parameters, and their influence on the coating weight is illustrated in Figure 2.7.

As can be seen from Figure 2.7 the interdependence of the coating parameters (running speed, solids content and viscosity) is not linear. The coating weight increases with the solids content, with the running speed, and with the decrease of the viscosity up to a maximum level, then decreases.

During the coating process disturbing phenomena may occur such as adhesive build-up on the machine, coagulum build-up on the machine and in the coated layer, turbulences in the coated adhesive layer, changes in the viscosity and foam formation. These phenomena depend on the rheology of the diluted system. Because of their practical importance, they will be discussed in Chapter 9.

Factors Influencing the Viscosity of Aqueous Dispersions. The following factors influence the viscosity of aqueous dispersions [72], namely: temperature; solids content; pH; dispersing agents; particle form, size, and size distribution; viscosity of the dispersing media; and shear forces.

In general the viscosity decreases with the temperature. The increase of the solids content implies an increase of the viscosity. The dependence of the viscosity on the pH is more complex, but generally it can be assumed that viscosity increases with increasing pH. The dependence of the viscosity on the dispersing agent is a function of the nature and concentration of the dispersing agent. A broad particle size distribution imparts low viscosity due to the mobility of the particles. Generally water-based dispersions contain a lot of different additives which influence the viscosity of the dispersing media. Shear forces produce in most cases a pronounced decrease of the viscosity (shear thinning). Too high a shear force may destroy the system producing coagulum. Too low an emulsion viscosity can cause wetting problems, foam problems, or adhesive bleedthrough when the emulsion is directly coated onto the face stock. The emulsion viscosity can also be controlled by either controlling the emulsion particle size, or by subsequent thickening/diluting.

3 RHEOLOGY OF THE PRESSURE-SENSITIVE LAMINATE

Pressure-sensitive adhesives are mostly used in a coated, bonded state enclosed within a laminate. Generally the adhesive acts as an intermediate layer between face stock and release liner, and finally, between face stock and substrate. Because of the relatively small thickness of the adhesive layer (Figure 2.8) its flow properties are strongly influenced by the properties of the delimiting solid surfaces. On the other hand, at least one component of the laminate (the face stock), but mostly both (face stock and release liner or substrate) are flexible, soft materials allowing no uniform distribution of the applied forces, and undergoing a pronounced deformation during converting or end-use. Thus one can suppose that the rheology of the bonded PSAs will be strongly influenced by the components, structure, and manufacture of the laminate.

Pressure-sensitive adhesives act like permanently liquid elastomeric layers enclosed between two thin, flexible, soft, but solid surfaces. The flow of the adhesive film is strongly influenced by the adhesive/surface interactions, and

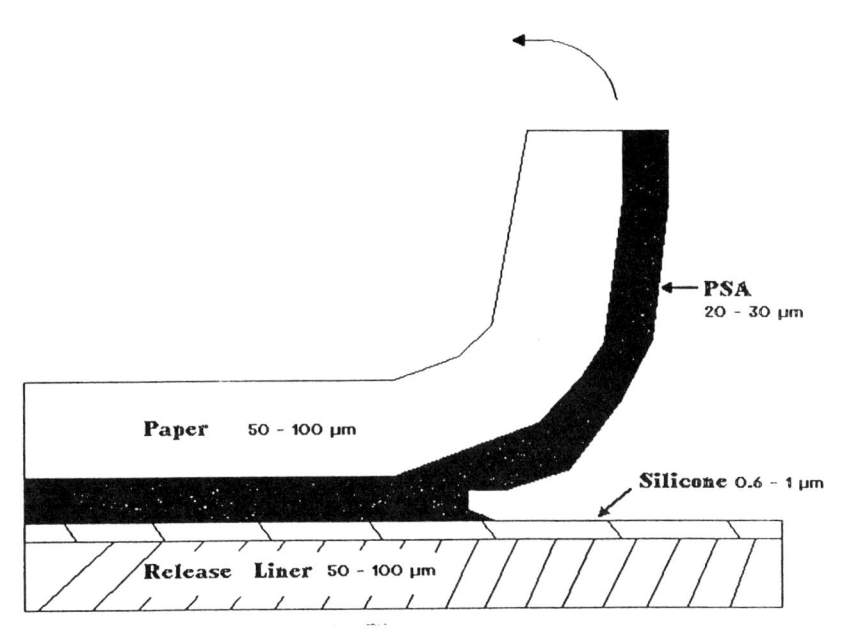

Figure 2.8 Geometry of a PSA paper laminate.

by the stress/temperature loading transmitted by and through these delimiting surfaces. The adhesive/face stock, adhesive/release liner, and adhesive/substrate interactions depend on the adhesive nature and on the characteristics of the solid components. One can conclude that the rheology of the adhesive in the laminate depends on its thickness, and the dimensions, respectively continuous/discontinuous nature of the laminate components, and on the laminate build-up. It is relatively easy to determine the rheological properties of the adhesive or of the materials used as face stock or release liner, but there is no means for direct measurement of the rheology of the pressure-sensitive laminate. Practically one can investigate the adhesive, converting and end-use properties of the coated (bonded) PSAs, and thus indirectly evaluate the rheology of the bonded PSAs. Both the uncoated adhesive and the solid components influence the rheology of the coated adhesive.

In Section 2 the most important features about the rheology of the adhesive are summarized. The influence of the solid laminate components on the rheology of the bonded adhesive will be described indirectly through the most important adhesive properties (i.e., tack, peel adhesion, and shear). Rheology depends both on the nature of a fluid and on its flow conditions; these are a function of the relative volumes of the fluid and, in the special case of a coated adhesive, on its coating weight. Therefore before examining the influence of the solid state components on the adhesive properties, the influence of the coating weight needs to be considered first.

3.1 Influence of Liquid Components of the Laminate

Earlier the rheology of uncoated PSAs was examined. The conclusions of this chapter may be taken as valid for the liquid component (adhesive) of the laminate. However, these considerations must be completed by the examination of the influence of the adhesive coating weight on the rheology of pressure-sensitive laminates.

The importance of the flow of the adhesive in achieving pressure-sensitive properties was outlined. Flow properties depend on the flow conditions, that is, fluid volume, rate, and interfaces. Pressure-sensitive adhesives are thin-layer coatings, where surface roughness and fluid volume may have a pronounced influence on the motion of the adhesive. Surface influences may be at least partially avoided by increasing the distance of the fluid in motion from the solid surface (i.e., by increasing the coating weight of the adhesive). Changes of the coating weight will influence the adhesive and converting properties of the PSAs.

Influence of Coating Weight on Adhesive Properties

The main adhesive properties of PSAs may be described as tack, peel, and shear, and each one of these properties depends in a different manner on the coating weight. As known from the labelling practice a certain coating weight is necessary in order to obtain a measurable tack. Different tack measurement methods show a different sensitivity towards the coating weight. On the other hand the nature of the solid components of the label also influences the relation tack/coating weight. The coating weight also influences peel adhesion. Generally the peel increases with the coating weight. This is a nonlinear dependence, and its relation depends on the substrate to be adhered to [74].

It was recognized in the first years of the development of PSAs that controlling the coating weight allows modifying the permanent/removable character of the adhesive. A decrease of the coating weight from 20–25 g/m^2 to 15–20 g/m^2 can yield a removable label [38]. Using less than 7 g/m^2 of dry adhesive labels removable from paper were developed [43,44]. Shear resistance tends to be inversely proportional to the coating weight. It is to be mentioned that because of the roughness of the solid components a minimum coating weight may be required in order to achieve a good shear strength.

Influence of Coating Weight on Converting Properties. Coating weight strongly influences the cuttability of the laminate. This dependence will be examined in detail in Chapter 7. Here, the influence of the coating weight on the cold flow of the adhesive will be discussed. The cold flow of the

Rheology of Pressure-Sensitive Adhesives

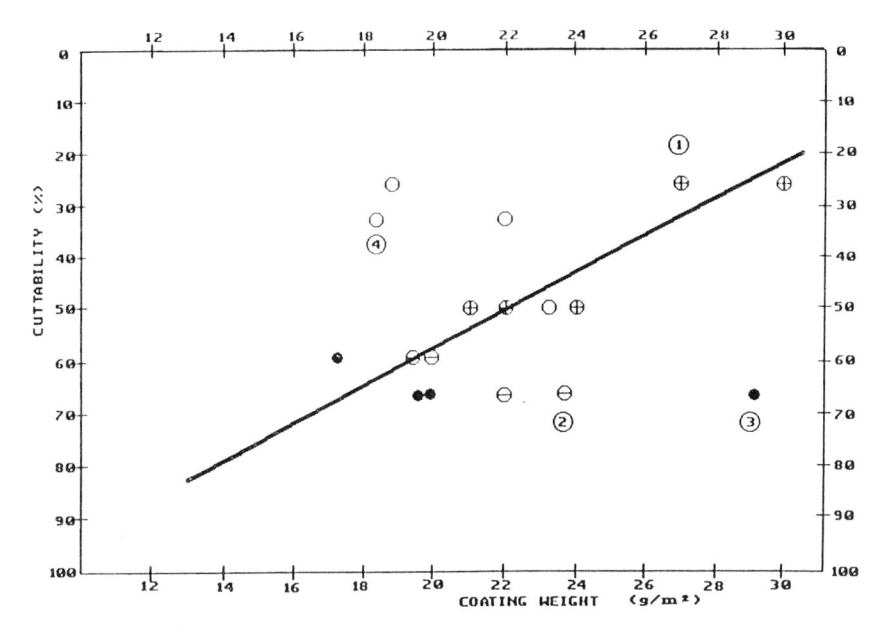

Figure 2.9 Dependence of the cuttability on the coating weight. Cuttability of soft PVC film labels: ① PSA with low PE-peel, medium tack, high cohesion; ② PSA with high PE-peel, low tack, low cohesion; ③ PSA with low peel, very low tack, high cohesion; ④ PSA with high peel, high tack, low cohesion.

adhesive in laminates is hindered by the marginal interaction of the adhesive with the "walls" of the laminate. Because of the local character of this interaction the middle layer displays free flow.

It is evident that the thickness of the free flowing layer depends on the coating weight and on the thickness of the anchored layer. The anchored layer depends on the coating technology and on the nature of the solid laminate components; the free flowing layer depends on the coating weight (CW) and coating technology:

$$\text{Cold flow} = f(\text{free flow}) = f(\text{CW, Coating Technology}) \quad (2.41)$$

Therefore, in a first approximation, the cuttability depends on the coating weight, as follows:

$$\text{Cuttability} = f\left(\frac{\text{Coating technology}}{\text{CW}}\right) \quad (2.42)$$

Figure 2.9 illustrates the dependence of the cuttability on the coating weight (i.e., the cuttability decreases with increasing coating weight).

The most important factors influencing the coating weight are the face stock material, the substrate to be adhered to, the permanent/removable character of the adhesive, and the coating conditions. Generally it may be stated that the optimum coating weight value depends on the chemical composition and end-use of the PSAs. The adhesive performance also influences the coating weight to be used. All these parameters and their influence on the coating weight will be examined in Chapter 6.

3.2 Influence of the Solid Components of the Laminate

The solid laminate components influence the distribution of the applied stresses on the adhesive (Figure 2.10) and the flow of the liquid adhesive in the laminate. The interactions between the PSAs and the delimiting surfaces act like a brake or resistance to flow.

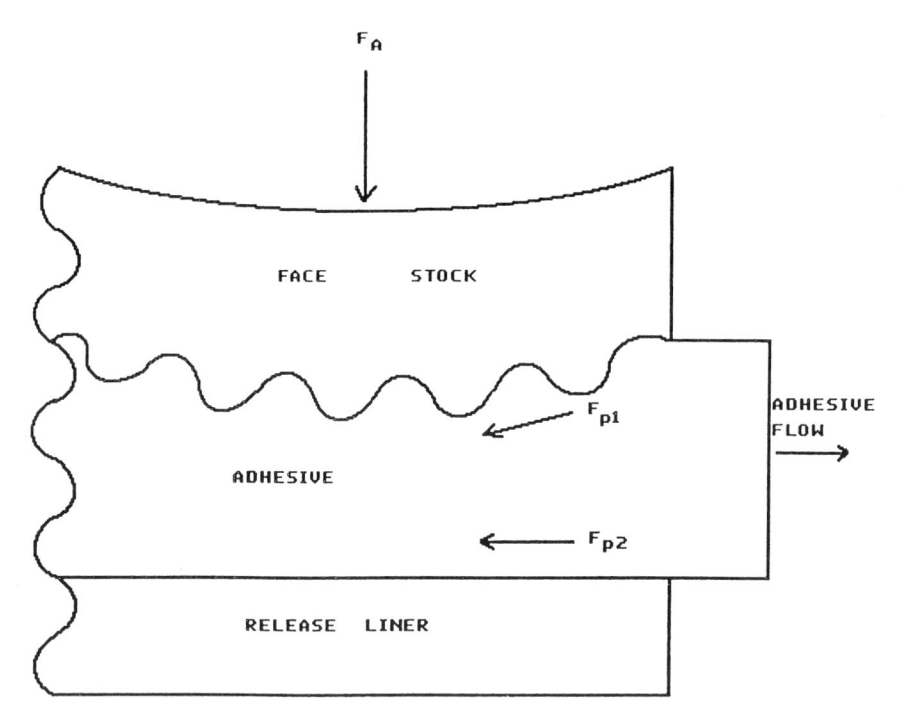

Figure 2.10 The influence of the solid state laminating components (face stock and release liner) on the active and passive forces in the adhesive's flow. F_A: Active forces, compression and impact stress on the laminate; F_{p1}: Passive forces, resistance against PSA flow, given by adhesive/face stock interaction; F_{p2}: Resistance against PSA flow, given by adhesive/release liner interaction.

Rheology of Pressure-Sensitive Adhesives

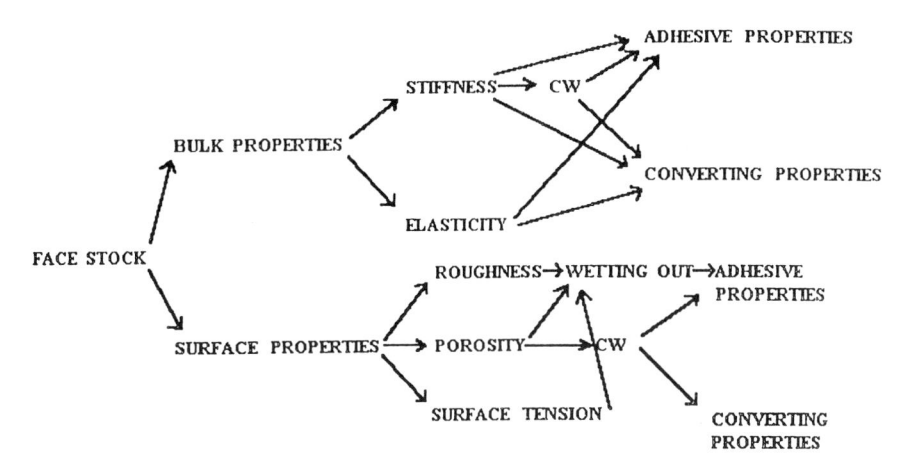

Figure 2.11 Face stock influence on the adhesive and converting properties. The influence of the stiffness, elasticity, roughness, porosity, and surface tension of the face stock directly and through coating weight on the adhesive and converting properties of the label material.

The stress distribution depends on the elasticity, hardness, and flexibility of the solid component (i.e., on its bulk properties) and dimensions. Controlling the flow is a function of the quality of the delimiting surfaces, that is, of the surface properties of the solid components. As solid component influencing the adhesive properties the adherent (i.e., the substrate) should be considered as well. In order to elucidate the influence of each one of the solid components of the PSA label (applied or not) a separate discussion of these components has to be carried out. First the role of the face stock in the design of the label will be examined in order to clarify its influence on the rheology of the adhesive. The influence of the face stock material on the rheological characteristics of the adhesive (modulus and viscosity) will be discussed while characterizing the face stock material by its own rheological properties (modulus). The best approach to describe these phenomena is their indirect examination through the adhesive and converting characteristics.

The properties of self-adhesive labelstock are primarily concerned with the printability of the face stock, the conversion characteristics of the laminate, and end-use adhesive performance. All these characteristics depend on the face stock material used. As shown in Figure 2.11, the face stock material influences the properties of the PSA label through its bulk and surface properties.

The bulk properties of the face stock material influence the adhesive and converting properties directly and indirectly. The indirect influence is exerted by the coating weight. The surface properties (roughness, porosity, and surface tension) influence the wetting out, but there exists an indirect influence of the adhesive on converting properties through the coating weight as well. The role of the porosity in the migration is well known. If the adhesive wets the adherent (face stock) surface, its performance is largely influenced by the bulk properties of the face stock material.

The enormous importance of the surface quality of the solid components of the PSA laminate on the properties of the PSA is illustrated by the recent rheological studies concerning high and low surface energy substrates. In case of low surface energy substrates the rheological approach of the properties is often not valid. The plasticity/elasticity balance of the face stock material, its stiffness and its capability to absorb energy, influence the adhesive and converting properties of the pressure-sensitive laminate.

Influence of the Face Stock on the Adhesive Properties of the Pressure-Sensitive Laminate

The dependence of the most important adhesive characteristics (peel, tack, and shear) on the bulk properties of the face stock material will be investigated. It should be pointed out that some of the tests of the adhesive properties of PSAs involve deformation of the face stock material (e.g., tack, peel); others (e.g., shear) are characterized by the deformation of the pure adhesive layer only. Thus the bulk properties of the solid components of the pressure-sensitive laminate influence more the tack and the peel of the adhesive. Peel adhesion depends on the face stock used as well as on testing conditions [6]. This strong dependence is due to the simultaneous deformation of the face stock material and adhesive during peeling. Also, some tack measurements induce face stock straining. In fact most of the tack measurements may be considered as peel, shear, or tensile strength measurements, and therefore they will be discussed in Chapter 6.

Face Stock Material and Peel Adhesion. From an energy point of view debonding is associated with energy dissipation. Energy dissipation during debonding is dependent on such parameters as interfacial adhesion, polymer rheology, rate of peel, temperature, angle of peel, web stiffness, and coating weight. Interfacial adhesion and web stiffness are characteristics of the face stock material. They also influence the angle of the peel. On the other hand, the peel rate, especially in the starting phase or onset of peeling, or during manual peel, strongly depends on the plasticity and elasticity of the face stock material (Figure 2.12).

Figure 2.12 presents the dependence of the peel on the face stock material for tackified PSAs dispersions with different tackifier levels. As can be seen from Figure 2.12, the value of the peel is quite different for a soft, plastic

Rheology of Pressure-Sensitive Adhesives 51

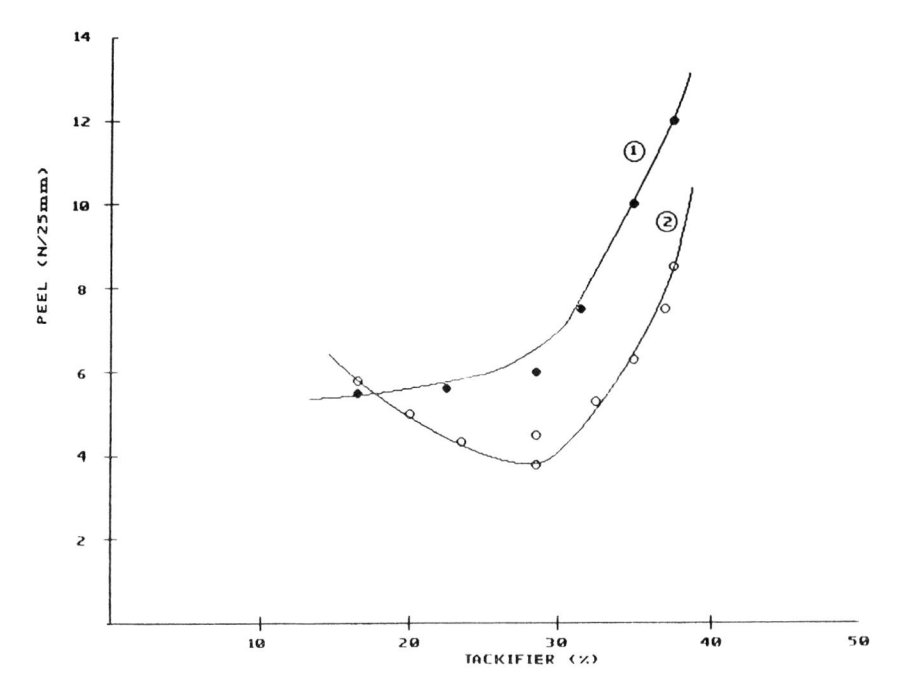

Figure 2.12 Dependence of 180° peel adhesion force on the nature of the face stock material. ① PVC; ② Aluminium.

face stock material. Adhesive formulations display quite different peel adhesion values when coated on different face stock materials.

Thus face stock flexibility and plasticity are the main bulk characteristics influencing the rheology and the peel. From release force measurements, it is well known that the peel force from a silicone release liner depends on the peel angle α_p and the peeling rate v:

$$\text{Peel force} = f(\alpha_p, v) \tag{2.43}$$

The peel angle is really a function of the flexibility f_{FS} of the face stock:

$$\alpha_p = f(\text{flexibility}) \tag{2.44}$$

Thus:

$$\text{Peel force} = f(\alpha_p, f_{FS}) \tag{2.45}$$

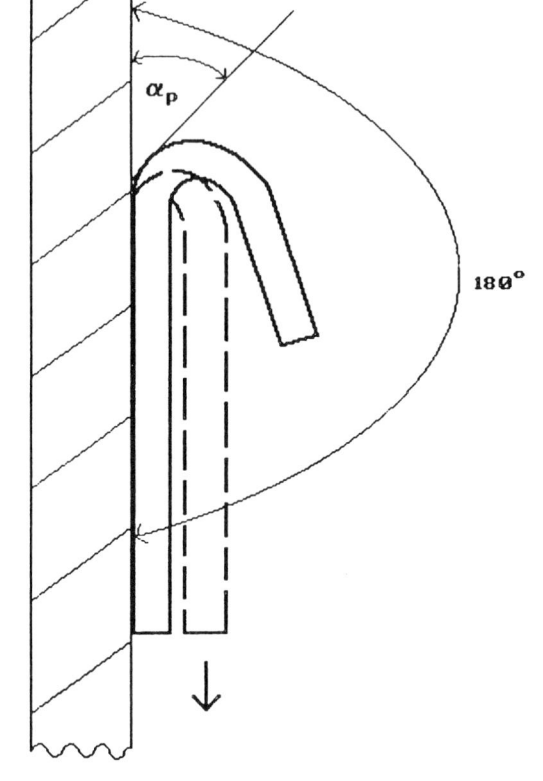

Figure 2.13 Theoretical and real peel angle during 180° peel measurement.

The peel angle is normalized (90° or 180°) but the real peel angle depends on the flexibility of the material (Figure 2.13), and its deviation from the theoretical angle is different for 90° and 180° peel tests.

As demonstrated by Wilken [75] for peel testing at 90° the delaminated length A of the peeled bond is given as a function of the modulus of the face stock E_f, the modulus of the adhesive E_a, the thickness of the adhesive coating h_a and the thickness of the face stock h_f:

$$A \sim \sqrt[4]{\frac{E_f}{E_a} \cdot h_a^2 \cdot h_f} \tag{2.46}$$

A more detailed description of the influence of the face stock elasticity (stiffness) on the peel is given in Chapter 6. Stress transfer to the adhesive,

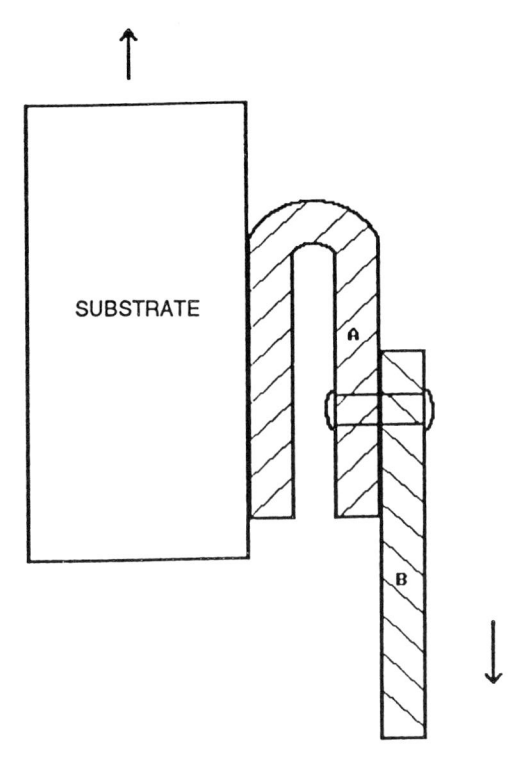

Figure 2.14 The influence of the elasticity/plasticity of the face stock on the peel. A paper label material (A) is being peeled through a 180° angle from a substrate, while the peel stress is being transmitted via different materials (B_i) with a different elasticity, and giving rise to different final peel forces: B_1: Paper (peel force: 7.4 N/25 mm); B_2: PET (7.7) B_3: Soft PVC (6.9); B_4: LLDPE (6.9).

and from the adhesive is influenced by its plasticity. The experimental data from Figure 2.14, illustrate this behavior.

Both phenomena—face stock stiffness and elasticity (resistance to elongation)—were studied by Gent and Kaang [46]. When adhesive labels (or tapes) are pulled away from a rigid substrate, the force required depends upon both the strength of adhesion and the resistance of the label to stretching [46]. The pull-off force F may be correlated with the work of detachment G_a and the effective tensile modulus E of the tape and its thickness t. The pull force is found to be neither proportional to G_a as might at first be expected, nor is it proportional to $(EG_a)^{1/2}$ as is found in many linearly elastic systems. Instead it is found to be proportional to $(EG_a^3)^{1/4}$, a result which emerges directly from the analysis as a consequence of the particular relation between the force F and the corresponding elastic displacement of the label where no further debonding occurs. It is to be noted that the values of detachment energy G_a obtained from pull-off experiments are generally lower than those obtained from peeling experiments. It was stated that:

$$F \sim \left(\frac{8G_a}{3}\right)^{3/4} (E \cdot t)^{1/4} \tag{2.47}$$

that is,

$$F = f(t^{1/4}) \tag{2.48}$$

A further prediction of the theory is that the product $F\theta$ (where θ is the detachment angle) will be independent of the stiffness of the tape (label). Differences in G_a from pull-off and from 90° peel measurements are attributed to additional energy losses (by peeling) due to the severe bending deformations imposed on the label. A possible cause is nonlinear elastic behavior of the label in tension. The effective tensile modulus E at small strains is greater than at large ones. The simplifying assumption of linearly elastic behavior is quite inadequate for film labels or for tapes which undergo large deformations as well as plastic yielding. The work to bend the label away from the substrate may be important too. In some circumstances this contribution can be both large and strongly dependent upon the magnitude of the peel angle [76].

Energy transfer from the face stock to the adhesive depends on the anchorage of the adhesive layer and on the existence of an energy absorbant layer on the face stock material. Both are surface dependent characteristics, thus the surface of the face stock material will influence the peel of PSAs (especially for removable and nonpolar adherent applications). Indirectly the anchorage of the PSAs and the existence of different adhesive layers on the surface of or in the face stock material depend on the migration of the adhesive or adhesive components in the face stock material. This phenomenon depends on the surface and bulk properties of the face stock material.

Influence of the Substrate. The nature of the substrate influences the peel adhesion, and thus the removability of the PSA label. Certain PSAs may be classified (depending upon the substrate) as removable or permanent adhesives. The most important parameters of the substrate that influence the removability of the PSA label are its flexibility, its elasticity/plasticity balance, and the anchorage of the adhesive onto the face stock. The peel does not depend on the deformability of the adhesive, but on the flexibility of the face stock and substrate [75].

The deformability of the adherents influences the stress distribution in the bond [35]. If the adherents are not deformable the adhesive would be stressed uniformly. If the solid components of the joint are deformable the stress distribution varies along the joint. Mathematically this phenomenon may be described by a factor n, where n is the ratio between maximal and mean stress (σ_{max}, σ):

$$n = \frac{\sigma_{max}}{\sigma} = \sqrt{\frac{\Delta}{2}} \coth \sqrt{\frac{\Delta}{2}} \qquad (2.49)$$

where Δ is given by the following correlation:

$$\Delta = \frac{G \cdot l^2}{E \cdot a \cdot b} \qquad (2.50)$$

where a is the thickness of the adherent, b is the thickness of the adhesive, G is the modulus of the adhesive, E is the modulus of the adherent, and l is the overlap length. The influence of the deformability of the adherents on the stress distribution by peeling has also been demonstrated [77]. The adhesive is under tension at the interface and under compression just behind it. The tensile stress is a function of the stiffness of the combined adherents and adhesive, as well as of the thickness of the system. The compressive load appears greatest at 180° peel between flexible and rigid adherents, less at 90° peel under the same circumstances, and the peel force is the lowest when measured between two flexible adherents.

Face Stock Material and Shear Strength. There are no experimental data available about the influence of the bulk properties of the face stock material on the shear resistance. Differences in shear strength of PSAs coated on different face stock materials are due to the surface of these materials.

Influence of the Face Stock on the Converting Properties of the Pressure-Sensitive Laminate

The most important converting properties of pressure-sensitive labels are the printability and the cuttability. The rheology of the adhesive strongly influences the cuttability of the label. Next the dependence of the cuttability on the bulk properties of the face stock material will be analyzed, and in particular the influence of the rheological characteristics of the face stock material. During the die-cutting operation, the mechanical strength of the face stock and release liner materials influence the distribution of the stresses from the cutting edge to the adhesive layer. Generally, this mechanical resistance is a function of the mechanical properties of the bulk material, and of the dimensions of the face stock and release liner:

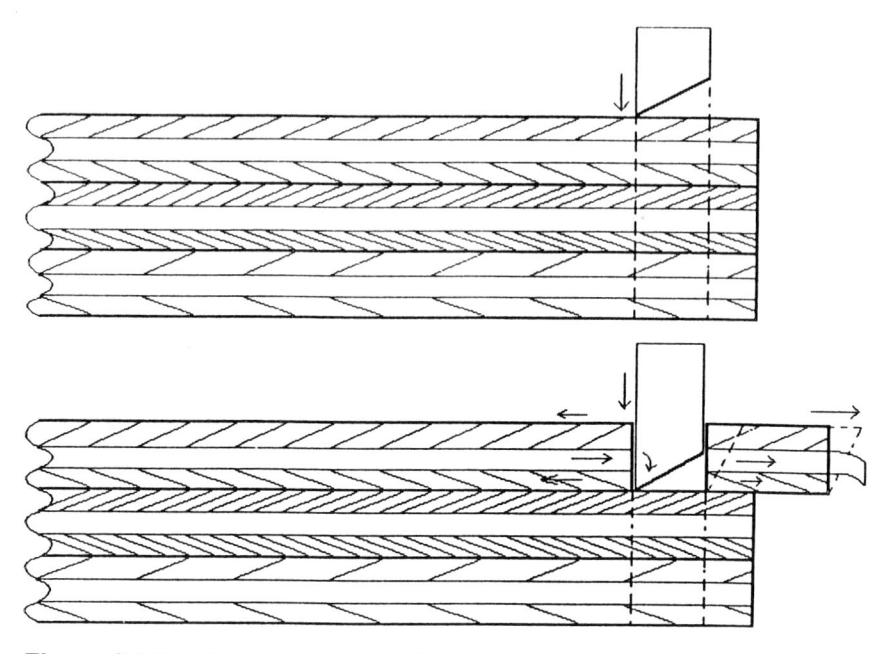

Figure 2.15 The motion of the laminate layers during cutting.

$$R_c = f(K_1, K_2) \tag{2.51}$$

where R_C is the mechanical resistance of the face stock material, K_1 denotes material characteristics and K_2 material dimensions. In some cases (sheet material) a certain number of superposed laminate sheets are cut during the same operation; thus the distribution of the force may be influenced by laminate slip (Figure 2.15). Therefore the laminate strength will depend on the motion of the laminate layers (face stock and release liner, respectively), with respect to one another (i.e., on the surface characteristics of the laminate: face stock, release liner) K_3:

$$R_c = f(K_1, K_2, K_3) \tag{2.52}$$

where K_3 is a surface-dependent factor influencing the laminate slip. As can be seen from Figure 2.16, the motion and cold flow of the adhesive will be a function of stress magnitude and distribution, but smearing of the cutting knife and of the cut (side, front) surface also depends on the porosity of this surface.

Thus the cutting behavior and the cuttability C of the laminate will depend on the mechanical properties of the bulk material (face stock, release liner) K_1, on the surface properties of these materials K_3, and on the dimensions

Rheology of Pressure-Sensitive Adhesives

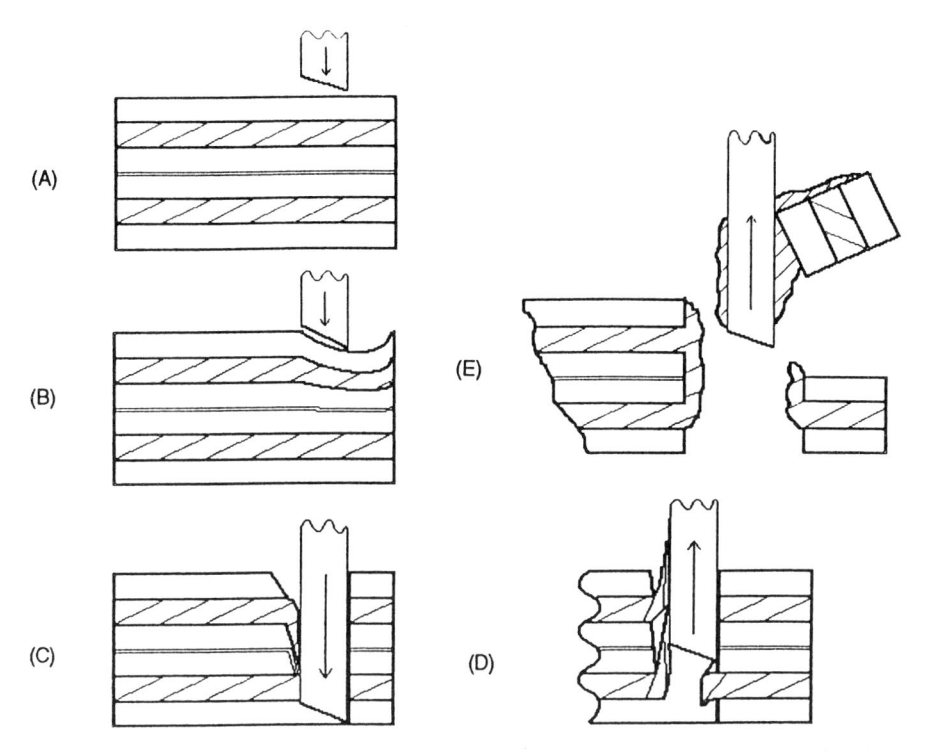

Figure 2.16 Flow and smearing of the adhesive during cutting. (A) through (E) are the different steps during the cutting operation.

(influencing the real value of the forces and the cut area) of the solid components of the laminate K_2:

$$C = f(K_1, K_2, K_3) \tag{2.53}$$

In order to identify the most important mechanical properties of the bulk material which may influence the cuttability, the nature of the forces and deformation during cutting have to be examined. Different kinds of stresses are present in a material as demonstrated, by creasing and folding paper board (Figure 2.17), [78,79].

One can conclude that the same stresses (i.e., tensile, compression, and shear) are working when cutting a pressure-sensitive laminate. In fact creasing the face stock constitutes the first operation during cutting. Material characteristics related to tensile, compression, or shear resistance of the face stock material will influence the cuttability. The distribution of the stresses, the

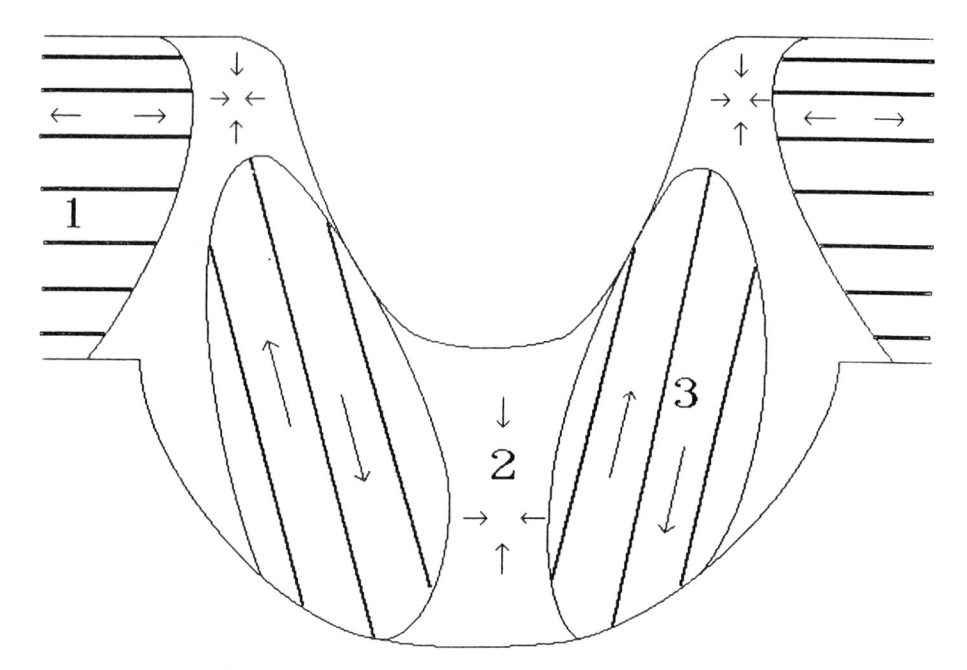

Figure 2.17 Stress distribution during creasing (cutting) of the face stock (label) material. 1) tensile, 2) compression, 3) shear stresses.

uniformity of the force acting on the laminate, the compression stress character in the first cutting step are determined by the rigidity of the face material (Figure 2.18).

The rigidity (stiffness) of the material is a function of its modulus. An incompressible liquid PSA may support a compressive stress well. In the first step of the cutting operation the force is acting perpendicularly on the whole laminate. This situation changes only through the concentration of the local stresses, and because of the weakness of the solid material, giving rise to the flexure of the face and shearing of the adhesive. Hence, the stiffness of the solid laminate components influences the stress distribution and the creep resistance of the adhesive. The value of the modulus of elasticity E of the face stock material (or/and release liner) influence the cuttability of the material. Thus:

$$C = f(E) \tag{2.54}$$

On the other hand, assuming that the cutting edge (knife) is acting on a partially supported material, flexure of the material occurs first (i.e., the

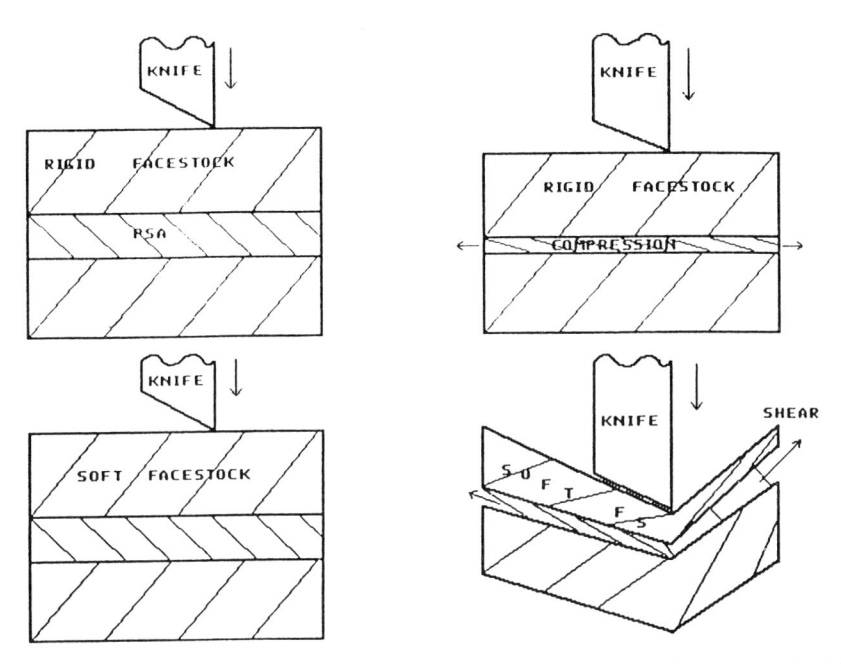

Figure 2.18 Stress distribution during cutting as a function of the nature (rigidity) of the solid components of the laminate.

flexural modulus of the material influences the cuttability). In a first approximation, the cuttability C depends on tensile, compression, flexural and shear modulus (E_t, E_c, E_f, E_{sh}) of the material and the value of these moduli may or may not differ from one another, as a function of the materials used (e.g., paper, plastic, or metal) and of the practical conditions:

$$C = f(E_i) \tag{2.55}$$

Cuttability also depends on the material dimensions. It should be mentioned that the stress distribution dependence on the thickness of the face stock material may vary according to the nature of the applied stress. It may be stated that cuttability improves with material thickness, that is:

$$C = f(E_i, h)^\alpha \tag{2.56}$$

where h is the material thickness and α is an exponent taking into account the improvement of the cuttability by the increased cut surface for a voluminous (thick) laminate component. For hygroscopic materials like paper, bulk properties may depend on the environmental humidity. Because of the fiber-like structure of the material, the surface texture may also change as a function of the humidity. Surface texture influences the smoothness (friction) of the surface and smoothness influences the cuttability. The surface friction of papers increases with increasing air humidity [80]. On the other hand, because the hygroscopic nature of the face paper, there is a continuous transfer of water between the adhesive and face material, influencing the composite character of the adhesive and its rheology (water-based adhesives). A more detailed analysis of the parameters influencing the cuttability will be presented in Chapter 7. Here the influence of the bulk properties of the face stock material on the distribution of the stresses acting on the laminate, and of the rheology of PSAs on the laminate will be discussed.

Influence of Surface Properties of the Face Stock Material

Figure 2.11 showed that the surface properties of the face stock material influence the adhesive and converting properties of the pressure-sensitive laminate, either directly or indirectly. The direct influence is based on the anchorage phenomenon at the surface of the face stock. Indirect influences are reflected by the porosity related adhesive penetration in the face stock and the changes in the coating weight caused by penetration.

Influence of Face Stock on Migration. It is difficult to store labelstock for any length of time, due to possible bleeding and/or oozing problems. A lot of problems exist with high gloss label material bleedthrough. Adhesive bleedthrough often interferes with the printing. Migration (penetration or bleeding) is due to the diffusion of the adhesive into and through the face stock material. This phenomenon depends on the properties of the adhesive, of the face stock material, on the coating technology and on the processing (storage, converting) conditions:

Migration = f (adhesive, face stock, coating technology, processing) (2.57)

Migration of the adhesive in the face stock material or of the additives from the face stock material in the adhesive produces changes in the composition of both materials and in their mechanical/rheological properties.

In many health care and other applications it is desirable to have a PSA film coated onto a porous web. Hot-melt PSAs offer a relatively easy route to coat PSAs onto fabric and nonwoven webs with a minimum of adhesive penetration into the web. The penetration of the adhesive in the porous web can be avoided with a release liner used as an intermediate web surface (transfer coating) for some processes, but during storage of the product diffusion in the porous web may occur. Evidently, migration (penetration) of the adhesive into the face stock depends on the adhesive's nature, but also on the face stock porosity. The dimensions of the pores strongly influence the penetration. It was demonstrated that pore diameters of 0.01–0.05 μ allow good penetration [81]. Theoretically, penetration under pressure and capillary-penetration in paper are different phenomena [82]. Penetration through pressure P_p would depend on the pressure p, the radius of the pores r, and the viscosity η:

$$P_p = f(p, r, \eta) \tag{2.58}$$

Capillary penetration CP (mainly water penetration) depends on the humidity of the paper H and its pH value:

$$CP = f(H, \text{pH}) \tag{2.59}$$

Porosity influences positively the anchorage of the adhesive through migration. Where pores, crevices, and capillaries are accessible in the surface of the adherent the adhesive will penetrate to some extent, and so increase adhesion. The capillary rise h is a function of the porosity [83]:

$$h = \frac{k \gamma_{LV} \cos \theta}{\xi R} \tag{2.60}$$

where R is the equivalent radius of the capillary, k is a constant, and ξ is the density of the liquid; h reaches a maximum when:

$$\gamma_{LV} = \frac{1}{2}\left(\gamma_c + \frac{1}{b}\right) \tag{2.61}$$

where γ_c is the surface tension of the liquid. One can conclude that the dimensions (radius) of the pores, the wettability of the material (surface tension), the humidity of the paper, and the pH of the paper are major factors influencing the migration. This is also confirmed by the following correlation [84]:

$$\frac{dL}{dt} = \frac{\gamma \cdot \cos \theta \cdot r}{4\eta \cdot L} \tag{2.62}$$

where dL/dt denotes the penetration rate, η is the viscosity of the liquid, r is the capillary radius, γ is the surface tension of the liquid, θ is the wetting angle of the capillary wall, and L is the penetration length. Generally, the penetration rate of the adhesive in the face stock depends on the capillary radius (i.e., on the paper density) [84].

Influence of Face Stock Porosity on Coating Weight. It is well know in the field of cold-seal adhesives that blocking is a function of the adhesive coating weight and of the face material. For a porous web (paper) a coating weight of about 6 g/m^2 is recommended; nonporous face stock (film or aluminium) need no more than 3 g/m^2. Similarly the dry coating weight (the weight of the adhesive applied per unit surface area) can vary substantially depending upon the porosity and irregularity of the face and of the substrate surface to which the face material is to be adhered. For instance higher PSA loadings are preferred for adhering porous, irregular ceramic tiles to porous surfaces, while lower adhesive loadings are required to manufacture tapes, films, and other articles from relatively nonporous, smooth surface materials such as synthetic polymer films and sheets. When the adhesive is applied to nonporous polymeric or metallic face materials, intended for adhesion to nonporous polymeric or metallic surfaces, adhesive coating weights of about 2 to about 20 kg of dry adhesive per 300 m^2 of treated surface are generally adequate [85]. A good adhesion for tapes manufactured from continuous sheet polymeric substrates can usually be achieved with dry coating adhesive weights of about 5–10 kg of adhesive per 300 m^2, while coating weights of about 10–20 kg of adhesive are usually employed for the manufacture of paper-backed tapes, such as masking tapes. In order to avoid a decrease of the coating weight due to the porosity of the face stock material or to prevent plasticizer migration, it is also possible to apply a barrier coating to the paper [86].

Influence of Face Stock on Adhesive Choice. Face stock sensitivity towards PSAs requires an adequate choice of the adhesive system. Shrinkage of soft PVC face stock requires the use of shrinkage resistant adhesive systems [87]. To this end crosslinked acrylics are generally employed. On the other hand, plasticizers in the vinyl film can have a pronounced effect on the adhesive properties during the aging process [88]; plasticizer-resistant adhesives have to be selected. These examples show that the face stock nature and its sensitivity towards certain adhesives may influence the choice of the adhesive and thus the rheology of the adhesive layer.

Influence of Face Stock on Coating Technology. In a first instance the mechanical and chemical characteristics of the face stock material determine the choice of the coating technology, the coating process (direct coating/transfer coating), use of cold/hot adhesives (solvent-based/hot-melt) or chemically agressive or inert adhesive systems (solvent-based/water-based), and coating machines (metering device, drying system, web transport system, etc.). The influence of the face stock material on the choice of the coating technology is illustrated in Figure 2.19.

As can be seen in Figure 2.19 the web is usually supported when making paper-based laminates via transfer coating, using a gravure roll as metering device, and solvent-based, water-based or hot-melt PSAs. For film coating a Meyer bar and/or gravure roll is suggested, usually using water-based PSAs.

Rheology of Pressure-Sensitive Adhesives

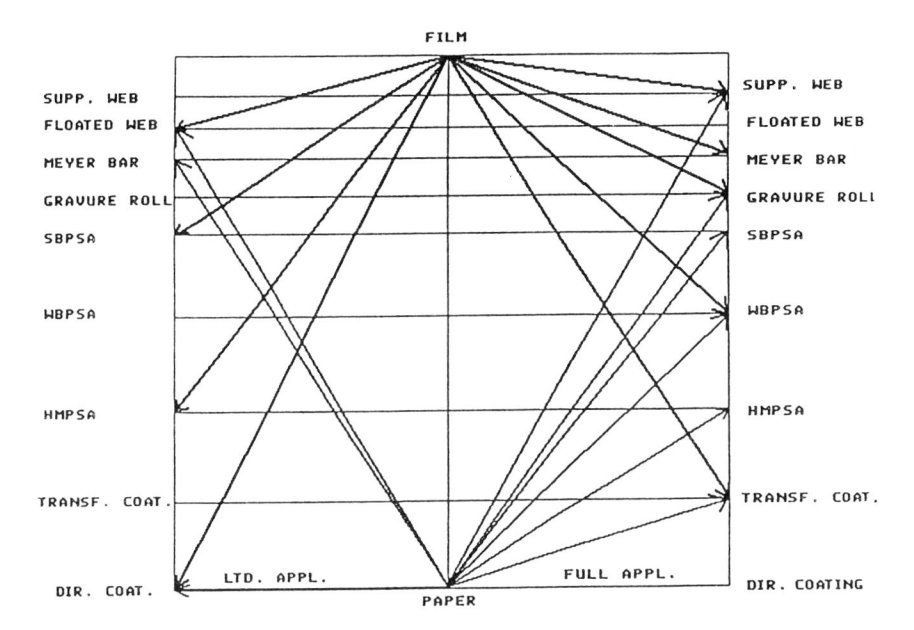

```
                                FILM
SUPP. WEB                                              SUPP. WEB
FLOATED WEB                                            FLOATED WEB
MEYER BAR                                              MEYER BAR
GRAVURE ROLL                                           GRAVURE ROLL
SBPSA                                                  SBPSA

WBPSA                                                  WBPSA

HMPSA                                                  HMPSA

TRANSF. COAT.                                          TRANSF. COAT.

DIR. COAT.     LTD. APPL.         FULL APPL.           DIR. COATING
                           PAPER
```

Figure 2.19 The influence of the face stock material on the coating technology. Dependence of the coating technique (direct/transfer), of the nature of the PSA (water-based, solvent-based, hot-melt) of the coating device and drying technology used, on the nature of the face stock material (paper/film).

Bonding (or anchorage) supposes the coating of a substrate (or face stock) with the liquid adhesive, that is, the spreading and wetout of the adhesive on the surface of the solid material. After wetting the surface the adhesive penetrates in the cavities of the solid material and physical and chemical contact points are formed. Both phenomena, bonding and debonding, depend on the quality of the surface.

Wetting Out as a Function of Face Stock Surface. Earlier the importance of the wetting out, and the parameters influencing it were discussed, considering in particular the liquid component (adhesive). Next the characteristics of the solid components influencing the wetout are examined. It was shown earlier that the bulk properties of the face stock material on the one side, and the surface properties on the other side are determinant for the properties of the pressure-sensitive laminate. The surface quality influences the wetting out of the adhesive. Wetting the solid surface by the adhesive is an initial necessary condition for the anchorage of this adhesive to the solid surface (e.g., face stock material) [36], and is a function of the surface tension

Table 2.2 Surface Energy Values of Common Face Stock/Substrate Materials

Material	Surface energy, dyne/cm	Reference
Poly(ethylene)	29.0–31.0	[35]
Poly(propylene)	28.0–30.0	[35]
Poly(vinyl chloride)	35.0–41.0	[35], [107]
Poly(vinylidene chloride)	40.0	[35]
Poly(styrene)	33.0–43.0	[35]
Polyamide (PA-6)	42.0–48.0	[65], [35]
Poly(methyl methacrylate)	33.0–44.0	[65]
Poly(ethylene terephtalate)	37.0–43.0	[65], [107]
Poly(tetrafluorethylene)	18.5	[35]
Stainless steel	45.0	[65]
Chrome plated steel	60.0	[65]
Glass	47.0	[65]
Aluminium	45.0	[65]

of the solid and liquid. Liquids, and for that matter solids, at the boundary between two phases (liquid/paper or liquid/solid) possess properties which are different from those within the mass of the liquid itself. This difference results from an inbalance between the attractive faces of the surface layer of molecules and the molecules contained within the liquid (or solid) [89]. The molecules on the surface layer compensate for charge imbalance by pulling themselves close together, thus creating the phenomenon of surface tension. In the most extreme case when the surface tension of the adhesive differs greatly from the initial surface tension of the film or paper (face stock or liner) fish eyes and cratering will result (when the face stock surface tension is less than the adhesive surface tension).

To obtain optimal wetout the surface tension of the adhesive must be equal to or less than the critical surface tension of the face stock or liner (a contact angle of 0°) (Figure 2.6). As can be seen in Figure 2.6, a high surface tension liquid beads up on nonpolar, low energy surfaces [65]. Table 2.2 lists the surface energies of materials commonly used as face stock and substrate.

Situations can also arise where the adhesive is absorbed on the surface of the face stock (e.g., film) by chemisorption, with a zero contact angle (good wetout). However, the mass of the liquid does not bond to this chemisorbed layer [89].

Except for hot-melt PSAs, PSAs are coated onto the face stock (or release liner) as liquids. The wetting out of liquid PSAs depends on the quality of the face stock surface, on the surface tension and roughness of the surface. The dependence of the wetting out on the roughness is a general phenomenon. The wetout time on cardboard (for an aqueous dispersion) may vary between 0.1–0.4 sec, depending on the roughness of the surface [82]. Moreover, Bikermann [90] stated that adhesion is due to the inherent roughness of all surfaces. Molecular forces of attraction cause an adhesive to wet and spread on the surface. But once achieved the mechanical coupling between the adhesive and the inherently rough surface is more than enough to account for a bond strength.

Influence of Face Stock on the Anchorage of PSAs

Interfacial adhesion is a prerequisite for bonding and debonding. Stable bonds are formed when there is a strong interaction between the adhesive and the face stock (or substrate). The different types of possible interactions include Van der Waals forces, polar interactions, electrostatic interactions, and chemical bonding. Effective bonding between the face material and the adhesive is critical to the performance of PSAs; given an adhesive with internal integrity, the adhesive properties exhibited by that system will be proportional to the affinity of the adhesive for the face stock material [42]. A useful approach for evaluating the compatibility of the adhesive and face stock is to measure the free energy of these surfaces. Theoreticians have formulated adhesion models based on the thermodynamic or reversible work of adhesion W_A. This is the change of free energy when materials are brought into contact, which represents the amount of energy released under reversible or equilibrium conditions to disrupt the interface. The expression is given by the following equation:

$$W_A = \gamma_{SV} \cdot \gamma_{LV} - \gamma_{SL} \tag{2.63}$$

In practical terms the face stock material and the adhesive should have comparable surface energy properties.

Solvent-based PSAs consist of a polymer dissolved in a solvent. The solvent-based PSA adhesion or anchorage on plastic films essentially concern the study of how polymers and solvents interact. In order to explain the phenomena which occur during the coating of solvent-based adhesives the nature of this interaction between polymers and solvents is to be examined. Solvent molecules are attracted to other solvent molecules. Similarly, polymer chains possess attractive forces within a chain, and from one chain to another. A polymer dissolves in a solvent, when the net effects of the solvent/solvent and polymer/polymer attractive forces are reduced. Hildebrandt [91] defined a solubility parameter δ as a measure of the internal attractive forces of the solvent, defined as heat of evaporation E_V per unit molecular volume V_M; δ is a measure of the energy density of the solvents:

$$\delta^2 = \frac{E_V}{V_M} \tag{2.64}$$

Hansen et al. [92,93] proposed an alternative to the single valued solubility parameter, namely:

$$\delta^2 = \delta_d^2 + \delta_p^2 + \delta_h^2 \tag{2.65}$$

where δ_d is the attractive molecular force from temporary dipole formations, δ_p is the attractive force from dipole/dipole interactions, and δ_h is the attractive force from hydrogen bonding. This concept allowed for the weighted effects of the highly polar and hydrogen bonding solvents. Polymers, unlike solvents, have a range of acceptable solubility parameters due to the effects of molecular weight distribution, areas of crystallinity, different monomer units on the same chains, etc. Hansen [92,93] developed a method for rating the solubility parameter ranges for polymers using the three-component parameter system. One can then define the solubility parameter range of a polymer in terms of a three-dimensional volume. Then any solvent or solvent blend with a solubility parameter that falls within the described volume will dissolve the polymer. For our purposes, an adhesive/solvent blend that falls within the solubility parameter area of a plastic film will develop good adhesion or anchorage. For the sake of simplicity, one can make a good approximation of the adhesive/substrate or face material interaction using a two-dimensional graph.

Face Stock Surface and Removability. The nature of the face stock influences the peel and thus the removability of a PSA label. Some PSAs may be classified (depending upon the face stock) as removable or permanent adhesive. The most important parameters of the face stock influencing the removability of the PSAs label are its flexibility and the anchorage (adhesion) of the adhesive to the face stock. The peel does not depend on the deformability of the adhesive only, but also on the flexibility of the face stock [75]. The influence of the bulk properties of the face stock on the peel, and the removability will be examined in Chapter 6. The influence of the surface properties of the face stock on the removability of PSAs will be discussed next.

An adequate anchorage of the PSAs imparts good removability. Some special adhesives display built-in release from untreated polypropylene. It means that the anchorage on treated polypropylene is higher than on the untreated one, and thus no need for release coating exists when coated on the corona-treated side of polypropylene tape [91]. The main criteria for removability are tack, peel, cohesion, and anchorage [43,60]. Energy dissipation during debonding depends on the interfacial adhesion [65]. Interfacial adhesion is a function of the surface quality. In the case of removable adhesives high rate peel forces must be dampened by high energy dissipation (i.e., good anchorage of the adhesive on the face material is needed). Adhesion to paper and stainless steel is greater than the anchorage to polyester film; therefore the bond fails due to a breakdown within the adhesive layer when paper is used; when polyester film is used it peels away leaving the adhesive layer virtually intact on the steel plate. On the other hand, peel characteristics change on plasticized PVC [95]. A thermodynamic approach considers adhesion between two materials as the result of two different interfacial interactions, namely a dispersive (I_d) and a polar (I_p) one [96]. The reversible adhesive work W_{ph} is described as the sum of these two types of physical interactions:

$$W_{ph} = I_d + I_p \tag{2.66}$$

In addition to these physical interactions there exist some chemical interactions as well. Both physical and chemical interactions depend on the quality, polarity, and chemical affinity of the face stock (or adherent) material. In general a peel energy measurement yields a greater value than the corresponding reversible work. On the other hand [96] peel energy depends both on macroscopic factors of energy dissipation $f(R)$ and on the molecular factors of energy dissipation $g(Mc)$:

$$P_F = W \cdot g(Mc) \cdot f(R) \tag{2.67}$$

where P_F is the peel energy and W is the reversible work of adhesion. Regarding the role of the macroscopic factors, Boutillier [96] showed the dependence of the dissipation energy (peel) on the polymer thickness. Polymer thickness implies coating weight, and as shown earlier, surface porosity influences the coating weight. These results refer to hot-melt PSAs, but one can admit their general validity for the PSA domain.

The influence of the anchorage on the peel is illustrated by the use of polyester (PET) as standard face stock material for PSA tests. Paper differs according to the manufacturing conditions and storage conditions. Therefore a dimensionally-stable material (e.g., PET) is used as standard face stock material in laboratory tests. Unfortunately, it is difficult to wet [97] and anchorage onto PET remains poor. Thus it was observed in many cases [98] that if the adhesive to be tested was applied on a polyester film, the film peels away, leaving the adhesive layer initially intact on the steel plate; however, if paper is used the bond fails due to breakdown within the adhesive layer (adhesive transfer). Generally, peel tests using PET face stock give lower

Table 2.3 Dependence of the Peel on the Face Stock Material

Coating weight, g/m^2	180° peel adhesion on PMMA at room temperature, N/25 mm	
	White	Transparent
2.0	0.89	0.62
2.5	1.04	0.87
3.0	1.19	1.02
3.5	1.34	1.17
4.0	1.50	1.32
4.5	1.65	1.47
5.0	1.81	1.62

Face stock: 70-μm, blown, low density polyethylene film, white and transparent.
Adhesive : Crosslinked water-based acrylic PSA.

values than the same tests using paper. The data from Table 2.3 illustrates the influence of the face stock material on the peel (removability).

For removable PSAs the adhesive must adhere better to the face stock material than to the substrate. This adherence or anchorage may generally be improved using primers [99–101]. The influence of the substrate surface quality on the peel was studied indirectly as the separation energy dependence on the surface tension of the substrate [15].

Converting Characteristics of the Pressure-Sensitive Laminate. It is evident that the surface properties of the face stock material influence its converting characteristics, as well as the converting characteristics of the pressure-sensitive laminate. Convertability of the face stock material means in a first approximation its coatability. Convertability of the pressure-sensitive laminate include its cuttability and printing ability.

The distribution of forces during the cutting operation depends on the slip of the laminate layers. The surface quality and smoothness of the face stock surface influence the cuttability (see Equation 2.52):

$$C = F\left(\frac{1}{\text{smoothness}}\right) \tag{2.68}$$

On the other hand surface smoothness influences the anchorage of the adhesive (i.e., its flow between the "walls" of the laminate). Cuttability increases with the anchorage of the adhesive and decreases with the smoothness of the face stock:

$$C = F\left(\frac{1}{S^a}\right) \tag{2.69}$$

where S denotes smoothness and a is an exponent taking into account the dependence of the cuttability on external slip surfaces (between various laminates) and internal flow surfaces (enclosed within a laminate).

Face Stock Geometry (Label Thickness) and Pressure-Sensitive Properties. From the labeling practice it is apparent that the label geometry influences the adhesive properties of the label. Generally it is difficult to find suitable PSAs for labels to be applied around sharp corners [102]. The label thickness influences the peel properties, whereas peel adhesion depends on the face thickness [103]. The influence of the thickness of the face stock material is given by the change in flexibility. The flexural resistance depends on the width b and thickness h of the sample (moment of inertia). Flexural resistance influences the peel angle. In practice however, the peel angle is different from the one theoretically supposed (90° or 180°C) [111]. The face stock (label) form (width) also influences the peel force [104], and peel adhesion increases linearly with the label width. It must be mentioned that the global geometry of the pressure-sensitive laminate must be taken into account for convertability tests (i.e., the whole thickness of the adhesive/solid component sandwich). Similar to the formula given by Whitsitt [105] for corrugated board construction, where the stiffness S is a function of the elastic modulus of the liners, t is the liner thickness, and H is the combined board caliper, one can write:

$$S = E \cdot \frac{tH^2}{2} \tag{2.70}$$

and thus the global laminate thickness influences the stiffness related properties (peel, tack, cuttability) of pressure-sensitive laminates.

Thickness changes of 4–5% for paper may give ± 15% changes of the flexural resistance [106]. There exists an exponential dependence between thickness and the flexural resistance (exponent = 2.5). The flexural resistance is not a material characteristic, but a result of the construction of the laminate. The effect of the face stock thickness on the peel was studied [32]; peel adhesion increased up to a limit with the thickness of the face stock (film). Especially at higher peeling rates the peel value increases with the thickness of the face stock film used [32].

3.3 Influence of the Composite Structure

The structure of the PSA label influences the rheology of the bonded adhesive. Generally the continuous/discontinuous character of the PSAs, the symmetry of the build-up, and the multilayer structure of the PSA sandwich determine the adhesive and rheological properties of the bonded PSA. As shown earlier the pull-off force of a label depends on the elastic modulus E, width w, and thickness t of the label (Equation 2.48). Experimental values of F were plotted against $N^{1/4}$ where N is the number of layers of tape (label) applied on top of another and pulled away together. The effective label thickness was proportional to N. In practice if the tape (or film label) face material stretches so that the detachment angle becomes increasingly large, then two or more film layers should be applied [46].

REFERENCES

1. W. Hoffmann, *Kautschuk, Gummi, Kunststoffe*, (9) 777 (1985).
2. S. W. Medina and F. W. Distefano, *Adhesives Age*, (2) 18 (1989).
3. A. Zosel, *Colloid Polymer Sci.*, **263**, 541 (1985).
4. *Adhesives Age*, (9) 37 (1988).
5. A. S. Lodge, *Elastic Liquids*, Academic Press, London and New York, 1964, p. 72.
6. D. Satas, *Adhesives Age*, (8) 28 (1988).
7. S. G. Chu, Viscoelastic Properties of PSA, in *Handbook of Pressure Sensitive Technology* (D. Satas, Ed.), Van Nostrand-Reinhold Co., New York, 1988.
8. J. Ferry, *Viscoelastic Properties of Polymers*, 2nd Edition, J. Wiley and Sons, New York, 1970.
9. A. W. Bamborough, 16th Munich Adhesive and Finishing Seminar, 1991, p. 96.
10. *Adhesives Age*, (11) 40 (1988).
11. J. P. Keally and R. E. Zenk (Minnesota Mining and Manuf. Co., USA), Canad. Pat., 1224.678/10.07.82 (USP. 399350).
12. A. S. Rivlin, *Paint Technology*, (9) 215 (1944).
13. M. Sheriff, R. W. Knibbs and P. G. Langley, *J. Appl. Polymer Sci.*, **17**, 3423 (1973).
14. P. J. Counsell and R. S. Whitehouse, in *Development in Adhesives* (W. C. Wake, Ed.), Vol. 1., Applied Science Publishers, London 1977, p. 99.
15. A. Zosel, *Adhäsion*, (3) 17 (1966).
16. W. Retting, *Kolloid Zeitschrift*, (210) 54 (1966).
17. A. Midgley, *Adhesives Age*, (9) 17 (1986).
18. K. MacElrath, *Coating*, (7) 236 (1989).
19. J. Class and S. G. Chu, *Org. Coat. Appl. Polymer Sci., Proceed.*, **48**, 126 (1989).
20. C. A. Dahlquist, Tack, in *Adhesion Fundamentals and Practice*, McLaurin and Sons Ltd., London, 1966.
21. C. A. Dahlquist, in *Handbook of Pressure Sensitive Adhesive Technology* (D. Satas, Ed.), Van Nostrand-Reinhold Co., New York, 1988, p. 82.

22. J. Kendall, F. Foley and S. G. Chu, *Adhesives Age*, (9) 26 (1986).
23. G. R. Hamed and C. H. Hsieh, *J. Polymer Physics*, **21**, 1415 (1983).
24. C. M. Chum, M. C. Ling and R. R. Vargas, (Avery Int. Co., USA), EP 1225792/ 18.08.87.
25. S. E. Krampe, L. C. Moore (Minnesota Mining and Manuf. Co., USA), EP 0202831A2/26.11.86, p. 20.
26. T. Matsumoto, K. Nakmae and J. Chosoake, *J. Adhesion Soc. of Japan*, (11) 5 (1975).
27. *Die Herstellung von Haftklebstoffen*, T1.-1; 2-14d; Dec. 1979, BASF, Ludwigshafen.
28. A. Gent and P. Vondracek, *J. Appl. Polymer Sci.*, **27**, 4357 (1982).
29. G. Fuller and G. J. Lake, *Kautschuk, Gummi, Kunststoffe*, (11) 1088 (1987).
30. M. Gerace, *Adhesives Age*, (8) 85 (1983).
31. G. Hombergsmeier, *Papier und Kunststoffverarb.*, (11) 31 (1985).
32. J. Sehgal, *Fundamental and Practical Aspects of Adhesive Testing*, Adhesives 85; Conference Papers, Sept. 10-12, 1985, Atlanta.
33. G. Meinel, *Papier und Kunststoffverarb.*, (10) 26 (1985).
34. *Adhäsion*, (10) 307 (1968).
35. R. Köhler, *Adhäsion*, (3) 66 (1972).
36. H. Merkel, *Adhäsion*, (5) 23 (1982).
37. R. Köhler, *Adhäsion*, (3) 90 (1970).
38. R. Mudge, "Ethylene-Vinylacetate based, waterbased PSA", in *TECH 12., Advances in Pressure Sensitive Tape Technology*, Technical Seminar Proceedings, Itasca, IL, May, 1989.
39. T. H. Haddock (Johnson & Johnson, USA), EP 0130080 B1/02. 01. 85.
40. *Kautschuk, Gummi, Kunststoffe*, (8) 758 (1987).
41. *Adhesives Age*, (12) 35 (1987).
42. M. A. Johnson, *Radiation Curing*, (8) 10 (1980).
43. EP 3344863.
44. *Die Herstellung von Haftklebstoffen*, T1.2.2; 15d, Nov. 1979, BASF, Ludwigshafen.
45. H. Müller, J. Türk and W. Druschke, BASF, Ludwigshafen, EP 0118726/02. 02. 84.
46. A. Gent and S. Kaang, *J. Appl. Polymer Sci.*, **32**, 4689 (1986).
47. *Coating*, (5) 11 (1971).
48. *Coating*, (3) 65 (1974).
49. *Kautschuk, Gummi, Kunststoffe*, (6) 556 (1977).
50. *Adhesives Age*, (10) 24 (1977).
51. J. B. Bonnet, M. J. Wang, E. Papirer and A. Vidal, *Kautschuk, Gummi, Kunststoffe*, (6) 510 (1986).
52. P. A. Mancinelli, *New Development in Acrylic HMPSA Technology*, p. 165.
53. *Adhäsion*, (9) 352 (1966).
54. D. Hadjistamov, *Farbe und Lack*, (1) 15 (1980).
55. *Kautschuk, Gummi, Kunststoffe*, (6) 545 (1987).
56. N. Hirai and H. Eyring, *J. Polymer Sci.*, **37**, 51 (1959).

57. A. K. Doolittle, *J. Appl. Physics*, **23**, 236 (1958).
58. J. D. Ferry and E. R. Fitzgerald, *Proc. 2nd International Rheology*, Butterworth, London, 1953, p. 140.
59. T. Timm, *Kautschuk, Gummi, Kunststoffe*, (1) 15 (1986).
60. H. A. Schneider and H. J. Cantow, *Polymer Bulletin*, (9) 361 (1983).
61. *Kautschuk, Gummi, Kunststoffe*, (9) 31 (1985).
62. A. N. Gent and A. J. Kinloch, *J. Polymer Sci.*, **A-2** (9) 659 (1971).
63. *Kautschuk, Gummi, Kunststoffe*, (8) 34 (1987).
64. A. Fau and A. Soldat, Silicone addition cure emulsions for paper release coating, in *TECH 12, Advances in Pressure Sensitive Tape Technology; Technical Seminar Proc.*, Itasca, IL, 3-5 May, 1989, p. 7.
65. A. C. Makati, *Tappi J.*, (6) 147 (1988).
66. D. W. Mahoney and J. W. Hagan, *PSTC Proceedings*, Pressure Sensitive Tape Council, Glenview, IL, 1982, p. 82.
67. *Coating*, (2) 39 (1970).
68. *Adhäsion*, (10) 273 (1975).
69. J. Hausmann, *Adhäsion*, (4) 21 (1985).
70. H. Hadert, *Coating*, (2) 35 (1985).
71. S. E. Orchard, *J. Oil, Lab. Chem. Assoc.*, **44**, 618 (1961).
72. C. P. Kurzendörfer, T. Altenschöpfer and H. J. Völker, *Henkel Referate*, (20) 33 (1984).
73. *Adhäsion*, (8) 29 (1982).
74. Polysar, *Product and Properties Index*, Arnhem, 02/1985.
75. J. Wilken, *apr*, (42) 1490 (1986).
76. A. N. Gent and G. R. Hamed, *J. Appl. Polymer Sci.*, **21**, 2817 (1977).
77. *Adhesives Age*, (6) 32 (1986).
78. J. D. Hine, *Verpackungsrundschau*, Technisch wissenschaftliche Beilage, 9-15 (1964).
79. H. Grosmann, *Verpackungsrundschau*, (4) 1799 (1986).
80. *apr*, (42) 1152 (1988).
81. *Adhäsion*, (2) 43 (1977).
82. E. Brada, *Papier und Kunststoffverarbeitung*, (5) 170 (1986).
83. J. Skeist, *Handbook of Adhesives*, 2nd Edition, Van Nostrand-Rheinhold Co., New York, 1977.
84. J. Wilken, *Papiertechnische Rundschau*, (4) 4043 (1988).
85. EP 0244997
86. E. Park, *Paper Technology*, (8) 14 (1988).
87. R. Lowman, *Finat News*, (3) 5 (1987).
88. R. A. Lombardi, *Paper, Film and Foil Converter*, (3) 76 (1988).
89. J. Pennace, *Screen Printing*, (7) 65 (1988).
90. J. J. Bikermann, *J. Colloid Sci.*, (2) 174 (1947).
91. J. Hildebrandt, J. M. Prausovitz and R. L. Scotty, *Regular and Related Solutions*, Van Nostrand-Rheinhold Co., New York (1970).
92. C. M. Hansen, *J. Paint Technol.*, (33) 104 (1967).
93. C. M. Hansen and K. Skaarup, *J. Paint. Technol.*, (39) 511 (1967).

94. Rohm & Haas, *Pressure Sensitive Adhesives*, Prospect (1986).
95. *Adhesives Age*, (3) 36 (1986).
96. H.K. Müller and W.G. Knauss, The Fracture Energy and Some Mechanical Properties of a Polyurethane Elastomer, *Trans. Soc. Rheol.*, 15 (1971) pp. 217-233.
97. J. Lin, W. Wen and B. Sun, *Adhäsion*, (12) 21 (1985).
98. British Petrol, *Hyvis*, Prospect (1985).
99. E. Djagarowa, W. Rainow and W. L. Dimitrow, *Plaste u. Kaut.*, (2) 100 (1970); in *Adhäsion*, (12) 363 (1970).
100. *Adhäsion*, (1/2) 27 (1987).
101. *Coating*, (4) 123 (1988).
102. Cham Tenero, 2nd International Cham-Tenero Meeting for the Pressure Sensitive Materials Industry, 1990, Locarno.
103. A. W. Aubrey, G. N. Welding and T. Wong, *J. Appl. Chem.*, (10) 2193 (1969).
104. J. R. Wilken, *apr*, (5) 122 (1986).
105. W. J. Whitsitt, *Tappi J.*, (12) 163 (1988).
106. G. Renz, *apr*, (24-25) 960 (1986).
107. Solvay & Cie, *Beschichtung von Kunststoffolien mit IXAN, WA*, Prospect Br. 1002d-B-0,3-0979.

3
Physical Basis for the Viscoelastic Behavior of Pressure-Sensitive Adhesives

It was shown earlier [1] that PSAs are required to bond rapidly to a variety of substrates under conditions of low contact pressure and short contact time. This characteristic feature of a PSA is called tack [1]. During bond formation contact on a molecular dimension between the adhesive and the adherent is established, in isolated points of the geometrical contact area, the number and size of which increase with the contact time through deformation and flow, as well as by wetting.

Tack is a characteristic of amorphous high polymers above the glass transition temperature (T_g) only. In terms of physical properties the T_g represents the temperature range through which the polymer changes from a hard, glassy state into a liquid, rubber-like state. In general a polymer with low viscosity (low T_g) will be able to wet a substrate surface and establish intimate contact with the adherent. An increase in T_g will lead to a stiffer polymer, decreased wettability and most likely a decline in adhesive properties. The adhesive polymer must however possess sufficient cohesive strength to resist further flow and subsequent deformation. The proper flow of the adhesive ensures quick coverage of the substrate through wetting, but a solid-like response to applied stresses. Efficient wetting implies a high deformability (i.e., a low modulus), whereas a solid-like response to stress means low deformability (i.e., a high modulus). Thus one can postulate:

$$\text{Wetting out} = f\left(\frac{1}{E}\right) \tag{3.1}$$

$$\text{End–use} = f(E) \tag{3.2}$$

It can be concluded that the viscoelastic flow properties of PSAs materials are characterized by the value of the elastic modulus E and the T_g.

Investigations [1] of the viscoelastic properties of many commercial labels and tapes showed that T_g and E at the application temperature are the most important features for a good PSA performance. The T_g may be measured from the peak of the tan δ curve (see Chapter 2). Several methods are used to determine the glass transition of a polymer material; the most commonly used ones are mechanical/rheological, calorimetric, dilatometric, and dielectric methods. The calorimeter method has gained wide acceptance; the most common instrument used is the differential scanning calorimeter (DSC).

1 THE ROLE OF THE T_g IN CHARACTERIZING PSAs

Adhesive bonding is only possible if the fluidity of the materials involved can ensure enough contact areas, in which chain segments would diffuse into each other. In general, the T_g is one of the most important parameters in order to determine the minimum usage temperature for polymeric materials; it is in fact the temperature below which large scale molecular motion does not occur. Below this temperature polymers become glassy and brittle, and develop a high modulus; above this temperature the polymer behaves like a more or less rubber-like elastic fluid. Its tensile strength and modulus decrease above the T_g (i.e., the solid begins to liquify) [2].

Polymers for use in adhesives must act like a continuous, tacky, and elastic layer. A continuous adhesive layer is formed by coalescence; tack is the result of a bonding (viscous flow, wetting out, elastical deformation) and debonding (deformation and failure) process, where elasticity and viscoelasticity refer to energy storage and dissipation. All these requirements imply fluidity of the adhesive at operating temperatures above the T_g.

On the other hand the internal strength of the adhesive (i.e., the cohesion) must be high. The break energy of the polymers depends on their molecular motion mechanism [3]. This molecular motion is a function of the T_g and of the molecular weight (MW). Thus, the T_g and molecular weight values are critical for the use of a polymer as an adhesive. This statement was confirmed by Midgley [4] and Foley et al. [1]; they demonstrated that the T_g, molecular weight, and molecular weight distribution (MWD) are the basic properties to characterize elastomer latexes used in adhesives. It will be shown later that the T_g is a function of molecular weight as well as of other chemical and macromolecular characteristics. The importance of the T_g as a characteristic of the molecular motion will be described in a detailed manner in Section 1.2. Also, it should be stressed that it is not the aim of the book to provide a detailed discussion of the chemical or macromolecular basis of PSAs, but rather an understanding of the adhesive process; there are some fundamental correlations between the physical basis and characteristics of PSAs which need to be discussed. First the value of the T_g, the factors influencing the T_g, and the adjustment of the T_g will be reviewed.

1.1 Values of T_g for Adhesives

Generally the desired value of the T_g for coatings (film-like polymers) depends on their application and end-use temperature, quality criteria and working mechanism. Paints and varnishes possess a T_g range of +3 to +16°C—this is a relatively high T_g range—that yield hard polymer surfaces [1]. Similarly dispersion-based offset printing inks show a minimum film-forming temperature (MFT) of +5 to +10°C (i.e., also a relatively high T_g range).

It should be mentioned that the required T_g of coatings also depends on the face surface. Coatings on soft face materials need a lower T_g. As an example, acrylic binders for absorbent monomers have a T_g below 0°C [5]. Adhesive coatings have to be softer than lacquer coatings or printing inks (their T_g is lower also), but it depends on the substrate in a similar manner. Contact adhesives display a T_g range of –10 to 0°C. These adhesives need fluidity only for film and bond forming. They must not display fluidity in the bonded state. Pressure-sensitive adhesives have to be liquid at room temperature and within the bond, thus they should have a T_g of about –15 to –5°C or less. In fact the T_g of some PSA base polymers is much lower (Tables 3.1 and 3.2).

The T_g of PSAs is generally the result of a formulation of a low-T_g base elastomer and high-T_g additives (see Chapter 8). As an example, the current T_g range of rubber/resin PSAs (–15 to –5°C) is a combination of the low

Table 3.1 Glass Transition Temperature for Different Adhesive Types

Adhesive type	T_g (°C)	References
Press (contact) adhesives	–5 to +15	[10]
Low energy curing adhesives	–40 to +30	[10]
Hot melt adhesives	+5 to +100	[10]
Wet adhesives	–40 to +15	[10]
Acrylic PSA	< –20	[49], [10]
Acrylic PSA	–40	[11]
Rubber/resin	–55	[4]
PSA generally	50–70°C below the use temperature	[49]

Table 3.2 Glass Transition Temperature Values for Base Materials for PSA Labels

Polymer	T_g (°C)	Applications	References
Polybutadiene	−85	[25]	Rubber/resin PSA
Polyethylhexylacrylate	−70	[25], [83]	Acrylic PSA
	−62	[10], [84]	—
Poly-n-triethylacrylate	−54	[25]	Acrylic PSA
	−46	[84]	—
Poly-i-butylacrylate	−21	[10], [84]	Acrylic PSA
Polyethylene	−30	[25]	Face stock or liner
Polyethylacrylate	−22	[25]	Acrylic PSA
	−14	[84]	—
Polymethylacrylate	+16	[84]	Acrylic PSA
	−10	[85]	—
Polyvinylacetate	+29	[25]	Acrylic PSA, EVA
Polystyrene	+110, +100	[25]	Hot-melt, face stock
Polymethylmethacrylate	+105	[28], [85]	Acrylic PSA
Polymethacrylic acid	+185	[25]	Acrylic PSA, thickener
Polyacrylic acid	+106	[25]	Acrylic PSA, thickener
Polyacrylnitryle	+100	[85]	Acrylic PSA
Polyvinylchloride	+85	[85]	Face stock
Poly SIS	−28	[86]	EB and HM on Kraton D-1320X basis
Poly [EVA-DIOM]	−25, −35	[86]	EVAc PSA
SBR latex	−17 (tan δ) 46% ST	[86]	WB PSA
	−18 44%	[86]	
	−46 25%	[1]	

Table 3.3 Changes in the Tg Due to Tackification—Aqueous Dispersions

	Components	
PSA concentration	Tackifier resin dispersion concentration, parts (w/w)	T_g (°C)
100	0	−54.7
85	15	−53.7
70	30	−38.7
55	45	−26.4
60	40	−17.1
60	40	−22.8
50	50	−27.8

T_g of the rubber—less than −20°C (Table 3.3)—and the high T_g of the tackifier resin (more than +10°C) as shown in Table 3.4.

In this case the loading level of both components is almost the same, or of the same order of magnitude. Another example shows that the relatively low amounts of high T_g additives used in the synthesis of low T_g base elastomers may have a decisive influence on the T_g of the product. Plasticizer free vinyl acetate homopolymer (VAc) dispersions display a T_g of +30°C, and build a hard, brittle film at room temperature. This is due (at least partially) to the protective colloid used. As known, in this case a polyvinyl alcohol (PVA) with a T_g of +70°C is used as protective colloid. The high value of the T_g of PVA has a negative influence in other cases of formulation of adhesive polymers too; PVA used as a thickener lowers the tack of PSA formulations. The value of the T_g for common PSAs have changed as a function of the raw material development. Some years ago, a typical PSA had

Table 3.4 Glass Transition Temperature and Ring and Ball Softening Temperature for some Tackifier Resins

Resin type	T_g (°C)	R&B softening temperature (°C)
Regalrez 1065	17	65
Regalrez 1078	23	76
Regalrez 1094	33	94

a T_g of about –40°C, actually common acrylic PSAs possess a T_g range of –40 to –60°C (see Table 3.1); the best performing styrene-butadiene-rubber latexes used for PSAs are those with a T_g of –35 to –60°C [6].

1.2 Factors Influencing T_g

The T_g is a function of chain mobility and flexibility. Chain mobility depends on internal interactions, the internal forces between polymer chains and segments. Intermolecular forces are a function of the nature and dimension of the polymer chains (i.e., of chemical composition, structure and molecular weight). The T_g depends mainly on the chemical composition and structure of the adhesive and the molecular weight of the base polymer. The chemical composition and structure of the polymers may be influenced by synthesis and formulation. As a general principle one can say that reduction of the intermolecular forces improves chain mobility and therefore lowers the T_g.

Dependence of T_g on Molecular Weight

Concerning the composition dependence of the T_g of compatible polymer blends, the parameters of an extended Gordon-Taylor equation are not only polymer specific, but also molecular weight dependent [7]. Chain mobility decreases with increasing molecular weight. The Fox-Flory equation relates T_g to the number average molecular weight M_n [8]:

$$T_g = T_g^\infty - \frac{C}{M_n} \tag{3.3}$$

where C denotes a constant. This dependence is not a linear function; the T_g dependence on molecular weight is more pronounced in the low molecular range. On the other hand PSA formulations are mainly mixtures of high molecular base elastomers and lower molecular additives. In some cases the base elastomer itself behaves like a mixture of two macromolecular compounds with different MW, due to a special sequence distribution. Therefore, it is more appropriate to discuss the dependence of the T_g on the molecular weight for the base elastomer and the lower molecular formulating additives separately.

Molecular Weight of the Base Elastomers. Because of the significant influence of the molecular weight on the T_g, a relatively low degree of polymerization DP is recommended for polymers used as adhesives (e.g., for polyvinyl acetate, the degree of polymerization amounts to several thousands); for PSAs the molecular weight should be lower. The polymer molecular weight has a significant effect on the balance of pressure-sensitive adhesive properties [1]. As discussed by Satas [9] shear is roughly proportional to molecular weight up to relatively high molecular weights at which the shear resistance drops off dramatically in some polymers. Tack is typically high at very low molecular weights and decreases gradually as molecular weight is increased. Peel adhesion typically exhibits a discontinuous behavior, increasing with molecular weight up to moderate molecular weight levels and than gradually decreasing as the molecular weight further increases. It is relatively simple to obtain low or standard molecular weight values for linear, uncrosslinked polymers. It is more difficult to limit the value of molecular weight for partially crosslinked natural or synthetic rubbers used as PSAs. As an example, for PSAs based on carboxylated styrene-butadiene (CSBR) latexes, the weight percentage of the polymer having a molecular weight of about 300,000 or more is employed as a relative measure of the effective molecular weight [10]. The molecular weight for PSA base polymers is situated in the range of 1000–2000 (M_n) [11]. As shown earlier [12] typically the molecular weight of solvent-based PSA polymers is limited to about 100,000, with a solids content ranging between 12 and 25%. At this molecular weight level the cohesive strength remains marginal and crosslinking appears necessary for better performance. High molecular weight polymers offer high shear and good thermal stability. An acrylic acid copolymer with a molecular weight of 70,000–80,000 is proposed for better wet adhesion properties [13]. Polyisoprene used for PSAs has a molecular weight of 4×10^3–5×10^6. Whereas the molecular weight of solvent-based acrylics is about 100,000, water-based ones may have molecular weight values as high as 1,000,000 [10].

Molecular Weight of Tackifier Resins. Tackifying resins used in common adhesive formulations possess a molecular weight ranging from 300 to 3000 where the T_g strongly depends on the molecular weight [14,15].

Dependence of T_g on Chemical Composition and Structure

The bonding/debonding characteristics of PSAs depend on their fluidity, that is, on the internal mobility of the polymer chains. Internal mobility is a function of the internal forces acting on a sterical (volume and structure) or polarity basis. Thus, in a first instance, polymer mobility may be regulated by changing the polymer (chain) composition through the choice of sterical/ polarity characteristics of the monomers. These changes are reflected by the value of the T_g. First the dependence of the T_g on the chemical composition will be reviewed.

Flexible Main Chain. The proper choice of the base monomers and/or those acting via polar forces permits adjusting the T_g. A detailed discussion of the interdependence of the base monomers will be given in Chapter 5. The influence of the base monomers on the T_g will be discussed first, going from general problems (polymers) to special ones (PSA).

Theoretically, the unpolar, low-volume chain of polyethylene (PE) should display a high mobility, and thus a low T_g. In fact PE has a medium to low T_g (- 30°C) but a very high tendency to crystallize. Therefore, it does not fulfill the criteria for tack, namely to be amorphous as referred to earlier. In order to avoid crystallinity and ordering of polymer chain segments on an ethylene basis, sterical hinder appears necessary, namely, that voluminous (branched) monomer units have to be inserted between ethylene units or inversely ethylene units should be inserted as "plasticizer" units between other vinyl, acryl, or other monomer units.

This is illustrated (for common plastics) by the plasticizing effect of ethylene in polypropylene [16]. In this case the T_g decreases from 262°K to 214–222°K. A similar, plasticizing effect of ethylene may be observed for common, nonpressure-sensitive adhesives [17]. The same T_g decreasing effect for polyvinyl acetate is produced by the addition of 32% butyl acrylate; adding 18% ethylene decreases the T_g to 0°C [18]. The T_g of polyvinyl acetate (PVAc) dispersions usually amounts to +30°C. Dibutyl maleinate dispersions have a T_g of –5 to +10°C, and vinyl acetate-ethylene dispersions have a T_g range of –10 to +5°C [19]. In the synthesis of PSAs cases exist where an adequate lowering of the T_g may be obtained by the use of ethylene only. Vinyl acetate dioctyl maleate copolymers with a dioctyl maleate content as high as 50% do not possess enough tack and peel [20]. The use of a monomer ratio of 15–30 wt% of ethylene with 35–50% 2–ethyl hexyl maleate, dioctyl maleate, or fumarate, introduces sufficient tack and peel. It should be mentioned that other than ethylene-based low-T_g materials show this crystallizing tendency as well. A striking example is given by the elastomeric sequences of rubbery block copolymers (hot-melt PSAs), where PE-like sequences are interrupted by butylene units.

Side Groups. Flexible main polymer chains with a low volume and unpolar side groups, yield a low T_g. Generally, in order to avoid intermolecular, segmental contact of the chains and their build-up into structures, sterical hindrance has to be incorporated. According to the length, bulk volume, and polarity of the side groups an increase or decrease of the T_g may be obtained. The effect of the side groups on the T_g can be illustrated in the field of the acrylics. Definite trends in physical properties of the polymers can be observed, when higher molecular weight acrylic esters are used; only a few acrylic monomers are adequate for PSAs. The most important of these are ethyl hexyl acrylate and butyl acrylate [9]. As the ratio of low molecular weight (short-branched) monomers is increased, the following changes in properties generally occur:

- Tackiness increases as T_g is lowered;
- Hardness decreases;

- Tensile strength decreases;
- Elongation at break increases.

According to the above scheme (at least theoretically), it is possible to classify the base monomers into softer (plasticizer) or stiffer (hardener) ones. The harder ones are short-side chain (bulky or polar) acrylates. As an example illustrating the effect of the side-chain length on the T_g, the ethyl acrylate/methyl methacrylate copolymers should be mentioned: 75% ethyl acrylate forms polymers with a T_g of −5°C, but the decrease of the soft monomer concentration to 25% leads to a T_g of +80°C.

The principal monomers used in emulsion polymerization can be divided into two main groups: those producing hard polymers such as vinyl acetate, styrene, and vinyl chloride, and those capable of softening hard polymers or producing soft polymers such as ethylene or butyl acrylate [20]. Generally soft polymer compositions must be hardened by polar monomers (see Chapter 5) which may interact physically or chemically, such as acrylic acid, glycidyl methacrylate, N-vinyl-pyrolidone, methacrylamide, acrylonitrile, and methacrylic acid [21]. Because of the very low T_g of ethyl hexyl acrylate, which is a principal monomer for PSAs, it is recommended to copolymerize with acrylonitrile (e.g., Acronal 80D, 81D), styrene, vinyl acetate, or methyl methacrylate in order to increase the cohesion (i.e., the T_g).

The level of the stiffening (hard) monomers should be kept as low as possible. As an example, the usual styrene content of CSBR latexes varies from about 40–70%; T_g values range from about −30°C to room temperature [6]. Generally, the concentration of the most commonly used carboxylic acids lies between 3 and 6%. It should be mentioned that the inclusion of voluminous side groups may be carried out by homopolymerization also (e.g., in the polymerization of butadiene the concentration of 1,2 butadiene units influences the T_g). Increasing the 1,2 butadiene content, the T_g increases also because of the hindrance of the internal rotation and lowering of the chain flexibility. The influence of the length of the side groups of the monomer on the T_g was studied in a detailed manner by Zosel [22] for common acrylic esters used as raw materials for PSAs. A strong dependence between side-chain length and the T_g of the PSAs was demonstrated, namely, the T_g decreases with increasing side-chain length.

Polar Monomers. Polar monomers are copolymerized with the base elastomer in order to improve the mechanical properties (cohesion) of the adhesives. Their inclusion increases the T_g. This may be explained by increasing dipole interaction, or enhanced hydrogen bonding. This effect was demonstrated by Satas [9]. Next the influence of the small scale parameters of the polymer structure, mainly the microstructure, the morphology, and the linear/crosslinked character of the polymer will be reviewed.

Influence of the Microstructure. Many variations can be made in the structure of a thermoplastic rubber (e.g., molecular weight), styrene content, the number of blocks in the chain (i.e., two, three, or multiblocks) and furthermore in their configuration (e.g., a linear polymer, or one that is branched or radial). Varying the structural parameters has a direct influence on their physical properties, hardness, and viscosity. As known the electron beam-crosslinkable S-I-S rubber has a branched molecular structure. Ethylene copolymers are directly synthesized in high pressure equipment, and the comonomer is randomly distributed along the polymer chain [23]. For copolymers containing large amounts of softening comonomer, the PE segments are short and the melting point and amount of crystallinity are depressed, reducing the heat resistance.

Graft or block copolymers are ideal as adhesives because their two-phase structure provides a broad operating range. They can be designed as structures containing microphases and separated into a tough, rubbery, low-T_g phase and a hard reinforcing phase (either a glass or a crystalline material).

Sequence Distribution and Length. Within the block copolymers used for hot-melt PSAs there are amorphous polymer segments with T_g less than room temperature (*RT*), and other, amorphous polymer segments with $T_g << RT$ [24]. The T_g of the thermoplastic component (polystyrene) influences the stiffness of the whole polymer. Its value may be changed, thereby modifying the length of the sequence. The T_g of polystyrene is a function of its molecular weight up to a value of about 20,000, after which the T_g remains equal at 100°C [25]. As an example, the molecular weight of the polystyrene block in S-I-S used for hot-melt PSAs is about 10^4, and its T_g is about 85°C [26]. Generally, the sequence length of the elastomer varies between 500–700 monomer units (with T_g at –70°C), where the styrene blocks have a length of 200–500 units [27]. According to these macromolecular characteristics, in S-I-S copolymers the polyisoprene (PI) phase shows a peak for the loss modulus in the temperature range of –40°C [28]. Another peak around 100°C is linked to the softening point of the pressure-sensitive domains. In a similar manner a separate T_g was observed for the hard phase (105°C) and for the soft phase (–45°C) in acrylic hot-melt PSAs [29]. In each case the high T_g end block will cause a high cohesion/stiffness at room temperature. The thermal stability improves by tackifying these elastomers with resins that have a melting point higher than the T_g of the polystyrene domain.

Sequence distribution is important because the additivity of the T_g is valid for random copolymers only. Foley and Chu [1] showed that the T_g of a random styrene-butadiene copolymer is approximately related to the styrene content, through the following equation:

$$\frac{1}{T_g} = \frac{w_1}{T_{g_1}} + \frac{w_2}{T_{g_2}} \tag{3.4}$$

where w_1 and w_2 are the weight fractions of each copolymer. The "inverse" additivity of the T_g is very important for adjusting the T_g (see Section 1.3).

Influence of Morphology on T_g. The nature and build-up of the polymer chain influences the polymer morphology. Intramolecular forces may promote crystallization. Crystallization reduces chain mobility, and thus increases the T_g. By crystallization or crosslinking through the main valencies appears a hindrance of the chain mobility [30]. Crystalline or crosslinked high polymers are not tacky, even above their T_g. This is an important characteristic not only for adhesives, but also for face stock materials, or substrates to be bonded.

Different adhesion levels can be achieved on high-density polyethylene (HDPE), which has high cristallinity, and low-density polyethylene (LDPE), which has low cristallinity. Because of its cristallinity, HDPE only displays one third of the contact area of LDPE. Thus different adhesion on polyethylene types with different crystallinity influences the anchorage of PSAs applied on face stock materials or substrates made from these polymers. On the other hand, crystallization phenomena may also influence the adhesive flexibility. As an example one can refer to the design of raw materials for hot-melt PSAs. Here the elastomer segment composition in a rubbery block copolymer must be designed to adequately suppress crystallinity by interrupting the polyethylene-like chains (with a butylene monomer unit) without at the same time enriching the polymer in butylene, in such a way that the T_g is substantially raised. The purpose of this restriction is to obtain a saturated olefin rubber block, with the lowest possible T_g and the best rubbery characteristics. Another possibility is to insert isoprene units. Each isoprene monomer addition to the growing chain incorporates two methylene and one propylene sequence (or one ethylene and one propylene for ethylene propylene rubber, EPR, or ethylene propylene diene rubber, EPDM, respectively). Although isoprene can also polymerize through the 3,4 carbons, the resulting hydrogenated structure which has an isopropenyl side group, does not provide any crystallinity or substantial increase of the T_g. In a similar manner, the increase of the acrylic acid content (decrease of the crystallization tendency) improves the peel values, through the lowering of the T_g [27].

Influence of Crosslinking. Steric hindrance by crystallization or crosslinking reduces chain mobility and thus increases the T_g. In the case of crosslinking the effect on T_g depends on the crosslink density. Common electron beam-curing polymers are not recommended for PSAs because of the high crosslinking density, which impart hardness and a tack free character. For PSAs it is recommended to have a T_g lower than $-20°C$ and a low crosslink density. Crosslinking agents in an adhesive influence the T_g and thus the minimum film-forming temperature. Woodworking glues having trivalent chrome nitrate as crosslinking agent (for better water resistance) display a MFT increase of 3–4°C.

Generally, the high temperature performance of hot-melt adhesives is limited by the T_g of the endblocks of the base polymers. Better temperature resistance may be achieved by crosslinking the polymers [31].

1.3 Adjustment of T_g

For practical use adhesives must be fluids (i.e., they must be applied at a temperature above the T_g). An end-use temperature higher than the T_g can be achieved by melting the adhesive (e.g., hot-melt PSAs) or by changing (at least temporarily) its T_g. Fluidity may be achieved by other ways, such as using a volatile plasticizer like a solvent (solvent-based adhesives), or a dispersing agent (water-based dispersions). These methods permit the application of the adhesive, but a continuous adhesive layer is formed above the T_g only. Thus adjusting the T_g appears required.

The composition of the adhesive is determined by polymer synthesis or most frequently through formulation: polymer synthesis regulates intermolecular forces by the nature and polarity of the chain building monomer units. Polymer formulation regulates intermolecular interactions by changing the internal structure of the adhesive polymer.

Polymer Synthesis

Earlier a short description of the parameters influencing the T_g was given. The macromolecular and chemical characteristics of these polymers (molecular weight, chemical composition, structure) can be adjusted through the polymer synthesis, by the choice of the raw materials and of the polymerization technology. In the case of solvent-based adhesives and hot-melt PSAs, the final composition of the adhesive layer is that of the bulk material (i.e., the polymer does not contain important amounts of other nonpolymeric additives).

In the case of water-based dispersions, the final adhesive composition may differ from that of the bulk adhesive, by the additives used, and/or the water included in the adhesive composition. Thus there are differences between the T_g of the synthesized, uncoated (bulk) polymer and the T_g of the applied adhesive. Water-based adhesives are almost always formulated offering the advantage of T_g adjustment by both procedures (changing the composition of the bulk or of the dispersed adhesive). As a function of the level of formulating agents, the T_g differs more or less from that of the base polymer. On the other hand, it was shown that for natural rubber latices, the T_g of natural rubber latex is effectively undistinguishable from that of the dry rubber separated from the same latex [32].

Adjusting the T_g by Formulating

In the polymer synthesis different copolymerization techniques of various monomers giving different chain flexibility and intermolecular forces (i.e., a different macromolecular order) lead to adequate, low-T_g PSAs. Another possibility to change the macromolecular order is the change of the intermolecular forces by the change of the "packing density" through the use of macromolecular or micromolecular diluting agents like tackifiers or plasticizers.

Use of Plasticizers. Plasticizers change the degree of order of the molecules, also changing their enthropy and enthalpy. The use of plasticizers for current thermoplastic materials (e.g., polyvinyl chloride, or PVC) is well known. Using high amounts of plasticizer it is possible to drastically reduce the T_g; as an example the addition of 40% plasticizer to PVC decreases its T_g to –40°C. Such a plasticized PVC used as face material for pressure-sensitive labels, possesses a T_g in the same temperature range as the PSAs applied on it which contributes to very good anchorage of the PSAs on the face stock material. Low stiffness and thus poor cuttability are the drawbacks associated with low T_g. The decrease of the T_g by a micromolecular plasticizer yields increased plasticity and thus better tack of the adhesive, thus plasticizers act like tackifiers.

Use of Macromolecular Tackifiers. In a manner different from plasticizers, the use of macromolecular tackifiers results in a higher T_g. In this case the better flow properties and improvement of the tack are due to the decrease of the modulus.

Earlier a special case to forecast the T_g for random styrene-butadiene copolymers was mentioned. Additivity of the T_g as described for this polymer is a general phenomenon. Generally the T_g is additive, that is, the T_g of polymer compounds is a function of the nature of the components and their ratio. Practically, this is only true for compatible polymers, and thus polymer compatibility also influences the adjustment of the T_g.

The dynamic mechanical properties of polymers are basically determined by their mutual solubility. If both are compatible and soluble in one another, the properties of the mixture are approximately those resulting from the random copolymer of the same composition. Many polymer composites however form two phases due to their natural insolubility. In that case the tan δ/temperature graph will show two peaks, which identify the T_g of each of the components. As an example, a good compatibility between natural rubber and glycerine ester of a hydrogenated colophonium was demonstrated by Zosel [22]; the T_g of the mixture increases for a higher resin content, but no second T_g is observed in the composition range of 80–200% rubber. For compatible adhesive components (most rubber-resin systems) an acceptable estimate of the tan δ peak temperatures can be made using the Fox equation [7].

For miscible polymer blends the Gordon-Taylor [33] equation may also be used [34]:

$$T_g = \frac{w_1 T_{g_1} + k w_2 T_{g_2}}{w_1 + k w_2} \tag{3.5}$$

where T_g, T_{g_1} and T_{g_2} refer to the peak temperature of tan δ for the blend and pure components; w_1 and w_2 are the weight fractions of the components and k is a polymer constant. For compatible polymers deviations from the additivity rule is less than 5°C [35]. It should be mentioned that different addition rules may give different results. The T_g for hot-melt PSAs calculated according to the Fox equation or by the mixing rule may yield different results [35]. On the other hand it was indicated that many miscible blend systems exhibit phase separation upon heating, caused by the existence of a lower critical solution temperature [36]. This incompatibility caused by heating may influence tack and peel adhesion of stored PSAs or the fluidity of PSAs during cutting.

Compatibility is a function of the molecular weight and this statement is very important for the tackification process of PSAs. Generally higher softening point resins are less miscible with the base polymer and any resin added after a certain saturation level is reached, does not contribute to raising the T_g of the system. Instead a second phase consisting of pure resin or a resin rich blend is formed. The existence of this incompatible phase becomes noticeable by the appearance of a second glass transition point. The structure of the low molecular weight resin is very important to its compatibility with elastomers, and consequently to its effect on the viscoelastic properties and performance as PSAs. A completely aromatic resin such as polystyrene exhibits poor compatibility with natural rubber, but is compatible with styrene-butadiene rubber (SBR). A cyclo-aliphatic resin such as polyvinyl cyclohexane is compatible with natural rubber and is incompatible with styrene butadiene rubber. An alkyl aromatic resin, such as poly-t-butyl styrene is incompatible with both elastomers.

1.4 Correlation Between the Main End-Use/Converting Properties of PSAs and T_g

Adhesive Properties and T_g

Room and elevated temperature cohesive strength, adhesion to various substrates, flexibility, etc., are all related to the film-forming ability of the adhesive (i.e., to its T_g) [37]. The viscoelastic properties of the polymer film determine its adhesion-cohesion balance (i.e., its adhesive and end-use properties). The value of the T_g determines the main adhesive properties. An adequate choice of the T_g allows to optimize the adhesive and end-use properties for a specific application area. Generally low T_g values ensure very high tack, whereas medium ones give an optimum peel combined with an acceptable cohesion. The T_g appears particularly useful when comparing the flexibility of several latexes, that is, the latex grade with the lower T_g can be expected to impart a softer bond at a given temperature and to remain flexible at lower temperatures (see Chapter 8 on the choice of latexes for general/deep freeze labeling). But care should be exercised when comparing latexes belonging to different polymer groups, as the nature of the polymer can also influence the flexibility. On the other hand a more complex situation arises by the synthesis of the new core shell latexes where particle structure is important. Here it is possible to identify the composite structure of the latex by determining the T_g and the MFT. The optical examination of the cross-section may show a random distribution of the soft and hard polymer phases, but unusual T_g values indicate the special structure. It must be stressed that T_g values of adhesive latexes are necessary but not sufficient to characterize these polymers. Particle characteristics (morphology) are also important.

Furthermore the T_g is not an absolute measure of an adhesive's suitability for use, but it is a good predictor. Once a material reaches temperatures below its T_g, the viscous component of its viscoelastic character is eliminated [38]. Therefore it can be explained why silicone-based PSAs suffer no delamination at –75 C, like acrylics do at –25°C. The T_g range of silicone-based PSAs is situated at –100 to –125°C, while organic PSAs possess a T_g range of –40 to –25°C.

The effect of the T_g on the peel adhesion at constant molecular weight was studied by Kendall and Chu [23]. They stated that the peel adhesion increases with increasing T_g (at constant tackifier resin level) and passes through a maximum. This maximum occurs at a T_g of about –55°C. In some instances maximum peel strengths are apparent at constant T_g and increasing resin level. Schrijvers [39] found that peel and tack increase with increasing T_g.

Converting Properties and Tg

Converting and laminating properties of the adhesive refer to its ability to be coated in order to form a continuous film on the face material. The fluidity required to coat the adhesive may be achieved using different procedures. Through the use of higher temperatures (hot-melt PSAs) or dispersing agents the viscosity of the adhesive can be reduced to a convenient level. Practically, the dispersed (or dissolved) adhesive will form a continuous film after drying, if its particles are soft enough to be deformed, to exhibit coalescence only.

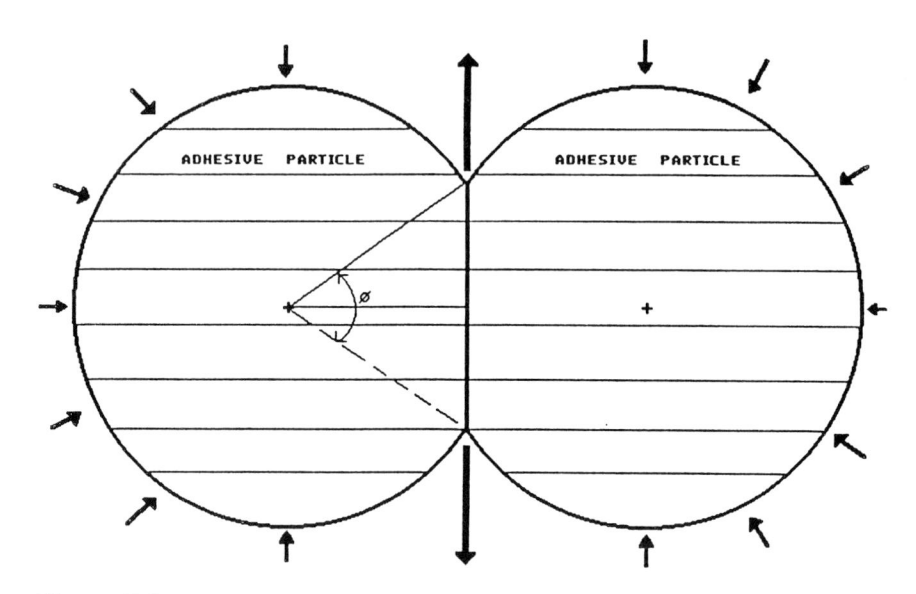

Figure 3.1 Coalescence of polymer particles.

Coalescence is only possible above the T_g. Near this temperature the coalescence is theoretically possible, but the viscosity of the polymer remains too high to achieve full coalescence. The coalescence depends on the viscosity of the bulk polymer in the polymer particles [40]:

$$\varphi^2 = \frac{3\gamma \cdot t}{\pi \cdot \eta \cdot d} \tag{3.6}$$

where φ is the contact angle of the particles (Figure 3.1), γ is the surface tension of the dispersion, η is the viscosity of the bulk polymer in the particles, d is the particle diameter, and t denotes time.

Pressure-sensitive adhesives are very low T_g polymers in a continuous fluid state with low viscosity. Thus PSA coalescence above the T_g is given at all temperatures. Therefore for these materials the definition of a minimum film-forming temperature has no sense nor importance. For adhesives other than PSAs the negative influence of the surface active agents (decrease of the surface tension, see above formula) and the much higher viscosity in the vicinity of the T_g leads to a higher MFT value than the T_g. Thus between the T_g of these adhesives and their MFT discrepancies exist [41], that is:

$$T_g \neq \text{MFT} \tag{3.7}$$

These discrepancies are due not only to the dispersion nature of the adhesive, but also to the test methods used for measuring the MFT; the MFT depends on particle size also. Excellent film-forming properties are obtained with small particle size dispersions. On the other hand, the particle size dependence of the MFT, makes it sensitive towards the face stock porosity. The MFT measured on porous surfaces (real conditions) is higher than the standard value measured on aluminum [42].

End-Use Properties and T_g

It is well known that different adherents require different adhesives (i.e., harder or softer formulations). Adhesives for vinyl substrates have a T_g value of $-10°C$ in order to obtain an optimum bonding [10]. For adhesion to wooden plates a T_g of $-3°C$ is required. Steel can be bonded optimally with an adhesive exhibiting a T_g of $+7°C$. The bonding technology will also influence the T_g. Wet laminating needs T_gs between -40 and $+15°C$, bonding by pressure requires a T_g range of -5 to $+15°C$.

Bond forming at high temperatures is possible for adhesives with a T_g range of $+5°C$ to $+100°C$, and low energy curing (UV, electron beam) requires a T_g range of -40 to $+30°C$. As discussed earlier PSAs generally display a T_g lower than $-20°C$.

2 ROLE OF THE MODULUS IN CHARACTERIZING PSAs

The T_g is an important characteristic of PSAs, allowing the selection of raw materials for PSAs applications. Its value defines the tack of PSAs; a low T_g is a prerequisite for tacky materials. On the other hand the T_g alone does not permit to obtain a real image of the adhesive performance. As an example, Zosel [22] showed that polyethyl hexyl acrylate (PEHA) and polyisobutylene (PIB) have about the same T_g values (-64 and $-60°C$, respectively), but PIB is less tacky because its modulus is 20–300 times higher than that of the PEHA. Thus it can be concluded that the knowledge of both T_g and modulus is necessary for characterizing PSAs. The interdependence between the creep compliance and tack was demonstrated by Dahlquist [43,44].

As discussed earlier (see Chapter 2) PSAs may be described as viscoelastic fluids, with the plastic, viscous part of the system characterized by viscosity (viscous flow occuring above the T_g) and the elastic deformation governed by the modulus. The T_g and modulus both characterize the adhesive flow and its mechanical response. During application of PSAs bonding occurs first; bonding implies adhesive deformability and wetting of the substrate. Deformability by wetting includes building of contact areas, anchorage and penetration. Thus:

$$\text{Bonding} = f(\text{fluidity}) = f(1/E) \tag{3.8}$$

$$\text{Anchorage} = f(\text{fluidity}^\alpha, 1/\text{fluidity}^\beta) = f[(1/E)^\alpha, E^\beta] \tag{3.9}$$

$$\text{Penetration} = f(\text{fluidity}) = f(1/E) \tag{3.10}$$

Ideally the most important property of an adhesive is the bond strength, or the debonding resistance. The bond strength increases with the increase of the modulus:

$$\text{Bond strength} = f(E) \tag{3.11}$$

This general statement was made for adhesives, using more or less complex mathematical correlations. In Chapter 2, Equation 2.12 showed that the flow limit is correlated to the value of the modulus [45].

Pressure-sensitive adhesives intended for similar applications also exhibit similar rheological properties [46]. The correlation between adhesive properties and dynamic shear storage modulus is quite good. The above given short examination of the dependence of bonding and debonding on the modulus clarifies the role of the modulus as a main parameter influencing the adhesive properties. Detailed theoretical considerations about the influence of the modulus on the rheology of the adhesion were provided in Chapter 2. Chapter 6 describes the dependence of adhesive characteristics on the modulus.

Dynamic mechanical analysis (DMA) currently is used in many application areas, especially in the adhesive industry. The relationship between rheological properties, as measured by DMA, and end-product performance was defined for many types of adhesives, particularly PSAs. The storage modulus G' is a measurement of hardness (Equation 2.9). At $G' = E + 0.8$ Pa and above, the hot-melt PSA is approaching the glassy region. A typical value of G' for a PSA at 23°C is between $1E + 0.5$ and $2E + 0.4$ Pa [47]. The loss modulus G'' is associated with energy dissipation; the greater the value related to G' the more flexible the PSAs. Tan δ is the balance of the viscous/elastic behavior; in practice it can be correlated to the cohesive strength of the adhesive; for hot-melt tan $\delta = 1$ is a limiting value. Above this value the hot-melt PSA can flow without induced deformation; below this value the hot-melt PSA has inherent cohesion.

2.1 Factors Influencing the Modulus

Generally the modulus is a material characteristic, but for materials in a composite structure it may depend on the structure and component characteristics of the composite. An example is that the elastic modulus of epoxy coatings is higher than that of the uncoated material [42].

Material Characteristics

For common plastics the modulus is not a constant as it depends on the molecular weight distribution (MWD) and on the manufacturing process [48]. This is generally valid for adhesives and PSAs in particular. Here the modulus is a function of the molecular weight and its distribution, the chemical composition, and the structure. It should be noted that in certain cases, when evaluating adhesive characteristics, that the creep compliance ($1/E$), not the modulus, and its dependence on the polymer need to be considered. As an example, special PSAs (e.g., skin adhesives) use grafted polymers; the number and composition of the attached polymeric moieties on the polymer, and the inherent viscosity (molecular weight) of the polymer are manipulated to provide an adhesive composition with a creep compliance value greater than 1.2×10^{-5} cm^2/dyn [49].

Modulus and Molecular Weight. Molecular weight is a parameter influencing the most important practical properties like tensile strength, elongation, modulus, T_g, viscosity, green strength, wettability, pot life, chemical resistance, and processability of the adhesives. Generally PSAs are polymer blends, thus the influence of the molecular weight on the modulus must be taken into account from the point of view of the mutual solubility (compatibility). A classical example of the modulus dependence on molecular weight is given by rubber/resin formulations. Here nonmilled rubber from dried natural latex shows a larger plateau modulus and a higher elastic modulus (G') than milled natural rubber. Milled rubber possesses a lower molecular weight because of the mechano-thermal degradation of the elastomer. For a CSBR latex, the increase of the molecular weight is accompanied by an extension of the plateau region of the dynamic modulus in the master curve towards higher temperatures [26]. Although the modulus increases with molecular weight, its adjustment by the molecular weight remains limited. It was shown [2] that the maximum of the modulus is situated at about 500–1000 atoms in the main polymer chain. Longer molecules do not give higher strength because their various segments act independently of each other. On the other hand an excessive increase of the molecular weight is not recommended because of the poor wetting out (low flow) and decrease of the ability of the adhesive to desorb energy (adhesive break). It was shown earlier that T_g related parameters (chain mobility) limit the molecular weight as well.

Another parameter is the molecular weight distribution, which may be characterized by dynamical rheology (i.e., the rheology is very sensitive towards MWD) [50]. This sensitivity is also valid for PSAs; in this case MWD influences the loss tan δ peak. McElrath [51] demonstrated that the molecular weight (and its distribution) of a tackifier influences the peak of the loss modulus and its location as a function of the frequency. For tackified formulations the loss tan δ peak is sharper for those resins with a narrow MWD. It is very broad and diffuse for the broad MWD obtained by blending two resin fractions [28]. In a diagram representing tan δ peaks for a resin-rich phase (on a log scale) the broad peak (large MWD, resin mixtures) does not equal zero. At zero, the loss modulus equals the storage modulus. Above zero there is more energy loss than storage. On the other hand, for tack it is important to dissipate energy during the debonding step in order to achieve high tack. Energy dissipation is higher for a broad MWD (i.e., for resin mixtures); this consideration is very important for the choice of the tackifier resins.

The influence of the molecular weight and MWD on the modulus should be taken into account for rubber blends too. For instance, it was shown [52] that mastication (used for natural rubber) will have a relatively small effect on long chain molecules and on high molecular weight elements. If they survive mastication to some extent, then their relative contribution to the elasticity will increase. Therefore in mixtures with fractions having different molecular weights, the influence of the high molecular weight factors is more pronounced. For tackified blends the molecular weight of the resin is also significant for the compatibility. In this case an optimum of the viscoelastic properties is linked to molecular weight below than 1000 with a narrow MWD. The dependence of the modulus on the mutual solubility was confirmed by Kendall, Foley, and Chu [1]. However in certain cases the use of low modulus polymers is dependent on coating technology. Low viscosity requirements imply low molecular elastomers as raw materials for hot-melt PSAs; therefore the toughness of hot-melt PSAs is also low [53].

Opportunities to adjust the molecular weight of the base polymers for adhesives allow the adjustment of the modulus as well. Theoretically the molecular weight may be modified by increasing the polymerization degree for the linear polymer, or by crosslinking. It was shown earlier that increasing the polymerization degree is limited because of the decrease of the flow properties (for processing and end-use purposes). Thus an adequate adjustment of the molecular weight should be achieved through crosslinking.

Chemical Composition. The modulus is a function of the chemical composition; polymers based on different monomers possess different modulus values. On the other hand minor compositional aspects such as sequence distribution also influence the modulus. Zosel [22] tested the dependence of the modulus of acrylates on the side-chain length. Polymethyl-, polyethyl-, *n*-butyl-, *i*-butyl-, and ethyl hexyl acrylate were studied in order to estimate the influence of the chain length (branching) on the T_g and E. Shorter side chains in acrylates impart higher modulus values, the $\overset{g}{E} = f$ (temperature) plot being moved to the right (higher T values) with a higher plateau. According to the theory of rubber elasticity the shear modulus G^A depends on the number of crosslinks V_c,

$$G^A = V_c \cdot R \cdot T \tag{3.13}$$

where R and T are the universal gas constant and the temperature, respectively. For tackified PSAs the addition of resin, the dilution of crosslinks, and the destruction of the elastomeric network decrease the modulus, but the latter increases again at higher resin loadings. The influence of the crosslinking through molecular weight increase and natural aging was discussed earlier. Here another special effect of the crosslinking on the modulus, due to the lack of interaction between gel and tackifier, will be examined. Rubber shows a highly reversible deformability combined with a low modulus (the modulus of unfilled rubber is 10^3 lower than that of plastics). Additionally rubber exhibits a technologically important property, namely the ability to undergo self-reinforcement [54] through formation of strain-induced crystalline structures; this imparts excellent tensile properties to rubber. Self-reinforcement by crystallization is a well-known possibility to improve the properties (modulus) of rubber-like synthetic polymers used as raw materials for hot-melt PSAs. In this case soft polyethylene butylene (PEB) segments (matrix) are reinforced by polystyrene segments at the end of the chain; this gives physical crosslinking [55]. Chemical crosslinking is characteristic for CSBR latexes, as they consist of a mixture of polymer species with linear chains, branched chains, and a crosslinked network. Many natural rubber solvent-based adhesive formulations also contain a crosslinking agent. The effect of crosslinking may be illustrated by the increase of G' (elastic modulus). The adhesive with more crosslinking possesses a higher modulus and higher shear holding power.

Classical rubber elasticity theory predicts that the plateau modulus G_N^* for a diluted polymer is proportional to the square of the volume fraction of the polymer. Scaling law interpretations by De Gennes [56] suggests that G_N^* should be proportional to the 2.25 power of the polymer volume fraction. Plots of SBR blends indicate higher values of 2.45–2.65 for the exponent. A possible explanation is that the SBR contains gels (see Chapter 5) which do not participate in the dilution by the resin. Therefore the resin concentration in the amorphous phase is higher than expected, resulting in an apparent higher power for the polymer volume fraction, and as a practical consequence for SBR tackification in a higher tack. The influence of the tackifier concentration on the modulus was studied by Foley and Chu [9]. They demonstrated that the minimum G' value (at 25°C) occurs at about 40–60% resin loading for rubber/resin systems. For a detailed examination of the modulus dependence on tackifiers, the reader is referred to Chapter 8.

Dependence of Modulus on Sequence Distribution. The sequence distribution exerts a strong influence on the modulus of block copolymers used as raw materials for hot-melt adhesives. By changing the sequence length, volume, and polarity it is possible to modify the modulus, as well as the adhesion/cohesion balance of the polymers; here polar styrene blocks reinforce these polymers [55]. It should be mentioned that a similar hardening effect of certain monomers was observed in some statistical copolymers as well [57]. For partially hydrolysed ethylene vinyl acetate (EVAc) copolymers the dependence of the modulus on the distribution of the hydroxy groups was shown [58].

However, the most important practical use of sequence regulation occurs in the synthesis of thermoplastic rubbers used as raw materials for hot-melt PSAs. These materials are comparatively low molecular weight polybutadiene and polyisoprene rubbers, intertwined with short polystyrene chains. When they are cooled from the melt the (incompatible) polystyrene end-blocks separate out and gather to form small hard domains of polystyrene. Since the polystyrene blocks are covalently bonded to the rubber, phase separation is restricted; this forces the copolymer in a special microphase morphology [59] (see Chapter 5). Special sequence distribution allows increasing the molecular weight without the loss of the tack. Thus segmented EVAc copolymers with very high molecular weight are adequate raw materials for hot-melt PSAs [60].

Softer, more tacky adhesives are obtained with polymers with a low polystyrene content such as Cariflex TR 1107 [61]. Some applications, however, require a high shear resistance (i.e., a higher polystyrene content). For lower viscosity hot-melt PSAs, which are easier to process at lower temperatures, S-I-S block copolymers with high melt flow rates containing a relatively high proportion of diblocks were developed. These low modulus polymers also promote better die cuttability of PSAs used in label stock applications. It can be expected that in the future S-I-S polymers will be designed with even more variations in the di-/triblock ratio, which will enable the manufacture of still softer hot-melt PSAs with a more aggressive tack. The higher modulus polystyrene-polybutadiene-polystyrene (S-B-S) block copolymers are mainly used in nontacky adhesives. For tacky hot-melt PSAs with better shear, low MW, low viscosity S-B-S grades with a high polystyrene content are used [4].

The size and morphology of the polystyrene domain in styrene block copolymers influence the cohesive strength of the system. However, if one increases the polystyrene domain by increasing the polystyrene content of the copolymer the storage plateau modulus increases, thus making the polymer considerably more difficult to tackify [62]. On the other hand it is well known that in tape applications the diblocks act as an elastically ineffective diluent in the formulated adhesive. A high diblock content limits the cohesive strength of the base polymer of the adhesive system. The tensile strength and the modulus decrease as the diblock content increases. For label stock adhesives the presence of diblocks is currently considered desirable for obtaining the required die cutting and skeleton waste stripping performance. The optimum level of required diblock depends upon the type and speed of the converting machine, the label stock face material used, together with the other components in the adhesive formulation.

Modulus and T_g

The tensile strength and elastic modulus decrease above T_g [63]. This general observation has a special implication for polymer blends. For certain polymer blends a phase separation temperature exists, the value of which depends on the T_g. A different dynamic-mechanical response is observed below and above this phase separation temperature (T_c). For temperatures below T_c a thermorheologically simple behavior is observed and the time/temperature (frequency/temperature) superposition principle can be applied [64]. For these temperatures master curves are obtained according to:

$$G_{\text{red}}(T_o, \omega) = \frac{G(T, \omega) \cdot T_o \zeta_o}{T_g}$$ (3.14)

where G_{red} is the modulus ($G'(x)$ or $G''(x)$) reduced to the reference temperature T_o. For polymer blends where no strong interactions occur between the polymers (i.e., one of the components acts as a low molecular weight solvent) the plateau modulus $G^o_{N_2}$ is given by the following equation:

$$G^o_{N_2} = G^o_N(\varphi = 1) \cdot \varphi_A^{2.25}$$ (3.15)

where φ_A is the polymer volume fraction. For phase-separated mixtures (and chemically crosslinked systems) an $\omega^{1/2}$ power law was observed (usually $G' \sim \omega^2$ and $G'' \sim \omega$).

Filler Effect on the Modulus. The use of fillers with elastomers yields increased stiffness, modulus, and rupture energy [65]. This effect of the filler is due to a hydrodynamic viscosity increase, an increased stiffness, and a lower T_g of part of the rubber phase at the interface with the filler, given by surface interaction. A covalent bonding of network segment to filler surface, and a tightening of the short chains between aggregates must also be considered. The filler particles which may be as small as 1.5×10^{-5} mm (about 30 polymer chain diameters) may also interface with the motion of the polymer molecules [66]. On the other hand, tackifier resins act partially as fillers. Exhibiting a surface free energy close to that of the elastomer the filler is perfectly wetted by the organic matrix and, as a consequence, good dispersions (i.e., lower viscosities) are obtained [65].

Dependence of the Modulus on the Tackifier Resins. It was shown earlier that the use of tackifier resins lowers the plateau modulus of adhesive blends. This effect depends on the nature and concentration of the tackifier. The dependence of the modulus on the tackifier concentration was shown by Sherriff [27] and Kamagata [67]. Zosel [22] demonstrated that the modulus reaches a minimum for 50% resin in the mixture (natural rubber/glycerine ester of colophonium). As discussed earlier the molecular weight and MWD of the tackifier resin strongly influence the modulus of the mixture. Generally the modulus increases by increasing the concentration and activity of the fillers [68]. Fillers increase the modulus proportional to their concentration [69]. The relative modulus increase was calculated as a function of the concentration. These correlations are calculated for fillers acting as chemically inert substances. Proactive fillers (such as metallic oxides or salts) may interact with the base polymer, actually crosslinking it. This addition of some oxide to carboxylated compounds dramatically reduces the pressure-sensitive nature of the adhesive [70].

Dependence of the Modulus on Other Factors

The modulus of adhesive joints depends on the adhesive's thickness [71]. In fact increasing the thickness of the adhesive layer decreases its modulus. This behavior may be explained by the special multilayer structure of the adhesive and the influence of the surface of the solid state components. The modulus of the laminate varies close to the interface with the adhesive. The width of this zone depends on the adhesive; for certain adhesives it varies between 0.04–0.2 mm [72].

It must be noted that the investigations on the influence of the modulus on the joint resistance, the dependence of the modulus, and adhesive joint resistance on the adhesive layer thickness were mostly carried out for thick wall sandwiches and adhesives other than PSAs. However it should be noted that the thickness (coating weight) of the adhesive would also exert a special influence on the adhesive resistance for PSA laminates. Increasing the coating weight increases the thickness of the mobile middle adhesive layer only (i.e., the cold flow of the adhesive). Another parameter influencing the modulus of plastics is their degree of orientation [47]. Actually there are no data available about a similar modulus dependence for PSAs. One should note the dependence of the modulus on the composite structure of the laminate or laminate components. Special attention must be paid to the influence of the humidity on the paper modulus.

The modulus of the paper depends on its chemical composition, temperature, physical structure and moisture content. When analyzing the elastic properties of paper, it is necessary to take into account the fact that paper is composed of both crystalline and amorphous cellulose [71], the latter under-

Physical Basis for the Viscoelastic Behavior of PSAs

going softening as the moisture content increases. The effect of moisture on the modulus of cellulosic materials is also affected by temperature. A moisture/temperature equivalence exists, that is, for lower moisture contents the softening effect occurs at higher temperatures. The decrease in relative modulus with increasing moisture content is different for papers of different crystallinity. Theoretically the modulus is a function of the number of hydrogen bonds (n) absorbing the applied strain:

$$E = k \cdot n^{1/3} \tag{3.16}$$

where k is the average value of the force constant and E is the elastic modulus of the paper. The decrease of the effective number of the hydrogen bonds results from moisture absorption, resulting in the decrease of E.

2.2 Adjustment of the Modulus

On the basis of the interdependence between modulus and adhesion-cohesion balance of PSAs there is a need to adjust the modulus value. As discussed earlier, the modulus is a polymer characteristic and depends on its chemical and macromolecular properties; thus the modulus should be modified by changing these properties. In a similar manner to the T_g, it is possible to change the modulus of the adhesives during the polymer synthesis or through formulation.

Synthesis

The choice of the monomers and the control of the sequence distribution and polymer structure allow an adequate regulation of the modulus. As an example, the T_g and modulus of the acrylic latexes can be adjusted by polymerizing monomeric acrylates at different rates [1]. On the other hand the choice of the adequate rubbery sequence for block copolymers for hot-melt PSAs allows the modification of the polymer stiffness. Jacob [73] pointed out that although more economical, there are only a few cases where the use of S-B-S blockcopolymers for hot-melt PSAs applications appears appropriate. This may be explained by the high modulus of these compounds, comparatively to S-I-S and by the higher stiffness and limited tackifying ability (compatibility).

Formulation

On the basis of a few adhesive groups/polymers, a large number of PSAs are compounded according to the very different requirements concerning the adhesion/cohesion balance. This is possible through the use of formulation. Formulation refers to the design of the adhesive properties through compounding. Designing of the adhesive properties involves mainly tackification (i.e., the increase of tack and peel) or, in a few cases, the improvement of the shear, cohesion (see Chapters 6 and 8).

Tackification imparts better flow properties and molecular mobility. This is achieved by the use of diluting agents, either micromolecular (plasticizer) or macromolecular (tackifying resins). Thermoplastic rubbers are unique in the way they bring together both rubber and plastic [74]. They comprise two inherently incompatible, but chemically connected segments which attempt to separate at service temperatures; this results in their having a morphological structure in which thermoplastic end-blocks such as polystyrene form domains holding together a rubbery network (e.g., polybutadiene; see earlier). Materials with fine basic morphologies can thus be produced.

In addition to these opportunities provided by changing the basic polymer structure, further modification of the material properties is possible during compounding. For example, by using low molecular weight compounding ingredients which demonstrate preferential compatibility with, or solvency for one or the other of the two phases in the base polymer, it is possible to change the effective ratio of the thermoplastic to rubber phase. More specifically it is often possible to move from a nonresilient morphology to a soft material by increasing the volume of the rubbery phase through the addition of a suitable oil, or by decreasing the amount of the polystyrene compatible aromatic resin in an S-B-S copolymer/oil/resin hot-melt adhesive formulation.

Crosslinking

Crosslinking yields a higher modulus and higher shear resistance. In the range of the crosslinked adhesives those with the highest modulus are the most crosslinked and have the highest shear and holding power [48]. Theoretically, on the basis of crosslinking data it should be possible to calculate the shear modulus of the polymer. Practically, in a few cases it is possible with the aid of crosslinking data to predict the gel point and the shear modulus for polyurethanes [75].

The theoretical basis for the dependence of the modulus on the crosslinking is given by the dependence of the modulus on the polymer chain network, on the number of crosslinked sites [57]. For small deformations (according to Flory [8], deformation means transition from a real network to a phantom network) there is a correlation between the modulus and the number of crosslinked points (entanglement):

$$G = 2C_1 + 2C_2 = (\nu_{CH} - \nu_C T_e) kT \frac{<r^2>}{<r_o^2>} \quad (3.17)$$

where G is the modulus, $2C_i$ is a function of crosslinking density, ν_e is the number of network points in the uncrosslinked polymer (polymer-specific content), and T_e is the trapping factor (i.e., the number of rigid network points); it is a function of molecular weight distribution and chemical crosslink-

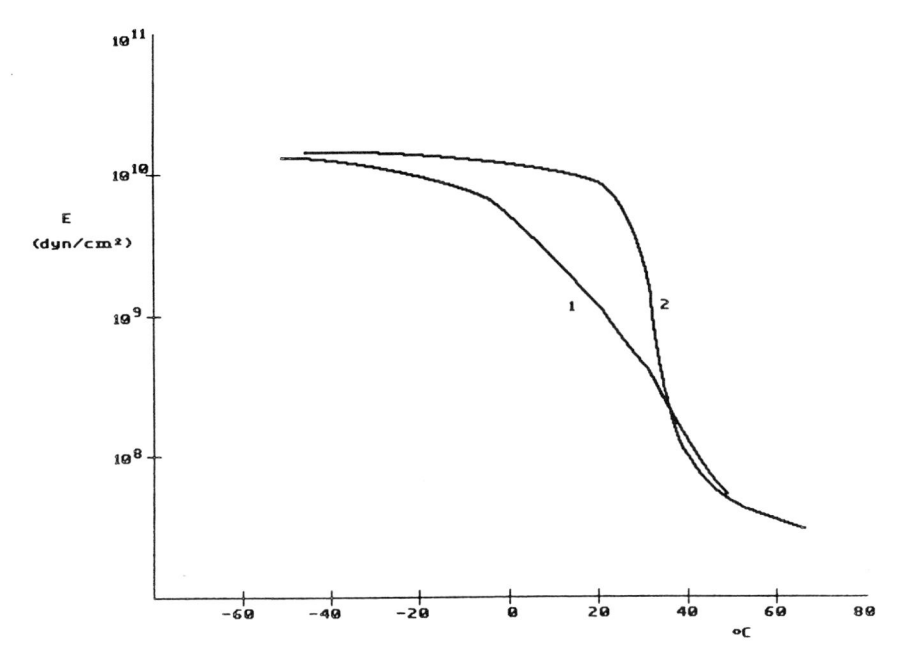

Figure 3.2 Modulus as a function of temperature.

ing density. Ideally, the chain length of the crosslinking agents also influences the modulus [76]; longer chain-crosslinking agents allow more elongation.

2.3 Modulus Values

The value of the modulus at a temperature above T_g characterizes the viscoelastic behavior of PSAs raw materials. The location of the rubbery plateau is very important. In order to obtain suitable adhesive properties at temperatures above T_g (room temperature or less), a low E plateau is needed (Figure 3.2).

The dependence of the modulus on the temperature and stress rate is well known. Thus it becomes very difficult to compare different polymers on the basis of single E data only. However, these values provide a first indication about the possible application field of the polymers. Pressure-sensitive adhesives are soft and mobile under certain conditions. Even the most crosslinked PSAs do not match any property of a thermoset plastic as the value of their modulus is low, and hence their films are not meant to be used without a face material. Table 3.5 summarizes some modulus values for polymers used for PSAs or related materials.

Table 3.5 Modulus Values for Materials Used for PSA Laminates

Material		Modulus value (N/mm²)			
Nature	Characteristics	MD	Mean	TD	Ref.
Polypropylene	BOPP, film	1400	—	2200	[77]
	BOPP, film	2400	—	4000	[77]
	BOPP, film	1500	—	2500	[77]
	BOPP, film	2000	—	2500	[77]
	BOPP, film	1200	—	2100	[78]
Ethylene vinylacetate copolymer	Bulky, grafted	—	280	—	[60]
	Bulky, 50–70% VA		200–1600		
Polyvinylchloride	Plasticized, film	67	—	61	—
Polyethylene	LD, film		200–500		[54]
Polyethylene	HD, film		700–1400		[54]
Polycarbonate	Film		2100–2400		[54]
Polystyrene	Film		3200–3250		[54]
Rubber	Unfilled		1–5		[54]
Polyvinylchloride	Hard, unplasticized, film		4000		[79]
Paper	—		2920–3700		[80]

A maximum tack implies a modulus of 1×10^5 to 2×10^4 Pa [81]. For a good wetout a modulus of about 3.10^5 Pa is required [82]. For different kinds of stresses different modulus values must be taken into account. Generally cold flow is characterized by the creep modulus E_c:

$$E_c(t) = \frac{\sigma}{\varepsilon(t)} \tag{3.18}$$

where σ is the stress (constant in time) and $\varepsilon_{(t)}$ is the time-dependent strain:

$$E = f(T, \text{stress}) \tag{3.19}$$

Generally for small deformations ($\varepsilon \leq 0.8\%$) the stress dependence of the deformation may be eliminated. Pressure-sensitive adhesives are subject to higher deformation levels, thus the stress dependence of the strain cannot be ignored and the value of the creep modulus is not a constant. On the other hand, for high deformation rates (e.g., peel or cutting) the dynamic modulus must be considered:

$$E' = 2G'(1 + \nu) \tag{3.20}$$

where v is Poisson's ratio. Dahlquist [43] has taken the $1/E$ value (creep compliance) as an index for PSAs, showing that for general PSAs this value should exceed 1×10^{-6} cm^2/dyn but a maximum tack requires a lower modulus (i.e., a higher creep compliance) [43,44]. The creep compliance for PSAs for medical and surgical applications should be about 1.2×10^{-5} cm^2/dyn, preferably 1.3×10^{-5} cm^2/dyn [25]; the higher the creep compliance, the bigger the adhesive residue left on the substrate. The fundamentals of creep compliance as they relate to polymeric materials and in particular to viscoelastic polymers are covered in [51].

REFERENCES

1. J. Kendall, F. Foley and S. G. Chu, *Adhesives Age*, (9) 26 (1986).
2. *Der Siebdruck*, (3) 69 (1986).
3. W. Retting, *Kolloid Zeitschr.*, (210) 54 (1966).
4. A. Midgley, *Adhesives Age*, (9) 17 (1986).
5. F. X. Chancler, J. G. Brodnyan and D. G. Strong, *Resin Rev.*, (3) 12 (1972).
6. A. Zawilinski, *Adhesives Age*, (9) 29 (1984).
7. M. J. Brenden, H. A. Schneider and M. J. Cantow, *Polymer*, (1) 78 (1988).
8. T. G. Fox and J. P. Flory, *Bull. Am. Phys. Soc.*, (1) 123 (1956).
9. D. Satas, *Handbook of Pressure Sensitive Technology*, Van Nostrand-Rheinhold Co., New York, 1982.
10. *Adhäsion*, (4) 13 91967).
11. EP 0212358, p. 3.
12. *Adhesives Age*, (11) 28 (1983).
13. C. T. Albright, EP 009087B1/23.12.87.
14. *Coating*, (7) 186 (1984).
15. L. Jacob, New Developments of Tackifier Resins for SBS Block Copolymers, 11th Munich Adhesive and Finishing Seminar, 1986, p. 87.
16. S. N. Gan, D. R. Burfield and K. Sogh, *Macromolecules*, (18) 2684 (1985).
17. E. Merz, *Double Liaison - Chimie des Peintures*, (226) 263 (1974).
18. R. P. Mudge (National Starch Chem. Co., USA), EP 0225541/11. 12. 85.
19. C. M. Chum, M. C. Ling and R. R. Vargas, (Avery Int. Co., USA), EP 1225792/18. 08. 87.
20. M. Bowtell, *Adhesives Age*, (12) 37 (1986).
21. P. K. Dahl, R. Murphy and G. N. Babu, *Org. Coatings. Appl. Polymer Sci. Proceedings*, (48) 131 (1983).
22. A. Zosel, *Adhäsion*, (3) 17 (1966).
23. G. Prejean, High Performance Adhesives from Ethylene Copolymers via Grafting, 15th Munich Adhesive and Finishing Seminar, 1990, p. 102.
24. G. Holden, E. T. Bishop and N. R. Legge, *J. Polymer Sci.*, (26) 37 (1969).
25. J. P. Keally and R. E. Zenk (Minnesota Mining and Manuf. Co., USA), Canad. Pat., 1224.678/19.07.82 (USP. 399350).
26. A. S. Rivlin, *Paint Technology*, (9) 215 (1944).

27. M. Sheriff, R. W. Knibbs and P. G. Langley, *J. Appl. Polymer Sci.*, **17**, 3423 (1973).
28. *Adhesives Age*, (9) 37 (1988).
29. *Adhesives Age*, (11) 48 (1985).
30. J. Class and S. G. Chu, *Org. Coat. Appl. Polymer Sci. Proc.*, 48, 126 (1989).
31. E. G. Ewing and J. C. Erickson, *Tappi J.*, (6) 158 (1988).
32. D. R. Burfield, *Polymer Comm.*, (6) 178 (1983).
33. G. R. Hamed and C. H. Hsieh, *J. Polymer Physics*, **21**, 1415 (1983).
34. J. M. Gordon, G. B. Rouse, J. H. Gibbs and W. M. Risen Jr., *J. Chem. Phys.*, (66) 4971 (1977).
35. A. Gent and P. Vondracek, *J. Appl. Polymer Sci.*, **27**, 4357 (1982).
36. G. Fuller and G. J. Lake, *Kautschuk, Gummi, Kunststoffe*, (11) 1088 (1987).
37. G. Hombergsmeier, *Papier u. Kunststoffverarb.*, (11) 31 (1985).
38. L. A. Sobieski and T. J. Tangney, *Adhesives Age*, (12) 23 (1988).
39. L. Schrijvers, *Coating*, (3) 70 (1981).
40. Solvay & Cie, *Beschichtung von Kunststoffen mit IXAN, WA*, Prospect Br 1002d-B-0,3-0979.
41. H. D. Cogan, *Off. Digest*, (33) 365 (1961).
42. *Adhäsion*, (5) 76 (1984).
43. C. A. Dahlquist, Tack, in *Adhesion Fundamentals and Practice*, McLaurin and Sons Ltd., London, 1966.
44. C. A. Dahlquist, in *Handbook of Pressure Sensitive Adhesive Technology* (D. Satas, Ed.), Van Nostrand-Rheinhold Co., New York, 1988, p. 82.
45. *Adhäsion*, (5) 100 (1984).
46. D. Satas, *Adhesives Age*, (8) 28 (1988).
47. D. P. Bamborough and P. H. Dunckley, *Adhesives Age*, (11) 20 (1990).
48. R. Köhler, *Adhäsion*, (3) 66 (1972).
49. S. E. Krampe, L. C. Moore (Minnesota Mining and Manuf. Co., USA), EP 0202831A2/26. 11. 86, p. 20.
50. G. Wu, *Polymer Mater. Sci., Eng. Proceed.*, (54) 239 (1986).
51. K. MacElrath, *Coating*, (7) 236 (1989).
52. S. Cartasegna, *Kautschuk, Gummi, Kunststoffe*, (12) 1188 (1986).
53. H. Kehler and W. M. Kulickje, *Chem. Eng. Tech.*, (10) 802 (1986).
54. U. Eisele, *Kautschuk, Gummi, Kunststoffe*, (6) 539 (1987).
55. D. Krügers, *Kautschuk, Gummi, Kunststoffe*, (6) 549 (1988).
56. P. G. De Gennes, *Macromolecules*, (9) 587 (1976).
57. L. V. Sokolova, O. A. Nikolaeva and V. A. Sersnev, *Vysokomol. Soyed.*, (**A27**) 1297 (1985); in *Kautschuk, Gummi, Kunststoffe*, (6) 563 (1986).
58. T. Matsumoto, K. Nakmae and J. Chosoake, *J. Adhesion Soc. of Japan*, (11) 5 (1975).
59. R. A. Fletscher, The temperature factor in compounding thermoplastic rubber based HMA, 11th Munich Adhesive and Finishing Seminar, 1986, p. 36.
60. A. Koch and C. L. Gueris, *apr*, (16) 40 (1986).
61. D. De Jaeger, Developments in Styrene Block Copolymers, 11th Munich Adhesive and Finishing Seminar, 1986, p. 96.

62. K. E. Johnson and Q. Luvinh, Dexco Triblock Copolymers for the Adhesive Industry, European Tape and Label Conference, Exxon, Brussel, 1989, Part 19.
63. S. W. Medina and F. W. Distefano, *Adhesive Age*, (2) 18 (1989).
64. R. Stadler, L. De Lucca Freitas, V. Kriegel and S. Klotz, *Polymer*, (9) 1643 (1988).
65. J. B. Bonnet, M. J. Wang, E. Papirer and A. Vidal, *Kautschuk, Gummi, Kunststoffe*, (6) 510 (1986).
66. G. L. Schneeberger, *Adhesives Age*, (4) 21 (1974).
67. R. Kamagata, K. Kosaka, *J. Appl. Polymer Sci.*, (15) 183 (1971).
68. B. Poltersdorf and D. Schwambach, *Kautschuk, Gummi, Kunststoffe*, (1) 43 (1988).
69. *Adhäsion*, (7) 242 (1972).
70. *Adhesives Age*, (10) 26 (1977).
71. L. Dorn and G. Moniatis, *Adhäsion*, (11) 32 (1967).
72. M. Schlimmer, *Adhäsion*, (4) 8 (1987).
73. L. Jacob, HMPSA a well performing, economic and environmentally friendly technology to manufacture tapes, Afera 93, Dresden.
74. D. L. Bull, Thermoplastic rubbers, the utilisation of their dual structure in compounding, Shell Chemicals, Thermoplastic Rubbers, Technical Manual, TR 8-9.
75. F. De Candia and R. Russo, *J. Thermal Analysis*, **30**, 1325 (1985).
76. H. Y. Tang and J. E. Mark, *Macromolecules*, (17) 2616 (1984).
77. Mobil Plastics, Europe, MA 657/06/90.
78. Courtaulds Films, Technical information, 01.04.1991.
79. P. Hammerschmidt, *apr*, (4) 190 (1986).
80. H. E. Krämer, *apr*, (31) 843 (1988).
81. *Adhesives Age*, (12) 35 (1987).
82. R. Köhler, *Adhäsion*, (3) 90 (1970).
83. R. Milker and Z. Czech, Solvent Based Pressure Sensitive Adhesives, 16th Munich Adhesive and Finishing Seminar, 1991, p. 134.
84. W. Hoffmann, *Kautschuk, Gummi, Kunststoffe*, (9) 777 (1985).
85. S. W. Medina and F. W. Distefano, *Adhesives Age*, (2) 18 (1989).
86. A. Zosel, *Colloid Polymer Sci.*, **263**, 541 (1985).

4
Comparison of PSAs

There is no special science dedicated to the study of PSAs and there is no special education geared to this particular field. Therefore specialists from other related fields work on the complex technical problems involved in the manufacture and use of PSA laminates. Pressure-sensitive adhesives are macromolecular compounds and thus one can suppose that a better understanding of this field requires the input of a polymer scientist or engineer. In fact certain mathematical correlations or approximations describing the dependence of the polymer behavior on the experimental nature and structure are also valid for PSAs. But the problem appears more complex because of the viscoelastic behavior of PSAs which depends on the experimental conditions; PSAs may obey correlations valid for plastics (viscous materials) or rubber (elastic materials).

Doing mostly empirical work, the PSA expert tries to act on a theoretical basis. This theory is based on the knowledge from processing plastics and elastomers, or from the more thoroughly investigated field (relative to PSAs) of general purpose adhesives. Therefore, a better understanding of the PSA practice requires a quintescence of the similarities/differences in the properties/rheology of plastomers/elastomers, adhesives, etc. Generally making a comparison of the rheological properties of plastomers/elastomers, PSAs, and of classical adhesives, one can observe certain similarities and differences. Similarities are given by the (limited) flow of plastomers/elastomers; the differences arise from the (partially) ordered structure of these materials.

1 COMPARISON OF PSAs WITH THERMOPLASTICS AND RUBBER

Similar to plastics and rubber, PSAs are macromolecular compounds; their properties depend on their chemical composition and macromolecular characteristics. But in a different manner from elastomers, plastomers, or common adhesives, the physical state and not the chemical composition plays a determinant role for PSAs. For plastics and rubber the fluid state is only required for processing purposes. For PSAs flow is a permanent need. Related to flow is the lack of crystallinity, the lack of orientation, the need for cold flow, the high rate of relaxation, and the ductility. It should be pointed out that in comparison to the flow of plastics or rubber during processing the flow of PSAs occurs isothermally. On the other hand the PSA fluid is always a composite, containing solids, liquids, or volatile components coming from the converting process (fillers, solvents, wetting agents, etc.). The level of additives is never zero, but remains relatively low (less than 20–30% by weight).

Generally the rheological similarity of PSAs, thermoplastics, and rubber are observed by examining the practical behavior (e.g., cuttability, re-adhering behavior, mechanical destruction, etc.) of the material. These properties, starting from similar practical processing or test conditions for polymers and from the parameters influencing the polymer behavior are discussed next.

1.1 Cold Flow

Pressure-sensitive adhesives must exhibit a good resistance to stringing, edge flow, and bleeding [1]. Even though PSAs are highly viscous, they nevertheless display cold flow [2]. The rheology of the adhesive determines its converting and end-use properties. In the industrial practice of laminate-converting, the storage (as a converting-related phase) and the cuttability of the laminate are affected by the flow of the adhesive. As far as the end-use properties are concerned the plasticity/elasticity balance is important. Both storage and cutting may be accompanied by edge flow (oozing) and bleeding of the adhesive. This phenomenon is due to the viscous flow of PSAs and occurs also at low (room) temperature.

The cold flow is a general characteristic of plastics. In order to understand its importance and the parameters influencing it, the flow of plastics and the similarities and differences between the cold flow of the plastics and that of the PSAs need to be examined. The cold flow of plastics is generally taken into account in the design of plastic elements (pieces) and where safety factors are used; when long-time, dynamical or thermal stresses are expected, lower values of the relevant material properties are used. The nature of the flow behavior does not change for amorphous plastics at higher temperatures, therefore their molten state and flow in the molten state may be taken as a simplified model for the flow of PSAs. The investigation of the parameters influencing the flow of plastics may yield information about the flow of PSAs during processing. The influence of the coating weight, shear rate, and the viscosity on the flow of PSAs will be reviewed based on the processing of plastics.

Coating Weight and Cold Flow

There are tests in polymer processing to determine the calendering ability of a polymer; these are based on the shear resistance of the material [3]. Here the maximal shear rate $\dot{\gamma}_{max}$ depends on the linear speed of the cylinder v, and the thickness of the layer h_o:

$$\dot{\gamma}_{max} = \frac{2v}{h_o} \tag{4.1}$$

that is

$$v = \frac{\dot{\gamma}_{max} \cdot h_o}{2} \tag{4.2}$$

Assuming a similarity between the material volume squeezed out during cutting of pressure-sensitive laminates and calendering of plastics (push-out of the plastic web) one can estimate that the rate of edge bleeding (i.e., the volume of the adhesive oozing out from the laminate) depends on its thickness (i.e., on its coating weight) [4], or:

$$\text{Bleeding} = f \text{ (coating weight)} \tag{4.3}$$

For polymers (plastics) the minimum pressure for flow can be estimated from the knowledge of the melt flow index (MFI); the MFI depends on the temperature and this dependence is a function of the temperature range used. Knowing the geometric parameters of the mold, temperature, and thermal properties of the melt, the minimum pressure requirement for flow may be calculated. On the other hand the thermal properties are polymer dependent (see Chapter 2, Section 1.5) [5]. For rubber blends there is a flow (creep) limit, tested with a Höppler consistometer [6], depending on sample volume V, modulus G and sample thickness h_L, according to the following equation:

$$T_o = \left(\frac{9 h_L^5 \pi G^2}{4 V^3} \right)^{1/2} \tag{4.4}$$

Viscosity and Cold Flow

During the extrusion of plastics the volume V of the extruded elastomer depends on its viscosity η_s [7]:

$$V = \left(i \cdot \frac{BH}{2} \cdot f_s \cdot v_2 \right) - \left(i \cdot \frac{BH^3}{12\eta_s} \cdot f_s \cdot \frac{dp}{dz} \right) \quad (4.5)$$

where i, BH, v, f_s, and p are polymer- and machine-dependent parameters. Assuming a similarity between the material squeezed out by extrusion and cutting of PSAs laminates, one can conclude that:

$$\text{Oozing} = f(\text{Viscosity}) \quad (4.6)$$

Thus, on the basis of Equations 4.1–4.5 describing the flow of plastomers during processing, one can suppose a similar dependence of the flow of PSAs on the viscosity, shear rate, and layer thickness; cuttability also depends on these parameters (see Chapter 2).

Other complex features of the polymer flow also show similarities with PSAs. An example is given by the melt flow viscosity of polymer blends. Here an "interfacial slip viscosity" is defined because of the observed minima in the viscosity. A multilayer structure is assumed for blends where the steady state shear viscosity of the blend at a constant shear stress may be written as follows [8]:

$$\frac{1}{\eta} = \frac{\varphi_1}{\eta_1} + \varphi_2 \eta_2 + \varphi_i \eta_i \quad (4.7)$$

where η_i are the viscosities of the adjacent layers and φ_i are the volume fractions. Measurements of η_i were lower by a factor of 10 with respect to the lowest viscosity of the mixed homopolymers. The minimum in viscosity-composition plots of the blends disappeared by the addition of selected block copolymers to the blend. Thus it may be concluded that interfaces with low frictional resistance may be formed and must be considered when modelling the behavior of polymer blends. In tackified PSAs similar viscosity-lowering phenomena occur, which may be partially avoided using polymeric additives.

Shear Rate as a Parameter for Cold Flow

Some rubber blends that exhibit good processability during pressing do not display good properties during injection moulding because of the different shear rates applied in both processes. Since the viscosity of these rubber blends is a function of the shear rate, the tests of processability have to be carried out at different shear rates according to the real shear conditions. Molten thermoplastics undergo high shear straining during processing; extrusion of polymer melts is carried out at shear rate values of 10^2 to 10^5 s^{-1}

[9]. At such a high rate, high shear processing may increase the polymer temperature and may decrease the polymer viscosity. Experimental work has developed a mathematical basis for this phenomenon. The amount of mechanical energy transformed into frictional heat (also called dissipation D) depends on the viscosity and shear rate [10].

$$D = f(\eta \cdot \dot{\gamma}^2) \quad (4.8)$$

where η and $\dot{\gamma}$ denote viscosity and shear rate, respectively. The increase of the temperature of a sheared polymer sample depends on the shear stress τ, the shear rate $\dot{\gamma}$, the density ρ, the heat capacity C_p, and on the shear time t. In a similar manner it may be assumed that cutting of pressure-sensitive laminates and the shear during cutting depends on the same parameters, namely, that the rate of the cutting and cutting time influence the bleeding and smearing of the adhesive during cutting, due (at least partially) to the increase of the temperature. Thus the equation given for the temperature increase of plastics will be valid for PSAs:

$$\Delta\theta = \frac{\tau \cdot \dot{\gamma} \cdot t}{\rho \cdot C_p} \quad (4.9)$$

where τ, $\dot{\gamma}$, C_p, and t are the above-mentioned parameters. In the processing of plastics the decrease of the viscosity at higher shear rates and the self-heating of the material is a positive phenomenon; for PSA-converting purposes however this phenomenon constitutes a disadvantage [11].

Flow Test of Plastics and PSAs

The shear rate must be examined by the test of the flow properties of the plastomers. The amount of the material yield V in a plastometer is given by Equation 4.10, where the shear rate $\dot{\gamma}$ plays an important role [12]:

$$V = \frac{\pi R^3 \dot{\gamma} T}{4} \quad (4.10)$$

where R and T are the universal gas constant and the temperature, respectively. Unfortunately in laboratory scale tests the melt flow of thermoplastic polymers is characterized by the melt flow rate (MFR), which depends on the molecular weight and molecular weight distribution measured under low shear conditions [13]. Thus no (or only limited) information about the real molding behavior may be obtained from the melt flow rate. Extrapolated to the field of PSAs, the MFR corresponds to a static, long cold flow time. It is interesting to note that like the MFR the shear of PSAs depends on the molecular weight. This statement is valid and important for tackified PSAs, where shear resistance may decrease exponentially with the level of low molecular weight tackifier resin. Like MFR values, static or low speed dynamical shear tests provide only limited information due to the high speed shear resistance (i.e., the cuttability of PSAs). In fact, the melt flow of polymers is both pressure (shear) and temperature dependent [14]. Therefore the same temperature dependence for holding power tests, or other low stress rate tests (e.g., Williams plasticity) may be assumed for PSAs. A similar temperature dependence of the flow characteristics may be observed in the evaluation of elastomers (e.g., hot-set test: tensile strength at high temperature and less than 75% elongation) [15].

Practically, there is a relation between the time/temperature dependence of the tested polymer characteristics and the processing behavior of the material. As an example Shenoy [15] demonstrated the dependence of the minimum pressure P necessary to fill a cavity during processing, on the polymer characteristics (i.e., $f(n)$, K, C and, n) and the melt flow index (MFI):

$$\frac{P_{min}}{C^{3n} R^{1+n}} = f(\eta) \frac{K \cdot \pi^n}{\mathrm{MFI}^n} \tag{4.11}$$

where η and K are rheological parameters; for the dependence of MFI on the temperature, the Arrhenius or Shenoy equation (Eq. 4.11) should be used. In a similar manner the dependence of the PSAs bleeding (or migration, especially for hot-melt PSAs) on the polymer rheological characteristics and temperature should be evaluated.

Generally there are many test methods available in order to investigate the temperature dependence of certain properties of plastics and elastomers (e.g., dependence of the tensile modulus on temperature [16], or of the critical tear energy [17]) which should be considered for similar measurements on PSAs. There are also similarities in the low temperature range and in low shear rate tests of plastics/PSAs. In the case of polymer deformation by elongation (tensile) the viscosity depends on the tensile rate (draw ratio) [18]:

$$\eta_c(t) = \frac{\sigma_{(t)}}{\dot{\varepsilon}} = \frac{\text{tensile strain}}{\text{elongation rate}} \tag{4.12}$$

The strain rate is low (below 1 sec^{-1}) and it enables the slow mechanical response of the material to be examined [19]. Therefore it may be assumed that tensile tests occur at similar low shear rates like static shear measurements (holding power). Such tests would not give any relevant data about high shear behavior (cutting) of PSAs. Table 4.1 illustrates the flow (cohesion) behavior of a PSA tested by tensile strength and statical shear (holding power). As can be seen from Table 4.1, the tensile strength of the adhesive is proportional to its shear strength but only for certain formulations.

Comparison of PSAs

Table 4.1 Test of the Adhesive Flow Via Tensile Strength and Statical Shear

Performance characteristics	Values for different formulations				
	1	2	3	4	5
Shear resistance (min)	900	200	50	170	30
Tensile strength (N/mm^2)	8.0	3.3	2.9	3.5	1.4
Elongation (%)	400	1020	1150	700	1550

Rheology of Plastics/PSAs

There is a special chemical basis of PSAs that provides the whole assembly of rheological properties (see Chapter 2). The main polymer characteristics (e.g., modulus and viscosity) may be controlled through chemistry. The basic knowledge of the interdependence polymer composition/structure and rheological parameters is known for plastomers/elastomers. Adhesive chemists have to apply these general methods to the special field of PSAs.

The most important molecular factors influencing the rheological behavior of polymer melts are the chemical composition of the monomers, the average molecular weight, the MWD, and the branching of the chains [20]. These same parameters also influence the rheology and the adhesion-cohesion balance of PSAs. Other more complex similarities also exist; the T_g is an example. The chemical composition of the monomer unit provides the flexibility and fluidity of the adhesive. Because of the volume dependence of the T_g substituted high-volume resins provide less tack. The chemical composition of the monomer makes crosslinking possible. The crosslinking density influences tensile strength and tension set [21]; in a similar manner it influences peel and cohesion of PSAs. In certain tackified rubbers, gel formed by crosslinking gives higher tack, but generally crosslinking leads to lower tack (nontackified mixtures). An increase in molecular weight by plastics may lead to incompatibility in plastic blends. In a similar manner tackifier resins with a molecular weight higher than 1000 cause incompatibility problems. On the other hand this behavior may explain the higher tackifying effect of low molecular liquid resins.

1.2 Relaxation Phenomena

For all elastomers, reinforced or not, the stress to be applied to obtain a given elongation is less important on a second elongation cycle [22]. This so called stress-softening or Mullinns effect [23] results from several mechanisms (i.e., not only the structure of the elastomeric network, but also the interactions of the polymer with the solid components of the laminate). The stress-softening effect should be taken into account for re-adhering adhesives or for the repeated (periodical) compression-yield-shear stresses when cutting pressure-sensitive laminates. The dependence of the peel force and rate for rubbery materials on the frequency of subsequent bonding and debonding (e.g., re-use, renewed application of the adhesive label), has also been described as a "memory effect". An applied label looses this memory effect after some time or when in a solvent. The memory as a function of the dwell time effect is associated with some rearrangements of the molecular structure of the rubber at the interface. This memory effect is important when re-adhering repositionable labels.

The role of the relaxation phenomena and their influence on the mechanical properties of plastics is well known. The dependence of the mechanical properties of plastic blends on the relaxation ability was demonstrated. Applying cyclical stresses to polymer/polymer mixtures leads to a dissipation of the energy in the elastical parts of the mixture; the tensile strength depends on the viscoelasticity and relaxational behavior. Similar phenomena exist for PSAs where a balance of elastic/viscous parts (given by a certain formulation) may allow a soft or hard debonding (see Chapter 2, Section 1.4) as a result of the relaxation.

1.3 Mechanical Resistance

The mechanical resistance of the plastics, rubber, and adhesives is quite different. As can be seen from the modulus values (see the Chapter 2), the modulus of the face stock materials (paper or plastics) differ by a factor of 10^3 from the modulus of rubber, or by 10^4 from the modulus of the adhesive.

Rubber is an ideal raw material for PSAs. Rubber displays auto-adhesion and shows strain-induced crystallization [24]. Amorphous rubber is very mobile during application (bonding) and strain-crystallized materials yield high shear. But PSAs are mostly mixtures of resin and rubber in a 1:1 ratio, thus it is yet not clear if the effects of the strain-induced crystallization can be extrapolated to PSAs. A more thorough examination of the main rheological and mechanical characteristics of plastics/elastomers and PSAs reveals a lot of differences in their properties. Generally the values of the modulus E, the tensile strength, and of the elasticity are much lower for PSAs than those of the elastomers (or plastomers). High modulus values as encountered with elastomeric block copolymers (e.g., 250–350 kg/cm^2), high elastic recovery values (of about 80%), or high elongation at break (800–1300%) cannot be obtained for PSAs. Their flow characteristics are very pronounced. There is a model rubber [DIN 53516] that is used to characterize rubber and it is possible to characterize the creep under compression of the elastomers in a "creep unit." If the deformation $\varepsilon \leq 0.8\%$ (see Chapter 3, Section 2.3), then for plastics one can write:

$$E \neq f(\sigma) \tag{4.13}$$

If the modulus does not depend on the stress value, the yield, compression, and flexural modulus are identical,

$$E_t = E_c = E_f \tag{4.14}$$

No model PSA exists and as deformation values for PSAs are higher than 0.8% and the modulus values for different kinds of stresses vary. On the other hand, not only the deformation value but also the mechanism of the deformation and destruction of plastics/elastomers differs from that of PSAs.

Plastics are viscoelastic materials and do not obey linear elastic fracture mechanics (LEFM) unless as an approximation where the plastic zone remains small [20]. For ductile fracture the F integral was suggested as a fracture criterion for large-scale plasticity. In a similar manner PSAs undergo ductile fracture; however in the design of removable PSAs the formula for the critical fracture toughness K_c given by Griffith [25] should also be considered:

$$K_c = \sigma_c \cdot a^{1/2} \cdot \psi \tag{4.15}$$

where ψ depends on the geometry of the specimen, a denotes the crack length, and σ_c is the critical value of the remotely applied stress at which the crack begins to grow.

It may be assumed that a similar correlation exists between a specimen geometry ψ_a, "leg" length q_a, a critical value of the applied stress σ_{ca}, and critical tear toughness K_{ca} for a removable PSA [26]:

$$K_{ca} = f(\sigma_{ca} \cdot q_a^\alpha \cdot \psi_a) \tag{4.16}$$

2 COMPARISON BETWEEN PSAs AND OTHER ADHESIVES

Common adhesives have to be fluid only during the coating process. Similar to PSAs, the flow of common adhesives during application is obtained through the aid of a dispersing agent (solvent or dispersing media) or by melting the bulk adhesive. After coating the adhesive bond is achieved by the increased viscosity of the adhesive (in comparison with the dispersed/molten one) and, in some cases, by the chemical interaction of the adhesive with the adherent. In special cases after bond forming, the adhesive itself undergoes a chemical transformation (e.g., crosslinking) that results in a higher adhesion/cohesion balance with respect to the uncoated adhesive.

This is possible by changing the viscous/elastic balance and increasing the value of the modulus. In this case the adhesive layer has similar rheological and mechanical properties to those of the adherent, and therefore debonding (without material destruction) or rebonding is no longer possible. The properties of such adhesive joints are similar to plastics or elastomers. Similarities with the rheology of PSAs are observed for the uncoated, or coated but unbonded adhesive only. On the other hand, the chemical basis of the common adhesives is almost the same as that of PSAs. The chemical composition of PSAs and common adhesives may be the same, although the role of the built-in fractional groups may be different. The presence of ithaconic and/or acrylic acid in vinyl acetate (VAc) terpolymers imparts excellent freeze-thaw stability to these emulsions. The same monomers in PSAs are used to crosslink and to impart special adhesion. The same statement is valid for the formulation. Plasticizers or tackifiers used in PSAs in order to achieve tack and peel give polyvinyl acetate (PVAc) adhesives better bonding "speed" [23].

Classical adhesives and PSAs differ in the reason why one uses plasticizers and tackifiers. For classical adhesives plasticizers and tackifiers modify the application viscosity and green tack, whereas for PSAs they modify tack and peel. The functional groups in classical adhesives serve mainly as crosslinking sites, while in PSAs they are adhesion promoters. Thus in hot-seal adhesives the resin improves the adhesion to difficult substrates, decreases the blocking resistance, and lowers the seal temperature [27]. Differences in the use of additives for water-based formulations may be noted also. Thus polymeric colloids for use in PSAs have a relatively low T_g (i.e., in the range of about –40 to –20°C), while the colloids for use in laminating adhesives have a T_g value within a range of –25 to –10°C.

REFERENCES

1. *Adhesives Age*, (3) 8 (1987).
2. H. G. Koch, *Adhäsion*, (11) 312 (1976).
3. J. Guillet and V. Verney, *Polymer J.*, (8) 773 (1984).
4. R. R. Lowman, *FINAT News*, (3) 24 (1987).
5. D. R. Saini and A. V. Shenoy, *Plastics and Rubber Proc. Appl.*, (3) 178 (1983).
6. B. Poltersdorf and D. Schwambach, *Kautschuk, Gummi,m Kunststoffe*, (5) 454 (1988).
7. H. J. Laake, *Kautschuk, Gummi, Kunststoffe*, (6) 506 (1989).
8. J. L. Jörgensen, L. D. Thomsen, K. Rasmussen and K. Soendergard, *Internat. Polymer Process*, (3-4) 122 (1988).
9. F. Kurihara and S. Kimura, *Polymer J.*, **17**, 863 (1985); in *Kautschuk, Gummi, Kunststoffe*, (3) 256 (1985).

10. H. Y. Tang and J. E. Mark, *Macromolecules*, (17) 2616 (1984).
11. H. Kehler and W. M. Kulickje, *Chem. Ing. Techn.*, (10) 802 (1986).
12. *Kautschuk, Gummi, Kunststoffe*, (9) 31 (1985).
13. S. Cartasegna, *Kautschuk, Gummi, Kunststoffe*, (12) 1188 (1986).
14. F. Grajewski, A. Limper and G. Schwarter, *Kautschuk, Gummi, Kunststoffe*, (12) 1188 (1986).
15. G. Bolder and H. Meier, *Kautschuk, Gummi, Kunststoffe*, (8) 715 (1986).
16. Y. K. Kawasaki and N. Watanabe, *Kautschuk, Gummi, Kunststoffe*, (2) 119 (1987).
17. T. Gehman, *Kautschuk, Gummi, Kunststoffe*, (6) 493 (1989).
18. T. S. Ng, *Kautschuk, Gummi, Kunststoffe*, (9) 830 (1986).
19. J. N. Hay, *Coaters Scientific Conference Communications, Surface Coating Evaluation and Performance*, 1988.
20. H. M. Laun, *Kautschuk, Gummi, Kunststoffe*, (6) 554 (1987).
21. G. Bolder and H. Meier, *Kautschuk, Gummi, Kunststoffe*, (3) 196 (1984).
22. J. B. Bonnet, M. J. Wang, E. Papirer and A. Vidal, *Kautschuk, Gummi, Kunststoffe*, (6) 510 (1986).
23. L. Mullins, *J. Rubber Res.*, (16) 275 (1947).
24. G. Hammed, *Rubber Chem. Technol.*, (5) 576 (1981).
25. A. Griffith, in *Coaters Scientific Conference Communications, Surface Coating Evaluation and Performance* (J. N. Hay, Ed.), 1988.
26. R. Dörpelkus, *apr*, (16) 456 (1986).
27. E. A. Theiling, *Kunstharznachrichten*, (1) 14 (1972).

5
Chemical Composition of PSAs

A typical PSA is derived from a film-forming elastomeric material such as styrene-butadiene rubber, butyl, silicone, nitrile and acrylic rubber. The elastomer provides flexibility, ease of tackification, and the desired bond strength when compounded with compatible tackifiers, pigments, plasticizers, waxes, and oils. Polymers for PSAs based on natural rubber represent 60% of the raw materials, block copolymers and water-based acrylics amount to 18%, and the remaining part is mainly covered by solvent-based acrylics [1]. Although new elastomers were identified as base materials for PSA applications, natural rubber is still the most preferred polymer.

From a chemical point of view classical PSAs are elastomers that exhibit viscoelastic properties due to their low T_g (–40 to –60°C) and modulus. Their macromolecular basis is built on long-chain polymers, with a certain degree of branching and with or without crosslinking. The balance between plastic and elastomeric properties is governed by the polymer nature and structure, and molecular weight. For classical rubber-based PSAs, structure and molecular weight adjustment occurs via crosslinking or mechano/chemical destruction of the network (e.g., mastication, tackification, plasticizing). Synthetic raw materials for PSAs can be tailored in order to achieve the chemical characteristics required to obtain the desired adhesive properties. A detailed study of the interdependence between the chemical basis and physical structure of PSAs was carried out by Krecenski, Johnson, and Temin [2].

Pressure-sensitive adhesives are amorphous, viscoelastic materials, and their flow behavior is characterized by the viscosity and the modulus (see Chapter 4). On the other hand a special structure and temperature dependence of the internal phase transitions serves as basis for the required rheology. Since the viscosity, modulus, and the glass transition temperature are dependent on the chemical composition, changes in the composition and/or in the macromolecular characteristics of the basic elastomers may influence the rheology of PSAs. From a theoretical point of view it is more accurate to discuss the influence of the compositional and macromolecular factors on the properties of PSAs separately. However, it is not within the scope of this book to present a detailed, purely theoretical approach to PSAs; therefore the chemical basis (composition and macromolecular characteristics) will be discussed together and, if relevant, related to the adhesive practice.

1 RAW MATERIALS

Early PSAs were based on organic solvent solutions of synthetic rubber that were tackified by "some type of resinous material" [2]. These so-called rubber-resin (RR) adhesives are still widely used. With regard to their chemical basis and application technology, PSAs may be classified in solvent-based systems, hot-melt adhesives, radiation-cured materials, and emulsions. Solvent-based systems may include rubber/resin adhesives, acrylics (thermoplastic and crosslinked), and silicones. Hot-melt PSAs are based on block copolymers and acrylics. The radiation (electron beam and UV)-cured materials can be either acrylic- or rubber-based. Finally emulsions (water-borne adhesives) comprise acrylics, natural and synthetic rubber latexes, and ethylene vinyl acetate dispersions.

The original compositions were heterogeneous physical mixtures of one or more elastomers (rubbers) with tackifiers and/or plasticizers. The second approach, essentially homogeneous polymers based mainly on acrylate esters, was introduced in the mid-1950s. Both techniques were refined and their usefulness extended as a result of extensive research and development on new raw materials and crosslinking methods, and both are currently available in water-borne and 100%-solids form as well as in the original solvent solution. Deficiencies persist, however, especially as the use of PSAs was expanded into new, more demanding markets. Particular needs exist for PSAs with good aging characteristics, with broader specific adhesion, higher cohesion and heat resistance, and reduced plasticizer-induced performance losses when used on flexible polyvinyl chloride (PVC) face stock materials. Some alternative compositional approaches have shown considerable improvement, especially for silicones, but these are too expensive for general use, as are urethane-based PSAs.

1.1 Elastomers

A main characteristic of PSAs is bond forming (tack). There are three fundamental criteria that must be met in order for an elastomer to exhibit tack [3]. The polymeric chain must come into molecular contact with the surface, thus the chains must interdiffuse across the interface and become entangled with one another. After bond formation the elastomer must display high cohesive strength to resist easy separation upon application of a force. This property distinguishes a tacky elastomer from a low molecular weight liquid. This combination of partially contradictory characteristics will be embodied by a viscous/elastic, rubber-like material. Therefore the most important component of a PSA formulation is the elastomer.

A partially crosslinked elastomer (i.e., natural rubber) was used as starting material for PSA formulations. Rubber-like synthetic materials, with a rubber-like (hydrocarbon-based) chemical composition were developed later and used with or without natural rubber. The development of soft and elastic polymer coatings (usually acrylic-based) opened up new ways in the synthesis of elastomers. Later multicomponent copolymerization of nonpolar monomers was developed, which yielded a new class of soft and elastic, thermoplastic elastomers. Heterocyclic compounds exhibiting a special bond stability function as the elastomeric backbone of the formulation (e.g., silicones). Thus there is a broad range of elastomers available for PSA formulations.

Uncrosslinked Elastomers

Diene-based natural and synthetic rubbers always contain more or less reactive side groups, or branching, and thus they always possess some crosslinked structure. The development of the acrylic-based elastomers allowed the synthesis of pure elastomers that are not crosslinked. Later, other comonomers, usually vinyl acetate and its derivatives, were copolymerized with acrylics mainly for economical reasons. Looking for a low-cost, low-T_g monomer forced the development of ethylene vinyl acetate elastomers and, related to these compounds, a new renaissance of the maleinate copolymers. Actually there is a broad range of uncrosslinked elastomers that are suitable as base materials for PSAs. It is evident that acrylics have the most sophisticated chemical basis and therefore may satisfy the extreme requirements concerning the different physical state (solvent-based, dispersed, or hot-melt PSAs) and end-use properties (permanent/removable, paper/film). On the other hand, the new ethylene multipolymers will win new fields of application due to their dual (technical/economical) advantages.

Natural Rubber as Base Elastomer for PSAs. Natural rubber was used almost exclusively as the base elastomer for PSAs in the early stages of PSA manufacturing. Mechanochemical destruction of the base elastomer (by mastication or dissolution) allows the modification of the molecular weight or molecular weight distribution. The good compatibility of natural rubber with different tackifiers or plasticizers allows an easy adjustment of the adhesion/cohesion balance. In some cases more cohesion is required; therefore blends of natural rubber with styrenic block copolymers are preferred.

Natural rubber is a well-known component of traditional rubber/resin systems, but its use (in latex form) in emulsion-based PSAs for permanent label stock applications is less widely appreciated [4]. Natural rubber latex can be used in SBR/tackifier resin systems to modify the balance of adhesive properties. High molecular weight natural latexes contribute to improved shear resistance, while the lower molecular weight ones enhance the tack (especially when natural latex is used at less than 10% of the polymer content in the formulation). Furthermore it was claimed that when SBR and natural rubber latexes are blended, a product with better aging characteristics results. Here the SBR latex crosslinks and hardens under oxidative aging, whereas the natural rubber latex undergoes chain scission and softens, whence a balance is achieved [5].

Natural rubber latex-based PSAs offer an advantage over solvent-based systems (e.g., milled smoked sheet) with regard to the molecular weight difference between the two. The high molecular weight portion of natural rubber is insoluble in solvents, therefore it cannot be used in solvent-based adhesives. Natural rubber must be milled to a Mooney viscosity of 53 or below to obtain complete solubility (at 30% solids) [6]. The natural rubber from dried latex shows a wider plateau of the modulus and a higher elastic modulus value at 100°C than the milled natural rubber [6]. The degradation does not change the tan δ peak temperature (T_g), but it does reduce the modulus at high temperatures. This modulus reduction relates to the low shear performance of solvent-based systems. An adequate rubber latex has a gel content low enough to allow good quick stick, but high enough to give good holding power when properly formulated [7]. A detailed discussion about the use of natural rubber per se (solid state) or in latex form will be presented in Chapter 8.

Synthetic Elastomers. Batch-to-batch consistency and staining are two problem areas and/or disadvantages of rubber/resin PSAs [8]. Molecular weight and molecular weight dispersion both influence rubber compatibility with tackifiers and the adhesion/cohesion balance. Unfortunately the molecular weight and its distribution depend on the natural rubber quality and on the processing parameters. Therefore the scattering of the adhesive performance values remains high for natural rubber-based adhesives. For a better reproducibility synthetic rubber-like products were tested. First pure hydrocarbon-based polymers (polidienes or diene-styrene copolymers) were synthesized; later acrylic and vinyl monomers were combined.

Synthetic, hydrocarbon-based elastomers, homo- or copolymers, and random- or block copolymers (two component or multipolymers), stereoregulated or not, are synthesized and used as raw materials for PSAs. Low-speed crystallization of polyolefins permits their use as semi-PSAs [9]. Macromolecular compounds based on dienes were used. Stereoregulated 1,4 *cis*-polyisoprene is recommended for special formulations [10]. On the other hand mono-olefins like atactic polypropylene are also used. The best results were obtained copolymerizing styrene with mono- or diolefins. The first products were designed for hot-melt PSAs; later carboxylated styrene-butadiene rubber (CSBR) dispersions were synthesized (see Chapter 8). In order to improve the adhesion of polybutadiene-based PSAs, these were modified with isopropyl dicarboxylate [11,12]. Other synthetic rubbers (butylene- and isobutylene rubber) were used as tackifier. The use of methyl rubber and polybutylene in adhesives was described in the literature [13]. Development of special products for specific end-uses, clarification of the structure/property relationships, and introduction of new products such as styrene-ethylene-butylene-styrene thermoplastic elastomers (TPE) and crosslinkable derivatives were documented [14].

Ethylene-, Propylene-, and Diene Copolymers. New, mainly amorphous ethylene/propylene/butene-(1)-terpolymers can be made tacky quite easily and they exhibit a low speed of crystallization. Propylene/hexene copolymers were also tested for hot-melt PSA applications [14]. With hot-melt adhesive formulations with a long "open time" one takes advantage of the slow crystallization. During cooling these products exhibit properties like a hot-melt PSA for a long period; afterwards, however, they are no longer tacky. They appear quite suitable for application by spraying. Hot-melt PSAs with an excellent price/performance ratio are based on a highly viscous, amorphous polyalpha-olefin (APAO), rubber, resin and a combination of plasticizers. The latter consist of a mineral oil and polyisobutylene (see Section 1.2). Amorphous polyalpha-olefins encompass a group of usually low molecular weight polyolefins which are obtained by coordination polymerization [15]. Atactic polypropylene (APP) is a byproduct in the production of isotactic polypropylene, using first-generation Ziegler-Natta catalysts. These are the distinctive product types of APAO made by direct reactor synthesis, namely homopolymers of propylene, copolymers of propylene and ethylene, and copolymers of propylene and 1–butadiene [15].

Single-Component PSAs. Proper selection of comonomers and synthesis conditions has led to the development of olefin copolymers which are pressure-sensitive [16,17]. Because these materials are neat olefin polymers, they offer several advantages over typical styrene block copolymer-based, tackified PSAs. These include the absence of oils which can bleed into the substrate, fewer skin sensitivity problems, a good thermal stability (< 10% viscosity decrease after 100 hr at 172°C in a forced-draft oven), low color and odor, the ability to be blended with most olefin compatible adhesive ingredients for common applications, and finally a lower density than conventional rubber-based hot-melt PSAs leading to a decreasing usage of raw materials. These raw materials are mainly amorphous terpolymers of alpha and omega olefins. A range of properties can be obtained using this new amorphous polyolefin (APO) technology. Examples include a T_g between -10 to $40°C$, a probe tack of 0–800 g, and a static shear ranging from << 1 to 30 hr. The 180° peel strength (PSTC–1), 90° quick stick (PSTC–5) and shear resistance (hr) amount to 2.5, 0.8, and 8, respectively.

Styrene Copolymers. Thermoplastic elastomers were known since the 1960s [18]. In 1960 Shell developed Kraton in the USA and in 1974 Cariflex; later, ethylene butylene copolymers (S-EB-S) were developed. The first styrenic thermoplastic elastomers were introduced to the market in 1964 [19]. The styrene block copolymers used today usually have a molecular weight in the range of 60,000–200,000 and a styrene content of 15–45% [18]. Since their introduction styrenic block copolymers have found their way into numerous and varied application fields in the adhesives industry. Their molecular structure allows these thermoplastic rubbers to be formulated and coated in solvent-free hot-melt systems which are widely accepted for a variety of flexible assembly adhesives and as well as PSAs for labels and tapes. Many variations can be made in the structure of these block copolymers which has led to a wide range of grades which differ in physical properties and are well suited for more specific applications. The most commonly used block copolymers for hot-melt PSAs still are the linear S-I-S grades with low polystyrene content, although interest is growing in the use of low viscosity S-B-S grades in combination with newly developed tackifying resins.

Generally block copolymers used for PSAs contain three blocks A-B-A [20] where A is an amorphous polymer with the T_g above room temperature (thermoplastic), and B is an amorphous polymer with T_g much lower than room temperature (in rubber, for example, A is polystyrene and B polydiene). Because of the incompatibility of polystyrene endblocks and elastomeric midblocks, there are two microdomains in the polymer; a discontinuous phase and an agglomeration of polystyrene endblocks is formed through Van der Waals forces. The first thermoplastic elastomers (TPE) were of the linear triblock form, S-B-S or S-I-S [21]. The key to their function is what was called a "spaghetti and meatballs" morphology. At an appropriate balance of end-block polystyrene molecular weight versus those of the elastomer midblock, a two-domain structure is formed which consists of polystyrene islands in a rubbery ocean. Polystyrene is not miscible in polydienes and the polystyrene blocks at the end of each polymer molecule associate with each other, forming an "associative crosslink" at room temperature which behaves like a covalent vulcanized bond. If the polystyrene content is lower than 50%, polystyrene domains are formed through the incompatibility of the polystyrene with the polybutadiene sequence.

In general, thermoplastic elastomers are block copolymers or polyblends with separate phases [19]. In these copolymers with an A-B-A structure, the A sequences contain 200–500 units and the B sequence 500–700 units [22]. The three types of commercially available block copolymers are the S-I-S, S-B-S, and S-EB-S types, where S, I, B, and EB represent blocks of polystyrene, polyisoprene, polybutadiene, and poly(ethylene/butene), respectively. The most commonly used rubber is of the S-I-S type, such as Kraton D 1107 [21]. The S-I-S types are widely used in hot-melt PSAs, but are too soft for use in nonpressure-sensitive hot-melt PSAs [23].

The molecular structure of S-I-S polymers allows them to be dissolved, formulated and applied in solvent, or to be formulated and coated in a hot-melt system at temperatures higher than the T_g of the styrene endblocks. Following evaporation of the solvent or cooling down of the hot-melt, the physical crosslink structure rebuilds to provide the desired strength and performance characteristics. As thermoplastic elastomers have a less stable elastic domain and a second T_g (identical with the melting point) they act like thermoplastics above this temperature. Unfortunately, the physical bonding is less stable [24]. Moreover, the nature of an S-I-S polymer prevents its use in PSAs which must withstand solvent or high-temperature exposure at or above T_g of the polystyrene endblock.

Recently new S-I-S polymers were introduced which can be crosslinked through electron-beam (EB) curing. This new radiation-sensitive polymer (e.g., Kraton D 1320 X rubber) shows good crosslinking with an electron beam dose of 5–7 Mrad [20,21]; it is a multi-armed S-I-S block copolymer with only 10% styrene content. Further development of the styrene block copolymers concerns the sequence distribution and the soft monomer unit.

Polystyrene-polybutadiene-polystyrene polymers are completely amorphous and therefore show limited compatibility with waxes; the polybutadiene mid-block is unsaturated and therefore exhibit limited stability during hot-melt processing. The commercial S-EB-S polymers have a much better stability during hot-melt processing because their mid-block is a saturated polyolefin rubber. The EB rubber mid-block of the commercial S-EB-S polymers is essentially a random copolymer of ethylene and 1–butene. In the commercial S-EB-S polymers the 1–butene content is controlled at a level high enough to make the EB copolymer mid-block almost totally amorphous, making the copolymer thoroughly rubbery and soluble in hydrocarbon solvents. In S-EB-S the saturated polyolefin confers excellent resistance to light, ozone, and to heat to the polymer.

Styrene block copolymers are usually synthesized via a three-step process. A new process produces styrene block copolymers via a two-step process using a difunctional initiator instead of a monofunctional one [19]. In this new technology the mid-block (e.g., the polyisoprene or polybutadiene) is built first and the polymer chain grows in two directions at the same time, with two reactive sites, one at each end. In the second step two polystyrene blocks of the S-I-S (or S-B-S) are built at the same time. As a result a pure 100% triblock polymer is formed.

The adhesive strength of adhesive tapes under humid conditions and the heat resistance of adhesives were increased by butadiene and methyl styrene copolymers with silicone units [25].

Synthetic Latexes. Synthetic styrene-butadiene latexes were first produced and sold into adhesive end-uses in 1946 [26]. Generally SBR latexes have a solids content of 41–54% with a butadiene content of 70–75% [27]. For PSAs the most effective SBR latexes are those with a styrene level of about 25–35% and a T_g between –60 to –35°C [28]. Low temperature synthetic butadiene rubbers (GR-S) with 30% styrene are soft and tacky; high temperature ones (with 45–80% styrene) are harder and less tacky [26]. The most important properties of SBR latexes used for PSAs include their molecular weight and molecular weight distribution, T_g, and styrene content [29].

Styrene-butadiene rubber can be made with or without carboxylic groups. Some of the more common vinyl acids used for the carboxylation are acrylic, ithaconic, and fumaric acids. They are usually present at a level of about 1.0–3.0%. When COOH groups are not present in the polymer chain, SBR are usually poststabilized with about 1–4% of surfactants; antioxidants and pH-adjusting agents are also added [26]. A carboxylated SBR latex is typically prepared through the emulsion copolymerization of butadiene, styrene, and a small amount (generally less than 5%) of a functional monomer of the vinyl carboxylated type (e.g., acrylic, methacrylic, fumaric, or ithaconic acid), either alone or in combination [29]. The CSBR dispersions possess small (typically 0.15 μm) particles. Butadiene can polymerize via two different modes of addition, namely the 1,2- or 1,4- configuration. Both configurations contain residual double bonds capable of further polymer growth, either branched or crosslinked. The resulting three-dimensional network structure improves the mechanical properties without apparently affecting the T_g.

The gel content may be adjusted through the ratio of styrene to butadiene and molecular weight. Tack increases with the butadiene content (50–70%), but changes little at higher levels. When measuring peel adhesion the failure mode is adhesive at lower butadiene levels and cohesive at high butadiene levels. Shear strength is higher as the styrene content increases. Above 30% gel content (70% butadiene content) the tack decreases rapidly. The usual range of hot carboxylated SBR latexes runs from about 40% styrene to ± 70% styrene; the normal range for PSAs latexes generally lies below 45% styrene. For many systems the T_g is fixed at –55°C for use in combination with an 80°C softening point resin [30]. The molecular weight and the T_g of carbox-

Chemical Composition of PSAs 127

ylated butadiene polymers constitute major variables that determine the pressure-sensitive properties.

Carboxylated SBR latexes consist of a mixture of polymer species having linear chains, branched chains, and crosslinked materials. The relative ratio of the species can be controlled by changing the concentration of molecular weight regulators, and by varying factors such as conversion, reaction temperature or polymer particle number [30]. In some cases, for the characterization of CSBR, the acid equivalent (carboxyl equivalent) for 100-g latex solids is given [7]. Carboxylated neoprene latexes have the following advantages [7,31,32]: mechanical stability, resistance to electrolytes, crosslinking at room temperature, thermal stability, good polyethylene (PE) adhesion, high shear, low peel, removability, temperature resistance, and conformance with FDA regulations (e.g., 175.105 for adhesives).

Aqueous acrylic and modified acrylic PSAs possess certain limitations, including poor adhesion to polyethylene and polypropylene surfaces. Attempts to improve their performance by the incorporation of tackifiers have met with only limited success [29]. Tackified carboxylated styrene-butadiene rubber latex systems, on the other hand, seem to be able to overcome the apparent inherent limitations of water-based systems and yet match the superior performance characteristics of the solvent-based rubber/resin systems. Until the 1980s, relatively few SBR latexes were available with the right characteristics necessary for PSAs. The most important factors in the design of CSBR for PSAs and their interaction with the other components of the system were discussed in detail [29]. When using CSBR latexes it is better to produce a low viscosity latex and control the final rheology of the adhesive through the addition of thickeners. A comparative evaluation of the formulation and end-use properties of CSBR and acrylic PSAs was summarized by Benedek [33] and will be discussed in Chapter 6.

Carboxylated Neoprene Rubber. Some years ago contact adhesives on neoprene basis were developed [32]. Later carboxylated neoprene latexes with better tack for PSAs applications were synthesized [31]. The formulation and end-use of carboxylated latexes will be reviewed in Section 2.9 of Chapter 8.

Acrylic- and Vinyl-Based Elastomers. The special structure of the diene monomers (e.g., butadiene, isoprene, etc.) leads to macromolecular compounds displaying rubber-like properties. In order to achieve viscoelastic properties the natural, synthetic diene-based elastomers must be combined with tackifiers. In order to achieve fluidity at processing temperature and a good adhesion/cohesion balance at room temperature, synthetic block copolymers (with a crosslinked structure) were synthesized. Both tackifying or internal crosslinking are used in order to achieve balanced PSA properties. The mostly nonpolar backbone of the hydrocarbon-based elastomers requires formulating in order to achieve a suitable adhesion/cohesion balance. The use of polar monomers allows the synthesis of single component, unformulated elastomers with an adequate viscoelastic balance. The most common monomers used for this purpose are acrylic and vinyl monomers. It is generally agreed that PSAs based on acrylics are the most important adhesives for the production of label stock. The last several years new classes of raw materials for PSAs (e.g., CSBR, ethylene vinyl acetate copolymers) were developed. There was a considerable interest recently in a comparative evaluation of acrylics and other raw materials in order to forecast the development trends.

Acrylics were used in PSA applications since 1928 [34]. In a manner different from natural and synthetic rubbers, hydrocarbon-based acrylates (AC) are synthesized from esters of polar, organic acids, mainly acrylic and methacrylic acid [35]. The comonomers are the acids or other nonacrylic monomers. The choice of the main monomer determines the T_g of the polymer and its characteristics, especially the tack. Introduction of comonomers allows the (re)adjustment of the adhesion/cohesion balance. Comonomers with reactive functional groups allow crosslinking and thus the improvement of the shear resistance. An advantage of the acrylics is their synthesis-based pressure-sensitive character (i.e., they possess pressure-sensitive properties without additional tackifiers and they display stable storage and aging properties). Moreover, acrylics are not sensitive towards oxydation. Their adhesion, cohesion and tack do not change after long aging periods at high temperatures [35]. Another advantage of acrylics is their lack of sensitivity towards UV light; acrylics are colorless and they do not change their performance characteristics during aging; they also display high transparency. Acrylics are resistant towards plasticizer migration; they do not contain oligomers and therefore they do not give rise to migration. Acrylics are polar, so they display a good adhesion towards polar substrates. Thanks to an adequate adhesion/cohesion balance acrylics usually display good wetout properties and die-cuttability.

Because of their wide chemical basis and well-known chemistry it is possible to polymerize acrylics in solution, dispersion or in the bulky state; there are solvent-based, water-borne, hot-melt, and radiation-curable acrylic PSAs. The relatively low molecular weight and linear, gel-free structure of solvent-based acrylics allows the use of a broad range of solvents, while the chemically inert nature of these solvents makes the use of different crosslinking agents possible. On the other hand the polarity, water insolubility, and high reactivity of the main acrylic monomers allow the synthesis of mechanically stable, protective colloid-free, grit-free aqueous dispersions with a high solids content. At the present time hot-melt acrylic PSAs are at the developmental stage; this is also the case for radiation-cured, solventless adhesives.

Polyacrylates are used mostly as solutions or dispersions. Solid state polyacrylate rubbers are also known [36]. They are produced by emulsion polymerization technology to form a latex, which is subsequently coagulated, washed, and dried in sheet form. Polyacrylic rubbers possess excellent physical characteristics. They are resistant to a wide range of chemicals and oils, have good low and high temperature resistance, and a high degree of resistance to weather, atmospheric oxidation, and UV due to the saturated backbone. Polyacrylates do not contain double bonds (as in diene rubber) [36]. They are basically saturated copolymers containing 95–99% backbone monomers consisting of alkyl, or alkyl and alkoxy acrylates such as ethyl, butyl, methoxyl ethyl, ethoxyl ethyl and 1–5% cure site monomers with chlorine or hydroxy functionality for vulcanization purposes. The main applications of polyacrylate rubbers are solvent-based and hot-melt PSAs (100% solids) pressure-sensitives. New solid and liquid polyacrylates with many functional groups (e.g., hydroxy, carboxy, epoxy, and isocyanate groups) for possible crosslinking were developed also. Physical characteristics of polyacrylic elastomers include resistance to temperature (–40 to 20°C), petroleum, synthetic oils, aliphatic hydrocarbons, oxidation, ozone, and UV.

Even though it is not the aim of this book to discuss the synthesis and monomer basis of PSAs, a short overview of the acrylic components follows. Long chain polyacrylates do not exhibit sufficient cohesive strength; the cohesion may be improved copolymerizing *n*-butyl acrylate, acrylic acid, and other polar comonomers like glycidylmethacrylate, N-vinylpyrolidone, methacrylamide, acrylonitrile, etc. [37]. The patent literature generally identifies alkyl acrylates and methacrylates of 4–17 carbon atoms as suitable monomers for PSAs [38]. The most commonly used monomers are 2–ethyl-hexyl acrylate, butyl acrylate, ethyl acrylate, and acrylic acid. A detailed list of monomers used for PSAs is given by Satas [38]; he discussed the use of more than 200 comonomers for PSAs. Monomeric acrylates or methacrylate esters of a nontertiary alcohol (having 1–14 carbon atoms) with the average number of carbon atoms between 4 and 12 may be used [39]. Additional monomers (e.g., ithaconic acid, methacrylic acid, acrylamide, vinyl acetate, *n*-butyl acrylate, etc.) can be included in minor amounts (e.g., a 74:20:6 iso-octyl acrylate:*n*-butyl acrylate:acrylic acid monomer ratio). Similarly other acrylates derived from alcohols with 4–14 carbon atoms may be included [40].

As discussed earlier in Chapter 2, definite trends in physical properties of the polymers can be observed, when higher molecular weight acrylic esters are used. As their ratio to low molecular weight monomers is increased, the following changes in properties occur, namely, the tackiness increases, the hardness decreases, the tensile strength decreases, the elongation increases, and the water absorbtion decreases.

The term *hard monomer* refers to a monomer which when homopolymerized yields a polymer with a T_g above –25°C, preferably above 0°C. Among such monomers are methyl acrylate, alkyl methacrylates (e.g., methyl methacrylate, ethyl methacrylate, and butyl methacrylate), styrenic monomers (e.g., styrene and methyl styrene), and unsaturated carboxylic acids (e.g., acrylic acid, methacrylic acid, itaconic acid, and fumaric acid). The term *soft monomer* refers to a monomer which when homopolymerized yields a homopolymer with a low T_g (i.e., less than $-25°C$). Examples are the alkyl acrylates (e.g., butyl acrylate, propyl acrylate, 2–ethyl hexyl acrylate, isooctyl acrylate, and isodecyl acrylate). Presently 2–ethyl hexyl acrylate and/or butyl acrylate are preferred [41]. Soft monomers, like the diester of fumaric acid (dibutyl fumarate) with 2–8 atoms in the ester group or alkyl acrylate with 2–10 carbon atoms (e.g., EHA), should be incorporated at 50–85% in PSAs [42]. Hard monomers like alkyl methacrylate (e.g., MMA) with 2–6 carbon atoms and unsaturated carboxylic acid (e.g., acrylic acid) with 2–8 carbon atoms should be incorporated at 0–10% in PSAs. As an ideal formulation for solvent-based and water-based PSAs, the following composition was given [43]: 70–90% soft monomers, 10–30% hard monomers, and 3–6% functional monomers.

Kuegler [44] discussed copolymerization of acrylic esters with a minimum amount of fumarate diesters. Copolymers containing octyl acrylate, ethyl acrylate, vinyl acetate and maleic anhydride were patented by National Starch and Chemical Corporation some years ago. These PSAs have a T_g range of -45 to $-65°C$ [45]. Emulsion acrylic PSAs from acrylic-based fumaric acid ester interpolymers may be synthesized when peel and tack values can be modified without affecting cohesive strength [42]. An improvement of the permanent adhesive character of the adhesives is brought by grafting polar groups on the polymer; these reactive species are able to cause chemical bonds between the polymer and the adherent [46]. Urethane acrylate oligomers may also be used as photosensitive raw materials for PSAs; their hardness may be modified by a reactive diluent (e.g., acrylate monomers) [47].

Natural rubber was used almost exclusively as the base elastomer for PSAs in the early stages of PSA manufacturing [48]. It remains difficult to synthesize high molecular rubber-like acrylics with a low softening point range, although polyacrylate elastomers were known for about 20 years. Acrylic hot-melt PSAs with an improved creep resistance at ambient temperatures and desireable melt viscosity at elevated application temperatures are prepared by copolymerizing acrylic and methacrylic acid and alkyl esters with 10–40 wt% of an acrylate- or methacrylate-terminated vinyl aromatic monomer-based macromolecular compound [49].

In recent years acrylic-based adhesives which possess a balance of high tack, high peel and high shear properties were prepared by coating a blend of monomers (such as isooctyl acrylate and acrylic acid) on a web, maintaining an inert atmosphere, and polymerizing the blend in situ. The commercial preparation of such products, however, requires a substantial investment in unconventional manufacturing equipment [50]. Currently low-voltage electron beam accelerators with high output are used [51].

The main competitors to acrylic monomers are the acrylic copolymers; they are macromolecular compounds with a relatively high (more than 10% wet/wet) content of a second, main comonomer like vinyl acetate (AC/VAc) or styrene (AC/S). Among the nonacrylic-based copolymers, one needs to mention the ethylene and vinyl acetate copolymers, the vinyl ethers (VE), and polyurethanes (PUR). Ethylene copolymers are elastomers with a second, main comonomer like vinyl acetate (EVAc), vinyl acetate and acrylics (EVAc AC), or maleic acid derivatives. In order to produce vinyl acetate copolymers maleic acid derivatives are usually copolymerized with vinyl acetate, in order to lower the T_g of the polyvinyl acetate/maleinate. Water-borne adhesives based on polyvinyl acetate were discussed in detail [52].

Ethylene Copolymers. In order to use ethylene polymers as raw materials for PSAs, their plastomeric character must be changed in an elastomeric one, and they must display a good compatibility with viscous components to achieve the unique viscoelastic behavior of PSAs. Through copolymerization branched structures are introduced into the ethylene linear backbone and thus made less compact. A level of 20% vinyl acetate makes PE practically crystalline, a level of 40% produces a fully amorphous material. Above 30% vinyl acetate, EVAc copolymers behave like elastomers [53]. Recognizing the potential advantages of EVAc for use in PSAs, a lot of research and development work was carried out in order to come up with practical, commercial systems.

A comprehensive treatment by Benedek [33] covered attempts to utilize EVAc dispersions as basis for formulating PSAs. Benedek compared tackified EVAc- and SBR-based PSAs with acrylates and concluded that with the exception of shear strength, the adhesive properties of EVAc and SBR PSA dispersions are inferior to those of acrylate dispersions. In ethylene vinyl acetate copolymers ethylene is the softening component; its softening effect is stronger than that of dibutyl maleinate. For an equivalent softness of EVAc copolymer films a level of 43% dibutyl maleinate, 45% butyl acrylate, 73% ethyl acrylate or only 25% ethylene is necessary. As little as 18% ethylene content decreases the T_g of polyvinyl acetate down to 0°C. For the same effect 20% vinyl maleate, 30–32% dibutyl maleinate, or 33–34% butyl acrylate are required [54]. The development of soft, tacky PSAs, starting from the hard vinyl acetate homopolymer is illustrated in Figure 5.1.

Polyvinyl acetate hompopolymers were used as adhesive dispersions since 1940 [55]. Later they were plasticized using alkyl maleinates and acrylates as comonomers. The first copolymers with ethylene were synthesized in 1960. Ethylene vinyl acetate (EVAc) copolymers were available for more than 25

GENERAL PURPOSE ADHESIVES	PAPER-BOARD PAPER-FILM BOARD-FILM FOIL-CHIPBOARD ADHESIVES	FILM-FILM ADHESIVES	PRESSURE- SENSITIVE ADHESIVES
→	→	→	→
PVAC (POLYVINYL ACETATE HOMOPOLYMER)	P(VAC-E) (VINYL ACETATE-ETHYLENE) COPOLYMER	SBR (STYRENE BUTADIENE) COPOLYMER ACRYLIC COPOLYMER VINYL ACETATE- ACRYLIC COPOLYMER	TACKIFIER
HARD, NONTACKY	SEMI-HARD, SEMI-TACKY FLEXIBILITY	SOFT	VERY SOFT, TACKY

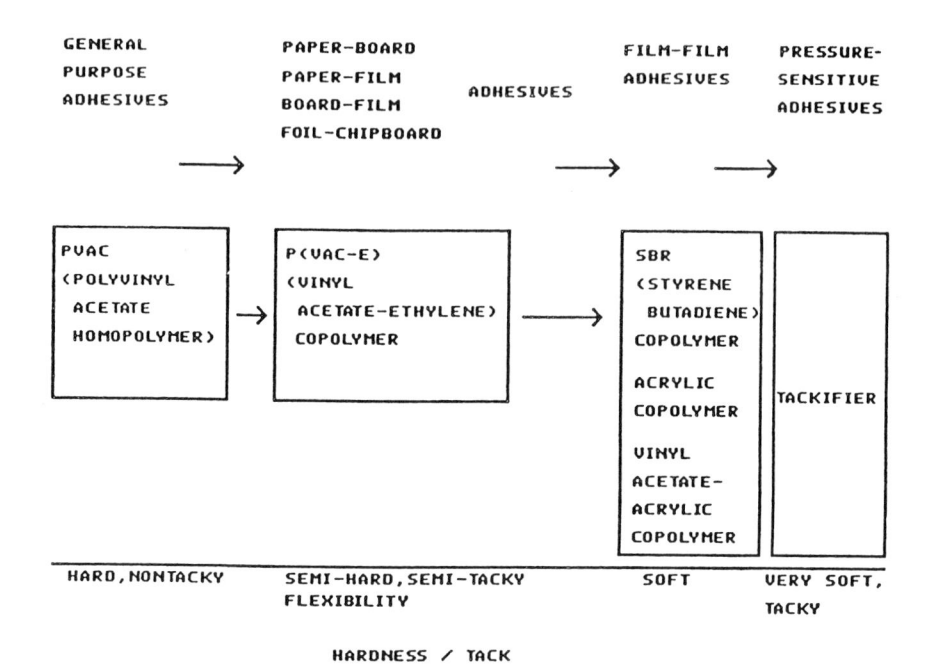

HARDNESS / TACK

Figure 5.1 Transition from hard, nontacky to very soft, tacky PSA.

years; their wide use in nonpressure-sensitive bonding applications is based to a large extent on improvements over other adhesives in those areas where traditional PSAs were deficient. In addition EVAc feed stocks were relatively low in cost and readily available [56].

Dupont introduced the first commercial products under the Elvax trademark in the early 1960s for use in hot-melt PSAs. These proved to exhibit excellent properties compared to the polyethylene based hot-melt PSAs and the aqueous adhesives such as dextrine and polyvinyl acetates. Water-borne EVAc for laminating adhesives were patented and introduced by Air Products in 1973. They quickly became the standard for many applications, particularly as replacements for externally plasticized polyvinyl acetate emulsions.

Polymerization of EVAc-containing polymers with intrinsic pressure-sensitive properties is a more basic approach. Ethylene vinyl acetate chemistry requires specialized, high pressure polymerization equipment and most of the work was carried out at companies having prior expertise. Wacker-Chemie GmbH was awarded the first U.S. patent [57]. Wacker's claims covered a copolymer blend of the following composition:

Chemical Composition of PSAs

Monomer	% by weight
Ethylene	10–30
Acrylic ester	29–69
Vinyl acetate	20–55
Methacrylamide	0.2–8
Others	0–12

The T_g range is specified at -20 to $-60°C$. The National Starch and Chemical Corporation is the other current holder of U.S. patents for EVAc-based interpolymer PSAs. A different approach was taken at National Starch; drawing on extensive EVAc experience in the packaging field, it was decided to maximize the ethylene content in the copolymer. A high EVAc content provides optimum specific adhesion and plasticizer tolerance when the adhesives were used on flexible PVC. While the original National Starch EVAc materials retained their peel value when aged on plasticized PVC face stock material, the cohesive strength was low. Development work resulted in a new U.S. patent [56], covering the following composition:

Monomer	% by weight
Vinyl ester of alkenoic acid	30–70
Ethylene	10–30
Di-2–ethyl hexyl maleate or di-*n*-octyl maleate (or the corresponding fumarates)	20–40
Monocarboxylic acid	1–10

Wacker Chemie GmbH continued developmental work and high solids vinyl acetate ethylene dispersions (with 60% solids) were synthesized [58]. The tackifying influence of maleic acid and maleic acid derivatives on vinyl acetate adhesives was recognized since the early development of vinyl acetate copolymers [59].

The first trials of incorporating polar monomers in ethylene (or ethylene vinyl acetate) copolymers were carried out with acrylates. Unfortunately, faster reacting acrylates (relative to ethylene or vinyl acetate) have a tendency towards block copolymerization with vinyl acetate. The disparity in the reactivity ratios between ethylene and acrylates with respect to vinyl acetate also limits ethylene incorporation into a terpolymer system. Maleates (or fumarates), on the other hand, do not readily homopolymerize, leading to alternating copolymers with vinyl acetate; this, coupled with a much more favorable reactivity towards vinyl acetate and ethylene, enables the production of terpolymers with an increased ethylene content. The two preferred comonomers among EVAc terpolymers for optimal pressure-sensitive performance are dioctyl maleate and 2–ethyl hexyl acrylate. The higher molecular weight of dioctyl maleate allows a lower mole fraction, permitting a higher ethylene content. Thus polymers commercialized by National Starch contain 20% or more ethylene by weight. Acrylic and ethylene copolymers may be incorporated in aqueous ethyl hexyl acrylate-based PSA systems [60].

Water-borne EVAc multipolymers, resulting from high pressure polymerization of ethylene and vinyl acetate with additional monomers, constitute a new class of PSAs. Their properties combine the most desired features of acrylic copolymers and tackified rubbers: high quick stick, resistance to oxidation and discoloration, clarity, adhesion to low energy surfaces, and compatibility with plasticized PVC material [61].

Acrylics exhibit lower peel adhesion and tack than ethylene-vinyl acetate dioctyl maleate copolymers. Tackified SBR, although exhibiting the tack of these polymers, generally display lower initial peel strength [62]. Generally the monomers are polymerized in an aqueous medium under pressure not exceeding 100 atm. The quantity of ethylene entering into the polymer is influenced by the pressure, agitation, and viscosity of the polymerization medium. Thus in order to increase the ethylene content of the copolymer, higher pressures are to be employed. A pressure of at least 10 atm is common [62].

As mentioned earlier, ethylene vinyl acetate copolymers were used since the beginning of the hot-melt adhesive technology as raw materials for hot-melt adhesives. The positive influence of the vinyl acetate comonomer units on the sealability of the ethylene-based plastics is well known for those familiar with the bonding technology of polymers. Ethylene vinyl acetate as a semi-crystalline polymer imparts internal strength, film-forming characteristics, and flexibility to hot-melt systems [63]. Unfortunately EVAc copolymers do not possess the agressive tack required for hot-melt PSAs. In the last decade some development was achieved through the synthesis of segmental EVAc copolymers, containing sequences with high vinyl acetate content and sequences with low vinyl acetate content [64]. These hot-melt PSAs possess the well-known UV and oxidation resistance of the ethylene vinyl acetate copolymers.

Other Elastomers. Vinyl ether polymers were used in solutions and dispersions for the formulation of PSAs, enjoying a much higher importance some years ago than today. On the other hand silicones, known for many years as dehesive materials only, are actually used as raw materials for PSAs.

Polyvinyl Ethers. Polyvinyl-, ethyl-, and isobutyl ethers were used some years ago as a main component of PSAs [65]. They are soluble in a broad range of solvents (e.g., ethanol, acetone, butyl acetate, toluene, and ethyl acetate); methyl ether also is soluble in water. This polymer has a good adhesion on polyethylene, is not sensitive towards atmospheric humidity, and therefore has been used on a large scale for roll stock materials. Because of their resistance to plasticizers, polyvinyl ethers (PVEs) were proposed as a tackifier for acrylic-based formulations for PVC [66]. Vinyl ether polymers used for PSAs are discussed by Müller [67] in a detailed manner. The water-soluble copolymer of methyl vinyl ethers and maleic anhydride is used as a thickener [68]. The most common polyvinyl ethers are based on vinyl methyl ethers, vinyl ethyl ethers, and vinyl isobutyl ethers [69]. The polymers are supplied either solvent free, as solutions, and as dispersions; the vinyl ether polymers may also be coated in the molten state [69].

Water vapor transmission of PVEs has the same value as for the human skin [66]; therefore it is recommended for medical tapes. Because of their water solubility PVEs are used for water-removable labels and as well as on humid surfaces [65]. Because of their compatibility with plasticizers PVEs are migration resistant and are recommended for PSAs coated on soft PVC. Polyvinyl ethers may also be used as tackifiers for acrylic dispersions, and polyurethane- and electron beam-cured PSAs; thus, polyvinyl ethers are used as tackifiers in crosslinked PUR-based PSAs and for electron beam-cured acrylic PSAs [70–72]. Polyvinyl ethers are mainly used in water-based PSAs in order to impart a hydrophilic character or as tackifier [69]. Unfortunately they are storage sensitive and have to be stabilized using protective agents. Polyvinyl ethers also are used as adhesives applied on PE [65,72].

Silicone-Based PSAs

Silicones are used as raw materials for PSAs as well as coated on release liners. In some cases silicone-grafted acrylic copolymers were used in order to achieve good removability.

Silicone PSAs are based on the combination of a "gum" component and a resin [73]. The gum component is a polysiloxane gum, such as methyl phenyl polysiloxane; the resin is also an organopolysiloxane. Specific resins are reacted with linear, high viscosity organopolysiloxane fluids [74]. Silicone polymers have inherently flexible siloxane backbones and a low T_g. Both contribute to excellent performance at low temperatures; they also offer good weatherability [74]. The performance properties of silicone PSAs can be designed by changing the resin/gum ratio [75]. A comparison was made between a typical dimethyl-based silicone and an acrylic-based organic PSA [75]; the silicone PSAs display good tack, peel, and shear at temperatures as low as −20°C.

Problems with silicone PSAs may arise concerning their shelf life and the availability of adequate release liners. While silicone PSAs have existed for over 15 years there was no adequate release technology available for them; their use was limited to industrial tapes. Several products on the market utilize dimethyl silicone in combination with diphenyl silicone PSAs. While diphenyl silicone PSAs have some advantages in specific applications, dimethyl silicone PSAs represent the bulk of silicone PSAs because of product cost. New release liner technology based on fluorosilicones was developed, providing a release surface for several types of silicone PSAs. A low energy surface does not suffice to offer release; this condition must be satisfied, but other conditions that favor a trend towards a low release energy are a contact angle greater than 30° and a negative spreading coefficient. The surface tension of the adhesive must be higher than that of the release coating. Fluorosilicones have critical wetting surface tensions of 15–18 dyn/cm or lower, but in general do not perform as release surfaces for PSAs because they do not display sufficient flexibility and ability to reorient their dimethyl groups like polydimethyl silicones. Crosslinked silicones may also be used as release layer for silicone PSAs [76]. The organopolysilicones may be produced directly by the hydrolysis of organic substituted halosilicones and depolymerization, or by hydrogenation of higher molecular weight organopolysilicones and heat treatment.

Silicone-Based (and Other) Release Coatings. The release surface will normally be provided by a layer of a silicone polymer or by a material providing similar release properties coated on the liner substrate (see Chapter 3). The usual solvent release coatings for PSAs are polyvinyl carbamates, acrylic ester copolymers, polyamide resins, and octadecyl vinyl ether copolymers. As an example, acrylonitrile stearyl acrylate copolymers, chlorinated polyolefins, and ethylene vinyl acetate copolymers may be used [77]. The aqueous release coatings (which are not silicone-based) presently marketed are based either on acrylic ester copolymers or vinyl acetate copolymers and are available in crosslinkable and self-crosslinking types. The self-crosslinking polymers have found application where heat resistance is required [78]. The various release coating technologies available on the market and their strengths and weaknesses were discussed by Craig [14].

The use of silicone as a release coating is based on its low surface energy (22–24 dyn/cm), where the surface energy of most adhesives averages 30–50 dynes/cm. This difference prevents the adhesive from wetting the silicone surface. The basic release liner is comprised of paper with a very thin silicone coating layer, adhering sufficiently to the adhesive to hold the laminate together, but enabling peeling off the release paper from the face material [79]. Silicone release resins are linear polymers or prepolymers in liquid form, with or without solvents. They are coated onto paper, film, or other backing materials and then cured, typically by heat and/or surface catalytic action to form solid, nontacky, crosslinked polymers in situ. The release resins are formed from halosilicones and consist predominantly of repeated units of the structure:

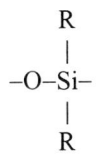

$$-\text{O}-\underset{\underset{R}{|}}{\overset{\overset{R}{|}}{\text{Si}}}-$$

where R is a hydrogen or a hydrocarbon radical, usually a lower alkyl or aryl (typically methyl) group, O is oxygen, and Si is silicone. The degree of polymerization is such as to produce a liquid linear prepolymer material with no significant crosslinking. It is believed that the currently known silicone linear release prepolymers consist of at least 95% repeating units of this structure with reactive end groups, but that small quantities of other modifying units may be present, if so desired [79]. For silicone-based PSAs, crosslinked silicone release liners are suggested [80]. Siloxane grafted vinyl copolymers may also be used as release layers [81]. Silicone coatings can be divided into two main groups, namely, thermally-cured and radiation-cured ones.

Thermal curing occurs through condensation or addition reaction. Thermally-cured silicone release systems consist of a reactive polymer, a crosslinker, and a catalyst. The radiation curing process is based on the chemical effect of radiation. Moisture curing silicone release coating technology is only at the developmental stage. More than 20 years of research were devoted to the concept of silicone addition curing systems (using a platinum-based catalyst) for paper release applications [82]. However it was not until the 1980s that this technique really began to develop after better control over the problem of inhibition had been achieved.

Although more expensive than the condensation system, additive curing offers several advantages [83]. It allows greater control over release characteristics. There is no postcure; once the coating is properly cured, there is little tendency to block. Differential release can be achieved using different polymers and additives, and, as there is no postreaction, the release force remains stable with time.

Catalyst System. With a tin catalyst system the curing is performed via a condensation reaction. The coating materials are supplied as two components which are crosslinked in the coating process; component A is a silicone polymer and component B contains the tin catalyst. The curing reaction acts as a postcure; on double-sided coatings, migration may occur which increases blocking [84].

With the platinum catalyst system the curing occurs via an addition reaction. The coating materials are supplied as two components which are crosslinked in the coating process; component A is a vinyl functional siloxane and acts as the catalyst, and component B contains a silane hydride functional polymer. For addition cure emulsions a platinum or rhodium component dispersed in a methyl vinyl siloxane polymer acts as a catalyst. The oxidation state of the metal, the nature of the binder, and the content of platinum will determine the level of reactivity of the catalyst and its shelf life stability. The base polymer consists of one or several dimethylpolysiloxanes which contain reactive vinyl groups. The portions of these groups, their number, environment, and accessibility will help to determine the reactivity of the system and the type of material obtained after cure, which constitutes the skeleton of the system [14]. For these systems a crosslinker composed of one or several methyl hydrogenopolysiloxanes may be used. In order to inhibit the reaction at room temperature stabilizers (e.g., acetylenic alcohol, acetylenic ketones, dialkyl maleates, and azodicarboxylates) may be used [14]. A tri- or tetrafunctional resin is used as a control release agent (CRA).

Radiation-Cured Systems. Ultraviolet- and electron beam-cured coatings are one component solventless systems. The process is based on the chemical effects of radiation; in the UV system it is necessary to use some heating energy [85]. Although beam-curing (UV and electron beam) silicones are available on the market this technology has not succeeded yet (apart from a few exceptions). The main advantages are the low space requirements and minimum substrate strain compared with thermal systems. The advantages/disadvantages of beam-curing will be discussed in Section 1.5 of Chapter 9.

Silicone coating technologies may be divided into three major categories, according to their application form [86]; they include solvent-based (addition and condensation products) silicones, emulsions, and solventless (thermal or electron beam-cured) silicones [85]. At least 15% of silicone coatings were carried out with solvent-free silicones in 1986 [87]. Other crosslinked compounds (e.g, polyurethanes) also may be used as base material for PSAs [88].

Crosslinked Elastomers

As discussed earlier tack, peel, and shear resistance of PSAs depend on the cohesive strength. However, enhancing the cohesive strength, greatly decreases the chain mobility needed to form a bond and hence the measured tack decreases. On the other hand, natural rubber can be processed to a relatively low molecular weight but, upon straining, it crystallizes, and hence resists separation. Induced strengthening mechanisms are favorable if a high tack is desired. Thus for high tack, the elastomers used as PSA components should have a low cohesion at low strain rates to facilitate bond formation, but a high cohesive strength at high strains to resist bond breakage.

The classical way to obtain a higher cohesive strength is to use crosslinkable base materials. This approach was initiated by the formulation of the old rubber/resin-based PSAs, where natural rubber with an inherent gel content was used as base elastomer. Later on chemically crosslinked or crosslinkable synthetic rubbers containing reactive functional groups (e.g., CSBR) were prepared. One other possibility was given by the synthesis of sequenced, physically crosslinked block copolymers (i.e., S-I-S and S-B-S). Acrylics are actually the most suitable PSAs raw materials on the market, but the most expensive ones. In this context ease of tackification of the HC rubber-based materials should be emphasized. Next a short description of the most important representatives of the crosslinked elastomers is presented.

Crosslinking of Acrylics and Their Copolymers. The crosslinking of acrylic-based copolymers (used for adhesives) has been discussed in many industrial papers [89]. The ability of acrylic monomers to polymerize with each other leads to many applications. These polymers can crosslink with epoxy resins, amines, reactive resins as well as alkyl urea derivatives [90]. Although the suitable polymers can contain functional, crosslinking monomers, such as N-methylol acrylamide, such monomers may release formaldehyde upon curing or cause the loss of tack and adhesion [91]. Thus the preferred polymers contain less than about 1% of N-methylol acrylamide monomer units.

In some polymer systems the unsaturated carboxylic acid is the effective ingredient for modifying polymer properties (e.g., cohesive strength); these modifications can be controlled through crosslinking with a suitable agent such as chromium acetate [42]. Multifunctional comonomers containing ethylene units, such as dialkyl maleate, trialkyl cyanurate, tetraethylene glycol dimethacrylate, etc., can be crosslinked by eliminating the in-saturation [92]. The crosslinking of polyvinyl acetate dispersions using metal salts was discussed by Homanner [93].

The mechanism of crosslinking for solvent-based acrylics was studied by Milker and Czech [94]. They used formulations with hydroxyethyl acrylate and acrylic acid as reactive monomers, and titanates and acetyl acetonates (Al, Ci, Ti, Zn, Zr) as crosslinking agents. As an example of crosslinking for a solvent-based acrylic PSA the following procedure is described [40]:

1. An organic solution of a PSA, a hindered phenol antioxidant, and a tackifying rosin ester are blended;
2. An organic solution of N,N'-bis-1,2–propenylisoftalimide crosslinker is added to the blend;
3. A thin layer of the adhesive solution is coated on a sheet backing.
4. The coated sheet is heated to remove the solvent and cure the polymer and crosslinker [95].

In another example an acrylic copolymer with N-methylol acrylamide as reactive comonomer was cured using ethyl acetoacetate diisopropoxyaluminium [96].

Crosslinking may be carried out using metal complexes, isocyanates, amino resins, polyamide, polyamine, epichlorhydrine resins, and other compounds [94]. Metal acetyl acetonates and orthotitanic acid esters are used together with alcohols as stabilizing agents. Crosslinking can only be obtained provided that the polymer chain is equipped with acid groups. The hydroxyl group alone does not contribute to crosslinking; however, in combination with COOH-groups, an unexpected synergism reflected in that higher shearing strength follows. The application of isocyanates remains limited in practice, since the pot life of PSAs being crosslinked with polyisocyanates is rather short. The special feature of amino resins used as a crosslinker is the low crosslinking speed at room temperature.

According to their chemical composition and to the nature of built-in reactive sites, quite different crosslinking agents may be used. For general use polyaziridines were proposed as crosslinking agents [92,97,98]. An organic solvent solution of N,N'-bis-1,2–propenylisoaftalimide may be used as crosslinker [40]. A chrome stearic acid complex (0.1 wt% of the dry adhesive) also was suggested [99]. Crosslinking via aluminium acetate/acrylic acid reactions were suggested for ethylene/maleinate copolymers [62]. Curing of silicone-based PSAs is made using benzoyl peroxide.

The additive trialkyl cyanurate is recommended for the electron beam crosslinking of polymers. This component allows a controlled adjustment of crosslinking density and improves mechanical characteristics. Macromolecular compounds like polystyryl ethyl methacrylate also may be used.

Crosslinking of Rubber Derivatives

Water-soluble polyamide-epichlorohydrine-type materials are effective as a crosslinking agent for CSBR latexes [100]. Although many SBR latexes are self-crosslinking special crosslinking agents can be used. When modifying by crosslinking it is common to add a solution of some suitable zinc compounds. A careful balance must however be considered as the addition results in decreased tack. It should be noted that addition of zinc oxide to carboxylated neoprene latex dramatically reduces the pressure-sensitive character of the adhesive [101]. A crosslinkable PSA composition was formulated on the basis of a butadiene acrylated rubber; it is crosslinked with aluminium propoxide or *sec*-butoxide [102].

Beam-Curing

The use of electron beam-curing in the paint industry was discussed by Linday [103,104]; the development of silicone release manufacturing via electron beam-curing was described by von Pilar [105]. Electron beam-curing may also be used for lacquers and laminating adhesives [106–108]. A great deal was achieved in this field through the activity carried out at the Fachhochschule München by Nitzl [109–112]. Proposals and guidelines for the formulation of beam-curable PSAs were given by Hinterwaldner [113]. Actually the most important use of beam-curing for PSAs concerns rubber-based adhesives. Different kinds of high energy beams (IR, UV, or electron beams) may be used as source of energy. The methods and possibilities of electron beam-curing for continuous web also were discussed by Holl [103]. Electron beam-cured formulations are composed of oligomers, prepolymers (50–100%), multifunctional monomers, and additives [114,115]. The most important oligomers used are polyester acrylics, epoxy acrylates, urethane acrylates, and polyether acrylates. Hexanediol diacrylates, trimethyl alyl propane triacrylates, tripropylene glycoltriacrylates, and pentaeritritol tetraacrylate are used as reactive diluting agents. The most important requirements for the polymers used as the electron beam-curing material for PSAs are a low crosslinking density and a T_g below 20°C [116].

A new class of polyaziridine derivatives giving a low crosslinking density were developed. Ethyl acrylate, butyl acrylate, glycidimethyl acrylate, and acrylic acid may be used as monomers for prepolymers [117]. The reactive diluents used to reduce the viscosity usually have hydroxyl groups as active sites. An example is hydroxylated caprolactam acrylate [118]. In some cases the addition of the reactive diluent not only reduce the viscosity of the uncured diluent composition but also increase the elongation of the cured coating [119]. The term *reactive diluent* generally refers to an unsaturated ethylene monomer which is miscible with the principal oligomers, which reduces the viscosity of the composition, and which reacts with the oligomer to form a copolymer.

Ewins and Erickson [19] studied the response of various resins and stabilizers to electron beam radiation. Their results show that proper selection of the formulating ingredients can minimize the required radiation dosage for the adhesive, while maintaining the desired balance of tack, shear, and solvent resistance. Possible formulations for UV-cured PSAs include acrylic oligomers; unfortunately their shear resistance is too low [120]. Ultraviolet-cured hybrid systems were discussed by Barisonek and Froelic [121]. In some cases vinyllactone is built-in as a sensitive comonomer [122]. Storage-stable latently-curable acrylic formulations based on triethylene glycol dimethacrylate and polymeric hydroperoxides were patented [123]. Liquid, saturated copolyesters with one terminal acrylic double bond per 3000–6000 molecular weight units are used in formulations for UV- and electron beam-curable PSAs [124]. Generally the initiator system for UV-cured PSAs may be free radical or cationic [125]. Water-based electron beam-cured coating systems were discussed by Loutz [126]. Acrylated polyester oligomers (with a molecular weight of 3000–8000) were tested as hot-melt PSAs [126,127].

In the USA two electron beam-cure silicone release coating machines have been installed in 1983 [103]. In Europe a plant at Tesa-Werk Offenburg (Beiersdorf) for curing of tapes was started [128] where classical adhesives are being used. The first release coating machine using electron beam-curing was installed in 1984 [129]. A comparative study of the costs for electron beam and water-based coating is given by Pagendarm [130]. Radiation-curable release coatings may be prepared with a low adhesive material of limited compatibility in the liquid, so that a thin layer migrates to the surface of the film [131]. A controlled release effect may be achieved using a different polydimethylsiloxane level [118]. Acrylic or epoxy derivatives of silicone may be used as active components. Generally polyfunctional acrylic or mercapto groups are used as reactive sites on the dimethylpolysiloxane [132].

1.2 Viscous Components, Plasticizers, and Tackifiers

Rubbers are elastic materials, exhibiting a pronounced self-recovering character. As PSAs are viscoelastic formulations, the change of the elastic character is possible by decreasing the plateau modulus.

A decrease of the plateau modulus is produced by the use of low modulus and/or low T_g formulating additives. The latter are viscous materials, providing a pronounced viscoelastic character to the rubber/additive blend. Their use as a reducing agent of the modulus is well known from PVC chemistry where plasticizers are used to transform hard and brittle materials into soft, elastic ones. Later on they were used to soften the polyvinyl acetate-based adhesives.

Plasticizers

The addition of plasticizer has an effect similar to that of a tackifying resin [38]. Plasticizers improve the ability of PSAs to flow under bond formation conditions, but reduce the higher frequency modulus, thereby improving tack and peel [133]. A detailed discussion of the plasticizers used in PSAs formulations is given in Chapter 8.

Tackifiers

Pressure-sensitive adhesives are usually blends of rubbers with low molecular weight resins. The resin is described as a tackifier if added to the rubber, as the tough dry rubber is converted into a product with PSAs properties. Different natural or synthetic resins are used as tackifiers. Tackifier resins were described in a detailed manner by Rich [123]. Colofonium derivatives, acid resins, resinates, and resin esters, and their chemistry, properties, applications, and suppliers were discussed by Jordan [134,135]. The properties of hydrocarbon resins are covered by Jordan [136,137] and the use of resins in the rubber industry was reviewed by Fries [138]; he describes the structure, properties, and use of different resins (hydrocarbon resins, rosins, and phenol-formaldehyde resins) in a comparative manner.

Among natural resins the rosin derivatives are the most widely used ones. Rosin is a thermoplastic acid resin obtained from pine trees. The rosin group consists of rosin, modified rosin, and derivatives. The most widely used rosin acids include abietic acid, neoabietic acid, primaric acid (neoprimaric acid), dehydroabietic acid, dihydroabietic acid, and tetrahydroabietic acid [138,139]. The storage stability and aging stability of the resins may be improved by hydrogenation, disproportionation, dimerization, and esterification [139]. There are different methods to extract the rosin. Gum rosin is obtained as oleoresin from the living tree. Wood rosin originates from aged pine, whereas tall oil rosin is obtained from tall oil which is a by-product of the paper industry. Different kinds of resins (e.g., disproportionated tall oil rosins, polymerized tall oil rosins, cured rosins, hydrogenated wood rosins, and polymerized wood rosins) were suggested [140,141]. The dimerized rosin acid content (35–60%) may yield harder or softer colofonium resins [142]. The range of resins formulating is illustrated in the Table 5.1. Pentaerithrytol esters of hydrogenated rosins with a ring-and-ball softening point of 104°C and an acid number of 12 are also used [8].

In 1985 hydrocarbon resins were by far the preferred tackifiers in PSAs formulations, representing about 80% of the market [143]. Hydrocarbon-based resins are low molecular polymers based on petroleum, pure monomers, natural terpenes, or coal derivatives [139]. The most important raw materials are aromatic C_9–C_{11} fractions, C_4–C_5 diolefins, styrene, alpha methyl styrene, vinyl toluene, dicyclopentadiene, alpha and beta pinene, d-limonene, isoprene, and piperylene. Generally hydrocarbon resins are classified as follows: aliphatic, aromatic, alkyl aromatic, and hydrogenated hydrocarbon resins. Low color, inert, low molecular weight, thermoplastic hydrocarbon resins produced from the terpene monomer alpha pinene with a ring-and-ball softening point of 112–119°C [8] were suggested. In some cases special chemically reactive synthetic resins like phenol-formaldehyde or melamine condensation products may be used.

1.3 Other Components

The most important single components in a PSA are the polymers (elastomers and tackifiers). Other micromolecular components like plasticizers, antioxidants, fillers, etc., are also included. Depending on the application technology carriers (solvents or water) may be required. Depending on the coating tech-

Table 5.1 The Range of Resins used for Formulation Purposes

Resin type	Trade name	Softening point (°C) Drop	Softening point (°C) R&B
Gum rosin	—	—	78
Wood rosin	—	81	73
Tall oil rosin	—	—	80
Polymerized rosin	Poly-Pale	102	95
Hydrogenated rosin	Staybelite	75	68
Pentaerythritol wood rosin	Pentalyn A	111	—
Glycerine-hydrogenated rosin	Staybelite Ester 10, Dermulsene DEG	83	
Pentaerythritol-hydrogenated rosin	Pentalyn H	104	—
Glycerine-highly stabilized rosin	Foral 85	82	—
Pentaerythritol-highly stabilized rosin	Foral 105	104	—
Hydroabietyl phtalate	Cellolyn 21	63	—
Olefin	Hydrolin	—	100–135
Cycloaliphatic hydrogenated olefin	Permalyn	—	85–135
Aliphatic petroleum hydrocarbon	Piccopale/Piccotac	—	70–122
Modified aromatic hydrocarbon	Hercotac AD	—	85–115
Dicyclopentadiene	Piccodiene	—	73–140
Mixed olefine	Statac, Super Statac, Wingtack	—	75–105
Alpha pinene	Piccolite	—	10–135
Beta pinene	Croturez, Piccolites	—	10–135
Terpene	Zonarez, Nirez 100	—	10–145
Alpha-methyl styrene-vinyl toluene	Piccotex	—	75–120
Alpha-methyl styrene	Kristalex	—	25–120
Styrene	Piccolastic	—	5–190
Terpene phenolic	Piccofyn, Dermulsene DT 75	—	100–135
Coumarone-indene	Cumar	—	100

nology wetting agents are used; converting and end-use properties of the label may require special additives. A detailed discussion of these components is given in Chapter 8.

2 FACTORS INFLUENCING THE CHEMICAL COMPOSITION

The most important means to adjust the chemical composition of a PSA are the synthesis of the base materials and the blending or formulation of the components. However, other factors may influence the chemical composition of the PSA, such as the physical state of the synthesized elastomer, the coating technology, and the solid state components of the laminate. As pointed out earlier a different end-use requires a different rheology; thus the end-use properties also determine the chemical composition of the adhesive.

2.1 Synthesis

The synthesis of the raw materials for PSAs provides a straight-forward way to modify the properties of PSAs through the chemical composition. Generally the choice of the monomers of the polymerization technology ensures an adequate change of the adhesive properties.

Previously the importance of the T_g and modulus for PSAs was discussed (see Chapter 3). The requirement for highly branched soft monomers (e.g., ethyl hexyl acrylate or butyl acrylate) as the main polymer backbone was pointed out earlier. On the other hand it may be desirable for various reasons to incorporate any of several modifying comonomers as part of terminally unsaturated vinyl monomeric portions of the adhesive system. For example, acrylonitrile imparts hardness and solvent resistance, t-butyl styrene improves tack, methyl methacrylate makes the adhesive harder, octyl vinyl ether softens the adhesive, etc. [8]. The choice of the comonomers in combination with special polymerization conditions influences the final chemical composition. This general statement is illustrated by the synthesis of special ethylene-maleinate multipolymers [142] or special styrenic block copolymers (Dexco triblock copolymers) [144].

The synthesis of the main polymer may influence the chemical composition of the adhesive through a built-in reactivity which can be activated at a later time. Functional groups are usually incorporated into the polymer for chemical reactivity reasons (e.g., crosslinking). This type of chemical reaction is desired to minimize the adhesive film's thermoplastic response and maximize its tensile properties. As an example, as the polymer contains copolymerized carboxyl groups, bond strength and other properties can be modified and improved by adding multivalent metal ions in a suitable form to promote ionic crosslinking [145]. Molecular weight adjustment during synthesis offers another way to influence the chemical basis and adhesive properties. Here solvent-based or emulsion-based polymerization of the same monomers provides another possibility to influence molecular weight. Emulsion kinetics differ from those of mass, solution, and suspension polymerization which show an inverse relationship between the rate of polymerization and the number average degree of polymerization [146].

Aqueous dispersions and solutions of the useful polymers can be manufactured by procedures known in the art to be suitable for the preparation of unsaturated olefin carboxylic acid ester polymers, such as acrylic ester polymers. For instance, aqueous polymer dispersions can be obtained by gradually adding each monomer simultaneously to an aqueous reaction medium at rates proportional to the respective percentage of each monomer in the finished polymer, then initiating and continuing the polymerization with a suitable polymerization catalyst. Promoters are free radical initiators and redox systems such as hydrogen peroxide, potassium or ammonium peroxydisulfate, dibenzoyl peroxide, lauroyl peroxide, ditertiary butyl peroxide, 2,2'-azobisisobutyronitrile, etc., either alone or together with one or more reducing components such as sodium bisulfite, sodium metabisulfite, glucose, ascorbic acid, etc.

The acrylate monomer may include an alkyl group with 1 to about 10 carbon atoms [147]. Suitable alkyl acrylate monomers are methyl acrylate; ethyl acrylate; N-propyl; isopropyl acrylate; butyl acrylate; 2–ethyl hexyl acrylate; and heptyl, octyl, nonyl, and decyl acrylate. The preferred alkyl acrylate comonomers are butyl acrylate and 2–ethyl hexyl acrylate. Acrylic acid used as a comonomer improves the cohesion [148]. Acrylic esters were copolymerized with small portions of monomers such as acrylic acid, methacrylic acid, ithaconic acid, acrylamide, methacrylamide, acrylic esters, vinyl esters, n-alkoxy-alkyl unsaturated carboxylic acid amides, half esters, half amides, amide esters, amides and imides of maleic anhydride, and the alkylaminoalkylene monoesters of maleic, ithaconic, or citraconic acids. If the average number of carbon atoms in the longest alkyl chain exceeds 12, the adhesive tends to become waxy and lacks sufficient adhesion. The useful PSA polymers contain a sufficient amount of one or more of the described functional monomers to increase the cohesive strength of the adhesive, relative to an otherwise identical PSA in the absence of such functional monomers. Detectable enhancement of cohesive strength is found in many polymers at functional monomer concentrations as low as 0.05 wt% [91]. Working with a butyl acrylate/vinyl acetate/methacrylic acid model terpolymer system, Brooks [149] used computer aided regression analysis to identify key factors influencing the properties and PSA performance of acrylic emulsion polymers. Experiments were designed to study the influence of the acid comonomer level, the chain transfer agent, surfactants, and buffer concentration on peel, shear, and tack values. Molecular weight, polydispersity, and gel content were also measured. The strongest influences were found with acid comonomer level and buffer concentration.

2.2 Formulation

Some adhesives typically are used without or with quite low levels of tackifying resins, but tackifying resins can contribute significantly to tack and peel [8]. In fact, formulation does not only mean tackification. Formulation is the general notion for a recipe containing all the components necessary to impart an adhesive the combination of the final (desired) adhesive, coating, converting, and end-use properties. Details about formulation are discussed in Chapter 8.

2.3 Physical State of PSAs

Generally the chemical composition of solvent-based, water-based, and hot-melt PSAs are quite different. As discussed earlier for solvent-based and water-based PSAs a formulating freedom is a given, that is, their composition does not have any limiting factors other than the T_g and modulus to be achieved. On the other hand hot-melt PSAs have to be fluid at the coating temperature, that is, their melting point has to be lower than their thermal resistance and, at the same time, they must display a low viscosity at the (high) coating temperature, but a high viscosity at the end-use temperature. This means that within the design of the chemical composition of hot-melt PSAs the processability of the adhesive should be considered. For this purpose conventional elastomers are not adequate. In a similar manner the high coating temperature requires aging and temperature-resistant tackifiers. In the case of water-based PSAs, the adhesive should possess water solubility or dispersability (i.e., the base elastomer should be designed so as to incorporate hydrophilic monomers).

Raw Materials for Hot-Melt PSAs

Twenty years ago only solvent rubber-based PSAs existed. Natural, high molecular weight rubber was used as base elastomer. In order to achieve better solubility and/or special adhesive properties, masticated or unmasticated rubber is preferred. Even a highly masticated natural rubber (with a low gel content) remains too viscous in the molten state in order to be coated per se, and has a too low temperature resistance/cohesion at the end-use temperature. For hot-melt PSAs low melt viscosity rubbers are needed; for these polymers, an acceptable cohesion level may be achieved only by physical crosslinking. Therefore thermoplastic elastomers are used [150]. In general hot-melt PSAs are based on a thermoplastic elastomer, an end compatible resin, a mid-block compatible resin, oil, and antioxidant [151]. The most commonly used block copolymers for hot-melt PSAs are linear S-I-S block copolymers with a polystyrene content usually ranging between 10 and 25%. They possess a relatively low modulus and are easier to tackify than S-B-S types. Current developments in the field of thermoplastic elastomers broadened the available raw material base for hot-melt PSAs. S-I-S block copolymers display better tack, compatibility, and aging resistance; they have a lower viscosity in hot-melt PSAs and remain soft after aging. Hinterwaldner [152] summarized the available raw materials used in hot-melt adhesives with a short description of the main backbone polymers (e.g., polyolefins and EVAc copolymers, polyesters, and polyamides). Dupont's Elvax, ethylene vinyl acetate, or EVAc copolymers were developed for use as a base resin in hot-melt PSAs. They are easily blended and have an excellent thermal oxidation stability. Ethylene vinyl acetates are used in hot-melt PSAs but do not provide much tack and peel. By copolymerizing ethylene and vinyl acetate with an acrylic or maleate monomer in water much enchanced properties are obtained [153]. A not statistical EVAc copolymer with a higher VAc content, molecular weight, and sequential structure was evaluated as raw material for hot-melt PSAs [152].

Although solvent and aqueous emulsion acrylic PSAs were in use for some time, it is only recently that hot-melt acrylics were developed. The wide variety of acrylic monomers available and their reactivity provide endless opportunities to tailor polymers to suit specific applications. Reactive hot-melt PSAs can be developed using solid and liquid functionalized polyacrylic elastomers, in combination with special hardeners or catalysts containing isocyanate for heat curing or UV curing with appropriate agents such as *p*-chlorobenzophenone or others.

Polyesters are also used as raw materials for hot-melt PSAs. Many epoxy- and carboxyl components can be used, hence the raw material base of polyester-based hot-melt PSAs is broader than that of block copolymers. Through synthesis or by a postfunctionalizing other reactive groups (e.g., carboxyl-, alkyl-, isocyanate-, acryl-, or alcoxy sylil groups) may be built into the polymer. Copolyesters are not dangerous to human health (some have FDA approval) and are insensitive to plasticizer migration [154]. The longer the chain length, the better the cohesion and adhesion, but the higher the melt viscosity.

Raw Materials for Solvent-Based PSAs

Solvent acrylic adhesives represent about 40% of the total solvent adhesive label stock production. Such systems are appreciated for their high performance level and quality, but remain the most expensive. They are increasingly being replaced by water-based adhesives but will likely remain preferred for demanding applications (e.g., outdoor usage, high humidity conditions). In the future, they may also be replaced by new hot-melt or reactive solventless systems. Rubber-based solutions still account for 55% of the solvent adhesives, but there is little research or development going on in this area.

Through the choice of the chemical base for solvent-based PSAs, the designer may use a broad range of chemically quite different raw materials. Theoretically all the base components suitable for molten or water-dispersed PSAs are soluble in organic solvents (i.e., they may be used for solvent-based PSAs as well). End-use and commercial considerations may play a determining role. Solution polymers generally have a narrower molecular weight distribution (MWD) than emulsion polymers; therefore the versatility of solution polymers allows a great deal of latitude [24]. Batch polymerization can theoretically yield a MWD approaching 1.0; in practice levels of 1.4–1.8 are more common. Continuous polymerization results in a MWD of 1.8–2.0 in most cases. The MWD along with the molecular weight exerts a very strong influence on processability.

Raw Materials for Water-Based PSAs

Among the various coating technologies water-based adhesives have grown from a small usage in the 1970s to the status of preferred technology, representing 63% of the whole label stock production in 1988. Solvent- and hot-melt-based adhesives are estimated at respectively 24% and 13% of the total production [155]. Taking into account the growing importance of the water-based PSAs technology a broad range of raw materials was developed. It should be mentioned that unlike solvent-based adhesives—where a rubber-based technology was consolidated and where only a late development in polymer synthesis (acrylics) made the manufacture possible of nonrubber-based PSAs—it was not possible to develop an aqueous technology based on natural or synthetic rubber latex in the field of water-based PSAs. The emulsions typically contain about 40–70% polymer when manufactured, whereas preferred latexes typically have a content of about 40–60 wt% polymer solids. The dispersed polymer particles can be of any size suitable for the intended use, although a particle size of at least about 120 nm is presently preferred. Most often the latexes have particles with a diameter in a range of about 120–1000 nm.

Acrylics as raw materials for water-based PSAs did not have any competitors for a long period of time. Only lately has the development of vinyl acetate copolymers led to a new, price attractive class of water-based dispersions as raw materials for PSAs. Some years ago water-based and hydrocarbon-based (rubbery) dispersions were synthesized on the basis of (carboxylated) styrene-butadiene copolymers. Natural rubber, CSBR, and acrylics make up the main base polymers for water-based PSAs [28].

Adhesives based on acrylic dispersions represent over 90% of the European production of water-based label stock adhesives. The success of water-based adhesives, particularly acrylics, is due to several factors. First water-based adhesives eliminate the use of solvents and their associated environmental concerns. Furthermore water-based adhesives display good coating characteristics with different types of coating equipment and offer good convertability to the converters. Blending with tackifier emulsions allows the development of products with specific adhesive performance characteristics, although adhesion on polyolefin substrates offers potential for improvement.

Raw Materials for Radiation-Cured PSAs

A major obstacle in the application of radiation curing technology for adhesives was the development of film-forming properties without adversely affecting the tack. Therefore multifunctional monomers were used in limited quantities in adhesives, because they minimize the segmental motion in the polymer and substantially reduce adhesive properties [156]. There are, however, multifunctional monomers with physical and functional characteristics that provide positive contributions to adhesive properties. Such a monomer is tetraethylene glycol diacrylate which displays enhanced polarity and molecular flexibility. When incorporated into the UV-curable PSA formulations as an adhesion modifier, a permanent type adhesive is obtained exhibiting 2.5–3.0 lb of peel adhesion with more than 2 hr of shear strength [156]. Substitution of tetraethylene glycol diacrylate with other multifunctional acrylates (e.g., pentaerithrytol triacrylate, hexanediol diacrylate, trimethylol propane) imparts a nonpermanent characteristic. A range of 100%-solid hot-melt PSAs on acrylic basis was developed [126,157].

Electron beam-cured adhesives are usually acrylate-based systems formulated from oligomers with reactive diluents in addition to the additives required to achieve the functional properties [158]. Generally prepolymers and reactive diluting agents are used [159], but the danger of phase separation (especially at concentrations lower than 10% of the diluting agent) remains [158]. On the other hand too high a radiation dose may lead to damage of the face stock and/or loss of tack, whereas too low a dose produces uncrosslinked monomers. The purpose of the reactive diluent monomer in the UV-curable adhesive is to provide a solubility medium for the resin system [156]. Such monomers used include 2-ethyl hexyl acrylate, hydroxyethyl methacrylate, cyclopentadiene acrylate, and tetrahydrofurfurol acrylate. Diluent monomers may also effect the PSA characteristics.

Low molecular weight polybutadienes (MW = 3000–5000) are able to make rigid resins more flexible and impart rubber-type properties. In addition to reactive side group vinyl unsaturation they possess a terminal functionality which can provide additional reactivity and/or property enhancement [156];

Chemical Composition of PSAs

they can contain amine, vinyl, carboxyl, and hydroxyl terminal functionalities, which can be incorporated in UV-curable adhesives at 15 wt%.

A variety of photoinitiator systems were tested for radiation-curing PSAs, including: benzophenone, isobutyl benzoin ether, diethoxyacetophenone, and benzoin ethyl ether [156].

Using the same formulation but a different radiation method (UV or electron beam) the adhesive characteristics are quite different [156]. The adhesive values for electron beam-curing are overall quite low compared to data acquired for the same formulations cured via the UV process. This is probably due to the fact that the electron beam is a much higher energy source than UV, thus promoting a higher level of crosslinking within the system.

Classical hot-melt formulations do not respond similarly when subjected to irradiation. To obtain optimum crosslinking with a minimum of radiation, the selection of suitable resins and plasticizers is necessary; for example, PSAs based on the newly developed S-I-S copolymers crosslink much faster with saturated resins than with unsaturated resins and plasticizers. Polybutadiene alone needs extremely high doses; blends with acrylic esters do not display acceptable PSA properties. Only built-in hydroxy and epoxy groups and double bonds (e.g., ethyl hexyl acrylates) give good results [159]. The best adhesive properties are achieved with a dose of 30–40 kJ/kg; unsaturated polyesters and monomers with alkylamino groups give good results for a radiation dose of 80 kJ/kg [159].

2.4 End-Use

High speed automatic label application requires moderate to high initial tack in order to prevent label delamination in the production cycle. Moderate to high shear properties are necessary on squeezable containers to allow the label to remain in contact with the bottle during deformation [160]. Good wetout characteristics are extremely important for proper application of clear label stock materials. See-through labels (no label look) also require water-like clarity of the adhesive system; therefore solvent-based acrylics usually are used. Some applications using opaque label materials require rubber-based systems. These few examples illustrate how very differently end-use requirements may influence the design of the chemical composition. Next the features of composition design for the most important PSA categories are discussed.

Raw Materials for Reel and Sheet Label Stock

Roll labels (usually on paper) are estimated at about 77% of the total label volume; sheet labels and other products represent 20% and 3%, respectively, of the overall production [155]. Quite different application (labeling) technologies are used for small labels and decals, for machine (gun) labeling and hand applied labeling. In most cases the adhesive tack and release force (peel force from the release liner) need to be closely monitored. Reel labels require an agressive tack, while sheet labels need more cohesion and less tack. Therefore the base elastomer for reel/sheet labels will be selected in a quite different manner.

Cohesive, low tack, high molecular weight polymers of short (side) chain acrylics, and ethylene vinyl acetate copolymers are adequate for such sheet applications. Soft, very tacky polymers of ethyl hexyl acrylate or butyl acrylate (i.e., long side-chain acrylics) are preferred for reel labels. These may be used as solvent-based or water-based adhesives. Competitive formulations for reel labels may also be designed on a hot-melt PSA basis. Water-based formulations are high tack, tackified acrylics, or CSBR-based, tackified PSAs. The use of special ethylene vinyl acetate maleinate multipolymers also was suggested [142]. The first generation ethylene vinyl acetate dioctyl maleate copolymers offered good tack and peel strength, broader specific adhesion than acrylates, and excellent PVC plasticizer resistance. Cohesive strength remains limited however, and the product is used mainly in roll label applications at moderate to high service temperatures.

Raw Materials for Film Labels

Clarity considerations together with a high UV resistance limit the choice of raw materials for film labels. Face stock/adhesive interactions also act as limiting factors. Therefore only a few materials appear suitable as PSAs on film face stock materials; the choice is restricted to the range of acrylics (usually solvent-based ones).

Raw Materials for Permanent and Removable Labels

As discussed in Chapter 2, removability implies a low peel level and no adhesion build-up. These criteria require low tack, high cohesion adhesives, and a good plasticizer compatibility. Because of the relatively low cohesion of hot-melt PSAs and of the high migration tendency of water-based PSAs (and their limited crosslinking) solvent-based PSAs were used as removable PSAs. The latest developments in water-based acrylics allow the formulation of removable PSAs on a dispersion basis. However care should be taken concerning the choice of the base elastomer. As an example it was observed that ethylene vinyl acetate copolymers used for film coatings interact with the soft PVC face stock material. Therefore aged peel values for removable formulations are higher than desired. A detailed description of the chemical basis of PSAs used for removable labels is given in Chapter 8.

As far as the pressure-sensitive adhesive component is concerned, the properties which contribute to peelability are principally limited tack, a low build-up factor, and a relatively soft adhesive. These properties can be achieved in a conventional way by omitting or only using a low level of tackifier, including tack deadeners such as waxes, fillers, or special agents, and by including plasticizers [27]. The chemistry of the polymer also has an effect but, apart from choosing among polymers which are commercially available, the chemistry is controlled primarily by the manufacturer of the polymer rather than the adhesive formulator. Small amounts of a highly polar monomer such as acrylic acid causes adhesion build-up. On the other hand small portions of acrylonitrile and methacrylonitrile improve removability [8]. An amount of about 30% natural rubber latex added to a vinyl acetate maleinate dispersion yields a removable PSA, without leaving any adhesive residues upon debonding [161].

Raw Materials for Other Requirements

Temperature. Low temperature peel adhesion for silicones increases to 12 pli at $-75°C$ (from $-25°C$); for organic PSAs the peel value is zero below $-25°C$ [75]. Hence silicone PSAs are recommended for low temperature conditions. Butyl systems should perform well at low temperatures, while acrylics are superior in a room temperature environment [162].

Water Resistance. Whereas wash-off labels were required on the market some years ago [163], conventional PSAs are hydrophobic in nature [99]. As such, they adhere poorly to wet or damp surfaces and generally cannot be effectively removed with water, even using automated cleaning procedures which employ detergents, warm water, or caustics.

Benedek [164] discussed the problem of the water resistance of PSAs in detail. The water resistance of water-based adhesives depends on the chemical composition of the base polymer and on the composition of the emulsion. The influence of the monomers and of the surface active agents used during the polymerization was described. As pointed out, formulating depends on the adhesion-water resistance balance. Water soluble (wash-off formulations) are limited by a narrow raw material basis and low adhesion levels. The practical requirement for wash-off labels were identified as the need to have a temperature-dependent water solubility and wet tack (i.e., adhesion on condensed water coated surfaces). A relatively simple formulation exhibiting water solubility can be designed by changing the ratio of water soluble/insoluble components in a common formulation (pseudosolubility). Time- and storage-independent wash-off formulations generally need more than 20 wt% special water-soluble components.

Adhesives based on polyvinyl methyl ether were among the earliest available water-soluble PSAs. While they showed fair adhesive properties and are used for limited applications, several deficiencies restrict their suitability, mainly that they did not absorb water rapidly enough to adhere to wet surfaces. Adhesives based on polyvinyl methyl ether, however, were selected for wash-off labels [99]. Partial esters such as polyvinyl methyl ether/maleic anhydride with nonionic surfactants of the nonyl phenol ethylene oxide adduct type have also been used. Polyvinyl ethers do not adhere to wet surfaces and maleic anhydride copolymers are very humidity sensitive. Other water-soluble polymers like polyvinyl pyrolidone (PVP) with water-soluble plasticizers, copolymers of acrylic acid and alkyl acrylates, and others also were used [165]. A need exists for an adhesive which has good adhesion to both wet and dry surfaces, whether cold or warm, whether polar or nonpolar, with a sufficient permanent character to be retained as tamperproof when combined with label papers, but yet easily removed when desired, with warm or cold water, or with detergents and alcohols used in commercial or domestic cleaning operations. Such a formulation may be given by acrylic-acid-based polymers with a broad MWD [99].

Water-soluble hot-melt PSAs are made from polyethylene oxide (MW = 50,000) and polyalkylene oxide (MW = 190–15,000) [166]. Other formulations contain a water-soluble ester of acrylic acid and polyethylene or polypropylene glycol [167]. One water-soluble PSA is made up of 2–ethyl hexyl acrylate or other tackifying monomers (optionally methyl methacrylate or other diluent monomers) and 7–30 wt% hydrophylic monomers selected from methacrylic acid (and salts thereof), hydroxy alkyl acrylates, and hydroxy alkyl methacrylates [168]. Other water-activateable or soluble hot-melt PSAs are based on vinyl pyrolidine/vinyl acetate copolymers [169].

Weather-resistant hot-melt PSAs are based on saturated thermoplastic rubbers; these contains no more than 15% styrene [170]. New generation acrylic water-based PSAs exhibit enough tack and peel after immersion in hot water [171]. Water-soluble surfactants remain in the dried film and give water-sensitive films. Ammonium salts of styrene-maleic anhydride lead to copolymers with improved water resistance [72].

In the area of health care tapes and disposable products, acrylate polymers are inocuous, in general, when exposed to human skin. Many health and other tape (label) applications exist where it is desirable to have a pressure-sensitive film coated onto a porous web. Acrylic hot-melt PSAs offer a relatively easy route to coating PSAs onto film or nonwoven webs, with a minimum of adhesive penetration into the web [172].

Others. Acrylic hot-melt PSAs have better initial tack after aging than classical, rubber-based ones [170]. They have a viscosity of 26–30,000 mPa·s at 180°C and display good peel adhesion values on different substrates [173]. A detailed comparison of the PSA properties based on different raw materials will be given in Section 2 of Chapter 6.

2.5 Coating Method

The coating method (direct or transfer coating) can influence the chemical composition of the adhesive. One can assume that direct coating on porous surfaces allows a rapid penetration of the adhesive in the face stock material. Evidently, low molecular compounds migrate preferentially (see Section 2.7 of Chapter 8). The viscosity of the coated adhesive and its components influences both the wetout and the bleedthrough. Therefore, for easy-to-wet, directly-coated face stock materials low viscosity dispersions may be used if bleedthrough is not an important issue. For difficult surfaces (release liners) or/and porous materials a higher application viscosity appears to be required. The viscosity of the PSAs to be coated may be controlled when formulating (e.g., by choice of the solvents used, solids content, thickness, pH, etc.). The viscosity of the bulk material (rheology) is composition and molecular weight dependent.

Chemical Basis for Directly-Coated PSAs. Each class of raw materials (e.g., natural or synthetic rubbers, acrylics, ethylene vinyl acetate copolymers, etc.) with or without tackifier may be used for direct coating of different face stock materials. In each physical state (molten, dissolved, or dispersed) materials may be used for direct coating onto paper face stock materials. Restrictions must be taken into account concerning the porosity/viscosity balance (usually for water-based PSAs) and the thermal sensitivity of special face papers. Similar to film coatings, this acts as a limiting factor when using hot-melt PSAs. Almost each rheological behavior (i.e., a high fluidity or a shear-induced viscosity decrease) may be tolerated for direct coating of paper or films.

In order to avoid penetration (i.e., bleedthrough of hot-melt PSAs) the application viscosity should amount to 17–20,000 mPa·s at 180°C [170]. Hot-melt PSAs are also suggested for coating onto nonwoven substrates [174]. Raw materials also differ as a function of the coating machine. As an example aqueous acrylic tape adhesives are designed as a function of the coater geometry [175]. The required viscosities are: for rotogravure, 75–100 mPa·s; for Meyer Rod, 300–800 mPa·s; and for knife-over-roll, 4000–8000 mPa·s. On the other hand foam generation is higher for Meyer Rod than for rotogravure, and even higher for knife-over-roll. A detailed discussion of the formulation differences as a function of the coating method is given in Section 1.1 of Chapter 9.

Chemical Basis for Indirectly (Transfer)-Coated PSAs. Indirect (transfer) coating implies good wetting at low surface tension, or high viscosity. Both are intrinsic properties of hot-melt PSAs or some solvent-based PSAs. Difficulties appear mostly with the use of water-based PSAs, where special wetting agents and thickeners must be used.

2.6 Solid State Components of the Laminate

Quite different plasticizer levels are required for the formulation of adhesives used for films (50%) or other materials (2–6%) because of the different flexibility and porosity of these substrates [176]. As illustrated by this example the solid state components of the laminate influence the choice of the adhesive components by their different affinity towards these components and by changing the adhesive composition through chemical interaction with the PSAs.

Pressure-sensitive adhesives for film coating must meet some special, face stock-related requirements. Pressure-sensitive adhesives for PVC coating should meet the general requirements for film coating in addition to some special PVC related requirements, as the requirements for soft PVC are quite different from those for hard PVC; similarly, filled (opaque) PVC possesses different properties from clear PVC. Sensitive film face stock materials need a nonagressive adhesive formulation; for example, soft PVC usually requires tackifier-free adhesive formulations. Stabilizers present in the PVC (e.g., Pb, Sn) produce oxidative degradation of the tackifier resin which is associated with the loss of the tack.

Table 5.2 PSA for Labels: Main Requirements as a Function of the Face Stock Material

	Face stock	
Required performance characteristics	Paper	Film
Adhesive properties	+ +	+ +
Converting properties	+ +	+ +
End-use properties		
Water resistance	– –	+ +
Shrinkrage resistance	– –	+ +
Ageing resistance	+ –	– –
Migration resistance	– –	+ +
Clarity	+ +	– –

+ + very important, + – important, – – not necessary

Table 5.3 Performance Characteristics of Different PSA as a Function of the Face Stock Material

Characteristics as a function of the face stock material		CSBR	EVAc	AC
Film	Clarity	+ +	+ +	+ +
	Aging resistance	– –	+ +	+ +
	Water whitening resistance	+ –	– –	+ –
	Shrinkage resistance	+ +	+ –	+ +
Paper	Migration resistance (bleeding)	– –	+ –	+ –

Porous, primerless papers as face stock material need nonmigrating adhesive formulations. Generally, migration of the low molecular components from the adhesive (e.g., plasticizer, tackifier, surface active agent, water, etc.) into the paper, or from the film (plasticizer) or paper (water, special chemicals like components of thermal papers) into the PSAs, may alter the chemical composition of the coated adhesive. These selected examples illustrate how the chemical composition may influence the solid state components of the laminate and vice versa. Not only the adhesive properties, but also the coating characteristics of the PSAs may be influenced by the solid state laminate components. As an example, carboxylated neoprene latex is preferred to other latexes because it provides better specific adhesion to aluminium foil. Tables 5.2 and 5.3 summarize the quite different requirements imposed on PSAs as a function of the face stock material.

REFERENCES

1. R. Dörpelkus, *apr*, (16) 456 (1986).
2. H. A. Krecenski, J. F. Johnson and S. C. Temin, *J. Macromol. Sci. Rev.*, in *Macromol. Chem. and Physics*, **C26**, 143 (1986).
3. G. R. Hamed and C. H. Hsieh, *J. Polymer Physics*, **21**, 1415 (1983).
4. *Adhesives Age*, (12) 36 (1986).
5. P. Green, *Labels and Labeling*, (11/12) 38 (1985).
6. J. Kendall, F. Foley and S. G. Chu, *Adhesives Age*, (9) 26 (1986).
7. J. C. Fitch and A. M. Snow Jr., *Adhesives Age* (10) 23 (1977).
8. EP 0213860, p. 2.
9. *Papier und Kunststoffverarb.*, (9) 48 (1990).
10. *Coating*, (18) 240 (1972).
11. J. C. Chen and G. R. Hamed, *Rubber. Chem. Technol.*, (2) 319 (1987).
12. G. R. Hamed and C. H. Hsieh, *Rubber Chem. Technol.* **59**, 883 (1986).

13. *Coating*, (12) 75 (1969).
14. C. Craig, *Adhesives Age*, (7) 31 (1988).
15. A. Sustici and B. Below, *Adhesives Age*, (11) 17 (1991).
16. P. McConell (Eastman Kodak, USA), US Pat. 4,072,812/07.02.78; in C. N. Clubb and B. W. Foster, *Adhesives Age*, (11) 18 (1988).
17. G. Trotter (Eastman Kodak, USA), US Pat. 4,264, 756/28.04.81; in C. N. Clubb and B. W. Foster, *Adhesives Age*, (11) 18 (1988).
18. C. K. Otto, *apr*, (16) 438 (1986).
19. E. G. Ewing and J. C. Erickson, *Tappi J.*, (6) 158 (1988).
20. G. Holden, E. T. Bishop and N. R. Legge, *J. Polymer Sci.*, (26) 37 (1969).
21. K. E. Johnsen and M. Dehnke, *Tappi, Hot Melt Symposium* 1987; in K. E. Johnsen and Q. Luvinh, *DEXCO. Tri-block Copolymers for Adhesives Industry*, European Tape and Label Conference, Exxon, Brussel, 1989, Part 19.
22. R. Koch, *Kautschuk, Gummi, Kunststoffe*, (9) 804 (1986).
23. D. St. Clair, *Adhesives Age*, (11) 23 (1988).
24. *Kautschuk, Gummi, Kunststoffe*, (1) 30 (1987).
25. V. M. Bokov, A. M. Medevedeva, L. P. Stogova and M. A. Smetanina, USSR Pat. 783. 326/30. 11. 80; in *CAS*, **94**, 122609s, 94,50 (1981).
26. *Adhesives Age*, (12) 35 (1987).
27. EP 0251672
28. A. Zawilinski, *Adhesives Age*, (9) 29 (1984).
29. *Labels and Labeling*, (11/12) 38 (1985).
30. R. Midgley, *Adhesives Age*, (9) 17 (1986).
31. Du Pont, *Bulletin*, DS 83/7 DGC/JP.
32. *Adhäsion*, (11) 30 (1983).
33. S. Benedek, *Adhäsion*, (12) 17 (1987).
34. R. Jordan, *Adhäsion*, (1/2) 17 (1987).
35. J. Andres, Haftklebstoffe in der Papier und Kunststoffverarbeitung. 4. PTS Klebstoff Seminar, PTS Vortragsband 03/84, p. 36.
36. M. Virin, in *TECH 12, Advances in Pressure Sensitive Tape Technology, Technical Seminar Proceedings*, Itasca, IL,May, 1989.
37. P. K. Dahl, R. Murphy and G. N. Babu, *Org. Coatings. Appl. Polymer Sci. Proceedings*, (48) 131 (1983).
38. D. Satas, *Handbook of Pressure Sensitive Technology*, Van Nostrand-Rheinhold Co., New York, 1982.
39. A. Zosel, *Adhäsion*, (3) 17 (1966).
40. EP 010046
41. Y. Sasaki, D. L. Holguin and R. Van Ham (Avery, USA), EP 0252717 A2/13. 01. 88.
42. C. M. Chum, M. C. Ling and R. R. Vargas, (Avery Int., Co., USA), EP 1225792/ 18. 08. 87.
43. R. Hinterwaldner, *Coating*, (4) 138 (1992).
44. H. Kuegler, USP 2,544,692/25. 01. 85.
45. Nat. Starch Chem. Co., USA, US Pat. 1,645,063; in *Coating* (7) 184 (1974).

46. M. Bernhard, Ausbluterscheinungen bei Hotmelts, 11th Munich Adhesive and Finishing Seminar, 1986, p. 9.
47. L. W. Wen and B. Sun, *Adhäsion*, (12) 21 (1985).
48. E. Lauretti, F. Mezzera, G. Antarelli and A. L. Spelta, *Gummi, Asbest., Kunstst.*, (16) 296 (1985).
49. J. A. Schlademann (Atlantic Richfield, USA), 0183386 A2/26.10.84.
50. J. P. Keally and R. E. Zenk (Minnesota Mining and Manuf. Co., USA), Canad. Pat. 1224.678/19.07.82 (USP. 399350).
51. *Coating*, (7) 238 (1989).
52. R. Pfister, *Coating*, (6) 171 (1969).
53. *Adhäsion*, (3) 13 (1974).
54. E. Merz, *Double Liaison - Chimie des Peintures*, (226) 263 (1974).
55. F. M. Rosenblum, *Adhesives Age*, (6) 22 (1972).
56. R. P. Mudge (National Starch Chem. Co., USA), US Pat. 4,753,846 (28.06.88).
57. Wacker, German Pat. 4,322,516 (1982).
58. *Adhäsion*, (12) 28 (1983).
59. *Adhäsion*, (2) 3 (1977).
60. G. R. Frazer (Johnson, USA), EP 259.842/16.02.88; in *CAS*, **21** (3) (1988); 109. 111731.
61. S. Cartasegna, *Kautschuk, Gummi, Kunststoffe*, (12) 1188 (1986).
62. R. P. Mudge (National Starch Chem. Co., USA), EP 0225541/11.12.85.
63. *Adhesives Age*, (11) 28 (1981).
64. A. Koch and C. L. Gueris, *apr* (16) 40 (1986).
65. *Coating*, (2) 45 (1969).
66. *Die Herstellung von Haftklebstoffen*, T1.2.2; 17d, Nov. 1979, BASF, p. 1.
67. H. W. J. Müller, *Vinyl Ether Polymers*, in *Handbook of Pressure Sensitive Technology* (D. Satas, Ed.), Van Nostrand-Rheinhold Co., New York, 1982, Chapter 14.
68. *Coating*, (4) 23 (1969).
69. H. W. J. Mueller, *Adhäsion*, (5) 208 (1981).
70. K. Hagenweiler and K. Scholz (BASF), DOS, 2.328.430/1973.
71. K. C. Stueben, *Adhesives Age*, (6) 72 (1977).
72. *Coating*, (8) 247 (1969).
73. PCT/US/85/02424/WO 87.
74. R. Goodwin Jr., US Pat. 2.857.736.
75. L. A. Sobieski and T. J. Tangney, *Adhesives Age*, (12) 23 (1988).
76. J. Pennace, *Screen Printing*, (7) 65 (1988).
77. G. Galli, L. Penzo and F. Pina (Manuli Autoadhesivi, Italy) US Pat., 4.725.454/16.02.88.
78. R. D. Gafford and G. R. Faircloth, *Adhesives Age*, (12) 24 (1987).
79. A. E. Bly, *Adhesives Age*, (10) 29 (1972).
80. J. Pennace and G. Hersey PCT/IPN/WO 87/03537/18.06.87.
81. L. H. Clemens, S. S. Kanter and M. Mazurek (Minnesota Mining and Manuf. Co., USA), US Pat., 4.728.571/01.03.88; in *Adhesives Age*, (7) 45 (1988).

82. A. Fau and A. Soldat, Silicone addition cure emulsions for paper release coating, in *TECH 12, Advance in Pressure Sensitive Tape Technology, Technical Seminar Proceedings*, Itasca, IL, May 1989, p. 7.
83. A. W. Bamborough, 16th Munich Adhesive and Finishing Seminar, 1991, p. 96.
84. I. Kesola, "Production of release paper", European Tape and Label Conference, Exxon, Brussel, 1989, p. 9.
85. *Coating*, (12) 366 (1986).
86. R. Thomas, *apr*, (16) 437 (1986).
87. H. Röltgen, *Coating*, (11) 400 (1986).
88. N. Toshio, N. Ken, A. Kazumi and F. Kazumide (Nitto Electric Co., Japan) Japanese Pat., C3 275 79/05.02.88; in *CAS*, **17**, (4) (1988); 109.39109m.
89. P. Kröger and W. Schimmel, Vernetzung von Copolymersaten auf Acrylatbasis, 11th Munich Adhesive and Finishing Seminar, 1984, in *Coating*, (2) 66 (1985).
90. M. Novak, *American Ink Maker*, (10) 44 (1983).
91. EP 0244997.
92. EP 0212358, p. 3.
93. A. Homanner, *Adhäsion*, (12) 16 (1982).
94. R. Milker and Z. Czech, Solvent Based Pressure Sensitive Adhesives, 16th Munich Adhesive and Finishing Seminar, 1991, p. 134.
95. O. Yoshinori, A. Kimihisa, A. Minoru and H. Takenao (Toa Gohsei, Japan), Japanese Pat. C3 89 345/20.04.88, in *CAS*, **21**, (10) (1988); 109:112327.
96. J. A. Fries, New Developments in PSA, in *TECH 12, Advances in Pressure Sensitive Tape Technology, Technical Seminar Proceedings*, Itasca, IL, May 1989, p. 27.
97. EP 3344863.
98. *Adhesives Age*, (12) 43 (1981).
99. C. T. Albright, EPA 0099087B1/23.12.87.
100. Hercules, *Bulletin*, OR-212C.
101. *Adhesives Age*, (10) 24 (1977).
102. British Pat., 1.063.324; in *Coating*, (4) 114 (1969).
103. P. Holl, *Verpackungs-Rundschau*, (3) 214 (1986).
104. E. Funk, *Paper, Film, Foil Conv.*, (8) 65 (1973).
105. G. Von Pilar, Einige Anmerkungen zur Diskussion über die Anwendung strahlenvernetzender Systeme im Bereich Trennpapiere, 7th Munich Adhesive and Finishing Seminar, 1982.
106. Hugo Klein, *Plastverarb.*, **37**, (4) 58 (1986).
107. Gerald Strohner, *Adhäsion*, (1/2) 24 (1985).
108. G.W. Drechsler, *Coating*, (7) 235 (1993).
109. K. Nitzl, 10th Munich Adhesive and Finishing Seminar, 1985, Fachhochschule München, in Hugo Klein *Plastverarb.*, (4) 58 (1986).
110. K. Nitzl, *Strahlungshärtende Massen*, Fachhochschule München; in H. Klein *Plastverarb.*, (4) 58 (1986).
111. H. Röltgen, *Coating*, (4) 102 (1985).
112. H. Röltgen, *Coating*, (5) 124 (1985).
113. H. Röltgen, *Coating*, (12) 311 (1985).

114. K. Nitzl, *Papier u. Kunststoffverarb.*, (5) 39 (1987).
115. K. Nitzl, *Adhäsion*, (6) 20 (1987).
116. R. Hinterwaldner, *Adhäsion*, (10) 24 (1985).
117. F. T. Birk, *Coating*, (9) 278 (1985).
118. R. Hinterwaldner, *Adhäsion*, (3) 14 (1985).
119. C. Y. Chieh (Morton Thiokol, USA), EP 0.147.142.
120. A. J. Morris, *J. Plast. Film Sheeting*, (1) 50 (1988); in CAS, *Adhesives*, **25** (1988); 109.191649v.
121. E. M. Barisonek and G. Froelic, *Adhäsion*, (11) 28 (1985).
122. W. J. Traynor, L. C. Moore, R. M. Martin and J. D. Moon, (3M, USA); US Pat., 4.762.982/23.02.88; in *CAS*, 17 (4) (1988); 109. 39101c.
123. R. D. Rich (Loctite Co.), EP 250.090/23.12.87 in *CAS*, 17 92) (1988); 109.398787.
124. H. F. Huber and H. Müller (Dynamit Nobel AG.); in 10th *Radcure*, 1986, *Conference Proceedings* 12/1-12/12.
125. R. Hinterwaldner, *Adhäsion*, (12) 15 (1985).
126. J. M. Loutz, *Coating*, (9) 328 (1987).
127. H. Bayer, *Adhäsion*, (6) 17 (1987).
128. *Coating*, (11) 292 (1982).
129. R. Hinterwaldner, *Coating*, (3) 73 (1985).
130. E. Pagendarm, 10th Munich Adhesive and Finishing Seminar, 1985, p. 42.
131. K. Brach (Design Cotg. Co., USA), US Pat., 4.288.479/08.09.81, in *Adhesives Age*, (12) 58 (1981).
132. R. Hinterwaldner, *Coating*, (6) 158 (1983).
133. S. W. Medina and F. W. Distefano, *Adhesives Age*, (2) 18 (1989).
134. R. Jordan, *Coating*, (8) 213 (1982).
135. R. Jordan, *Coating*, (10) 278 (1982).
136. R. Jordan, *Coating*, (12) 335 (1982).
137. *apr*, (16) 462 (1986).
138. H. Fries, *Gummi, Asbest, Kunststoffe*, (9) 454 (1985).
139. L. Jacob, New Developments of Tackifier Resins for SBS Block Copolymers, 11th Munich Adhesive and Finishing Seminar, 1986, p. 87.
140. EP 0.141.504.
141. *Coating*, (9) 27 (1972).
142. P. Mudge, Ethylene-, vinylacetate based, waterbased PSA, in *TECH 12, Advances in Pressure Sensitive Tape Technology, Technical Seminar Proceedings*, Itasca, IL, May 1989.
143. M. Bowtell, *Adhesives Age*, (12) 37 (1986).
144. K. E. Johnson and Q. Luvinh, European Tape and Label Conference, Exxon, Brussel, 1989.
145. A. B. Bofors, Swedish Pat. 14.308/1972.
146. *J. Adh. Coat. Technol.*, (364) 323 (1986).
147. EP 0.130.050.
148. *Adhäsion*, (2) 15 (1973).
149. T. W. Brooks, *Tappi J.*, (9) 29 (1984).
150. D. De Jaeger, 11th Munich Adhesive and Finishing Seminar, 1986, p. 87.

151. *Adhäsion*, (4) 28 (1985).
152. R. Hinterwaldner, *Coating*, (4) 110 (1991).
153. L. Schrijver and W. Schimmel, 11th Munich Adhesive and Finishing Seminar, 1986, p. 44.
154. *European Adhesives and Sealants*, (9) 3 (1987).
155. J. Lechat, *The Pressure Sensitive Labelstock Market in Western Europe*, Finat World Congress, Monaco, 1989.
156. W. C. Perkins, *Radiation Curing*, (8) 8 (1980).
157. *Coating*, (3) 68 (1985).
158. R. Kardashian, *Adhesives Age*, (6) 38 (1986).
159. J. R. Seidel, 4th PTS Adhesive Seminar, Munich, 1984, p. 49.
160. J. M. Casey, *Tappi J.*, (6) 151 (1988).
161. *Vinnapas*, Eigenschaften und Anwendung, Wacker, München (1976), p. 11.
162. M. Gerace, *Adhesives Age*, (8) 85 (1983).
163. *apr*, (5) 228 (1987).
164. S. Benedek, *Adhäsion*, (4) 25 (1987).
165. *Coating*, (10) 309 (1969).
166. *Coating*, (10) 180 (1975).
167. *Adhäsion*, (10) 80 (1966).
168. A. Kenneth (Allied Colloids), EP 0.147.067/29.11.83.
169. *Adhesives Age*, (12) 53 (1982).
170. R. Hinterwaldner, *Coating*, (7) 176 (1980).
171. *Coating*, (3) 360 (1985).
172. *Tappi J.*, 67 (9) 104 (1983).
173. *Coating*, (11) 316 (1984).
174. Wolfgang Graebe, *Papier u. Kunststoffverarb.*, (2) 49 (1985).
175. *Adhesives Age*, (11) 28 (1983).
176. *Coating*, (4) 11 (1974).

6
Adhesive Performance Characteristics

1 ADHESION-COHESION BALANCE

Pressure-sensitive adhesives possess adhesion, required for bonding and debonding, and cohesion necessary against debonding. Adhesion is characterized by tack and peel, whereas cohesion is described by shear resistance (and partially by peel). The special balance of these properties, the adhesion/cohesion balance, embodies the pressure-sensitive character of the adhesive. For any PSA, tackified or not, both tack and peel adhesion can be considered as the end result of two distinct processes [1], namely, a bonding process and a bond breaking process. The efficiency of the bonding process is related to the adhesive's ability to exhibit viscous flow. In order to achieve peel adhesion the bonding stage involves some dwell time. During this time the adhesive must flow in the absence of any externally applied forces. The more liquid-like the behavior of the polymer under these conditions, the more pronounced the degree of bond formation. The debonding process involves a more rapid deformation of the adhesive mass. The polymer's resistance to deformation at higher strain rates becomes very important; the higher this resistance, the higher the force which must be applied to separate the adhesive from the adherent (i.e the peel resistance). Therefore, high tack, high peel strength adhesives should exhibit good flow at low strain rates, but good resistance to flow at higher strain rates [2].

1.1 Tack

In Chapter 2 a short definition of the tack was given and the interdependence tack/adhesive rheology was discussed; this chapter describes the external aspects of the tack. Tack of PSAs is not an exactly defined, physical characteristic; it may be defined as a separation energy. Tack is a function of T_g (i.e., Tack > 0 if $T > T_g$). The tack is inversely proportional to the elasticity modulus [3]. On the other hand, the modulus depends on the branching, as does the T_g.

Tack is the resistance offered by an adhesive film to detachment from a substrate; it is the measure of the stretchiness of an adhesive, or the ability to form an instant bond when brought into low pressure contact with a substrate to which the adhesive is to adhere [4]. Tack, and the methods used to measure it were discussed by Johnston [5]. The concept of tack remains difficult to define. It was called tack, wet tack, quick stick, initial adhesion, finger tack, thumb tack, quick grab, quick adhesion, and wettability. The Pressure-Sensitive Tape Council prefers quick stick [6] and defines it as "that which allows a PSA to adhere to a surface under a very slight pressure." The American Society for Testing and Materials [7] defines tack as the force required to separate an adherent and an adhesive at the interface shortly after they were brought rapidly into contact under a light pressure of short duration.

Measurement of the Tack

Tack is measured either by touch or quantitatively by means of a loop tack tester. Loop tack, defined as the "quick stick" tack value, is the force required to separate at a specific speed, a loop of material (adhesive surface facing outwards) which was brought into contact with a specific area of a standard surface (in the absence of significant pressure). Tack also may be measured by the rolling ball or rolling cylinder method (for a detailed description of the methods see Chapter 10).

Tack Measured as Coefficient of Friction. It is believed that the resistance to rolling motion of a ball on a PSA reflects tackiness of the adhesive, because the motion must be closely related to bonding and debonding processes, which occur simultaneously at the contact surface. This way of expressing tack is useful in some practical cases, but the physical meaning of the measured values is not necessarily clear. It is more scientific to express the tack in terms of a rolling coefficient of friction which depends on the physical properties of the materials [8]. The rolling coefficient of friction can be determined experimentally from the "pulling cylinder" method more easily than from the "rolling ball" method, and one can theoretically calculate the rolling coefficient of friction by making some assumptions concerning the deformation and failure mode of PSAs.

The velocity of the rolling ball changes at every moment and, at the same time, the rolling coefficient of friction varies as a function of velocity. Assuming that F is the static frictional force, N is the normal force, θ and I are

Adhesive Performance Characteristics

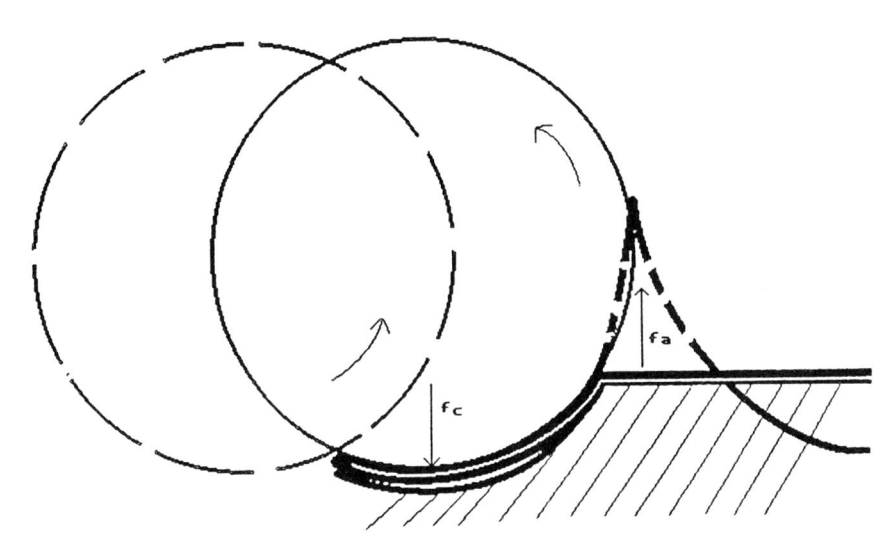

Figure 6.1 Rolling friction caused by compressive deformation of the substrate (adhesive) and by its extension deformation during rolling of the ball.

the angle of rotation and moment of inertia of a cylinder, respectively, f is the rolling coefficient of friction of the PSA, P and Mg are the force components in parallel with and perpendicular to the rolling cylinder, respectively, and if the cylinder is pulled at a constant velocity, P is given by the following equation:

$$P = \left(\frac{f}{R}\right) Mg \qquad (6.1)$$

The rolling coefficient of friction f of a viscoelastic material can be written as the sum of two terms:

$$f = f_c + f_a \qquad (6.2)$$

where f_c represents rolling friction caused by the compressive deformation of the substrate material and f_a that caused by adhesion or extensional deformation of the substrate, which are shown in Figure 6.1.

In the case of PSAs f_a must be much larger than f_c. If compressive deformation of the adhesive is neglected, the strain of the adhesive ε at θ is expressed as [8]:

$$\varepsilon = \left(\frac{R}{h}\right)(1 - \cos\theta) \qquad (6.3)$$

and the rate of strain $\dot{\varepsilon}$ is:

$$\dot{\varepsilon} = \left(\frac{R\theta}{h}\right)\sin\theta = \left(\frac{v}{h}\right)\sin\theta \tag{6.4}$$

where h and v are the original thickness of the adhesive layer and velocity of the cylinder, respectively. For a mechanical model, the stress σ generated by the elongation of the adhesive can be expressed as a function of θ, where θ_f is the angle at the moment of failure. Assuming that the moment caused by the extended part of adhesive is equal to $f_a Mg$ (according to the definition of the rolling coefficient of friction) f_a may be expressed as a function of the cylinder characteristics (i.e., width b, and radius R, moment, and stress):

$$f_a = \frac{R^2 b}{Mg} \int_0^{\theta_f} \sigma(\theta) \cos\theta \sin\theta\, d\theta \tag{6.5}$$

Graphs of f_a versus log v are different for different viscoelastic models and failure criteria [8]. Mizumachi calculated the f_a/V functions for a single

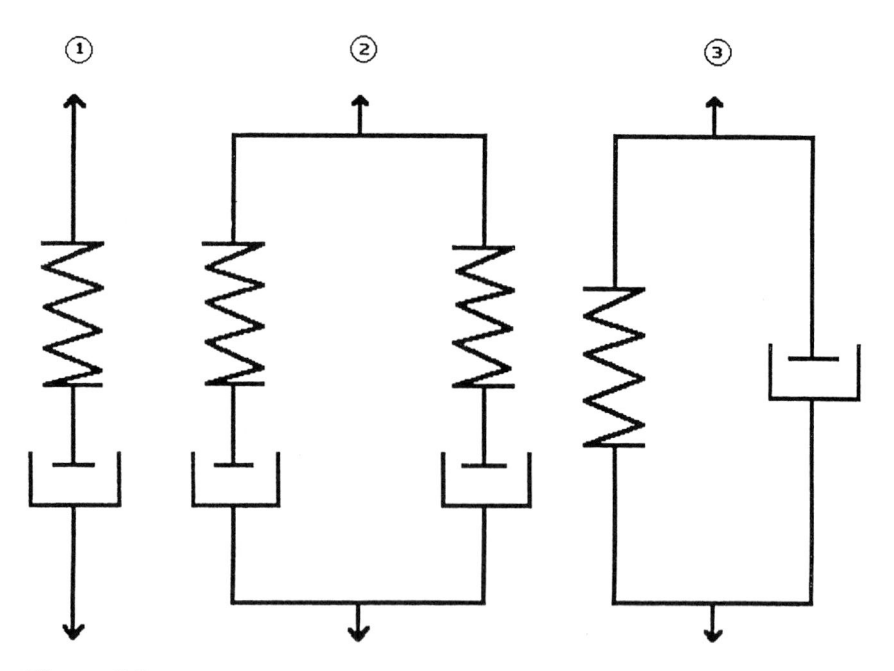

Figure 6.2 Viscoelastic model elements for PSA. 1) A single Maxwell element; 2) Two Maxwell elements in parallel connection; 3) A single Voigt element.

Voigt element, a single Maxwell element, two Maxwell elements, multiple Maxwell elements, and multiple Maxwell elements in parallel (Figure 6.2).

In some cases a dependence of f_a on the modulus E, viscosity η, velocity V, adhesive thickness, and cylinder radius was found. It would be reasonable to expect that a curve of f versus log V for a PSA generally exhibits a peak value or values and that f becomes small when the velocity becomes extremely high or extremely low [8]. For the adhesive practice it is very important to summarize the dependence of the tack on the adhesive rheology (modulus/viscosity) and on the coating weight CW:

$$\text{Tack} = f(E, \eta, CW) \tag{6.6}$$

The influence of the rolling ball or cylinder geometry on the measured tack value may be derived from the general conditions of the adhesion, friction, and wear of elastomers [9]. Equilibrium conditions and the kinetics of adherence between a single hard spherical asperity and a smooth surface of a soft elastomer were studied using fracture mechanics concepts. Conditions for the appearance of reattachment folds and detachment waves were established as a function of several parameters; namely, the normal applied load, the sliding or rolling speed, and the radius of the curvature. The radius a_H of the contact area is given by:

$$a_H = \frac{(PR/K)}{3} \tag{6.7}$$

where P is the normal applied load, R is the radius of the ball, and

$$K = \frac{4E}{3}(1 - v^2) \tag{6.8}$$

where E and v are the modulus and Poisson's coefficient of the adhesive. The contact area is given by two different fields, one of compression in front of the ball and another of tension (smaller than that of compression) behind the ball [10]. The friction force of the ball depends on the temperature and on the velocity of the ball, and this dependence is similar to that of the loss modulus on the frequency and temperature (i.e., it may be described by the Williams-Landel-Ferry equation) [11]. The friction is given by the propagation of the small debonding "waves" (Figure 6.3), which are related to the peeling off of the ball,

$$VT = nvF \tag{6.9}$$

where v is the velocity of the ball, T is the local resistance against the motion of the ball, V is the relative speed of the ball, and F is the peel force. As can be seen from Equation 6.9 rolling ball measurements can be formulated as

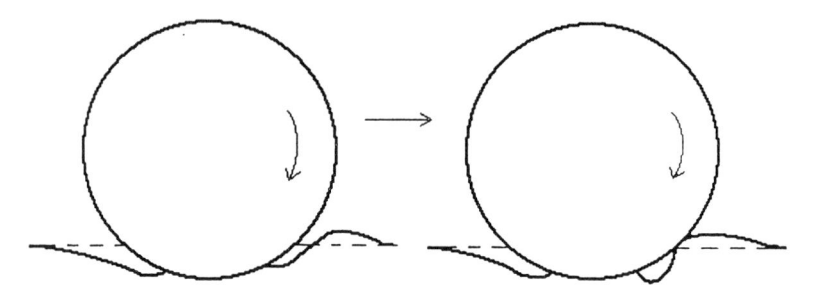

Figure 6.3 Friction given by the propagation of small debonding "waves."

peel measurements. It is evident that peel adhesion will depend on the relaxation of the adhesive material. The friction is the result of the competition between the entrainment of the adhesive by the ball and the interfacial relaxation due to the molecular motion. This interfacial relaxation depends on the rheology of the PSA, but at the same time it depends on the thickness of the adhesive and its anchorage onto the face stock material. If this holds, the nature of the face stock surface has a strong influence on the tack measured by the rolling ball. This conclusion is very important because of the quite different dependence of the tack measured as loop tack on the face stock material. In this case the flexibility of the surface plays a determining role on the measured tack value. It should be mentioned that, on the basis of our knowledge about peel adjustment with the aid of the face stock surface (primer), it is quite normal to have the same dependence for tack measurements by rolling ball, where as shown earlier, peeling phenomena occur.

The examination of the friction force allows a correlation between tack and shear resistance. The friction force F_f is the sum of two components [12]:

$$F_f = F_{adh} + F_{def} \tag{6.10}$$

where F_{adh} is the adhesion component and F_{def} the deformation component. The latter is a function of the internal cohesion of the material, therefore tack will be a function of the cohesion as well.

Tack Measured as Cohesion. Hamed and Hsieh [13] measured tack as cohesion. In the debonding step tack is a function of the cohesion; the upper limit of the tack of an elastomer is its cohesive strength. If during bond formation, complete contact and interdiffusion occur, then the interface will be modified, and the measured tack must be identical to the cohesive strength of the elastomer [13]. Hamed and Hsieh proposed "relative tack" (tack divided by cohesive strength) as a measure of the completion of the tack bond. If the relative tack equals zero, then no tack bond was formed. Relative tack is rate dependent [13]; Hamed and Hsieh measured tack as the average peel force per unit width. At low test speeds the ratio of tack to cohesive strength is less than the same ratio at higher test rates [13]. At sufficiently low stress rates the short interdiffused chains have time to relax, relieving the applied stress. On the other hand, the cohesive strength is controlled by the longer chains, which cannot slip past one another very easily. At higher strain rates even the short molecules may have little time for the molecular rearrangements needed to relieve stress; hence bond strength will increase rapidly as there is insufficient time for failure by chain slippage and chains may rupture.

Tack and Adhesive Fracture Energy. Adhesion and tack of polymers are not fundamental material properties like the modulus or the viscosity; they strongly depend on the test methods used and the measurement conditions [3]. Adhesion performance is characterized by the adhesive fracture energy. Zosel [3] developed an instrument which measures the fracture energy and allows the adjustment of the most important parameters during the bonding and debonding processes, such as contact time, contact pressure, rate of separation, and temperature. The (bonding/debonding) force versus time plot of the instrument can easily be transformed into a stress-strain curve, giving the tensile stress as a function of the tensile strain during bond formation and separation. An appropriate measure of adhesive bond strength and tack is the energy of separation per unit area of surface (i.e., the adhesive failure energy) [14–16]. This apparatus resembles the probe tack tester and a very high correlation between both methods was found.

Peel Test Used for Estimating the Tack. A modified 90° peel test (now known as PSTC test Method 15 for Quick Stick) is used for tack measurements [17,18]. Kendall [19] indicated that a similar method had been used as the Krech Test.

Tack Measured as Plasticity. The measurement of the tack is not as reproducible as the measurement of the Williams plasticity [20]. Practically tack is measured as quick stick, loop tack or rolling ball tack. None of these methods is sufficient to characterize tack alone (see Chapter 10). Therefore the tack index was defined as [21]:

$$\text{Tack Index} = \frac{A + B + 50(17 - C)}{3} \tag{6.10}$$

where A is the value measured in a 90° quick stick test, B is the value of a loop tack test (g/cm), and C is the value of the rolling ball tack in cm (up to a maximum of 15 cm). This index represents an attempt to take into account the three measures simultaneously.

Table 6.1 Tack of Unformulated PSA on Different Chemical Basis. Rolling Ball Tack of Water-Based Acrylic, CSBR, and EVAc PSA

Coating weight, g/m^2	Rolling ball tack, cm		
	AC	CSBR	EVAc
22	1.5	5.8	17.0
40	1.5	5.8	8.5

AC, V 205 (BASF); EVAc, 1360 (Hoechst); CSBR, 3703 (Polysar).

Tack Level. The main adhesive properties (tack, peel, and cohesion) depend on the state of the polymer (pure or formulated), the coating technology (direct/transfer), the coating weight, and the face stock. Of course the most important parameter remains the chemical composition of the polymers. Thus estimating the tack level supposes a comparison of PSAs on the basis of different polymers. Acrylics are polymers of long-chain alkyl-acrylates. There are only a few commercially available low-T_g acrylates (e.g., ethyl hexyl- and butyl acrylate). High tack implies a high level of these fluidizing comonomers. The obtained polymers are tacky, but they lack the necessary cohesive strength. Hence one can conclude that common (nonformulated) acrylic PSAs are not tacky enough and that special acrylic PSAs are not cohesive enough. On the other hand, the tack of pure acrylic PSAs is lower than that of classical (tackified) rubber-based PSAs, but higher than that of other untackified adhesives (except polyvinyl ethers):

Tack: AC (unformulated) > SBR (unformulated) > EVAc (unformulated)

(6.11)

The superior tack level of unformulated acrylic PSAs is illustrated in the Table 6.1, comparatively to water-borne EVAc or CSBR. It can be seen that acrylic PSAs possess a better tack at any given coating weight.

Factors Influencing the Tack

Tack is influenced by the nature and amount of the adhesive (coating weight) and face stock material; as shown earlier test methods and conditions also influence the measured tack value.

Influence of the Nature of the Adhesive. Tack may differ according to the chemical composition or state of the adhesive. Pressure-sensitive adhesives that have different chemical bases (e.g, natural/synthetic rubber, acrylics, vinyl acetate copolymers, maleinates) exhibit different tack levels. On the other hand, PSAs within the same class of monomers (e.g., acrylic PSAs) may display different tack, depending on the specific monomer characteristics, chain structures, and molecular weight. However, most PSAs are formulated and the formulating additives (tackifiers, plasticizer, etc.) change the tack. Water-based PSAs need special formulating additives. Thus the nature of the uncoated adhesive (solvent-based, water-based or hot-melt) will also influence its tack. Generally, additives such as tackifiers (resins or plasticizers) improve the tack; their influence depends on their compatibility. As an example, polybutenes added as a solution in toluene generally improve the rolling ball tack, whereas the use of Acronal 81 D causes a marginal decrease in tack [22].

Influence of Molecular Weight. Tack is a function of the molecular weight. Zosel [3] demonstrated that the tack of polyisobutylene increases with the molecular weight, up to a maximum for MW = 5.10^4. Midgley [23] noted a slight tendency for tack to depend on increasing molecular weight. The useful range of molecular weight in which good PSAs can be identified is limited. The upper limit of molecular weight corresponds to adhesive failure caused by a lowered ability of the adhesive to absorb energy before the interfacial bond is broken or by poor wetting because of the lowered flow of the adhesive. For solvent-based acrylic PSAs, the viscosity is dependent on molecular weight, and thus limited by it. For water-borne acrylic PSAs there is no technological molecular weight limit, but there is an optimum molecular weight for superior tackification.

At a sufficiently short contact time the bond strength (tack) is due primarily to the interdiffusion [13]. The self-diffusion rate is inversely proportional to the second power of the molecular weight; thus tack decreases with increasing molecular weight. On the other hand, adhesive diffusion in the face stock material influences the coating weight and tack depends on the coating weight. Therefore the molecular weight will have a contradictory effect on the tack. Relatively low molecular weight base elastomers and relatively high molecular weight tackifier resins should be used for a good tack.

The dependence of the tack on the molecular weight is demonstrated by the aging of S-I-S rubbers [24]. In this case chain scission occurs and tack increases. In general, resins with a softening point below about 50°C impart tack, but give poor cohesive strength, while resins with a softening point above 70°C give good cohesive strength but poor tack [25]. Tack properties at a given tackifier level are superior with the lower melting point tackifier [26]. For a natural rubber/rosin ester mixture a resin with a lower melting point (e.g., Staybelite E 10, melting point of 75°C) imparts the maximum tack at a higher level (52%) than a higher melting point resin (e.g., Pentalyn H, melting point of 95°C) at a 33% loading level. At the same time, for maximum peel adhesion, an even higher tackifier level (70%) is needed.

For CSBR tackification, increasing the melting point of the resin improves the quick stick of the adhesive [27]. New, high temperature tackifiers for formulating PSAs were proposed [28]. Narrower MWD resins contribute to a higher tack than broad MWD resins [29]. A detailed description of the resin characteristics and the correlation resin melting point/tack is presented in Chapter 8.

Effect of Crosslinking. Crosslinking yields high shear and lower tack properties [30]. As demonstrated for formulations containing an electron beam crosslinkable rubber, the tack is good and remains fairly constant for the Polyken probe and loop tack, regardless of the applied electron beam dose. However, the rolling ball tack values tend to become slightly poorer as the radiation dosage is increased. Generally, crosslinking reduces the chain mobility with increasing T_g, thus decreasing the tack. As an example, crosslinkable water-based acrylic PSAs (e.g., Acronal 29 D, 30 D) suffer a loss of tack after thermal crosslinking.

The tack of silicone PSAs depends on their coating weight and curing [31]:

$$\text{Tack} = f\left(\frac{\text{Coating weight}}{\text{Curing}}\right) \quad (6.12)$$

Lower peroxide levels and a greater adhesive thickness increase the tack. High peroxide levels increase crosslink density and cohesive strength [generally 0.5–3.0% benzoyl peroxide (BPO) is used]. The modification of the tack by crosslinking is a function of the materials used, and of practical experience. Despite the reactivity of Cymel resins, the tack of the modified neoprene latex remains quite excellent [32].

Tack of Uncoated PSAs. Surface-active agents in water-based formulations reduce the tack [23]. This influence appears less pronounced for the other adhesive characteristics. With the exception of a slight change in rolling ball tack, little difference was found between the same adhesives cast from water- and solvent-based formulations. In fact, the different tack level of solvent- or water-based PSAs may be observed mainly at low coating weight (undercoating). At higher coating weight the tack level of water-based acrylic PSAs approaches that of the solvent-based ones (Table 6.2). The obtained tack level also depends on the nature and amount of the surface-active agents [33]; fillers (e.g., silica) also decrease the tack [34].

Table 6.2 Tack of Solvent-Based Acrylic PSA Versus Water-Based Acrylic PSA as a Function of the Coating Weight

	Rolling ball tack (cm)	
Coating weight (g/m^2)	Solvent-based PSA	Water-based PSA
8.0	4.30	6.9
10.0	3.20	6.0
12.5	2.80	5.2
15.0	2.40	4.0
17.5	2.20	3.5
20.0	2.10	2.8
22.5	2.00	2.3
25.0	1.95	2.0
27.5	1.30	1.9

Influence of the Coating Weight. The thickness of the adhesive layer influences the flow conditions. Therefore the coating weight will influence the tack; in general the tack is directly proportional to the coating weight. The exact nature of this dependence also is influenced by the physical state of the adhesive. Figure 6.4 illustrates the dependence of the tack on the coating weight for a solvent-based and water-based adhesive, respectively, namely, that tack increases with the coating weight of the PSA.

Solvent-based adhesives display a higher tack at lower coating weights. The dependence of the tack T on the coating weight is not linear; tack depends on the nature of the adhesive, face stock, and substrate,

$$T = f(CW^\alpha) \tag{6.13}$$

where $\alpha \neq 1$.

Influence of Time/Temperature on Tack. Probe tack values increase with the applied load; even after a 100 sec dwell time the tack value is still increasing, indicating that optimum contact has not yet been achieved [35]. Bates [36] proposed that tack could be described by an equation of the Arrhenius type such that the tack energy T

$$T = A \cdot e^{mx} \tag{6.14}$$

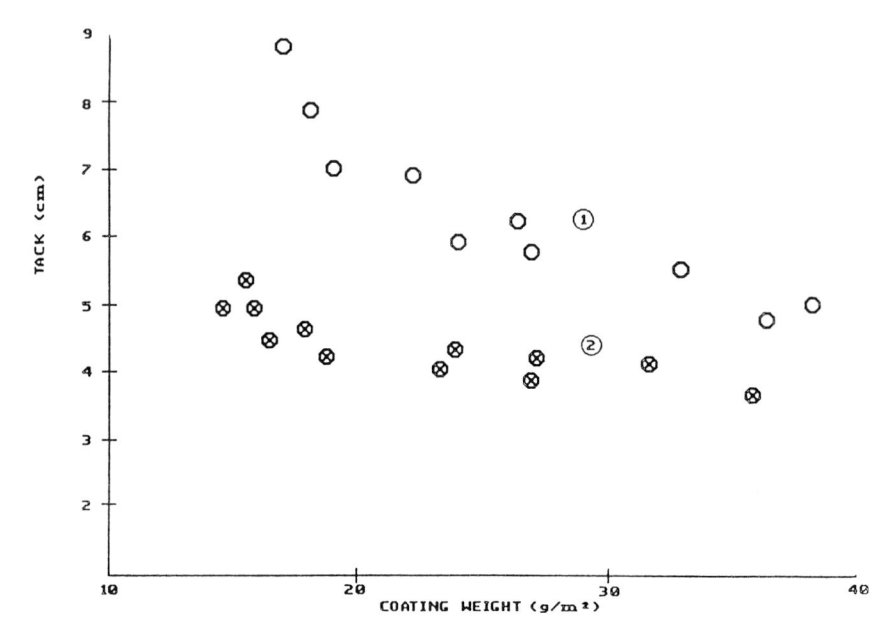

Figure 6.4 Dependence of the tack on the coating weight. Rolling ball tack as a function of the coating weight for 1) water-based and 2) solvent-based PSAs.

where A and m are constants which involve the effect of load and dwell time, and x is the separation rate of the probe.

Zosel [3] studied the tack dependence of polyacrylates on the temperature and found that the tack increased with the temperature going through a maximum at a given temperature as a function of the acrylate structure. Branching produces a shift of the modulus towards lower temperatures, that is, branched acrylates display the same modulus at lower temperatures.

Influence of Test Methods and Conditions. Different test methods give rise to different tack values. On the other hand, the labeling practice points towards a strong influence of the labeling conditions (mainly of the temperature) on the tack. Chapter 2 gives a detailed description of the tack dependence on the experimental conditions (time/temperature).

Influence of Face Stock Material on Tack. As observed when measuring the tack with different methods the nature and dimensions of the face stock material influence the value of the tack. Loop tack measurements are sensitive to the face stock flexibility, while rolling ball tack is influenced by the anchorage of the adhesive on the face stock (i.e., by the face stock sur-

Table 6.3 The Influence of the Face Stock Material on the Tack

Coating weight (g/m^2)	Loop tack (N)	
	Soft PVC	Paper
22	7.3	5.7
24	5.8	5.5
43	3.3	3.2
46	2.5	3.8
56	3.3	4.2

A water-based EVAc PSA was used.

face). Wetout on the face stock material influences the tack of the pressure-sensitive laminate; however, experimental results show some discrepancies; Toyama [37] found maximum tack for a surface tension of the face equal to the surface tension of the adhesive. On the other hand, Sherrif [2] and Counsell [38] found that tack increases with the surface tension of the face material. The surface influences the tack values measured as loop tack (Table 6.3) [39].

Tack characterizes the short bonding time and debonding behavior of the adhesive. During bonding the pressure-sensitive laminate must conform to the substrate in order to establish a maximum contact area. A high debonding resistance is favored by a large contact area. Thus the flexibility (i.e., the bulk properties) of the face material will strongly influence the tack of the PSAs. This behavior is illustrated by the sensitivity of the tack test methods to the flexibility of the face stock material.

In rolling ball tack tests the adhesive does not merely contact the ball, but climbs up the back of the ball just prior to release, very much like a low angle peel. The more flexible the face, the worse the situation becomes, especially if the self-adhesive tape specimen is not firmly secured prior to the test [19]. Kaelble [40] showed that there is an area of compression just behind the peel front. This results in automatic pressure application of the tape to the test panel no matter how minimal the initial application pressure. A polyethylene (PE) film, unlike other film tapes, does not have a tack plateau but continues to climb in tack value [5]. At maximum load the PE film contributes to the value of tack obtained due to its extremely high extensibility.

Tack Values. In the adhesive practice there is a need for reference values of the tack since the formulator and end-user have to compare the obtained tack values with their target performance. This remains a difficult question

Table 6.4 Tack Values for PSA on Different Chemical Basis, Measured with Different Methods

Chemical basis			Rolling ball tack (cm)	Loop tack (N/25 mm)	
Base polymer	Supplier	Chemical nature		Glass	PE
1360	Hoechst	EVAc	17.0	6.9	3.4
FC88/PC80	UCB	AC	7.0	13.1	7.6
FC88	UCB	AC	7.5	13.3	8.0
V205	BASF	AC	2.0	6.0	4.2
V208	BASF	AC	> 30.0	7.9	7.9
3703	Polysar	CSBR	25.0	5.3	2.7
3703/V205	Polysar/BASF	AC/CSBR	20.0	6.2	1.8
3703/V205	Polysar/BASF	AC/CSBR	6.5	8.4	7.6

Tackified emulsions were used.

because tack values depend on the measurement methods and conditions, the face stock material used, and the coating weight. Tack values obtained for adhesives with a different chemical basis or physical state differ. However, for those skilled in the formulation and use of PSAs, there are some target values of the tack which need to be achieved. Table 6.4 summarizes some tack values for PSAs with different chemical bases and measured with different methods.

Improvement of the Tack

On the basis of theoretical considerations and practical experience one can assume that hot-melt and solvent-based PSAs possess a higher tack than water-based ones (cf. the lower molecular weight for the base polymers of hot-melt PSAs and no poor tack additives for solvent-based PSAs); soft base polymers (flexible, branched chain) yield better tack. Low molecular weight polymers and uncrosslinked polymers (or with a low degree of crosslinking) display higher tack levels. Tackified polymers possess better tack as do plasticized polymers. Finally, higher coating weights result in higher tack levels (see Figure 6.4).

The coating weight also influences the peel adhesion and shear (see Sections 1.2 and 1.3), as well as other characteristics like wettability (see Chapter 2). Therefore, before the discussing other adhesive characteristics, a short examination of the coating weight as a quality parameter is presented first.

Role of Coating Weight and Parameters Influencing It

Benedek [41] summarized the influence of the coating weight on the PSA label quality and its importance for the screening and testing of PSAs. The coating weight influences the geometry and the quality of the PSA layer. Its influence on the quality of the PSA layer is indirect, via drying [42]. The coating weight influences the drying of the adhesive layer. The drying rate R is given as a function of the mass transfer coefficient c_m, the heat transfer coefficient c_h, and the latent heat of evaporation L [43] (i.e., as a function of mass and heat transfer):

$$R = f\left(\frac{c_h \cdot s \cdot \Delta t}{L}\right) = f(c_m \cdot s \cdot \Delta p) \qquad (6.15)$$

where

$$\begin{aligned}
\Delta t &= t - t_s \\
\Delta p &= p_s - p \\
t_s &= \text{surface temperature} \\
p_s &= \text{vapor pressure at } t_s \\
p &= \text{partial pressure} \\
s &= \text{surface area}
\end{aligned}$$

For the drying of aqueous adhesive layers, one can assume that the drying rate R also depends on the layer thickness h, concentration c, Δc, and specific gravity φ [42]:

$$R = \frac{h \cdot \Delta t \cdot \Delta c}{\varphi \cdot L \cdot h} \qquad (6.16)$$

For solvent-based adhesives, the drying rate depends on the diffusivity D, concentration c, and layer thickness h [42]:

$$R = \frac{\pi \cdot D \cdot \Delta c}{4h^2} \qquad (6.17)$$

Generally, the drying rate is inversely proportional to the layer thickness (i.e., to the coating weight CW):

$$R = f\left(\frac{1}{CW^\beta}\right) \qquad (6.18)$$

As shown in Figure 6.5 for low coating weight values, the exponent β is smaller than 1, whereas for high coating weights, the exponent should be larger than 1 (i.e., for drying PSAs with a high coating weight, the layer thickness is the most important rate determining parameter). The dependence

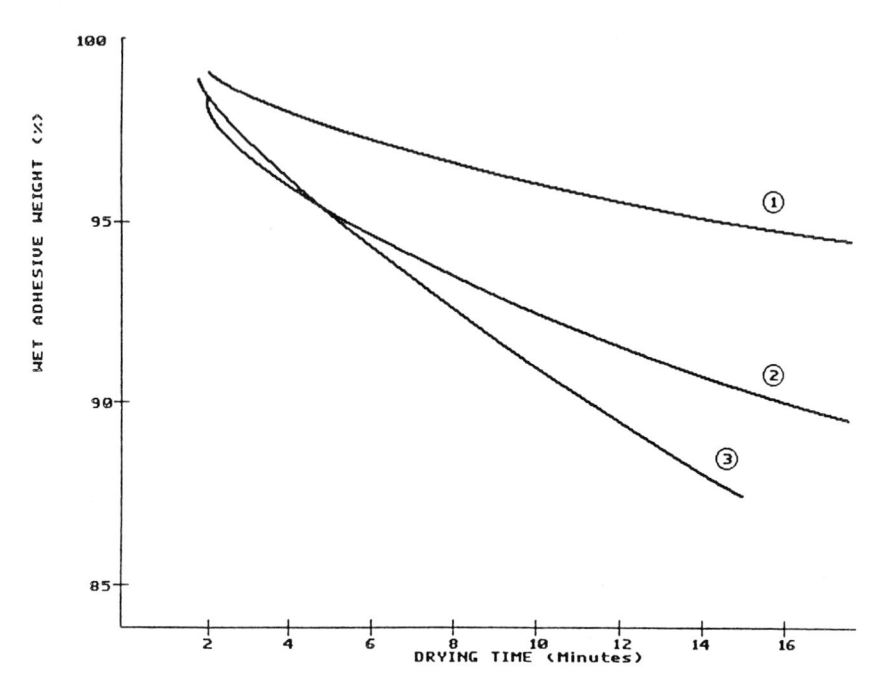

Figure 6.5 Dependence of the drying speed on the coating weight. Weight loss (%) as a function of the time, for different coating weights, (wet adhesive thickness, μm: 1) 60; 2) 100; 3) 200) during drying of a water-based PSA.

between the wet coating weight, solids content, and drying time was studied by Hansmann [43].

The strong influence of the coating weight on the drying has an indirect effect on the adhesive characteristics. The equilibrium water content of the paper and water-based PSAs layer depends on the drying degree, (i.e., on the coating weight). On the other hand, residual humidity affects tack, peel, and shear (see Section 1.2).

The direct influence of the coating weight on the peel and shear will be discussed in Chapter 10. As a general statement, the coating weight has a positive effect on tack and peel, and a negative one on the shear and drying. In a first approach tack T and peel P are proportional to the coating weight, shear S and the drying speed R are inversely proportional. Thus:

$$T = f(CW)^{\gamma} \tag{6.19}$$

$$P = f(CW)^{\delta} \tag{6.20}$$

$$S = f\left(\frac{1}{CW}\right)^{\varepsilon} \tag{6.21}$$

$$R = f\left(\frac{1}{CW}\right)^{\zeta} \tag{6.22}$$

In summary the dependence of the adhesive properties AP on the coating weight CW can be represented as follows:

$$AP = f\left(\frac{CW^{\alpha+\delta}}{CW^{\varepsilon+\zeta}}\right) \tag{6.23}$$

namely, the coating weight influences the versatility of the adhesive in a complex manner. Evidently, the development of the PSA chemistry leads to better adhesives (i.e., with a higher tack and peel adhesion) at a lower coating weight.

The most important factors influencing the coating weight are the face stock material, the substrate to be adhered to (adherent), the permanent/removable character of the adhesive, and the coating conditions. In general it can be stated that the optimum coating weight value depends on the chemical composition and on the end-use of the PSAs. On the other hand, the coating weight must be correlated to the uniformity of the coated layer and to the coating (film-forming) quality, depending on the combination of the following parameters:

- Machine characteristics: speed and sense of rotation of the metering cylinders, concentricity of the rolls, gap widths, properties of rolls (e.g., surface finish, stability of material, deflection), configuration, number and combination of rolls, hydraulic pressure, and temperature control;
- Adhesive characteristics: cohesive strength, flow, and rheological properties of the adhesive.

Chemical Composition and Coating Weight. According to their chemical nature, synthesis, or formulation, different PSAs possess a different adhesion-cohesion balance. Tack and peel may be improved by increasing the coating weight. Thus, the chemical nature of the PSAs will influence the required coating weight.

Influence of Coating Conditions on the Coating Weight. The most important parameter is the coating technology (i.e., the direct/indirect nature of the coating operation). Direct coating supposes depositing the adhesive on the face stock and thus adhesive bleeding through the porous surface remains possible. Hence the measured coating weight could be lower than the one being deposited. A detailed analysis of the influence of the coating conditions

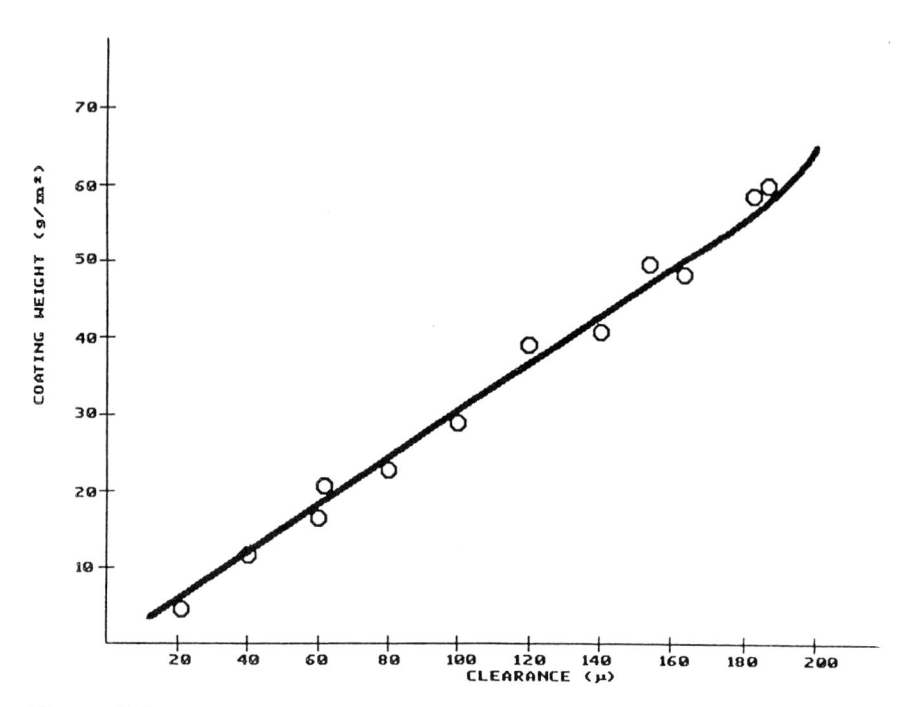

Figure 6.6 The dependence of the coating weight on the clearance (blade opening.) Unfilled, water-based PSA formulation.

(coating technology) on the coating weight will be given in Section 1.2 of Chapter 9.

A lot of parameters influence the coating weight [41]. For laboratory tests it is important to describe the dependence of the coating weight on the knife opening (clearance) and viscosity. For a coating device with a blade, the pressure difference Δp depends on the viscosity η, width of the blade b, web velocity V_w, and blade angle θ as follows [44]:

$$\Delta p = \frac{4\eta \cdot V_w}{b \cdot \theta} \tag{6.24}$$

For the simplest case of a parallel blade, the blade angle may be replaced by the blade opening (i.e., the distance between the blade and the cylinder). In a first approximation a linear correlation is found between coating weight and blade opening (Figure 6.6).

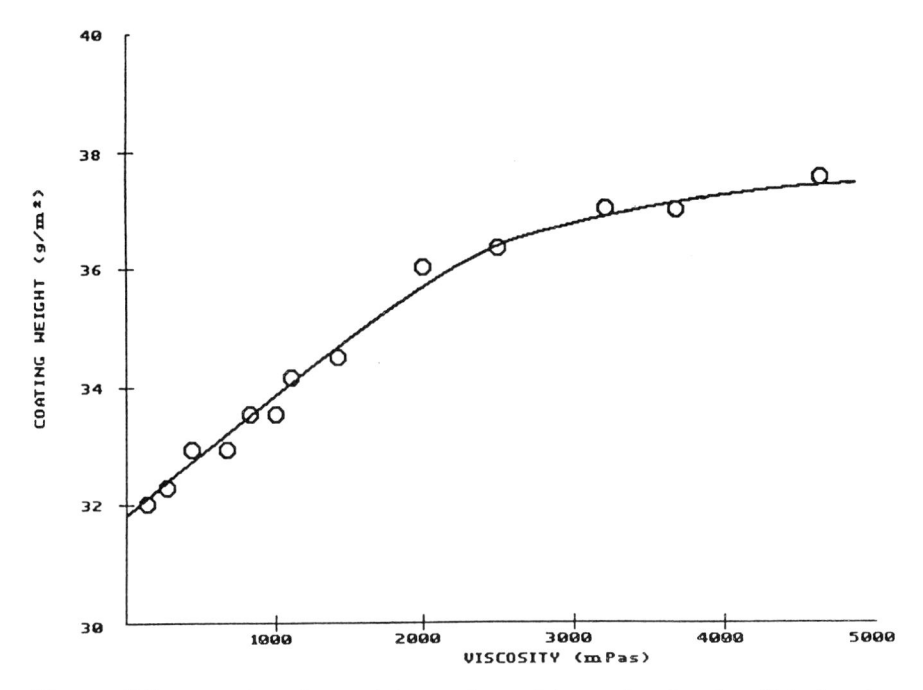

Figure 6.7 The dependence of the coating weight on the viscosity. Water-based, unthickened PSA formulation.

On the other hand, the dependence of the coating weight on the viscosity appears more complex. The coating weight or layer thickness H_w for a Newtonian fluid depends on the surface tension S_T, on the blade opening d, viscosity η, and specific gravity φ [45]:

$$H_w = f(S_T, d, \eta, \varphi) \tag{6.25}$$

There is a complex dependence of the coating weight on the viscosity (Figure 6.7). Unthickened dispersions behave quite differently from thickened ones; thickened dispersions give a higher coating weight at higher viscosities than unthickened ones (Figure 6.8).

The viscosity also depends on fillers and pH; adjusting the coating weight with fillers (viscosity) or through the pH also is possible. Table 6.5 demonstrates that filled dispersions do not display a linear dependence of the coating weight on the blade opening.

Surfactants may change the viscosity of the aqueous dispersions and hence the coating weight. Storage time and temperature also influence the viscosity

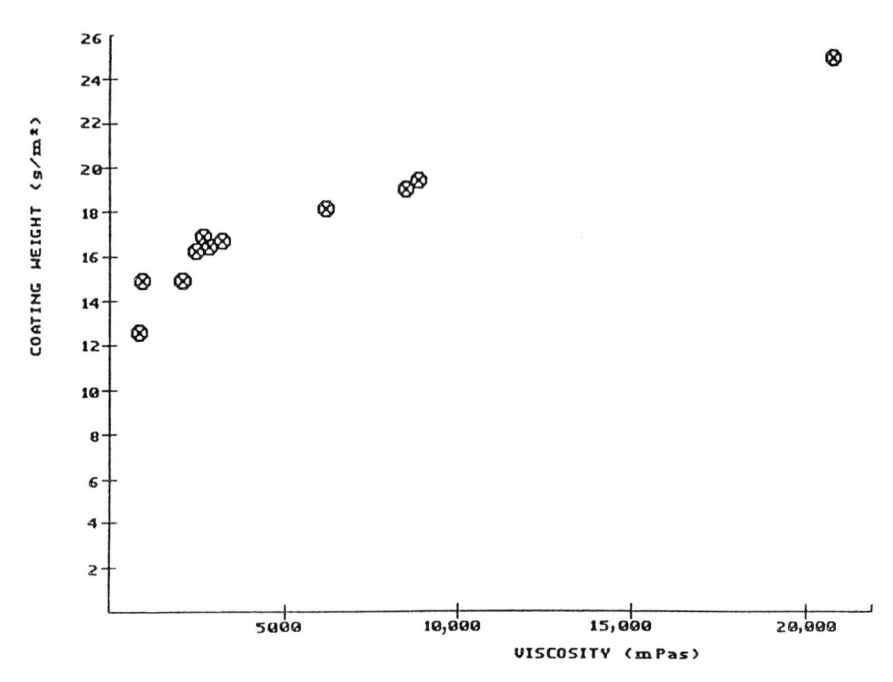

Figure 6.8 Coating weight dependence on the viscosity. Water-based, thickened PSA formulation.

Table 6.5 Dependence of the Coating Weight on the Clearance (Blade Opening) for Water-Based Acrylic PSAs Containing Filler

Clearance	Coating weight (g/m^2)				
	Filler content (parts/100 parts, wet weight)				
	0	5	10	20	30
60	25.5	20.0	33.0	44.0	29.0
100	43.0	36.0	40.0	38.0	45.0

and thus the coating weight. In fact practical coating conditions differ quite a bit from laboratory ones. Shear thinning and thickening and flow under different centrifugal forces may influence the coating weight. Laboratory coating conditions may be regarded as static whereas industrial conditions are dynamic. Here the web velocity and shear conditions are the most important parameters of the coating weight for a given adhesive and coating device geometry.

Practically, during coating, in order to keep the coating weight tolerance within maximum ±5% the viscosity of the emulsion must be kept at a constant level [46]. For an emulsion PSA on a reverse gravure coater, it is ideal to have a high solids content with a low viscosity. Low viscosity means a high coating weight; conversely, high viscosity means a low coating weight. The minimum level of the viscosity also depends on the surface tension of the adhesive; common viscosities for reverse gravure range from 17–24 DIN sec.

Pressure-sensitive adhesive solutions or dispersions that have different chemical bases may have a different solids content and viscosity. Thus, for the same viscosity and wet coating weight, a quite different dry coating weight and tack level may be obtained. This is illustrated by Figure 6.9.

For aqueous dispersions the coating weight also depends on the porosity of the web (paper). Therefore direct or transfer coating will result in different coating weights (i.e., for the same coating weight the thickness of the adhesive layer on the coated side will differ) (Table 6.6).

The obtained coating weight depends on the type and characteristics of the metering device used. A gravure cylinder with a line screen of 40, 70 μm depth, yields a coating weight of 5–7 g/m^2 [47]. When coating a 50%-solids adhesive with adequate viscosities, the rheology of the system is such that to achieve a 3–4 g/m^2 solid coating weight, a 35-μm gravure cell is suggested, while 2–2.2 g/m^2 solids requires a 22-μm deep gravure cell.

For a metering cylinder, the coating weight will depend on the blade opening between blade and cylinder, the machine speed, and the film transfer from the cylinder to the substrate [48]. Machine speed influences the coating weight; higher velocities yield higher coating weights [49].

Influence of the End-Use of the Label on Coating Weight. The end-use of the label requires PSAs with a permanent or removable character. On the other hand, the end-use of the label determines its application and labeling technology. This also depends on the adhesion/cohesion balance. Labeling guns require very tacky adhesives; for decals a high tack is less important. Peel is a function of the coating weight. In Chapter 2 it was shown that removability depends on peel adhesion and on the coating weight. According to the permanent/removable character of the designed label, one can use a higher or lower coating weight.

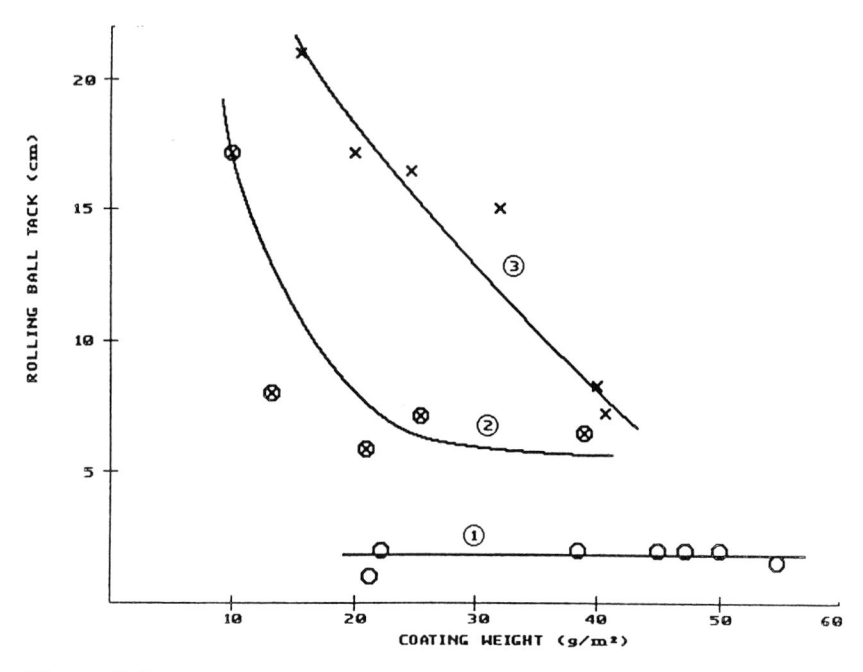

Figure 6.9 Tack dependence on the coating weight. Rolling ball tack as a function of the coating weight for water-based PSAs on different chemical bases: 1) CSBR tackified; 2) CSBR; 3) acrylic.

Table 6.6 Dependence of the Coating Weight and Peel on the Coating Method

PSA	Coating weight (g/m^2)	Coating method		Peel on glass (N/25 mm)
		Direct	Transfer	
1	18.0	x		17.0
	20.0		x	17.0 PT
2	19.5	x		21.0
	21.0		x	21.0 PT
3	15.0	x		12.0
	21.0		x	19.0 PT
4	10.0	x		7.0
	15.0		x	9.0

Water-based acrylic PSA coated on paper face stock material.

Influence of the Face Material on the Coating Weight. The dry coating weight (i.e., the weight of the dry adhesive applied per unit surface area) can vary substantially depending upon the porosity and irregularity of the face material and of the substrate surface to which the face stock is to be adhered. For instance, higher adhesive loadings are preferred for adhering porous, irregular ceramic tiles to porous surfaces, while lower adhesive loadings are required to manufacture tapes, films, and other articles from relatively nonporous smooth-surfaced materials such as synthetic polymer films and sheets. When the adhesive is applied on nonporous polymeric or metallic face materials intended for adhesion to nonporous polymeric or metallic surfaces, adhesive loadings of about 5–50 lb of dry adhesive per 3000 ft^2 are generally adequate [50]. Adequate adhesion in tapes manufactured from continuous sheet polymeric substrates can usually be achieved with dry adhesive coating weights of about 10–20 lb/3000 ft^2, while coating weights of about 10–20 lb/3000 ft^2 of adhesive are usually employed for paper-backed tapes, such as masking tapes. The influence of the face surface on the coating weight is linked to its porosity and deformability.

Influence of the Substrate on Coating Weight. The adhesive flow is infuenced by the nature of the contact surface (liner as adherent). Chemical, physical, and mechanical interactions with the solid state component may hinder the adhesive flow. Therefore the chemical nature (e.g., affinity and polarity) of the surface and the physical nature (e.g., porosity and roughness) will influence the interaction of the adhesive with the substrate. The adhesive flow influences the bonding and debonding characteristics.

The coating weight determines the real thickness of PSAs and the thickness of the adhesive layer which can flow freely. Thus the coating weight will depend on the nature of the substrate. As a practical example, adhesion onto HDPE is generally lower than adhesion on LDPE [51]; therefore the coating weight should be increased for adhesion to HDPE.

Dependence of Coating Weight on Adhesive Quality. The coating weight may also depend on the quality of the adhesive. Under certain conditions the emulsion polymerization of water-soluble monomers may yield suspensions (i.e., the monomer is homopolymerizing) with bulky undispersed polymer particles remaining (coagulum, grit). This phenomenon is especially encountered with acrylonitryle copolymers; Acronal 81 D may contain so much coagulum that a coating weight of at least 100 g/m^2 is recommended (i.e., more than the diameter of the largest solid particle in the fluid adhesive layer) [52].

Coating Weight Values. There are a lot of factors influencing the value of the coating weight. On the other hand, the development of the PSA chemistry has led to adhesives with improved tack and peel adhesion (i.e., with

Table 6.7 Coating Weight Values of Industrial Labelstock

Adhesive characteristics					
Nature		Adhesive layer			
Chemical	Physical	Coating weight (g/m^2)	Coating thickness (μm)	Face stock	Refs.
Acrylic	Hot melt	—	25	Paper	[26]
Acrylic	Water-based	25	—	Paper	[54]
Acrylic	Solvent-based	20	—	Paper	[54]
—	—	30	—	—	[55]
—	—	26–29	—	—	[56]
Acrylic	Water-based	27–29	—	Film	[4]
Acrylic	Water-based	25	—	Paper	[57]
Acrylic	Water-based	25	—	Paper	[58]
CSBR	Water-based	25–28	—	Film	[22]
Acrylic	Water-based	—	20	Paper	[59]
Acrylic	Hot melt	—	25–150	Paper	[60]
Acrylic	Solvent-based/ Water-based	—	25	—	[25]

lower coating weight). Some years ago coating weight values of 30–100 gm^2 were recommended [53]; actually 15–20 g/m^2 are more common. Table 6.7 gives some typical values of the coating weight used for labels on different face stock materials and for different end-uses. Techniques for measuring the coating weight and tolerances are covered in Chapters 8, 9, and 10.

1.2 Peel Adhesion

Peel adhesion is the force required to remove a PSA-coated flexible material from a specified test surface under standard conditions (e.g., specific angle and rate). It gives a measure of adhesive or cohesive strength, depending on the mode of failure [61]. Johnston [35] stated as an analogy that the adhesive consists of many little springs, as depicted in the Maxwell and Voigt models to simulate viscoelastic behavior, and that these springs range from strong to weak. The former contribute to the adhesion (peel), the latter contribute to tack. In fact, the measured peel energy includes a significant contribution from energy dissipation within the bulk of the polymer. Thus for a given extensible polymer, the higher the work of adhesion (due to the interaction at the surface), the higher the dissipation energy (due to polymer deformation) and the higher the peel energy [62].

Measurement of the Peel Force

Peel is measured as the force required to remove a PSA-coated material which was applied to a standard test plate, at a specific rate and angle. In some cases not only the value of the peel force, but also the failure mode (deformation or destruction of the components of the pressure-sensitive laminate or of the substrate) has to be examined in order to obtain a correct appraisal of the removability of PSAs.

Factors Influencing Peel Adhesion

Bonding plays an important role in the peel resistance. Parameters influencing the adhesive flow during the bonding step will influence the peel adhesion. In a manner similar to tack, peel is influenced by the PSA nature and geometry, and by the nature of the face stock material. Test methods and conditions modify the peel value as well. More pronounced for the peel force are the influence of the dwell time and of the adherent. Wilken [36] studied the influence of different parameters on the measurement of the peel force and stated that peel adhesion depends on the nature of the substrate, the peeling rate and angle, and on the sample dimensions; on the other hand, peel does not depend on the cleaning agent used or on the number of rolling passes used to laminate the sample onto the test plate [36].

Influence of Adhesive's Nature on Peel. Similar to tack, peel resistance is a function of the chemical nature and macromolecular characteristics of the base polymers. On the other hand, and in contrast to tack, peel adhesion increases with the cohesion (i.e., with the molecular weight) up to a limit. A quite different adhesion/cohesion balance is required for good peel adhesion onto polar or nonpolar substrates. Regarding the influence of the adhesive's nature one should differentiate between the chemical composition of the base elastomers, the chemical composition of the formulated adhesive, and the technology dependent composition of the coated PSAs; one should also note the influence of aging, the chemical composition of the PSA, and solid state components of the laminate, because of the environmental and reciprocal interactions between the adhesive and the neighboring components (e.g., face stock and release liner).

Influence of the Chemical Composition of the Base Elastomer on the Peel. The anchorage of the adhesive on the substrate is governed by a physical and a chemical component. Chemical interactions between the adhesive and the substrate result in adhesion build-up; the chemical composition of the base elastomers strongly influences this adhesion build-up. Small portions of highly polar monomers (e.g., acrylic acid) produce adhesion build-up. On the other hand, PSAs with a vinyl polymeric backbone and with grafted polysiloxane moieties produce initially repositionable labels [64].

The inclusion of polar carboxylic groups in the adhesive may influence its peel value and change its affinity towards polar surfaces [65]. In a similar manner such groups influence the peel of the adhesive through water sensitivity. Built-in carboxyl groups may produce a change in the peel value from special substrates through pH changes [65]. Introducing polar, functional groups into ethylene vinyl acetate copolymers improves their adhesion [66]. The contribution to adhesion strength by these groups is evaluated by surface energy measurements and by peel strength measurements (energy dissipation due to polymer deformation).

Influence of the Chemical Composition of the Formulated Adhesive on Peel. The effect of the formulating additives on the peel will be reviewed in Section 2.9 of Chapter 8. Some of these influences are caused by the chemical interaction of the viscous and elastical components of the formulation (e.g., elastomer, tackifier, plasticizer) and exist in a freshly coated adhesive layer, but others are due to the interaction of adhesive/environment or adhesive/laminate components. Changes in the peel value given by adhesive/environment interactions may be irreversible (aging) or reversible, like the influence of the atmospheric humidity on the peel of tackified (with aqueous resin dispersions) or untackified water-based PSAs.

Tackification remains the most pronounced formulating effect on the peel. Peel adjustment through tackification will be discussed in a detailed manner in Chapter 8. Here it should be mentioned that for removable (low peel adhesives) no tackifier or a low level of tackifier should be used. More than 50% tackifier will make the adhesive nonpeelable [67]; fillers also influence the peel adhesion. Generally, they improve the modulus, but decrease the contact area and diffusion rate. Therefore over 10–15% filler concentration (on wet/wet basis) decreases the peel [65]. In hot-melt PSA formulations peel values sharply decrease with the use of fully saturated mineral oils [68].

Effect of Crosslinking on Peel Adhesion. The 180° peel strength of PSA formulations based on crosslinkable rubber (e.g., S-I-S, electron beam-curable) remain fairly constant over the entire range of electron beam doses [30]. Generally, crosslinking exerts a more complex influence on the peel. A low degree of crosslinking improves the cohesion and thus the peel, whereas a high crosslinking degree lowers the tack and the peel. As an example one can consider the increase of the peel for Cymel crosslinked neoprene latex and the dramatic reduction of the peel upon adding zinc oxide to the formulation [32].

Adhesive Performance Characteristics

Influence of the Molecular Weight of the Adhesive on the Peel. The influence of the molecular weight of the polymer components of the PSAs on the rheology of the PSAs was examined in Chapter 3. The rheology of the adhesive influences its adhesive characteristics; therefore, the peel also is a function of the molecular weight [69]. The increase of the modulus with the molecular weight reinforces the peel; however, it reduces the chain mobility and diffusion rate, and decreases the peel [70]. Physical adhesion can develop from adequate wetting at a well-defined interface or by diffusional interpenetration of segments across the interface, when this is thermodynamically possible. The fracture energy G of a joint follows a relation of the form [71],

$$G = f(Dt)^\alpha \tag{6.26}$$

where D is the diffusion coefficient, which is inversely proportional to the molecular weight, and t is the time. For adhesive formulations tackified with butyl rubber, peel adhesion increases with the molecular weight of the tackifier. On the other hand, when tackifying CSBR higher melting point resins decrease the adhesion on polyethylene.

Influence of the Adhesive Geometry on the Peel Adhesion. Adhesive geometry includes the dimensions and form of the adhesive layer; the former depends on the coating weight. In general peel adhesion is a function of the thickness of the adhesive [72] and, like tack, peel is very sensitive towards the coating weight. Generally, at low coating weights, the peel increases with the coating weight, implying a debonding through adhesive failure.

Concerning the influence of the coating weight on the peel resistance, the peel force P depends on the peel stress s, sample width b, layer thickness h, and creep modulus E_c as follows [39]:

$$P = \frac{s^2 \cdot b \cdot h}{E_c} \tag{6.27}$$

As can be seen in Figure 6.10 there exists a complex dependence of the peel on the coating weight; generally there is an inflection point in the plot. Up to this point peel increases rapidly with the coating weight; after this point peel depends less on the coating weight. The shape of the plot depends on the adhesive and face material.

In order to avoid quality consistency deviations caused by a variation of the coating weight, the coater should use a coating weight above this critical value. It should also be mentioned that in the upper domain (over the critical value), the peel dependence on the coating weight is less dependent on the

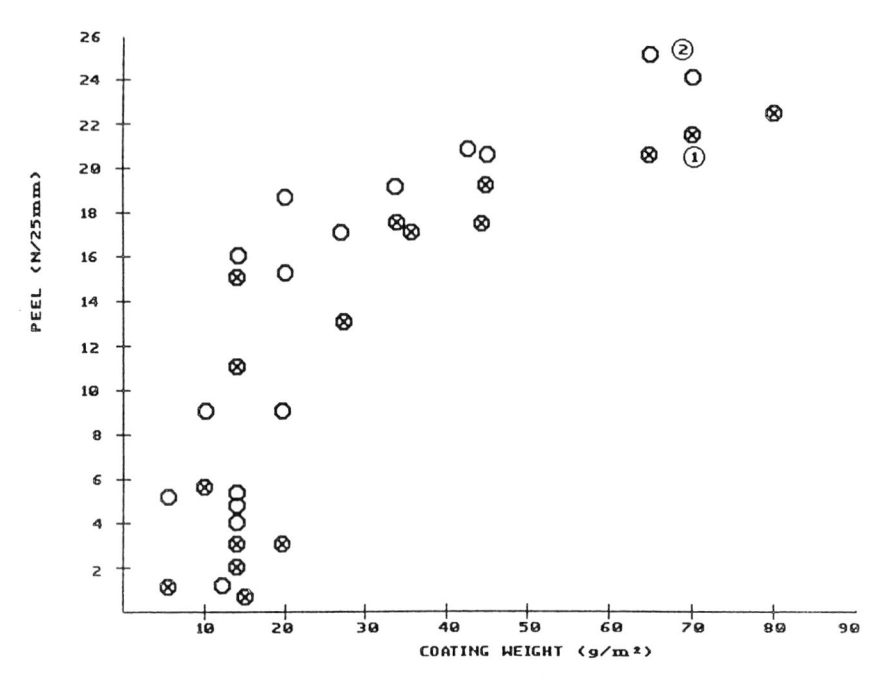

Figure 6.10 Dependence of the peel on the coating weight. 1) Peel on polyethylene; 2) Peel on glass.

face stock. Generally, the dependence of the peel on the coating weight may be formulated as follows:

$$P = f(CW^b) \tag{6.28}$$

where $b \neq 1$ and possesses quite different values in the over and under critical domain.

As discussed earlier, tack also is measured as peel adhesion. Tack increases with the coating weight. Figure 6.11 illustrates the simultaneous increase of tack and peel with the coating weight; however, if the debonding occurs by break in the adhesive layer (cohesive failure) the peel does not depend on the coating weight [73]. The peel strength depends on both the thickness of the adhesive layer and of the face stock material [74]. Peel increases with increasing thickness of the adhesive. This increase of the peel with the adhesive thickness is illustrated in Table 6.8. The influence of the thickness of the adhesive layer on the peel was also studied by Schlimmer [75], who found that peel adhesion does not depend on the layer thickness.

Adhesive Performance Characteristics

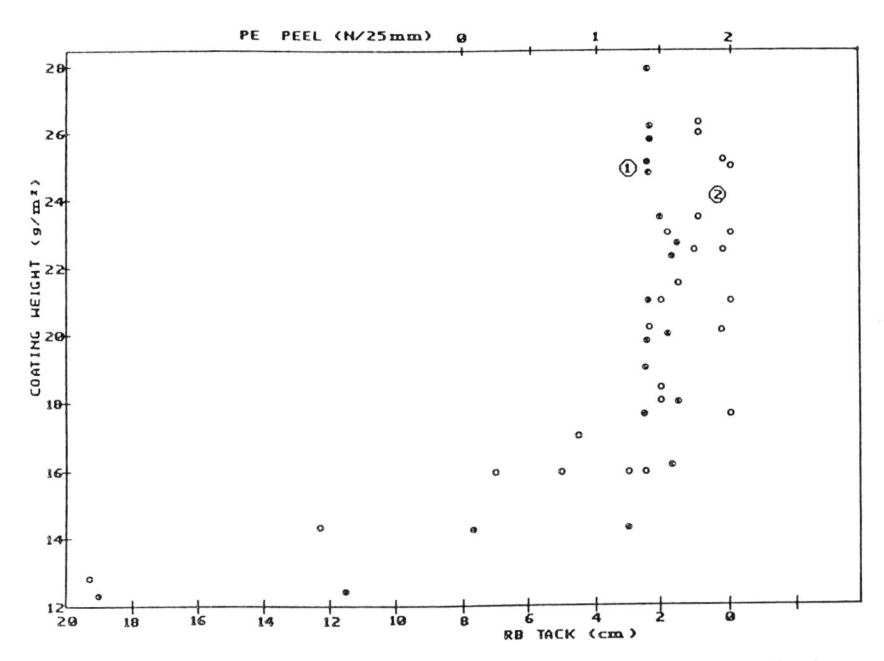

Figure 6.11 The influence of the coating weight on the tack and peel. The dependence of 1) the rolling ball tack and 2) of the peel on polyethylene on the coating weight for a water-based acrylic PSA formulation.

Table 6.8 Dependence of the Peel on the Thickness of the Adhesive Layer

Wet adhesive layer thickness (μm)	Coating weight (g/m^2)	Peel (N/25 mm)	
		PVC	Paper
60	22	7.3	—
	24	—	5.6
70	24	5.8	—
	25	—	5.5
100	43	3.3	—
	46	—	3.2
130	46	2.5	—
	47	—	3.8
160	56	3.3	—
	67	—	4.2

A 200-mPa·s EVAc water-based PSA was coated on soft PVC via a 80 g/m^2 paper (transfer coating).

Table 6.9 Dependence of the Peel on the Coating Weight (Peel on Glass and Polyethylene as a Function of the Coating Weight)

Coating weight (g/m^2)	Peel (N/25 mm)	
	Glass	PE
9.4	17.0 PT	15.0
18.8	17.0 PT	16.0
33.5	22.0 PT	14.0
39.6	25.0	17.0
61.8	25.0	19.0

A tackified water-based acrylic PSA was used.
PT = Paper tear

The use of large amounts of adhesive leads to a relatively strong bond between the adhesive and the target substrate, thus making the product less peelable. Generally, one finds that lower coating weights make peelability of the product in use easier to achieve, but make a uniform coating more difficult to obtain. For paper coating with acrylic PSAs the coating weight amounts to at least 12 g/m^2, and usually 14–20 g/m^2 depending on the nature of the adhesive. The influence of the coating weight on the peel was studied by several authors [5,69,74,76]. The 90° peel adhesion test (FTM 2) results in a higher (usually more than double) peel adhesion from PE than the 180° peel test [27] (see Section 3.1 of Chapter 10). For the 90° peel a nonlinear dependence of the peel on the coating weight was established; for 180° peel adhesion a linear dependence of the peel force on the coating weight is given.

Peel dependence on the coating weight also is a function of the substrate, face stock, and adhesive. For silicone-based adhesives, in most cases a coating weight-independent adhesion value on steel is obtained at coating weights over 4–6 mil. Table 6.9 shows that peel adhesion increases with the coating weight; after a certain value of the coating weight (which depends on the adhesive and substrate) peel adhesion increases at a slower rate and eventually reaches a plateau.

Dependence of the Peel on the Shape of the Adhesive Layer. In general defects in an applied continuous PSA layer should be avoided for common adhesive coatings. However, there are special coatings with a discontinuous character of the adhesive layer, mostly in order to avoid the build-up of high peel resistance. On the other hand, the structure of the continuous adhesive layer also influences the peel resistance [73]. See Chapter 2 for a more detailed discussion of the influence of the adhesive form on the peel.

Influence of Coating Technology on Peel. Direct coating achieves better anchorage of the PSAs on the face stock, and therefore better removability. For optimum anchorage, sheet-stock PSAs are often directly coated on the face material, rather than transfer coated via the release liner. The state of the adhesive during the coating operation also influences the peel. Water-based PSAs contain different water-sensitive and hygroscopic components; hence they may contain less or more humidity as the laminate. The residual water content acts like a plasticizer and reduces the peel [51].

Influence of Face Stock Material on the Peel. The rheology of the PSA depends on the solid state components of the laminate. Peel is a function of the rheology of the adhesive and it also depends on the face stock material. Both the bulk properties of the face stock material and its surface quality affect the peel resistance. Peel is determined not only by the deformability of the adhesive, but also of the face stock [77]; the stiffness (flexural modulus) of the face stock influences the peel angle and thus the peel value [73]. The effect of the face material becomes part of the result; as an example one can consider a 180° peel adhesion test, where work is being done not only to separate the adhesive from the applied surface but also to bend the face stock through 180° [5]. The same adhesive at the same coating weight can display a variety of peel values when coated onto different face stock materials. Depending on the substrate, adhesives may be classified as removable or permanent. The same substrate material may have different surface properties; for example, yellow plastic items may be labelled less easily than white ones because of the fillers used [78].

The bulk characteristics of the face stock material, its stiffness, and plasticity/elasticity affect the transfer of the debonding forces (see Chapter 2). On the other hand, the chemical composition of the face stock material influences the interaction between the adhesive and the face stock material. This influence should be taken into account especially for PVC foils [73]. The geometry and dimensions of the face stock can affect the peel resistance as well.

Mechanical effects in the peel test were studied by Kim et al. [79]. The peel strength measured by the peel test method is a practical adhesion (an engineering strength per unit width) and does not represent the true interfacial adhesion. The measured value is a combination of the true interfacial adhesion strength and forces required for plastic and elastic deformation of the adhesive film and substrate. The major controlling factors in the peel strength are thickness, Young's modulus, yield strength, the strain hardening coefficient of the adhesive film, compliance of the face material, and interfacial adhesion strength. The mechanical characteristics of the face stock material, its flexibility and elasticity (plasticity) influence the transfer of the debonding energy to the adhesive layer (see Chapter 2).

The decrease or damping of the peel force is caused by the deformation (elongation) of the face stock material under tension, changing the direction or distribution of the force (change of the debonding angle), and by the stiffness of the material; these phenomena decrease the peel force [5]. Thus, one can write:

$$\text{Peel} = f\left(\frac{1}{\text{Stiffness, Plasticity}}\right) \quad (6.29)$$

The stiffness depends on the flexural modulus E_f, whereas elongation depends on the elongation modulus E_e of the face stock. Thus:

$$\text{Peel} = f\left(\frac{1}{E_f, E_e}\right) \quad (6.30)$$

This equation is confirmed by practical data from the labeling industry. Stiff aluminium foils used as face stock material exhibit a lower peel value. Similarly, opaque soft PVC gives a lower peel value than clear (softer) PVC.

A detailed study of the peel as a function of the experimental conditions (e.g., peel angle, peeling rate, and the properties of the adherents) was carried out by Wilken [63], who found that peel adhesion increases with the modulus of the face stock and decreases with the modulus of PSAs. For some adhesives, in spite of the weak adhesion, there is a possibility of achieving high peel strength values if the adhesive film exhibits a low modulus [80]; this means that:

$$\text{Peel} = f\left(\frac{E_{\text{facestock}}}{E_{\text{adhesive}}}\right) \quad (6.31)$$

A theoretical minimum peel is observed for a 180° peel angle [63]. In reality, minimum peel values were obtained for lower peel angles. These discrepancies may be explained by the low flexibility of the face stock material [63]; however, gradually increasing face stiffness and thickness decreases the peel angle and so the peel adhesion [5]. On the other hand, the extra work needed to bend the face material will raise the observed adhesion value. In order to overcome this, a "reverse adhesion test" can be performed by mounting the tape to be evaluated, adhesive surface up, to a standard test panel, to eliminate the backing effect. The 180° peel adhesion values (underlayered) show a linear dependence on the thickness of the face material [5]. Table

Adhesive Performance Characteristics

Table 6.10 Peel Adhesion Values as a Function of the Face Stock Material and Peel Angle (Peel Force, N/25 mm)

PSA	Face material		Peel angle (°)	
	Paper	Film	90°	180°
Tackified water-based acrylic	x	—	8.3	25.0 PT
Water-based acrylic	—	x	8.0	12.0
Tackified water-based CSBR	x	—	8.4	25.0 PT
Rubber-resin solvent-based	x	—	12.5	23.0
Rubber-resin solvent-based removable	x	—	4.6	13.0

6.10 contains typical 180° and 90° peel adhesion values for labels with different face stock materials and peel angle.

As can be seen from Table 6.10, the stiffness of the face stock exerts a strong influence on the peel values. The dependence of the peel value on the peel angle and dwell time is shown in Table 6.11. In fact, not only the stiffness but also the resistance to elongation influence the peel value (Table 6.12).

In these measurements, the transfer of the peel force was achieved with the aid of a different material (i.e., the face stock had a combined paper/ material construction) where the paper layer was contacting the adherent and

Table 6.11 Dependence of the Peel on the Peel Angle and the Dwell Time (Peel Value, N/25 mm)

PSA		Face material		Peel angle (°)		Dwell time
Chemical nature	Physical state	Paper	Film	90°	180°	
Tackified acrylic	Water-based	x	—	8.3	25.0 PT	0
Acrylic	Water-based	—	x	8.0	12.0	0
CSBR acrylic	Water-based	x	—	8.4	25.0 PT	0
Rubber-resin	Solvent-based	x	—	12.5	23.0	0
Rubber-resin	Solvent-based	x	—	4.6	13.0	0
Acrylic	Water-based	—	x	5.7	5.2	0
Acrylic	Water-based	—	x	6.0	12.0	15 sec
Acrylic	Water-based	—	x	10.0	14.0	24 hr

Table 6.12 Dependence of the Peel on the Bulk Properties of the Face Stock Material. The Influence of the Mechanical Properties of the Face Stock Material (Maximal Tensile Strength of a PE Film) on the Peel Force

Ultimate tensile strength (N/cm)	Peel (N/cm)
8.0	0.13
9.0	0.23
10.0	0.30
11.0	0.40
12.0	0.54
13.0	0.60
14.0	0.70
16.0	0.87
18.0	1.07

A nontackified acrylic PSA was used. 180° peel adhesion on stainless steel was measured.

the material layer (paper, film, etc.) was subjected to tension. Thus by using different force transferring materials the peel force was more or less dampened according to the plastic/elastic character of the material. Soft, plastic materials (e.g., film) exhibit less peel strength than resistant ones (e.g., paper) (Table 6.12). As an example for the peel dependence on the face stock stiffness or rigidity, aluminium-laminated paper is removable without primer (using an adequate PSA), and a solvent-based silicone release liner should be used in order to get an appropriate release force.

Peel strength increases with the thickness of the face stock material, according to a power law dependence [74]. With gradually increasing face material thickness one needs to consider two additional features that affect the result, namely, the rapidly decreasing peel angle, which will drop the observed peel value, and the extra work needed to bend the face material, which will raise the observed peel adhesion value [5]. As an example PET face film shows a drop in peel adhesion from around 2 mil onwards, aluminium shows a continual rise, and polyethylene film shows little change or a slight decrease. To overcome this, a "reverse adhesion test" can be performed [5]; a standard PET film strip is applied and rolled down to standardize the face material effect. A plot of adhesion versus face thickness can then be drawn which can be extrapolated (for different PET thicknesses) indicating the value of adhesion without the added influence of the face stock.

When a laminate is being debonded, there are at least four possible modes of failure, including: adhesive failure with debonding of the adhesive from the web, adhesive failure with debonding of the adhesive from the substrate, cohesive failure or splitting of the adhesive itself, and paper/film tear or failure of the web itself. The failure mode is influenced by the surface quality of the face stock [81,82]. Clay-coated papers exhibit paper tear if the surface strength of the layer is not sufficient [83]; water-based dispersions of polymethyl acrylate can be used for removable paper labels, but they are not removable from PVC and laquered surfaces [84]. The substrate nature influences the peel and removability; for example, a two-year warranty was given for Fascal 1800, except from PMMA, nitrolack, and polystyrene surfaces.

In an ideal case, the correlation between the anchorage of PSAs to the face stock material A_F, cohesion of the adhesive C, and adhesion of the PSAs on the substrate A_S should be valid [73]:

$$A_F > C > A_S \qquad (6.32)$$

In the case of permanent adhesion A_S should be higher than the mechanical resistance of the face stock material or of the substrate. For removable adhesives, A_S should be less than the minimal forces producing a deformation or deterioration of the face stock material or of the substrate. Energy considerations imply an adequate anchorage of the PSA onto the face stock material, but this will depend on the surface quality of the face stock material. In fact the surface characteristics of the face stock material influence the anchorage of the PSAs on the face stock material through the wetting of the surface and the anchorage of the coated, wetted adhesive on this surface. The first step in the bonding process is the wetout of the surface by the adhesive; surface roughness and energy affects this wettability. Wettability is an initial prerequisite to high adhesion. Adhesion is at a maximum at zero interfacial energy [74].

Influence of the Release Liner on the Peel. The surface and bulk properties of the release liner influence the peel resistance of the PSA. The structure, nature and thickness of the adhesive layer affect the peel value [73]. Silicone transfer to PSAs, contamination of the adhesive layer through crosslinking catalysts, or mechanical damage of the PSA layer (as a replica of the release layer) may negatively influence the peel resistance.

In a manner similar to the face stock the bulk properties of the liner influence the peel of a label from the release liner. When increasing the thickness of the release liner, the high speed release values are approximatively doubled when the liner is removed from the label. This confirms that the nature of the release liner (probably its stiffness) has a considerable influence on the high speed release force.

Table 6.13 Peel Dependence on the Laminate Structure. Stiffness and Peel Values for Mono- and Bilayer Laminates, with the Same Face Stock and Different Adhesives

	Laminate structure		Characteristics	
	Face stock	PSA	Stiffness (mN/cm)	Peel on glass (N/25 mm)
First layer	Soft PVC	Untackified acrylic	—	—
Second layer	Paper	Tackified water-based acrylic	698	17.4
		Tackified water-based acrylic CSBR	480	17.8
		Tackified water-based acrylic	515	16.0
		Tackified solvent-based rubber-resin	483	10.5

The untackified acrylic adheres to the test substrate.

Influence of Label Laminate Construction on Peel Adhesion. The energy absorption during peeling decreases the value of the peel force. Therefore, multilayer-labels, with more deformable liquid adhesive layers and with the ability to absorb the debonding energy by viscous flow, show lower peel values than single-ply labels (Table 6.13).

The peel resistance depends on the rheology of the adhesive layer R and on the peel force F (see Chapter 2). The rheology of the adhesive layer depends on its thickness h. In the case of multilayer labels, the adhesive thickness is the sum of the different layers or:

$$h = \sum h_i \qquad (6.33)$$

Therefore the increased layer thickness improves the flow properties of the adhesive (i.e., its energy absorption capacity). Thus it is assumed that the peel is inversely proportional to the number N of the adhesive layers,

$$\text{Peel} = f\left(\frac{1}{N}\right) \qquad (6.34)$$

On the other hand, increasing the layer number increases the stiffness of the laminate (i.e., it changes the debonding angle) and thus decreases the peel too (see the influence of the face stock on the peel). Therefore, it is more

Adhesive Performance Characteristics

Table 6.14 Dependence of the Peel on the Number of Layers

Laminate characteristics		Peel (N/25 mm)	
Number of layers	Stiffness (mN/cm)	On glass	On PE
2	18	4.3	2.9
4	60	3.9	2.6
6	100	3.6	2.6
8	152	3.7	1.5

Water-based acrylic formulation. Film face stock material.

accurate to assume a pronounced effect of the number of layers through viscous flow (higher adhesive thickness) and peel angle (higher laminate stiffness), or:

$$\text{Peel} = f\left(\frac{1}{N^\alpha}\right) \tag{6.35}$$

where $\alpha > 1$. As shown in Table 6.14, this hypothesis is confirmed by experimental data. Increasing the number of layers increases the stiffness of the laminate and at the same time decreases the peel value. Increased stiffness requires more energy for the flexure of the face stock material (i.e., the peel force increases). Thus, the peel will depend on the number of layers according to a more complex correlation.

$$\text{Peel} = f\left(\frac{N^\beta}{N^\alpha}\right) \tag{6.36}$$

where $\alpha >> \beta$. A multilayer film-based label may increase the energy absorption by the viscous flow of the solid state components (face stock included) (i.e., the peel decreases again). Thus, one can assume that:

$$\text{Peel} = \frac{N^\beta}{N^\alpha \cdot N^\gamma} \tag{6.37}$$

where α takes into account the adhesive slip and the change of the debonding angle, β denotes the increase of the mechanical resistance with the number of layers, and γ accounts for the deformation of special film-based face stock materials. For multilayer filmic labels the dependence of the peel on the number of layers may be summarized by the following equation:

Table 6.15 The Influence of the Number of Layers on the Peel Adhesion of Film Laminates

Number of layers	Stiffness (mN/cm)	Peel on glass (N/25 mm)
2	39	23.5
4	236	19.5
6	408	12.0
8	644	8.0

Water-based PSA coated on soft PVC.

$$\text{Peel} = f\left(\frac{N^\beta}{N^\delta}\right) \tag{6.38}$$

where $N^\delta = N^\alpha \cdot N^\gamma$ (i.e., $N^\delta > N^\alpha$). It may be assumed that the peel decrease with increasing number of layers is more pronounced for film labels than for paper labels. This hypothesis is confirmed by the data in Table 6.15. Paper laminates show a less pronounced peel decrease with increasing number of layers (Table 6.16). In a similar way paper layers in hybrid (paper/film) labels impart increased peel adhesion.

In some experiments the influence of the stiffness deformation resistance of multilayer laminates was studied. In these tests a film label was reinforced with paper labels coated with different PSAs. The distribution of the peel forces depends on the deformability of the film face stock, or on the slip between the labels (i.e., the adhesive nature). As seen from Table 6.16, reinforcing the film label with an elongation resistant material (paper) improves the peel value. On the other hand, the peel value of multilayer laminates increases with the internal rigidity (no motion) of the sandwich (i.e., with the

Table 6.16 Peel Dependence on the Number of Layers for a Multilayer Paper/Film Laminate

Laminate construction						Laminate stiffness (mN/cm)	Peel on glass (N/25 mm)
PSA	Film	—	—	—	—	46	8.4
PSA	Film	PSA	Film	—	—	130	6.4
PSA	Film	PSA	Paper	—	—	245	8.0
PSA	Film	PSA	Film	PSA	Paper	483	7.8

adhesion between the laminate face components). The slight anchorage of a rubber-resin adhesive on soft PVC dramatically decreases the peel value of the sandwich. Control of the viscous flow can be achieved by superimposing adhesive layers [85].

Influence of Experimental Conditions on Peel. The surface tension, surface structure (roughness), and thickness of the substrate influence the peel resistance [73]. Test conditions such as dwell (contact) time, contact pressure, debonding rate, debonding angle, and temperature also affect the peel resistance. At lower speed, viscous flow is the predominant influence on adhesion (peel value), while at higher speed, the elasticity determines the adhesion; at lower speed, peel adhesion increases [73]. Some of these parameters were discussed earlier (see Chapter 2). Here only the influence of the dwell time will be discussed.

Concerning the stress rate and its influence on the peel, an increase of the temperature allows increased molecular mobility resulting in increased tack and reduction in shear resistance. A high rate of stress is equivalent to a drop in temperature [5]. Adhesive transfer may occur at higher unwinding force and rate; faster peeling rates may transfer the adhesive or break the label.

Physical and chemical interactions between the adhesive and the substrate result in a build-up of the adhesion (i.e., the peel increases with the dwell time) [73]. This is a general behavior observed for PSAs with different chemical compositions (e.g., acrylics or neoprene latices) [36]. The build-up of the peel depends on the chemical composition and time [64,65,86]. Because of the slow flow of the adhesive, adhesion tests require at least 0.5 sec of dwell time [87]. The surface of a PSA can be affixed to a substrate to which it will be adhered, but once affixed, adhesion builds to form a strong bond. Removable adhesives can be formulated so as not to display adhesive build-up, that is, their adhesion to the substrate does not increase to the point where the label (tape) cannot be removed cleanly, even after exposure to heat in a drying oven. Adhesion build-up requires a long time [5]. Different peel values are obtained when testing immediately after application (PSTC 1), after a 20-min dwell time (ASTM D-1000) and after a 24 hr dwell time. The adhesion can still be climbing after 30 days. Peel adhesion is a function of the dwell time [88]; after 10, 30, 60, 180 min, and 24 hr dwell times the following peel values were obtained: 4.0, 4.6, 5.3, 5.5, and 8.0 N/2 cm. Different methods specify different dwell times: AFERA 4001 requires 10 min, PSTC 1 and FINAT specify 20 min and 24 hr, respectively [78]. In order to test adherent fiber tear, tape is applied to cardboard with a 2-lb roller, immediately pulled off at a 90° angle, and rated for fiber tear.

When increasing the temperature the viscosity of the adhesive decreases and its anchorage to the substrate will be improved [64]; on the other hand, adhesion increases and thus peel reaches a maximum for a given temperature. At low temperatures adhesion decreases and the peel adhesion decreases as well [65]. One should take into account the poor peel adhesion properties of most current PSA labels at low temperatures (deep-freeze labels). The rate of separation and the temperature are related to the adhesive strength in a complicated manner, and these factors can dramatically influence whether or not adhesive or cohesive failure will occur. Peel decreases with the temperature [3]. Cohesion and autoadhesion of styrene butadiene elastomers were studied by Hamed and Hsieh [13], through peel tests at different peeling rates and temperatures.

Humid environments can adversely affect bond strength. Soaking conditions, even high humidity, often cause irreversible loss of adhesive strength or failure.

Peel Force from the Release Liner. A special case of peeling is the separation of the label from the release liner (i.e., labeling). The peel force to separate the PSA from the release liner depends on the peeling rate and angle, and adhesive/release nature [89]:

$$\text{Peel force} = f(V, \alpha) \tag{6.39}$$

where V and α denote peeling rate and angle. Separation rates of 200 m/min are common during labeling. At these rates condensation silicone release layers display an increase of the peel. In the FTM 10 test method a much lower test rate is used, while the FINAT method is designed for a rate of 10–300 m/min with the first test apparatus built in 1984. For testing purposes a sample of 2.5×70 cm or 5×70 cm was proposed; samples have to be stored under pressure (70 g/m^2, 20 hr, 23°C) before testing. Performance characteristics after aging should be tested upon storage at 40 and 70°C, respectively [89], because the release nature and age also influence the peel. There are some new materials that are supposedly insensitive to release nature and age (e.g., Nacor 80). Acrylic PSAs require a separate study for each formulation with regard to the level of the release force and its stability. For current water-based acrylic PSAs on solventless release liners, release forces average 0.05–0.06 N/25 mm (FTM 3). Interaction between water-based acrylic PSAs can increase this value to more than 0.1–0.15 N/25 mm (e.g., Vinacryl 4512).

Influence of the Adherent on Peel. It is recommended that quite different materials like glass, metal, plastics, wood, coated paper, cardboard, textiles, nonwovens, and surfaces coated with alkyd, PUR, acryl, epoxy, nitro, and powder lacquers, and plastisols be used as contact surfaces for testing purposes [84]. Crystal glass is proposed as substrate by FINAT, steel by AFERA, and glossy stainless steel by ASTM or BWB-TL [90]. These materials are used in order to eliminate the influence of the adherent on the peel.

Table 6.17 The Influence of the Adherend Surface (Substrate) on the Peel. Peel Values of PSAs with Different Chemical Bases, as a Function of the Substrate Used

Chemical composition		Peel (N/25 mm)		
Base components	Supplier	Glass	Cardboard	PE
1360	Hoechst	12.0 PT	2.0	11.0
V205	BASF			
80D	BASF			
CF52	A&W			
FC88	UCB	15.0 PT	11.0 PT	13.0
PC80	UCB			
CF52	A&W			
FC88	UCB	18.0 PT	11.0 PT	12.0
CF52	A&W			
V205	BASF	18.0 PT	—	15.0
80D	BASF			
CF52	A&W			
V205	BASF	14.0 PT	10.0	13.0
V208	BASF			
80D	BASF			
CF52	A&W			

If the only factor governing the peel force were the total extent of contact (i.e., the number and type of interactions across the interface) then the scale of the contact zone would not influence the measured peel strength [13]. In general this is not true, particularly for soft vulcanized elastomers, whose strengths are a measure of dissipative energy losses that occur while they are being peeled away. Adhesive strength to paper and stainless steel is greater than to polyester film [56]. Therefore the bond fails due to breakdown within the adhesive layer when paper is used. When polyester film is used as face material it peels away, leaving the adhesive layer virtually intact on the steel plate. The influence of the adherent is illustrated by PSA labels applied on PVC (where the peel value changes in plasticized PVC) or by removability from fragile substrates (e.g., light-weight papers) [91]. Peel and tack should be tested on different substrates (e.g., cellulose, polyamide, glass, cardboard, metal) [92]. Temperature resistance, the lowest application temperature, and the suggested operating end-use temperature should be determined. The influence of the adherent surface on the peel is shown in Table 6.17.

Other Factors. Peel adhesion does not depend on the number of passes with a standard application roller and it does not depend on the solvent used to clean the test plates [63]. On the other hand, peel increases with the application pressure and temperature [73].

Rheology is frequency dependent and strain frequency influences the peel; repeated, cyclical peeling off reduces the peeling force (see Chapter 3). Readherence onto a paper substrate decreases as the number of peeling trials grows. In a lot of experiments the peel of an experimental PSA from newspaper as substrate was tested [93]. The following data of the peel force were collected as a function of the number of peel trials:

- Peel from paper, first trial: 110 g/12 mm;
- Peel from paper after 50 peel trials: 85 g/12 mm;
- Peel from paper after 100 peel tests: 75 g/12 mm.

The dependence of the peel on the cohesion is very complex. For some adhesive formulations based on natural rubber and SBR a direct proportionality between peel and cohesion was established [94].

Improvement of Peel Adhesion

Improving the peel means optimizing it in such a way that it meets the end-use requirements. Regulating means increasing or decreasing the peel value and changing the character of the bond rupture related to the requirements of a permanent (paper tear) or removable (adhesion break) bond.

Permanent/Removable PSAs. Pressure-sensitive adhesive coated materials are functionally divided into two broad classes: permanent and removable [52]. The first of these is represented by the so-called permanent materials in which the properties of PSAs are selected so as to form an adhesive bond with the target substrate which is strong and, apart from degradation, does not weaken significantly with time. The second broad class consists of "removable" or "peelable" PSA-coated materials in which the PSAs form an adhesive bond, of functionally adequate strength with the target, but which after an extended period of adhesion (typically days to months) can be peeled away from the substrate, without damaging it, without leaving any residue of PSA on the target substrate, and without the adhesive coated material tearing itself apart. To be peelable PSA-coated materials require a combination of properties. The PSAs must form a bond of adequate strength with the target substrate, but which does not subsequently either dramatically weaken, in which case the adhesive-coated material might fall off the substrates, or significantly increase in strength, as this would tend to limit peelability. From these considerations it follows that the adhesive should possess an adequate cohesive strength in order to minimize any tendency of the adhesive film itself to rupture, as well as a high bonding strength to the face material (anchorage) onto which the adhesive is coated. It is clear that the formation of a strong bond between a PSA and its face material and a relatively much weaker bond between adhesive and target substrate are in conflict. Consequently, many of the peelable PSAs currently on the market do not fully meet these contradictory requirements; effectively, they are only less permanent than typical permanent PSA-coated materials. The reduction of the peel is a relatively simple procedure which (at least theoretically) can be achieved by increasing the viscous component of the PSA, that is, the component converting the high-speed bond-breaking force in permanent adhesive deformation (viscous flow). Unfortunately this viscous flow may exhibit adhesive break (i.e., after bond breaking adhesive residue is left on the adherent). Such a peel reduction is inadequate to obtain a convenient removable PSA.

Summarizing, a limited peel value and a high shear strength are required for removable adhesives [59]. A typical value for removable pressure-sensitive tapes is 10 N/25 mm [95]; lower values (2.5–5 N/25 mm) are required for labels. Shear values of at least 100 min at room temperature were suggested [95]. In fact these values should be considered as guidelines only since paper tear occurs at 20–25 N/2.5 cm. Removable PSAs should exhibit a peel adhesion value of 1.5 N/2.5 cm [95]; 1–6 N/2 cm are suitable values for removable labels [97]. In general, the same removable water-based acrylic PSAs for labels may be used for tapes [80]. Removable PSAs labels are a special type of permanent labels, where low peel force levels, as a function of dwell time, and a pure adhesive break from the substrate are assumed. Removable behavior depends on energy dissipation which influences the debonding (Figure 6.12).

The adhesive (its bulk properties, formulation, nature, structure, and molecular weight), laminate components, laminate geometry, and debonding conditions influence the removability. During peeling tests at least four possible modes of failure may occur [98]:

1. Adhesive failure (primary): debonding of the adhesive from the face material (i.e., not removable);
2. Adhesive failure (secondary): debonding of the adhesive from the substrate (i.e., removable);
3. Cohesive failure: splitting of the adhesive itself (i.e., not removable);
4. Web or film tear: failure of the face material itself, hence not removable.

Energy dissipation during debonding depends on such parameters as interfacial adhesion, polymer rheology, rate of peel, temperature, angle of peel, face stiffness, and coating weight.

Several requirements have to be fulfilled in order to achieve good removability [99]. These include self-adhesion on the substrate (tack), peel adhesion (in order to assure bonding), cohesion in order to avoid legginess or stringing,

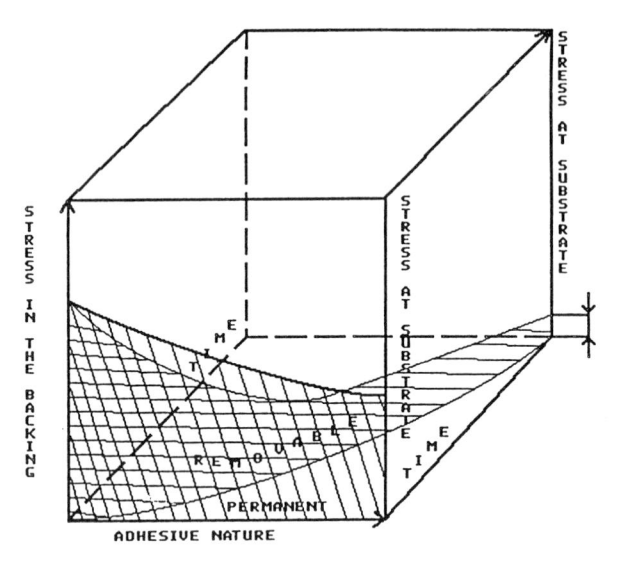

Figure 6.12 Schematical presentation of stress dissipation as a function of the time for permanent and removable PSAs.

and superior adhesion to the face stock rather than the substrate (anchorage). For suitable peelable labels adhesive failure from the substrate is desired [84]. As discussed earlier removability implies lower stresses, and lower stresses are possible using stress-restricting polymers, fillers or a primer coating [100].

Peel depends on the nature and geometry of the bonding components (i.e., the self-adhesive label and substrate). For a given adhesive, face stock, and substrate, the interaction of the adhesive with both solid state components of the joint will depend on the interface adhesive/solid state component. The type of face material is governed by the end-use of the label; therefore the labelmaker should optimize the interface between the face stock and the PSA. The most important parameters of the face stock influencing the removability of the PSAs label are its flexibility and the anchorage (adhesion) of the adhesive to the face stock. Effective bonding between the face and the adhesive is critical to the performance of the PSAs; for example, given an adhesive with internal integrity, the adhesive properties exhibited by that system will be proportional to the affinity of the adhesive for the face material. Face stock interactions bring the definition of "permanent" and "nonpermanent" PSAs much more into focus by emphasizing the fact that if a permanent bond is formed between the label and the substrate, the failure mode will be face oriented and the result is 100% transfer of the adhesive to the substrate. Conversely, if the bond to the substrate is nonpermanent, then a clean release from that substrate is encountered and the adhesive remains on the face material.

Use of Primers. Primers are necessary when using removable PSAs [101, 102]; they modify the face stock-PSA interaction. Polyolefin face stock materials are generally top coated to improve printability; in a similar manner primer coatings are applied on polyolefin face stock materials to increase the anchorage of the adhesive [103]. PVC-based tapes also need a primer coating [104]. Soft PVC should be primed when used with rubber-resin-based PSAs [105]. There are a few studies describing the use of primers in order to obtain removable adhesives [106–108]. The use of the primer generally decreases the peel value [109]. Primers are generally used to promote a stronger bond between PSAs and face materials, to absorb stress, and to strengthen the face stock, especially when paper labels are concerned.

There are very different recipes for primers. As in the case of adhesive base elastomers, primer coatings use more crosslinking agents, harder resins, resin solutions, or self-crosslinked elastomers. A primer coating for polyethylene face stock material typically is composed of a chloroprene rubber (20–60 pts, 60% chlorine), an EVAc copolymer (40–80 pts, with 25% VAc), and chloropolypropylene (1–15 pts with 25–50% chlorine) [109]. While stearic acid is effective in preventing adhesive build-up, it leaves a greasy deposit referred to as a "ghost" which could stain the substrate; the metallic stearates do not do this [110]. Contact cements (polymers with high cohesive strength and a strong tendency for two surfaces of the polymer to adhere to one another when placed in contact) are preferred as primers [67]; isoprene polymers and polychloroprene are particularly good as contact cement. More about primer formulations can be found in Chapters 7 and 8. One approach to use a key coat is including particulate matter, so as to make the surface of the coating rougher, and provide a better key for PSAs; primers provide a bond between adhesive and face stock [111,112]. Primed face stock ensures more uniform peel values and thus better removability (Table 6.18).

In order to achieve a maximum anchorage of the adhesive to the face material, a silicone release coating may be used as primer for silicone adhesives [31]. In making PSA tapes, polyacrylamide may be used as the binder between the hydrophilic cellulose and the hydrophilic butadiene-styrene latex. The primer also improves the anchorage for natural rubber-based PSAs [113]; the primer modifies the separation energy of the surface [3].

Primers are somewhat flexible and generally have high anchorage to the face material. Thus, they can absorb impact stresses without adhesion failure [100]. There are removable PSA formulations with a primer coating that contain a contact cement with the same contact cement in the PSA as well [67]. The flexible sheet label has on one side a primer coating firmly bonded

Table 6.18 The Influence of the Primer on the Removability of PSA Labels

Substrate	Removability	
	With primer	Without primer
Steel	2	3
Copper	2	2
Aluminium	1	3
Glass	2	4
PE	2	5
HPVC	2	3
PMMA	2	2
PA	2	5
PET	1	5
Average	1.8	3.4

Removability was evaluated subjectively. The following removability indices were used: 1 - very low peel, no residues; 2 - low peel, no residues; 3 - low peel, legging; 4 - medium peel, no residues; 5 - high peel, no residues.

to the face stock, which includes a contact cement, and above the primer coating is a peelable PSA (including the same contact cement). The contact cement serves to firmly bond the primer coating and the adhesive coating together. Contact cements are a class of adhesives which function by exploiting the properties of certain polymers, which exhibit a high cohesive strength and a natural tendency for two surfaces of the polymer to adhere when placed into contact. Examples include polymers like isoprene, natural rubber, polychloroprene, and one class of acrylic polymers. A low T_g is a property common to polymers which are used in contact cements.

The stress absorption capacity is a function of the primer's own softness. As shown in Table 6.19, adhesives with a different "hardness" (peel) act as primer coating in different manners. The soft (removable) ones do not cause any peel decrease; the peel adhesion decrease is proportional to primer stiffness. In these experiments different adhesives were used as primer (first) coating. The second layer (i.e., the PSA contacting the substrate) has always been the same (a permanent, tacky PSA). The necessity of stress absorption (i.e., energy absorption) is illustrated by the data of Table 6.20. It can be seen from Table 6.20 that the peel value of the removable adhesive layer increases with the

Adhesive Performance Characteristics

Table 6.19 The Influence of the Softness of the Primer on the Peel Value, Removability of PSA

Primer	Peel (N/25 mm)
Soft hot-melt PSA	20.0
Removable rubber-resin solvent-based PSA	23.0
Hard hot-melt PSA	24.0
Permanent rubber-resin solvent-based PSA	24.0 PT
None	24.0 PT

A tackified water-based acrylic PSA was used.

Table 6.20 Influence of the Primer Structure/Geometry on the Peel Value/Removability

	Removability			
	Relative thickness of the primer layer (%)			
Substrate	100	75	50	0
Steel	2	1	1	3
Copper	2 C	1	1	2
Aluminium	1	2	2	2
Glass	2	1	1	3
PE	2	2	2	5
PMMA	2	1	1	2
PET	2	1	1	5
Average	1.85	1.3	1.3	3.1

Removability was examined subjectively. The following removability indices were used: 1 - very low peel, no residues; 2 - low peel, no residues; 3 - low peel, legging; 4 - medium peel, no residues; 5 - high peel, no residues; C - cohesive failure.

coating weight of the primer coating. A primer coating weight equivalent to 5–10% of the permanent adhesive weight is able to change the peel of removable PSAs.

The peel force will be influenced by the material (PSA) in the noncontact area as well [13]. When the elastomer is more viscous, the strength will be less affected by flaws (noncontacted zones) compared to an elastomer that responds more elastically. As discussed above, primers may improve stress distribution. Another stress reducing possibility is the choice of the adequate polymer (PSA) formulation or the use of adequate fillers [100].

The primer coat typically is present on the coated face stock in an amount ranging from 2–20 and more often 5–10 g/m² [67]. The lower limit is determined by the requirement to ensure complete coverage of the coated face material, with a good coating pattern, and the upper limit is determined primarily by costs as there appears to be no significant benefit in increasing the coating weight of the primer beyond the limit indicated [67]; generally 2–5 g/m² of primer is used [59]. For tapes with an adhesive layer of 16–20 μm, the thickness of the primer coating should be 1–1.5 μm [114]. The thickness of the primer (manufactured via electron beam) from a monomer blend should be less than 5 μm, preferably less than 1 μm, and ideally less than 0.5 μm. Indeed it is believed that thicknesses approaching a single molecular layer would function efficiently [115].

The use of polyaziridine as a primer shows the effect of the crosslinking on the removability of PSAs [67]. Peel adhesion of the crosslinked PSA (after 24 hr) is 0.51 kg/cm versus 0.67 kg/cm for the uncrosslinked one. The crosslinking of PSAs shows similarities with the use of lightly crosslinked rubbers in thermoplastics for controlled soft-phase dispersions.

Re-adhering adhesives permit removal of the coated face material and allow re-adherence to another surface. Paper substrates which were precoated with removable, re-adhering adhesives are commercially available. One commonly known product of this type is marketed under the trade name "Post It" as note pads [116]. In order to achieve a good removability, many different procedures are used, including the use of hard (T_g > 10°C) suspension particles, crosslinking the matrix, and low coating weights (7 g/m²) [99].

Stress-Resistant Polymers. Stress-resistant polymers are those which develop a controlled crosslink density or which are internally soft [85]. This means that crosslinking agents or plasticizers may improve the removability. Like craze-deformation for plastic-elastomeric blends, PSAs are subject to energy dissipation via deformation work; thus shear deformation is necessary [100].

Flexibilizers. Polymers exhibit two kinds of fracture mechanisms: ductile and brittle. Ductile fracture involves overall yielding of the specimen and since the volume of yielded material is large, this failure mechanism absorbs a great deal of energy; however, it is slower than brittle fracture. Removable PSAs undergo ductile fracture in that they absorb all the break energy.

A number of superimposed adhesive layers having different gradients of shear/creep compliance can meet the requirement of releasable adhesion to plastic surfaces such as polyethylene [85]. Flexibilizers are plasticizer compounds which react with the polymer; because of their volume they are able to keep the chain segments apart. These plasticizers increase tack, but lower cohesion [100]. Adding a 15% butyl dispersion to a water-based PSA formulation may improve its removability properties [117]. Flexibilizers, like plasticizers introduce creep, which is a function of molecular weight and plasticizer loading, or

$$\text{Creep} = f\left(\frac{\text{Plasticizer}}{\text{Molecular weight}}\right) \tag{6.40}$$

A high molecular weight and a low plasticizer level reduce creep. Plasticizers may be incorporated in the peelable PSAs to soften the adhesive and thus improve peelability. Care should be exercised as some plasticizers can have a tackifying effect on adhesive polymers and this may limit the amount to be used; however, it is possible to use quite large quantities of plasticizers, even up to 50 wt% on the solid adhesive, but more often they are used at 10–20 wt%. Generally, plasticizers impart a fast tack increase, a slower peel increase, and very fast decrease of the cohesive strength [118].

Fillers. Stress reduction by fillers (i.e., reducing peel adhesion) is well known in the tape industry (e.g., 0.5–10% filler, particle size 150 μm, in rubber-based PSAs) [119]. Peel adhesion decreases in formulations containing fillers [65]. Rubber dispersions also may be used as fillers for water-based PSAs [117].

Strengthening of the Face Stock. Paper face materials can lead to paper tear, when peeling off the label. Reinforcing the fiber structure of the paper face stock with a primer can strengthen the surface layer in order to avoid fiber tear. High strength face stock materials (e.g., rubber latex-impregnated papers) are often used. In some cases the primer strengthens the face stock material [113].

Reduction of the Contact Surface. Peel adhesion depends on the surface area between the label and the adherent [63]. Reducing the contact surface between the label and the adherent improves peelability or removability. There are different methods to reduce the contact area between the adhesive and the substrate, including: reducing the coating weight of the adhesive, increasing the stiffness of the face stock (or substrate), applying a discontinuous coating on the face stock material, and removing contact points or areas between adhesive and substrate.

Discontinuous adhesive coatings (i.e., ungummed areas) reduce the contact area since in some cases, it suffices to coat only a minor portion of the face stock surface with adhesive [120–123]; a number of adhesive and nonadhesive zones may be used [116, 124]. There are several patents concerning the use of filled PSAs that decrease the adhesive/adherent contact area. Some use glass beads as filler [125]; microspheres also may be used [126]. The reduction of the contact surface may be achieved by overlapping the contact surface of the PSAs with powdery materials [127]. As an example, adhesive tapes applied for sealing windows and bonding side mouldings onto cars use UV-cured acrylic adhesives containing 2–15 parts per hundred (phr) hydrophylic silica-glass microspheres [128].

Coating Weight Adjustment. As far as the PSA component is concerned the properties which contribute to peelability are principally a limited tack, a low build-up factor (lack of age hardening), and a relatively soft adhesive. These properties can be achieved conventionally, by omitting or only using a low level of tackifier, or by including tack deadeners, such as waxes, and by including plasticizers [67]. The most important parameter is the adjustment of the coating weight. A way to reduce the peel (i.e., to get a removable PSAs) is the reduction of the coating weight. Reducing the coating weight from 20–25 g to 15–10 g/m^2 allows removability for PSAs formulations that are based on acrylic, vinyl ether, and plasticizer. For PSAs coated on paper the coating weight for repositionable labels may be as low as 7 g/m^2. In this case an initial peel value of 110 g/12 mm (180° peel at 300 mm/min) is enough to resist a vertical application (5 hr debonding method) [99].

Shear Strength for Removable Adhesives. Permanent adhesives generally possess a higher cohesive and shear strength than removable ones. On the other hand, several high-shear adhesives used for tapes also are proposed for removable labels (e.g., Ashland, Aroset 2530-W-50). For removable adhesives a shear value of 50–1000 min at 70°C [129] or 50–1000 min at 75°F is suggested [50]. The minimum level of shear for a removable PSA also depends on the coating weight. From a good removable PSA no adhesive transfer can be accepted.

Problems Concerning Removability. The most frequently encountered problems for removable PSAs are too high peel forces, the change of the peel force in time (build-up), adhesive residue left on the substrate, low tack, and pronounced bleeding, smearing, and migration. It is difficult to find removable labels which can be removed without leaving any adhesive residues behind. Generally, the migration or bleedthrough of removable PSAs is more pro-

Table 6.21 Typical Peel Values

Adhesive nature			180° peel on steel		
Chemical	Physical	End-use of the laminate	Value	Units	Ref.
Acrylic	Hot-melt PSA	Label	90–105	oz/in	[26]
Acrylic	Solvent-based	Transfer tape	> 7	lb/in	[130]
—	-	Highly agressive label/tape	60–100	N/100 mm	[162]
SIS	Hot-melt PSA	Packaging tape	10–18	N/25 mm	[163]
SBS	Hot-melt PSA	Double-sided carpet tape	8.5	N/cm	[164]
—	Hot-melt PSA	Permanent label	69–77	oz/in	[165]
—	Hot-melt PSA	Removable label	11–14	oz/in	[165]
Acrylic	Water-based	Aplication tape/film	1.3–5.0	N/25 mm	[166]
Acrylic	Wate-based	Protective film	100–200	cN/cm	[167]

nounced than for permanent PSAs [73]. This will be discussed in more detail in Chapter 8.

Peel Adhesion Values

Improved peel adhesion requires a higher tackifier loading than tack. Higher melting point resins are suggested. Permanent and removable labels/tapes, labels for hand-applied labeling or gun labeling, labels for room temperature or high temperature use, labels for use on plastics or metals, etc., have quite different peel adhesion values. Table 6.21 summarizes some of the performance characteristics specified by manufacturers of PSA laminates.

1.3 Shear Resistance (Cohesion)

Pressure-sensitive adhesives possess typical viscoelastic properties which allow them to respond to both a bonding and a debonding step. For permanent adhesives the most important step is the debonding one; the adhesive should not break under debonding (mainly shear and peel) forces (i.e., permanent adhesives must provide a higher level of cohesive or shear strength than removable adhesives). This is an inverse requirement with respect to Dahlquist's criterion (see Section 2.3 of Chapter 3) for a minimum value of the compressive creep compliance to achieve tack. Creep (fluidity under low forces) results in edge bleeding, migration, and poor die cuttability; high resistance to fluidity allows low instantaneous flow (i.e., low tack). The cohesion (shear) is important as an index for label-processing characteristics (bleeding, migration, die-cutting) and also end-use properties.

Shear resistance is measured as a force required to pull the pressure-sensitive material parallel to the surface to which it was affixed with a definite pressure. Practically one measures the holding time under standard conditions. Unfortunately there are no standard values of the time until shear failure occurs.

Measurement of the Shear Resistance

The shear resistance of PSAs may be measured statically or dynamically. Currently, static shear test methods use a constant load at longer test times; they generally show poor reproducibility and need very long measurement times. Better results are obtained with the hot shear test, where the cohesion of the sample is measured at an elevated temperature. Dynamic shear tests measure the cohesion of the sample in a tensile tester under increasing load (force). If a hot shear test is carried out in such a way that the test temperature is gradually raised, and if the temperature at which the bond fails is taken as a characteristic value, specialists speak about a dead load hot strength test or shear adhesion failure temperature (SAFT). There also are shear test methods that measure neither the time until debonding nor the temperature of the debonding but rather the deformation of the sheared sample (i.e., the slip of the sample after a given time is measured) [130].

A combined peel and shear strength of the adhesive can be tested (e.g., 20° hold strength of the adhesive to corrugated board) [50]. The important shear properties of adhesives are the shear strength or the shear stress at failure, the shear modulus or the shear stress/shear strain ratio, the creep modulus or the shear stress/shear strain ratio at time t, the shear stress or the load/bond area ratio, and the shear strain or the shear slip/adhesive thickness ratio [131]. A detailed description of the shear measuring systems is included in Chapter 10. Here the importance of hot shear measurements only will be highlighted. They exhibit a lower discrepancy of the test values and allow extrapolation of the data for long end-use periods.

Room temperature shear values display variations ranging from 100–1000%. Hot shear values normally do not vary more than 100–200%. On the other hand, better die-cuttability requires hot shear values to be higher by an order of magnitude (1000%); hence, a real improvement of the cohesion can only be evaluated via hot shear measurements. Figure 6.13 illustrates the measuring errors for the most commonly used hot shear test methods (relative to room temperature shear tests); the preheated plate method is better. Note that the value of shear or hot shear alone is not important, but rather it is the interdependence between them.

Adhesive Performance Characteristics 215

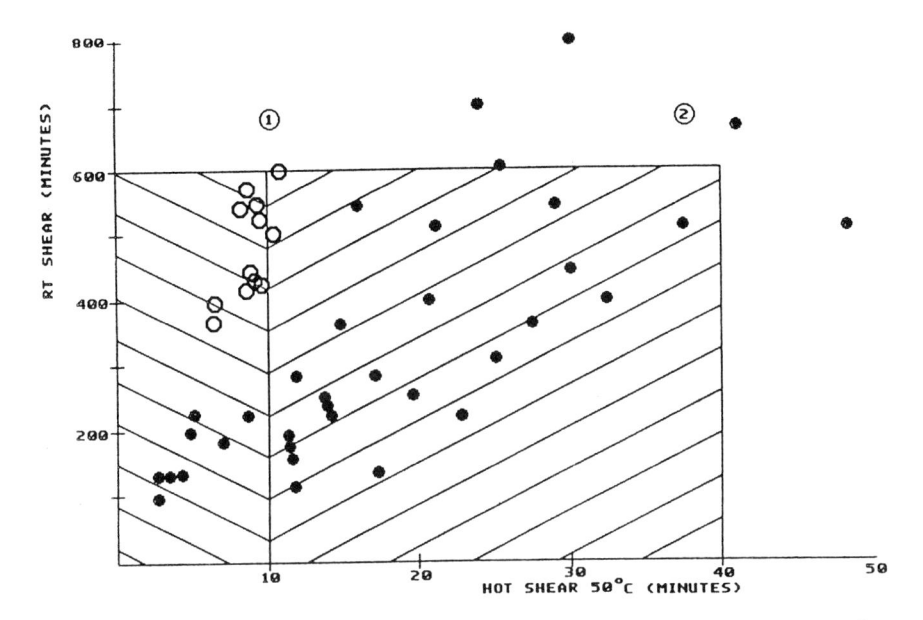

Figure 6.13 Interdependence: room temperature shear/hot shear. 1) Test using heated pannel; 2) Test using climatized atmosphere.

At times the hot shear values may be converted into room temperature data. Using the Williams-Landel-Ferry time/temperature superposition shift factors developed for SIS triblocks the 80°C data may be extrapolated to yield a shear holding time at 25°C of 387 days for a pure triblock versus 86 days for a commercial product.

The cohesion is a real measure of the internal structural resistance of the polymer. Generally, the mechanical properties of a polymer depend on its cohesion. Different methods exist to determine the mechanical resistance of a polymer, such as the measurement of the tensile strength, elongation at break, etc. Generally, adhesives are used as thin layers, where the adherents undergo shear deformation during debonding and delamination. Therefore, the measurement of the shear was accepted as a criteria for the cohesion.

Cohesion also can be measured by the Williams plasticity [132]. A measure of the Williams plasticity is given by the thickness of an adhesive pellet (mm) after 14 min compression at a fixed temperature and under a fixed load. Peel adhesion measurements also provide information about the cohesion. Shear may be also measured as tension. For polymers obeying Hooke's and Newton's laws, shear is directly proportional to tensile strength [133].

Interdependence of Adhesive/Cohesive Properties. Generally, the practical use of PSA labels requires an adequate adhesion/cohesion balance, that is, a certain agressivity of the adhesive, characterized by instant tack, a time-dependent final adhesion (bonding strength) characterized by peel and shear, according to the method of debonding (i.e., the end-use stresses). The fast, instantaneous flow required for high tack does not allow less flow and lower deformation when debonding (i.e., shear is inversely proportional to tack and peel is partially so). On the other hand, the adhesion/cohesion equation is more sophisticated. A good peel adhesion supposes a certain flow and a low to medium cohesion, but a high peel needs high cohesion when debonding. There is an internal relationship between peel adhesion and shear strength (Figure 6.14). For PSA specification and current production control purposes both values should be measured. For Acronal V 205 minimum and maximum hot-shear values (70°C, coated on PET) are 25 min and 100 min, respectively, which correlates with a peel from steel of 6–8 N/25 mm.

Factors Influencing Shear Resistance

Like tack and peel adhesion, shear is influenced by the characteristics of the adhesive and of the laminate, as well as by the debonding conditions.

Nature of the Adhesive. Special built-in functional groups, crosslinking, and high molecular weight can ensure a high shear and low tack properties. The use of harder monomers also improves the cohesion. For example, a higher acrylic acid level in solvent-based acrylic PSAs yields higher shear

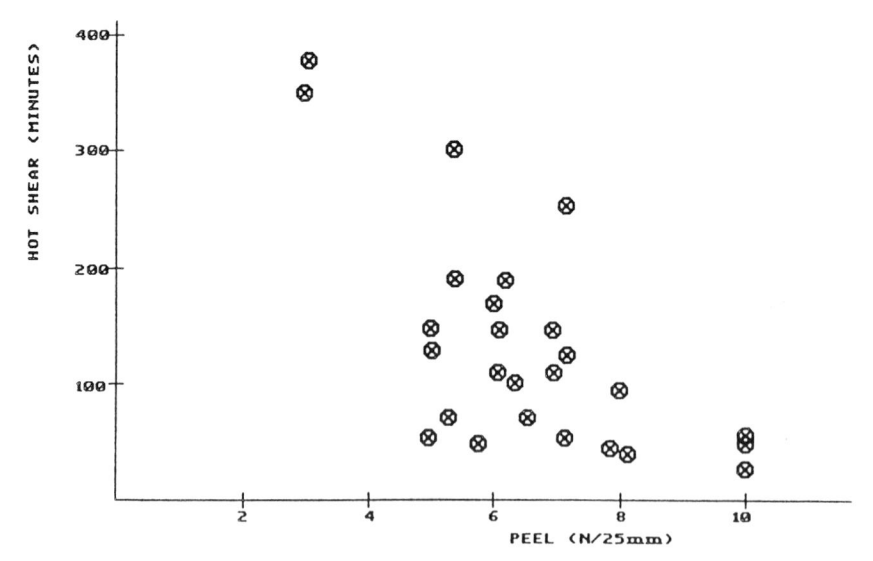

Figure 6.14 Interdependence: hot shear/peel. Tackified water-based acrylic PSA.

values [134]. Cohesive strength is independently controlled by the hard monomer content and increases as the concentration of this hard monomer increases [4]. Cohesive properties of a polymer can be affected by introducing polar groups that interact by forming secondary bonds; such interactions are classified as hydrogen, dipole-dipole, and dipole-induced bonding. Monomer-reinforced copolymers are suitable as PSAs, since they possess high shear strength [135]. Minor amounts of unsaturated olefin carboxylic acids and/or sulfoalkyl esters of such carboxylic acids improve the cohesive strength of PSAs; it is preferred that the polymer contains at least about 0.1 wt%, usually about 0.1–10 wt% of these components [50]. The higher the percentage of iso-octyl acrylate, the tackier the adhesive; conversely, the higher the percentage of acrylic acid, the higher the shear properties [129].

Permanent adhesives generally provide a higher level of cohesion shear strength than removable adhesives. Rubber-resin adhesives provide a compromise between high cohesive strength for conformance to curved containers and good quick stick required in automatic labeling [136]. Acrylic adhesives provide a high level of adhesion to surfaces that are difficult to adhere to. However, most acrylic-based adhesives do not have a cohesive strength equal to normal rubber-resin systems and may display cold flow. Showing up as a halo of adhesive around the die-cut label, cold flow can inhibit efficient labeling. All adhesives used in automatic labeling systems require a level of quick stick that will provide for efficient adhesion with a minimal application pressure.

Shear Dependence on Molecular Weight. Cohesion increases with the molecular weight. The ultimate tensile strength of a polymer reaches a maximum at about 500–1000 main chain atoms. Longer molecules do not impart higher strength because their various segments act independently of each other. The influence of the molecular weight on the shear resistance is illustrated by the changes of the shear resistance due to adhesive aging (i.e., destruction and depolymerization). S-I-S block copolymers undergo oxidative degradation at elevated temperatures by a mechanism which leads predominantly to scission of the polymer chains. This leads to a fall in molecular weight and a resulting decrease in viscosity and holding power [137]. The composition of the adhesive (i.e., its nature and molecular weight) influences the shear; shear measured as SAFT is directly dependent on the softening point (i.e., the molecular weight of the resin) [138]. The dependence of the shear on the molecular weight is illustrated by the manufacture of hot-melt PSAs. When high shear mixers are used the degradation depends on the mixing environment. In contact with open air, degradation occurs rapidly and polymer breakdown is observed. A dramatic reduction in melt viscosity and holding power results from even immediate degradation [139]. Polymer vis-

Table 6.22 Shear Dependence on the Molecular Weight

Molecular weight of the base elastomer	Shear, min at temperature (°C)		
	50	60	70
M_1	20	2	1
1.5 M_1	420	120	60
3 M_1	> 1200	> 1200	780

A CSBR was used as base elastomer.

cosity and shear are interdependent [140]. The latter depends on the molecular weight and its distribution. A broad MWD polymer may have a lower cohesive strength than one with a narrow MWD and lower molecular weight. The dependence (increase) of the cohesive strength as a function of the molecular weight was described by Schrijvers [141]. The shear resistance is improved by the increase of the melting point of the resin [27]. Gordon et al. [142], using different Escorez (manufactured by Exxon Chemical) resins with different softening points, list the SAFT data of 60/40 adhesive formulas. The SAFT improves with increasing softening point of the resin. In a similar manner shear increases with the increase of the ring-and-ball softening point of Bevitack (Bergvik) resins used as tackifier for hot-melt PSAs [143]. For tackification of Neoprene Latex 102, the 32°C holding power increases if the 49°C softening point (SP) resin is replaced by a 72°C SP resin. Table 6.22 illustrates the increase of the shear resistance with the increase of the molecular weight.

Influence of Crosslinking on the Shear. Crosslinking affects polymer properties in a manner similar to molecular weight increases but in a more pronounced way [30]. Crosslinking imparts high shear and lower tack [56,129]. The phenomenon is well known from the field of plastics where shear is proportional with the degree of crosslinking for polyethylene (i.e., the peroxide concentration) [144]. Similarly, increasing the amount of the resin in a silicone adhesive increases its crosslink density and results in greater cohesive strength and higher peel adhesion values [145]. Physical crosslinking has the same influence as chemical crosslinking. The beneficial influence of a 100% pure triblock in hot-melt PSA formulations is demonstrated in creep testing. Here the PSA is subjected to a dead load shear, and if every molecule of the rubber is able to fully participate in the associative crosslinking shear, bond holding times are excellent.

Table 6.23 Influence of the Crosslinking on the Shear

Adhesive nature		Crosslinking agent		Shear			
Chemical	Physical	Nature	Level (%)	Value	Units	Temperature (°C)	Ref.
Acrylic	Solvent-based	AZ	0	0.1	hr	RT	—
Acrylic	Solvent-based	AZ	1	1.0	hr	RT	—
Acrylic	Solvent-based	—	—	1	min	70	[9]
Acrylic	Solvent-based	—	1	5000	min	70	[9]

AZ, azinidine; RT, room temperature

Reactive resins with methylol groups in the fluid state are efficient tackifiers but also crosslink with the neoprene rubber used as elastomer, and lead to high strength bonds [25]. Electron beam-cured crosslinkable rubbers are used for PSAs also; the SAFT increases for these polymers with increasing gel content, but the relationship between gel content and SAFT is a gradual slope [30]. A polymer gel content greater than 0.8% is required to increase the high temperature holding power above 300 min. Crosslinking increases the modulus; therefore crosslinked adhesives have the highest shear holding power (Table 6.23).

Unfortunately crosslinking is associated with the loss of tack (and peel) and thus it may not be suitable for PSA modification, except under strictly controlled conditions (e.g., solvent-based adhesives). Ulrich [146] addressed the need to improve both the adhesion and the heat resistance at 70°C of an acrylate/acrylic acid adhesive when applied to low energy surfaces; this can be achieved by employing a tackifying resin for better adhesion and by crosslinking the tackified PSA for better heat resistance at 70°C. The new class of acrylic hot-melt PSAs that contain reactive methacrylic end groups that can be crosslinked, imparts good temperature stability [140]. Conventional natural rubber adhesives do not have the required temperature resistance for masking tape [118]; to impart the required degree of heat resistance to the adhesive, it must be crosslinked [147].

Shear Dependence on Composite Status. Generally, the composite structure of the PSA layer contains technological (polymerization or formulating) additives necessary for converting the liquid adhesive, and additives for improving the convertability of the laminate. Additives related to the water-based nature of the adhesive (e.g., surfactants, defoamer, and water) generally decrease the cohesion level of the adhesive. On the other hand, converting additives (fillers) improve the cohesion level. Thus,

$$\text{Cohesion} = f\left(\frac{\text{Fillers}}{\text{Water, Surfactants}}\right) \tag{6.41}$$

The resistance of separating liquid-bonded plates depends on the layer thickness and viscosity (see Eq. 2.14). Concerning the influence of the coating weight on the shear, the flow (creep) limit F_L may be formulated as a function of the adhesive modulus, elastic work, and the dimensions of the sample. Shear (defined as the limit of the slip) also depends on the layer thickness. Experimental data reveal a complex dependence of the shear on the coating weight. Above a minimum coating weight level the shear values decrease with the coating weight. Therefore the following dependence between coating weight and shear S can be written:

$$S = f\left(\frac{1}{CW^d}\right) \tag{6.42}$$

where $d \neq 1$. Since the coating weight influences the drying rate, and vice versa, therefore the shear strength depends on the drying as well (Figure 6.15).

Generally, the thickness of the adhesive layer influences the strength of the adhesive joint [75]:

$$\gamma = \tan \gamma = \frac{v}{D} \tag{6.43}$$

where γ is the shear resistance, v is the displacement of the component of the joint, and D is the thickness of the adhesive layer. The strain rate is given by the following equation:

$$\dot{\gamma} = \frac{d}{dt}\left(\frac{k}{D}\right) \tag{6.44}$$

On the other hand, the mechanical resistance of adhesive joints depends on the length of the overlap [148]. For adhesives obeying Newton's and Hooke's law, the shear is inversely proportional to the thickness of the adhesive layer (i.e., the coating weight) [133]. Shear really depends on the adhesive's intrinsic cohesion and on the adhesive/label geometry (i.e., shear resistance measurements should be carried out with well-defined samples).

Influence of Substrate on Shear. The influence of the substrate on sheer is illustrated in Figure 6.16 [114].

Adhesive Performance Characteristics 221

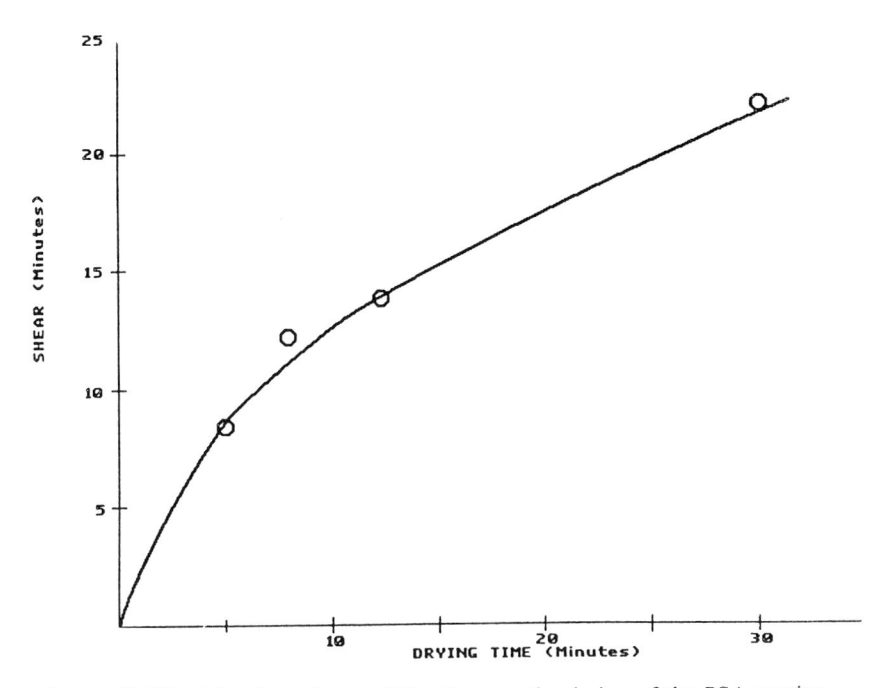

Figure 6.15 The dependence of the shear on the drying of the PSA coating.

Shear Resistance and Face Stock Material. The evaluation of the shear strength should be carried out relative to the face stock used. Unmodified PSAs are used mostly for film coating, while tackified PSAs are used mainly for paper labels. Consequently, film and paper laminates display different shear levels. Shear values for the same adhesive differ when coated on different face stock materials. Common, untackified acrylic PSAs display less cohesive strength when coated on paper than on polar films.

The shear resistance of the joints depends on the surface tension of the substrate [149]. A surface treatment (i.e., an increase of the surface tension) may improve shear characteristics. Table 6.24 illustrates the dependence of the shear resistance (measured as SAFT) on the face stock surface.

Because of the static nature of the shear tests in general, the deformability, the elasticity of the face stock material is less important, but may nevertheless influence the test results. Therefore shear tests are carried out with nondeformable, dimensionally-stable face stock materials (like paper or PET) or by reinforcing the plastic label on the back side (see Section 3.1 of Chapter 10). The applied coating technology influences the coating weight and smoothness of the adhesive layer, and thus the shear resistance.

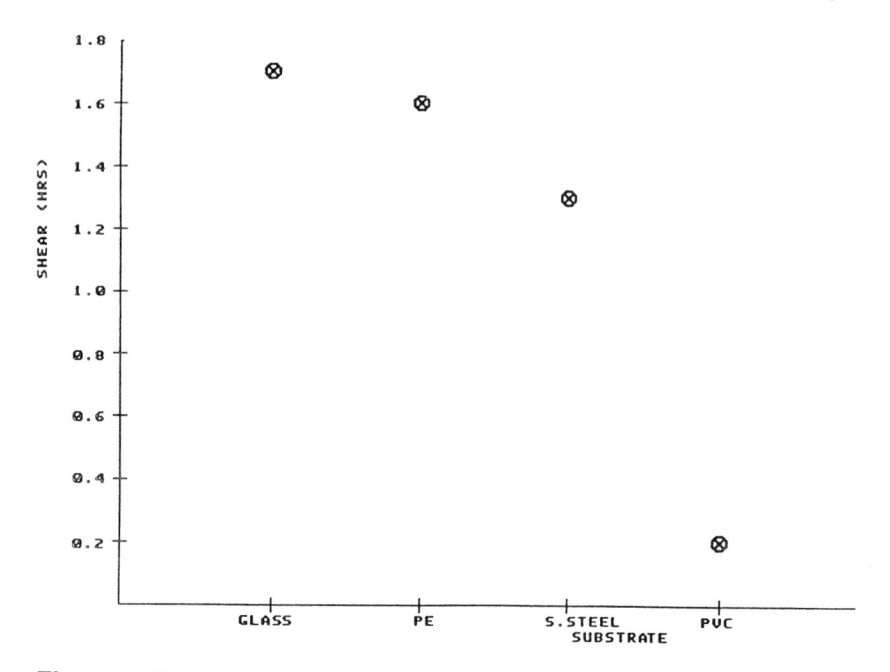

Figure 6.16 The influence of the substrate nature on the shear. Shear values for samples on glass, polyethylene, stainless steel, and PVC as adherend.

Table 6.24 Dependence of the Shear on the Surface Nature of the Face Stock Material

PSA formulation	SAFT values (°C)	
	PET	Paper
A	90	60
B	60	45
C	70	55
D	75	60

Improving Shear Resistance

The values of the shear resistance depend on the cohesion of the adhesive and on the laminate geometry. Cohesion may be improved with hard monomers and by increasing the molecular weight (crosslinking). The laminate geometry influences the adhesive flow. Upon improving the anchorage of the adhesive to the face stock the adhesive bond fails via cohesive break (i.e., within the adhesive layer). Therefore a low coating weight (an optimum as a function of the roughness of the solid state components of the label) will improve the shear resistance. Cold flow may be limited by use of fillers; in some cases the use of reactive fillers improves the shear resistance also. The homogeneity of the adhesive strongly influences the resulting shear strength (see Chapter 8).

Shear Values

There are many data available about shear strength. Unfortunately most of them are static shear values at room temperature. For these the method used (i.e., the sample, dimensions, weight, adherent) influences the measured absolute values. On the other hand, and in a quite different manner from tack and peel values, shear test results for PSAs with different chemical bases do not allow a real comparison of their cohesion-related properties (e.g., cuttability); they are used more as an internal check of the adhesion/cohesion balance for small changes in a given formulation.

Shear values obtained for hot-melt PSAs are generally lower than those for solvent-based PSAs (Table 6.25). Dispersion-based PSAs may exhibit lower or higher shear values as a function of the formulation. High shear, high molecular weight solvent-based acrylic PSAs are superior in shear to rubber-resin based PSAs; CSBR exhibits higher shear than common acrylic PSAs. Crosslinking improves the cohesion even more.

2 INFLUENCE OF ADHESIVE PROPERTIES ON OTHER CHARACTERISTICS OF PSAs

2.1 Influence of Adhesive Properties on the Converting Properties

Wettability depends on the adhesion-cohesion balance; high shear, low tack adhesives exhibit better wettability than soft, tacky ones. Printability of the laminate depends on its stiffness, porosity and surface characteristics; all these properties depend on the stiffness, anchorage, and migration of the adhesive (see Section 2.2 of Chapter 7 and Chapter 8).

Table 6.25 Typical Shear Values

Chemical basis			Physical state			Note	Shear resistance				Ref.
Acrylic	Rubber-resin	Styrene rubber-resin	Solvent-based	Water-based	Hot-melt		Value	Unit	Temperature (°C)	Method	
—	—	x	—	—	x	Electron-beam cured SIS	300	min	95	PSTC 7	[30]
x	—	—	—	x	—	—	300	min	RT	PSTC 7	[30]
x	—	—	—	x	—	Crosslinked	>20	hr	RT	PSTC 7	—
x	—	—	x	—	—	Self-crosslinked	260	hr	RT	—	[150]
—	—	x	—	—	x	—	4.6	d	RT	—	[151]
—	—	x	—	—	x	Kraton GX/Regalrez 100/150	>5000	hr	RT	1"×1", 1000g	[152]
x	—	—	—	x	—	—	5–10	min	RT	—	[153]
x	—	—	—	x	—	—	0.5–2	min	70	—	[153]
x	—	—	x	—	—	Crosslinked	>200	min	RT	—	[153]
x	—	—	x	—	—	Crosslinked	20–100	min	70	—	[153]
—	—	x	—	—	x	—	100	hr	88	—	[154]
x	—	—	x	—	—	—	>16	hr	RT	0.5"×0.5", 1000 g	[26]
x	—	—	x	—	—	—	0.7–2	hr	RT	0.5"×0.5", 500 g	[4]
x	—	—	x	—	—	—	83	hr	70	1.27×2.54 cm, 1000 g	[155]
x	—	—	x	—	—	Removable	16–83	hr	70	—	[129]
—	x	—	x	—	—	—	4–5	hr	RT	—	—

RT, room temperature

2.2 Influence of Adhesive Properties on End-Use Properties

The dependence of the removability on the adhesive properties is well known. High peel adhesion and peel build-up do not allow removability. On the other hand, low shear may cause legging or adhesive residue on the substrate after peeling.

Labeling properties depend on the label adhesion to the release liner (i.e., on the release or peel force). For gun labeling they also depend on the tack and peel. Repositionable adhesives need low initial peel adhesion, whereas tamper-proof ones require a high initial peel adhesion level.

2.3 Influence of Peel Adhesion

The influence of the peel on the end-use properties is evident. For permanent/removable or tamper-proof labels the peel value remains the main property. In a similar manner peel influences the labeling process, where the release force (peel from the release liner) has to be different for labels not used in dispensers (where a high release force would not be of great consequence), and those used in dispensing guns (where a high peel force of the label from the release would indicate poor conversion properties, or poor labeling ability). The build-up of the peel resistance as a function of the time is very important for the design of removable (no build-up), repositionable (slow build-up), and tamper-proof (fast build-up) labels. Different peel levels are required for labels with a multilayer construction (i.e., similar to application and masking tapes). The peel level has to be quite different for smooth, polar surfaces (e.g., glass or steel) or rough ones (e.g., cardboard).

2.4 Influence of Shear

Low cohesion limits the applicability of the label/decal PSA grade with respect to PVC film shrinkage both on the release liner during storage and when applied to a substrate, with regard to wing-up (flagging) on curved surfaces (mandrel hold), stringing and edge ooze during converting. Shear strength for film laminates is important for removable adhesives in order to avoid the build-up of adhesive residues; for permanent adhesives cohesion ensures good die-cutting properties and avoids oozing and bleedthrough through the face material.

3 COMPARISON OF PSAs ON DIFFERENT CHEMICAL BASES

It is evident that on the basis of their chemical composition and rheological behavior PSAs with different chemical bases can display quite different properties. Next a comparative examination of PSAs on a different chemical basis is presented. First the classical rubber-resin solvent-based PSAs will be compared to acrylic PSAs.

In a separate discussion the advantages of water-based acrylic PSAs over other water-based PSAs is examined. Although the whole range of properties is covered, it remains clear that the most important properties of a PSA label are those correlated with the adhesive characteristics.

3.1 Rubber-Based Versus Acrylic-Based PSAs

Table 6.26 summarizes the most important differences (i.e., the general adhesion properties, the special adhesive properties and the stability) between rubber-resin or acrylic-based PSAs. As can be seen from Table 6.26, rubber-resin adhesives possess very good general adhesive properties, but remain defensive in shear. Acrylic PSAs display better specific adhesion properties.

Table 6.26 Comparison of Rubber-Resin PSA Versus Acrylic PSA

Performance characteristics	Chemical basis of the PSA			
	Acrylic		Rubber/Resin	
	Solvent-based	Water-based	Solvent-based	Hot-melt
General				
Tack	Fair	Fair	Good	Good
Peel	Fair	Fair	Good	Good
Shear	Good	Good	Good	Fair
Specific				
Cuttability	Good	Fair	Good	Insufficient
Resistance to migration	Good	Fair	Fair	Insufficient
Aging				
Thermal stability	Good	Good	Fair	Insufficient
UV stability	Good	Good	Fair	Insufficient

Adhesive Performance Characteristics

Generally, the advantages of acrylic systems include chemical and water resistance, UV and oxidative stability, heat resistance, tack at varying temperatures, and the balance of adhesive and cohesive properties. Acrylic-based PSAs offer a unique combination of performance advantages relative to hydrocarbon-based rubber-resin adhesives, and are used extensively in end-use markets that demand excellent color and clarity, weatherability, durability, and plasticizer migration resistance (i.e., overall versatility of the adhesive). Acrylic PSAs do not exhibit yellowing and display good chemical resistance [156]. Rubber-resin adhesives provide a compromise of high cohesive strength for conformance to curved containers and good quick stick required in automatic labeling, but they exhibit a varying batch-to-batch consistency and can cause staining problems [147]. Staining and softening usually result when using rubber-based PSAs on PVC substrates. Low energy face stock surfaces and soft rubber-based PSAs bond fairly well (anchorage), but they are difficult to convert, whereas acrylic PSAs convert well, but exhibit low bond strength (anchorage). Properties that are correlated with the chemical instability of the rubber-resin basis (e.g., temperature and light resistance) and the pronounced liquid character of the rubber-resin mixture (e.g., cold flow and shear) remain inferior for rubber-resin PSAs as compared to acrylic PSAs.

3.2 Acrylics and Other Synthetic Polymer-Based Elastomers

In order to make a competitive evaluation of the properties of acrylics used as raw materials for PSAs, the advantages and disadvantages of acrylics as well as their main technical features and adhesive properties should be examined. This study deals with the evaluation and examination of the adhesive and end-use properties of acrylic-based PSAs as compared to other PSAs. First tack, peel adhesion, cohesion, and aging resistance are used as examination criteria. The main adhesive properties (tack, peel, and cohesion) depend on the state of the polymer (pure or formulated), the coating weight, and on the face stock used.

Tack

Common acrylic PSAs are not tacky enough and special acrylic PSAs are not cohesive enough. Therefore acrylic PSAs need compounding or tackifying. The tack of unformulated acrylic PSAs is lower than that of tackified rubber-based PSAs (their main competitor) but higher than that of the other untackified adhesives, except polyvinyl ethers (see Equation 6.11).

The better tack level of unformulated acrylic PSAs, compared to other unformulated PSAs (e.g., EVAc and CSBR) is illustrated in Table 6.27. These data show that acrylic PSAs possess a better tack at any coating weight level.

Table 6.27 Comparison of Acrylic PSAs Versus EVAc and CSBR PSAs. Tack and Peel of Unformulated WB PSA on Different Chemical Basis, as a Function of the Coating Weight

Coating weight (g/m²)	Chemical Composition									
	AC				EVAc				CSBR	
	Peel (N/25 mm)		RB tack (cm)		Peel (N/25 mm)		RB tack (cm)	Peel (N/25 mm)		RB tack (cm)
	Glass	PE			Glass	PE		Glass	PE	
10	—	—	—		—	—	—	10.0 PT	3.0	17.0
15	—	—	—		10.0 PT	1.0	18.5	10.0 PT	5.0	8.0
20	22.0	20.0	1.5		10.0 PT	1.0	17.0	12.0 PT	12.0	5.5
25	27.0	20.0	1.5		—	—	—	16.0 PT	14.0	4.9
30	—	—	—		14.0 PT	1.0	17.0	—	—	—
35	—	—	—		13.0 PT	1.0	15.0	—	—	—
40	21.0	14.0	1.5		15.0 PT	1.0	7.0	21.0	14.0	6.0
50	25.0	15.0	1.5		—	—	—	—	—	—
55	35.0 PT	17.0	1.5		—	—	—	—	—	—
60	35.0 PT	22.0	1.5		—	—	—	—	—	—

AC: Acronal V205 (BASF); EVA: 1360 (Hoechst); CSBR: 3703 (Polysar)
RB: rolling ball

Table 6.28 Tack as a Function of the Tackifier Level

Tackifier level, parts per 100 PSA (wet weight)	Tack (quick stick; g/cm)
20	270
40	570
60	710
80	800
100	880

Quick stick measured for tackified water-based PSA (25 g/m^2 on PET).

In order to evaluate the tack of formulated acrylic PSAs, the following examination criteria will be considered: ease of tackification, the tackifier level, and the tack level.

Ease of Tackification/Compatibility. Acrylic PSAs display good overall compatibility with common tackifiers (plasticizers or tackifying resins). Both solvent-based and water-based tackifying resins can be fed into water-based acrylic PSAs. It is not possible to add molten tackifiers to water-based acrylic PSAs.

Tackifier Level. A level of 30–40% (by wet weight) tackifier resin is commonly used in formulated acrylic-based PSAs [67,118] and a tackifier level of 10–20% suffices for tack improvement of acrylic-based PSAs (Table 6.28). This is very important because adhesive formulations for soft films should contain a minimum resin level in order to avoid resin/plasticizer interaction.

Tack Level. In general, the ease of tackifying and the obtained tack level depend on the chemical composition, structure, and molecular weight of the acrylic PSAs. With respect to the chemical composition and structure of the polymer, there exists a broad range of possibilities. For solvent-based acrylic PSAs viscosity is dependent on molecular weight; thus the molecular weight constitutes a limiting factor. For water-borne acrylic PSAs there is no technological molecular weight limit, but the molecular weight should be limited because of compatibility reasons. There is an optimum molecular weight in order to achieve good tackification. Table 6.29 illustrates the low tackifying response of high molecular weight acrylic dispersions.

Table 6.29 Tack Dependence on the Molecular Weight

	Chemical composition of the PSA (parts, wet weight)				
Compound code	3703	20094	CF 52	V205	
Supplier	Polysar	Syntho	A&W	BASF	RB tack
Nature	CSBR	Acrylic	Tackifier	Acrylic	(cm)
	32	33	35	—	25.0
	32	23	35	—	20.0
	32	13	35	—	6.5
	32	—	35	33	2.5

Influence of a high molecular weight acrylic PSA on the tack of a CSBR-based, tackified water-based PSA formulation.
RB, rolling ball

Acrylic-based PSAs need lower tackifier levels than CSBR and EVAc copolymers. The tack level of tackified acrylic PSAs is superior to that of tackified EVAc, but lower than the tack of tackified CSBR, namely:

Tack: CSBR (formulated) > AC (formulated) > EVAc (formulated)

(6.45)

as far as the tack properties are concerned. As discussed in Chapter 2, adhesive properties also depend on the face stock material. A short examination of the performance of acrylic-based PSAs in comparison with other adhesives coated on paper and film, follows next.

Permanent/Removable Paper Label Applications. Generally, removable adhesives exhibit low tack. Common acrylic-based removable PSAs are untackified formulations. Rubber-based, solvent-based, or water-based removable formulations, generally are tackified and thus they are tackier than acrylic-based ones (Table 6.30). On the other hand, the removability assumes a primer coating and/or a low coating weight.

Film Application. SBR has low light and aging stability and that EVAc has a limited water resistance. These disadvantages make them inadequate for film coating applications. On the other hand, both need tack improvement (i.e., a tackifier). The use of soft PVC as the main face stock material implies no or a low tackifier level. Thus, only acrylic-based PSAs meet the whole range of quality requirements for foil coating; they show the best unformulated tack, another important feature for film-coating adhesives (Table 6.31).

Table 6.30 Rolling Ball Tack (cm) of Rubber/Resin-Based Formulation Versus Acrylic PSA. Dependence of the Tack on the Coating Weight

Coating weight (g/m^2)	Chemical composition of PSA	
	Solvent-based rubber/resin	Water-based tackified acrylic
8.6	4.0	6.4
18.4	2.1	2.4
28.0	1.9	1.9

Adhesion to Polar/Nonpolar Surfaces. Acrylic-based PSAs are the best unformulated materials in the whole range of required peel forces for polar substrates. Unfortunately, the peel of untackified acrylic PSAs on polar surfaces is not sufficient (Table 6.32). The peel of other PSAs on nonpolar substrates is low too, except for the new class of ethylene–vinyl acetate–maleinate and ethylene–vinyl acetate–acrylic copolymers [157].

Peel of Tackified Acrylic PSAs

Tackification improves adhesion to polar and nonpolar surfaces. Polymers like SBR and EVAc need more tackifier than acrylic PSAs (Table 6.33). A detailed discussion about tack and required tackifier level is included in Chapter 8. Formulations with less than 30 wt% tackifier do not provide enough peel on polyethylene for common acrylic formulations (Table 6.34). On the other hand, there are some new acrylic PSAs which display adequate peel adhesion on polyethylene at low (20 wt%) tackifier loading levels (Table 6.35).

Proposals from tackifier suppliers include the use of 25–50% tackifier resin for agressive, soft, acrylic dispersions on an ethyl hexyl acrylate basis (Figure 6.17). Other base elastomers require higher tackifier levels.

Figure 6.18 and Table 6.36 illustrate the low tackifying response of CSBR-based PSAs as compared to acrylic-based PSAs. For common water-based PSA formulations with standard adhesive performance characteristics, EVAc and SBR require higher tackifier loadings than acrylic PSAs (Figure 6.18). In a similar manner tackified acrylic PSAs show a more pronounced response to tackification than ethylene vinyl acetate copolymers (Table 6.37).

Generally, higher peel adhesion is obtained with less tackifier for acrylic PSAs; the absolute value (level) of the achieved peel also is higher (Table 6.38). One can conclude that the peel of tackified acrylic PSAs on polar or nonpolar surfaces is superior:

$$\text{Peel: AC (Tackified)} > \text{CSBR (Tackified)} >> \text{EVAc (Tackified)} \quad (6.46)$$

Table 6.31 PSA used for Film Coating

Basis	Chemical composition			Adhesive characteristics	
	Code	Supplier	Parts (wet)	Peel (N/25 mm)	RB tack (cm)
Acrylic	V205	BASF	100	0	5.0
	V205	BASF	70	15.0	2.5
	V208		30		
	PC80	UCB	100	6.0	3.5
	FC88	UCB	100	8.0	3.5
	ECR555	Exxon	100	5.0	1.0
	FC88	UCB	80	—	3.5
	PC80		20		
	FC88	UCB	70	10.0	6.5
	880D	BASF	30		
	V205	BASF	80	11.0	4.5
	20084	Syntho	20		
	FC88	UCB	70	11.0	4.5
	2313	Nobel	30		
EVAc	1360	Hoechst	100	4.0	6.0
	1360	Hoechst	70	—	3.0
	1370		30		
Acrylic EVAc blend	913	UCB	50		7.0
	1360	Hoechst	50		
	FC88	UCB	30	10.0	3.5
	V205	BASF	30		
	1360	Hoechst	30		
EVAc acrylic	EAF60	Wacker	100	—	10.0
	VU895	Wacker	100	11.0	2.5
EVAc maleinate	Nacor 90	National	100	7.0	8.5

Adhesive Performance Characteristics

Table 6.32 Typical Peel Values of Untackified PSA on Polyethylene

Code	Supplier	Nature	Peel (N/25 mm)
V205	BASF	Acrylic	20.0
80D	BASF	Acrylic	3.8
85D	BASF	Acrylic	10.3
1360	Hoechst	EVAc	5.0
EAF60	Wacker	EVAc	7.0
3703	Polysar	CSBR	7.0
20094	Syntho	Acrylic	0.2

Instantaneous peel was measured; coating weight 20 g/m^2.

Table 6.33 Required Tackifier Level for AC PSA Versus CSBR PSA. The Dependence of the Peel and Tack on the Tackifier Level.

Tackifier level (parts/100 wet weight)	AC formulation			CSBR formulation		
	Peel (N/25 mm)		RB tack (cm)	Peel (N/25 mm)		RB tack (cm)
	Glass	PE		Glass	PE	
20	20.0 PT	15.0	2.5	14.0 PT	10.0	3.0
33	22.0 PT	17.0	2.5	15.0 PT	11.0	2.5
43	22.0 PT	20.0 PT	2.0	17.0 PT	16.0 PT	1.5
50	22.0 PT	19.0	1.5	19.0 PT	14.0 PT	1.5

AC: A 80/20 wet weight parts blend of Acronal V205/80D; CSBR: 3703 (Polysar); Tackifier: CF52 (A&W); PT: paper tear; RB: rolling ball.

Table 6.34 Required Tackifier Level for AC PSA Formulations. Peel Dependence on the Tackifier Level (Peel Adhesion on PE Film and Plate).

Tackifier level (% wet weight)	Peel (N/25 mm)			
	PE film (coating weight, g/m^2)		PE plate (coating weight, g/m^2)	
	10	20	10	20
16	6.0	2.0	23.0	6.0
20	15.0	5.0	20.0	7.0
24	20.0	5.0	15.0	5.0
29	17.0	8.0	20.0 PT	5.0
33	15.0	8.0	21.0 PT	20.0 PT

Table 6.35 Peel on Polyethylene for Low Level Tackified Special WB PSA

	Chemical composition, parts (wet weight)				
Code Supplier Nature	5650 Hoechst AC	ECR-576 Exxon AC	Exoryl 2001 Exxon AC	SE-351 A&W Tackifier	Peel on PE (N/25 mm)
	100	—	—	—	16.5 PT
	100	—	—	20	19.5 PT
	—	100	—	—	17.5 PT
	—	100	—	20	18.0 PT
	—	—	100	—	5.2
	—	—	100	20	15.4

Adhesive Performance Characteristics 235

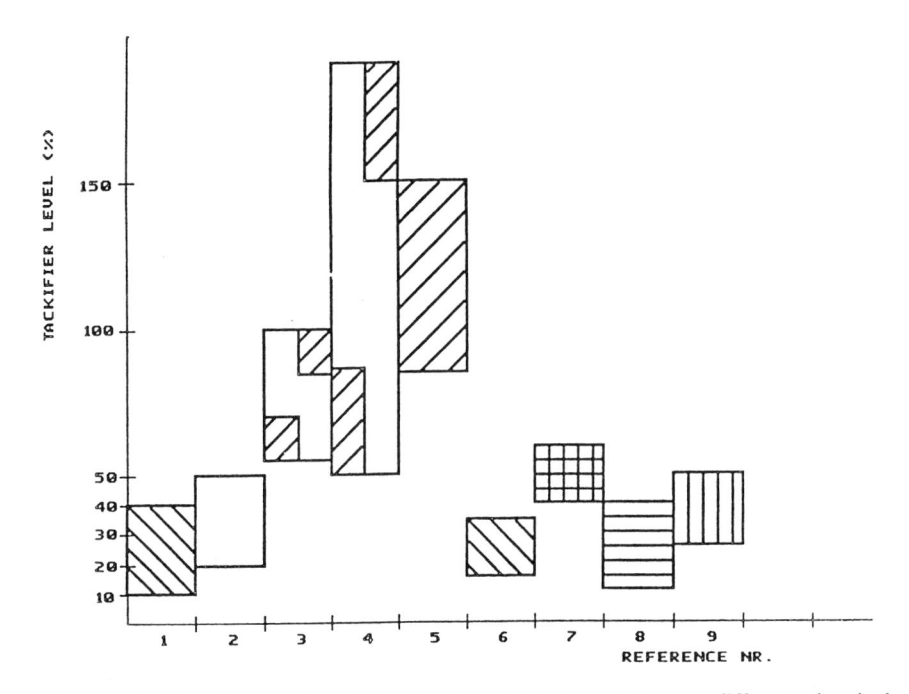

Figure 6.17 Suggested tackifier level for PSA formulation on different chemical basis. 1) acrylic; 2) Vinylacrylic; 3) SIS; 4) SIS; 5) SIS; 6) acrylic; 7) SBR; 8) Vinylacrylic; 9) EVAc.

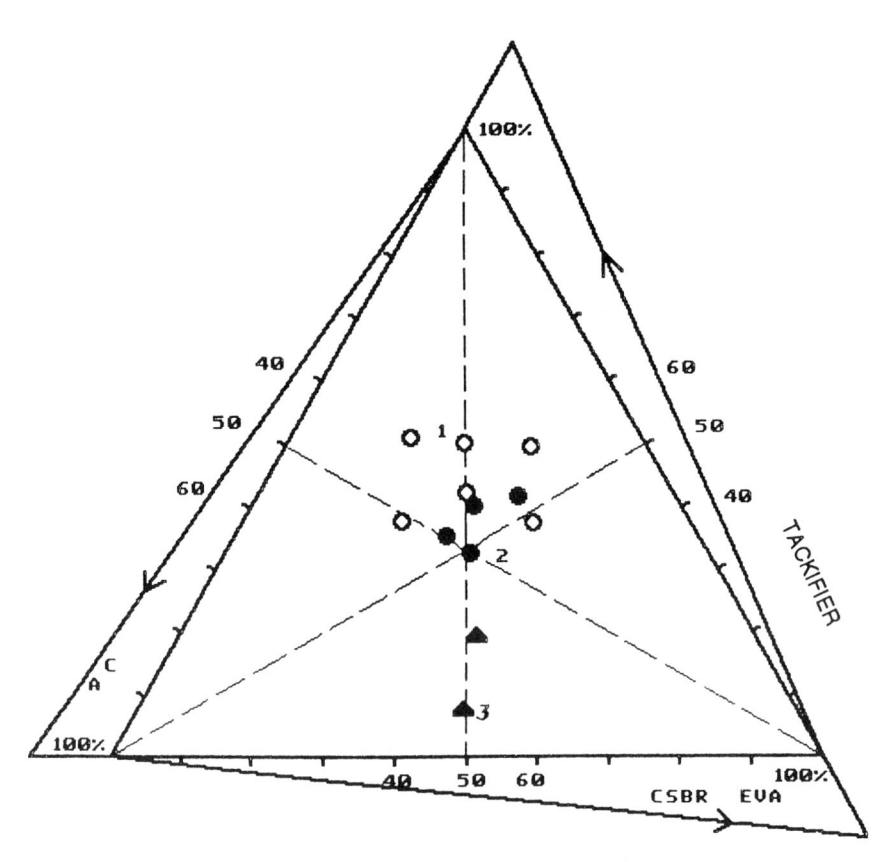

Figure 6.18 Ternary composition diagram. Common PSA formulations for 1) CSBR, 2) EVAc, and 3) acrylic formulations. Water-based tackified PSA.

Table 6.36 Tackifying Response of a Water-Based CSBR PSA Formulation

Chemical composition (parts, wet weight)		Adhesive characteristics		
		Peel (N/25 mm)		
PSA 3703 (Polysar)	Tackifier CF52 (A&W)	On glass	On PE	RB tack (cm)
50	50	19.0 PT	14.0 PT	1.5
57	43	17.0 PT	16.0 PT	1.5
67	33	15.0 PT	11.0	2.5
80	20	14.0	10.0	3.0

Cohesion of Acrylic-Based PSAs

Except for some special adhesives with improved low temperature (and room temperature) fluidity (e.g., deep freeze adhesives), common unformulated acrylic PSAs possess enough intrinsic shear strength to be converted and processed without problems. Other polymers (e.g., EVAc, CSBR) have better cohesion (measured as shear) than acrylic PSAs, but their shear-correlated properties like cuttability are not superior. As an example for the upper limit crosslinked solvent-based acrylic PSAs exhibit shear levels in excess of 260 hr [150]. On the other hand, soft, high tack acrylic PSAs show 15 min shear times. Table 6.39 summarizes characteristic shear values for PSAs on different chemical bases; it shows that acrylic PSAs display the best shear values in

Table 6.37 Tackifying Response of a Water-Borne EVA-Based PSA Formulation

Chemical composition (parts, wet weight)				Adhesive properties	
Code Supplier	1360 Hoechst	1370 Hoechst	CF52 A&W	Peel (N/25 mm)	RB tack (cm)
	50	50	100	7.0	> 30
	50	30	100	3.0	> 30
	70	30	100	15.0 PT	20
	50	50	70	6.0	> 30
	50	30	70	3.0	> 30
	50	30	50	8.0	> 30

Table 6.38 Peel/Tackifier Level Correlation for Water-Based PSAs on Different Chemical Bases

PSA Code	Supplier	Nature	Level	Tackifier level	Peel (N/25 mm)
3703	Polysar	CSBR	50	50	19.0 PT
			66	33	15.0 PT
			77	23	14.0 PT
Nacor 90	National	EVAcM	50	50	20.0 PT
			66	33	20.0 PT
			77	23	18.0 PT
FC88	UCB	AC	50	50	16.0
			66	33	8.5
			77	23	7.0
V205	BASF	AC	40	50	19.0
80D	BASF	AC	10		
V205	BASF	AC	49.5	33	10.5
80D	BASF	AC	16.5		
V205	BASF	AC	57.75	23	9.0
80D	BASF	AC	19.25		
3703	Polysar	CSBR	25	50	27.0 PT
80D	BASF	AC	25		
3703	Polysar	CSBR	35	33	25.0 PT
80D	BASF	AC	35		
3703	Polysar	CSBR	38.5	23	23.0 PT
80D	BASF	AC	38.5		
1360	HOE	EVAc	50	50	15.0
			66	33	12.0
			77	23	6.5

The concentration (level) of the components is given in parts wet weight; Tackifier: CF52 from A&W.

Table 6.39 Shear Values as a Function of the Physical State and Chemical Nature of the PSA

Chemical characteristics								Shear resistance			
Chemical basis			Crosslinked		Physical state						
AC	RR	EVAc	Yes	No	SB	WB	HM	Value	Unit	Temperature (°C)	Method	Ref.
x				x	x	x		0.7–2	hr	RT	0.5"×0.5", 500 g	[4]
x				x		x		5–10	min	RT	—	[153]
x				x		x		>200	min	RT	0.5"×0.5", 500 g	[153]
x				x		x		40–90	min	RT	—	—
x				x	x			1	min	70	—	[155]
x			x		x			5000	min	70	—	[155]
x			x		x			>10000	min	RT	0.5"×0.5", 500 g	[26]
x			x		x			260	hr	RT	—	[150]
x			x		x			>200	min	RT	—	[153]
x				x			x	500–4500	min	RT	—	[26]
	x				x			3	hr	RT	0.5"×0.5", 1000 g	[158]
	x				x			4.6	d	RT	—	[151]
	x				x			50–400	min	RT	—	[159]
	x			x			x	300	min	95	—	[30]
	x			x			x	100	hr	88	—	[154]
		x				x		1	hr	RT	—	[160]
		x				x		1	hr	RT	—	[161]

AC: acrylic; RR: rubber-resin; EVAc: ethylene vinyl acetate; SB: solvent-based; WB: water-based; HM: hot-melt; RT: room temperature.

Table 6.40 Influence of the Tackification on the Shear of Water-Based PSA on Different Chemical Basis

PSA	Supplier	Nature	Tackifier level						
			20	25	30	35	40	45	50
V205	BASF	AC	41	38	36	33	38	—	25
80D	BASF	AC	39	37	34	32	22	—	25
Hot shear			9	7	—	7	4	—	—
RT shear			100	80	55	39	27	—	55
3703	Polysar	CSBR	40	—	—	—	30	—	25
V205	BASF	AC	20	—	—	—	22.5	—	19
80D	BASF	AC	20	—	—	—	7.5	—	8
Hot shear			—	—	19	22	13	—	—
RT shear			250	—	—	—	25	—	14
3703	Polysar	CSBR	40	37.5	35	—	30	22.5	25
80D	BASF	AC	40	37.5	35	—	30	22.5	25
Hot shear			> 1500	> 1200	240	—	2	2	2
RT shear									
3703	Polysar	CSBR	—	—	70	—	—	55	50
Hot shear									
RT shear					1300			500	368

Shear data given in minutes. Hot shear was tested at 50°C /minutes. An acid rosin was used as tackifier.
RT: room temperature; AC: acrylic

the range of unformulated polymers with similar tack. This statement is very important for film label applications, where new ethylene-maleinate multipolymers display very good nonpolar peel adhesion, but low cohesion.

Shear Strength of Tackified Acrylic PSAs

Common tackifier levels lead to a decrease in cohesion. This is a function of the nature of the acrylic, the nature of the tackifier, and the mixing ratio. Shear loss is less pronounced for CSBR and EVA. Table 6.40 illustrates the variation of the shear through tackification for different classes of raw materials. Unfortunately higher shear does not lead to better converting properties for CSBR/EVAc.

Table 6.41 Aging Stability of PSA on Different Chemical Bases

Adhesive			Adhesive properties				
			Initial		After aging		
					Thermal treatment		
Code	Supplier	Nature	Peel (N/25 mm)	Tack (cm)	Peel (N/25 mm)	Tack (cm)	Light exposure
2382	Syntho	AC	15.0	2.5	14.0	4.5	1
2343	Syntho	VAc-AC	14.6	2.5	24.0 PT	2.5	—
1338	Syntho	CSBR	20.4	5.5	15.0	8.0	5
10066	Syntho	S-AC	22.0	7.5	20.0	10.0	5
1247	Syntho	CSBR	—	—	—	—	5
EAF60	Wacker	EVAc	25.0 PT	4.0	25.0 PT	6.0	1
V205	BASF	AC	25.0 PT	2.5	25.0 PT	4.0	1

1 = good, 5 = poor; AC, Acrylic; VAc, vinyl acetate; CSBR, carboxylated styrene-butadiene rubber; S-AC, styrene acrylic; EVAc, ethylene vinyl acetate

Aging Resistance

For common PSAs generally, time and temperature stability of the adhesive/laminate at room temperature is required. There are special end-uses where high temperature stability or low temperature performance is necessary. Table 6.41 summarizes data concerning UV and temperature stability of different PSAs.

Because of their bonding ability (accentuated flow) common acrylic PSAs show only a limited temperature stability in PSA labels. High molecular weight or partially crosslinked acrylic PSAs display an improved thermal stability; partial crosslinking may be obtained with special built-in monomers (e.g., acrylonitrile). Therefore some common ethyl hexyl-based acrylic PSAs with acrylonitrile as comonomer will display higher than normal shear values. Because of the temperature resistance of these nitrile-based crosslinks on especially high hot shear values are obtained.

Laminate Properties

Some aspects of the laminate properties are dealt with in Chapters 2 and 8. The adhesive/solid state component interactions influence the stability of the laminate properties. First, these interactions may be restricted to mutual interactions between adhesive and paper-based face stock material. The most important mutual interaction takes place between the porous, rough face stock material and the fluid adhesive (i.e., a physical interaction between paper and the PSA); this is known as migration.

Migration Resistance. The migration resistance of untackified acrylic PSAs remains superior to other formulations, with solvent-based ones better than water-based dispersions. Tackified acrylic PSAs show better migration resistance than competitive raw materials:

Migration: AC (Tackified) >> EVAc (Tackified) >> CSBR (Tackified)

(6.47)

There are some acrylic PSAs with lower migration resistance (e.g., removable and deep freeze PSAs). A high emulsifier loading decreases the migration resistance of water-based PSAs.

Anchorage. Anchorage is the phenomenon of adhesion between face stock material and adhesive. For fiber forming, textured supports (e.g., paper) this is a physical phenomenon also. Physical interaction and flow of the adhesive are the main components for the anchorage of PSAs onto rough, textured substrates. Common, tacky, unformulated acrylic PSAs display good anchorage on paper, even if transfer coated. Hard, low tack acrylic PSAs do not exhibit enough anchorage on paper. Removable, unformulated acrylic PSAs give less anchorage on unprimed surfaces than tackified formulations. Anchorage also is a function of the emulsifier level; therefore solvent-based acrylic PSAs display better anchorage than water-based ones. Untackified EVAc or CSBR exhibit poorer anchorage than acrylic PSAs; untackified acrylic PSAs give better anchorage than competitive materials. Tackification improves the anchorage. Tackified acrylic PSAs and tackified rubber display better anchorage than tackified EVAc. With respect to the chemical interaction between acrylic PSAs and paper, untackified acrylic PSAs do not interact chemically with normal paper. A similar behavior is observed for competitive base polymers. Tackified formulations may interact with special (e.g., thermal) papers, due to the tackifying resin.

Most plastic films are compounded materials (i.e., they contain additives). Therefore there are possibilities for an interaction between micromolecular components of the film (e.g., PVC) and the adhesive. Plasticizer and emulsifier from the film could migrate into the adhesive. On the other hand, monomers, oligomers, and surface agents from the adhesive can migrate in the face material, causing stiffening of the face stock (or its shrinkage) as well as the loss of the adhesive properties. Because of their low UV stability rubber-resin adhesives cannot be used for films. EVAc copolymers may interact with some film materials and cause shrinkage. Thus only acrylic PSAs are recommended for film coating. In a similar manner the anchorage of unformulated acrylic PSAs on nonpolar, untreated film is better than the anchorage of its competitors (except for ethylene maleinate multipolymers and Nacor 90).

Acrylic-based PSAs exhibit varying shrinkage on soft PVC. There are some leading acrylic PSAs dispersions that show shrinkage values of 0.3–0.9% (Table 6.42). A detailed discussion of these characteristics is presented in Chapter 10.

It is evident that adhesives interact with the release liner as well. Acrylic PSAs with different formulations exhibit different release levels and release level stability. Ethylene vinyl acetate maleinate copolymers display less sensitivity towards the nature and age of the release liner. Unformulated acrylic PSAs display the best die-cuttability. In the range of formulated products, crosslinked, tackified, solvent-based adhesives show the best die-cuttability (see Chapters 8 and 9). Generally, acrylic formulations require less surface active agents for stabilizing and wetout, thus the humidity content of the laminate is lower. Therefore acrylic-based laminates are superior compared to CSBR/EVAc-based laminates with respect to lay-flat, die-cuttability, and water resistance.

Special Features of Water-Based Acrylic PSAs as Compared to Other Water-Based PSAs

Water-based PSAs show different characteristics according to their chemical nature and manufacturing technology. The properties of the aqueous dispersions as well as those of the coated layer and of the laminate vary widely.

In general, water-based acrylic PSAs possess a higher solids content and a higher solids content/viscosity ratio than common CSBR/EVAc dispersions; they also respond better to diluting. Water-based acrylic PSAs exhibit a lower surface tension than common EVAc or CSBR dispersions. Current surface tension values range from 34–40 dyn/cm, whereas EVAc dispersions show values above 45 dyn/cm. Rubber-based dispersions have a surface tension between 45–60 dyn/cm. Furthermore water-based acrylic PSAs show higher coagulum (grit) than CSBR dispersions, although they display a better mechanical stability than CSBR. Water-based acrylic PSAs generate less foam formation than CSBR and EVAc. Finally, water-based acrylic PSAs show better wetout properties than EVAc- and CSBR-based PSAs, and can be converted at higher speeds than CSBR and EVAc dispersions.

Special Features of Solvent-Based Acrylic PSAs as Compared to Solvent-Based Rubber PSAs

Solvent-based acrylic PSAs have some special characteristics compared to classical rubber-resin PSAs. First, a higher solids content (more than 20–30%) is possible and the solids content/viscosity ratio can be adjusted exactly.

Table 6.42 Shrinkage Values for PSA-Coated PVC (Different PSA)

	PSA		Test conditions		Shrinkage (%, Face Stock)				Test method		
	Nature				White PVC		Transparent PVC				
Code	Status	Chemical basis	Temperature (°C)	Time (days)	Longitudinal	Transverse	Longitudinal	Transverse	Mounted	Unmounted	Ref.
01	WB	AC	70	7	0.20–0.30	0.15	—	—	x	—	—
02	WB	AC	70	7	0.50	0.40	0.85	0.60	—	x	[168]
03	WB	AC	70	7	—	—	0.85	0.50	—	x	[168]
04	WB	AC	70	7	—	—	1.70	1.20	—	x	[168]
05	WB	AC	67	7	—	—	6.0	—	—	x	[168]
05	WB	AC	67	1	—	—	17.0	—	x	—	[168]
06	SB	AC	67	7	—	—	8.0	—	—	x	[168]
06	SB	AC	67	1	—	—	11.0	—	x	—	[168]
07	WB	EVAc	70	7	—	—	0.50	0.40	x	—	[169]
07	WB	EVAc	70	7	—	—	2.60	1.40	—	x	[169]
08	WB	AC	67	7	—	—	0.40	0.40	—	x	[170]
09	WB	EVAc	70	7	—	—	2.0–3.0	—	—	x	[160]
10	WB	AC	70	7	—	—	0.5–1.0	—	—	x	[160]

WB, water-based; SB, solvent-based; AC acrylic;; EVAc, ethylene vinyl acetate

Usually special solvents are used and both solutions or dispersions can be produced. In general fewer or no plasticizers and antioxidants are used; finally crosslinking is possible.

Modification of Adhesive Properties

Adhesive properties of the PSAs can be influenced through polymer synthesis, formulation, and converting.

Polymer Synthesis. There exists a broad range of different functional acrylic/methacrylic monomers which can be used for PSA synthesis. Acrylic PSAs have the broadest monomer basis. Some of the monomers impart adhesive properties, others are used as emulsifier or stabilizer monomers. A butyl acrylate (BuAc) base provides a higher cohesion level than ethyl hexyl acrylate (EHA); other nonacrylic comonomers also can be used. For example acrylonitrile imparts hardness and solvent resistance, styrene and alpha-methyl styrene impart firmness and improve peel adhesion, t-butyl styrene improves tack, methylmethacrylate makes the adhesive firmer, vinyl acetate improves adhesion to certain plastic surfaces, etc. Compared to acrylic-based PSAs, rubber- and EVAc-based copolymers possess a narrower raw material basis.

EVAc-based PSAs must incorporate more than 18% ethylene, while SBR-based PSAs contain less than 25–35% styrene. Thus, their formulating freedom remains limited. Most vinyl acetate copolymers need protective colloids as their particle size and particle size distribution is limited; they have a predetermined viscosity and solids content/viscosity ratio, and a built-in water resistivity. EVAc-based PSAs do not have crosslinked structures, whereas acrylic-based PSAs may have them. Styrene-butadiene rubber-based PSAs must always have a gel/sol ratio as their properties (peel, shear, migration) depend on this.

A pressureless well-known polymerization technology is available for acrylic PSAs. Styrene-butadiene rubber need medium pressure and EVAc require high pressure reactors. For solvent-based acrylic PSAs a crosslinking technology was developed; for water-based acrylic PSAs branching and grafting procedures were established.

Polymer Formulations. A general compatibility exists between different acrylic PSAs, as well as between acrylic PSAs and other polymers. However, no similar compounding freedom exists for rubber- or EVAc-based polymers. Plasticizers and tackifiers can be used for acrylic PSAs and other polymers. Crosslinking by built-in comonomers, metal complexes, or PUR is possible for acrylic PSAs; EVAc have no similar crosslinking possibilities. Attempts were made to crosslink CSBR using polyisocyanates.

Converting. Theoretically there are three converting technologies for acrylic PSAs: hot-melt, solvent-based, and water-based. Practically, only a few experimental hot-melt acrylic PSAs exist (e.g., experimental hot-melt EVAc-based PSAs). Rubber-based PSAs also can be converted using all of the above technologies.

The coating technology is a function of the adhesive, face stock, release nature, and of converting, processing and end-use requirements. Both acrylic-, rubber-, and EVAc-based PSAs allow direct or transfer coating. Acrylic solutions are coated directly on paper and special films, and indirectly on soft PVC. Solvent-based rubber PSAs are directly or transfer coated on paper. Water-based acrylic PSAs are directly or transfer coated on films. Through direct/transfer coating, anchorage, die cutting, water resistance, and release can be adjusted.

High viscosity (and high solids) water-based and solvent-based acrylic PSAs are available, as well as low viscosity (ready-to-use) ones. Primers provide better anchorage and less migration, and increase the removability. Rubber-based PSAs need primers; acrylic PSAs can be converted with/without primer. Common acrylic dispersions need more wetting agents for transfer coating than for direct coating. They need more wetting agents on solventless silicones than on solvent-based silicones. Wetout depends on the siliconized paper. The age of the siliconized liner influences the wetting also. Anchorage of the coated adhesive also depends on the age of laminate.

Special Features of Acrylic PSAs

Next the main attributes of acrylic PSAs are highlighted. From the point of view of the end-user acrylic PSAs provide the best balance between adhesive and cohesive properties, the best independence from the face stock and substrate, the best aging properties (e.g., oxidative stability, heat resistance, UV stability), and the best tack at varying temperatures. In addition, they exhibit the best machinery properties (adhesion/release balance), as well as FDA and BGA approval. The advantages of acrylic PSAs for the converter include a broad range of raw materials, a broad compability with other polymers, and different tackifying possibilities (e.g., high polymers, resins, plasticizers). They accept a broad range of tackifying resins, and display at low tackifier level a more pronounced tackifier response. They are generally suitable for different face stocks and for permanent and removable applications in rolls and sheets. Acrylic PSAs possess excellent converting properties (e.g., wetting out, drying speed, low foam, high mechanical stability) and allow many different ways for adjusting peel adhesion and shear.

The disadvantages of acrylic PSAs for the end-user are relatively low adhesion on untreated, nonpolar surfaces, relatively low die-cuttability if formulated for high tack and coated on light weight laminate components, and relatively low water resistance for water-based PSAs. Disadvantages of acrylic PSAs for the converter include the lack of available acrylic hot-melt PSAs and the lack of or limited compatibility with rubber-based adhesives. Solvent-based acrylic PSAs are superior to water-based PSAs.

Although acrylic PSAs are the most expensive PSAs, they are used with increasing dynamism. There are only a few manufacturers formulating monomers for acrylic PSAs (mainly for EHA and BuAc). The most important component for acrylic PSAs are the tackifier resins. In this field there is an abundant supply of new and classical products.

Formulating Ease and Costs. Acrylic PSAs are multipolymers, so their properties can be varied using different monomer feeds and polymerization technologies. Solvent-based technology is more expensive but has the advantage of crosslinking (other quality and price level). Water-based technology presents the advantage of high solids, seed and branching, and particle size distribution control. Depending on the end-use requirements, it is possible to synthesize or buy acrylic PSAs for compounding, resulting in more or less expensive products. Acrylic PSAs can be tackified using a medium level of tackifier loading only; they have a reserve of adhesive properties and theoretically they can be modified with fillers.

The price of the laminate depends on the following components: the face stock, the release liner, the adhesive, and the choice of the converting (laminating) technology.

Paper. Tackified acrylic PSAs show good anchorage and low migration (solvent-based or water-based), thus they also can be used with inexpensive nonsurface-treated paper. On the other hand, water-based acrylic PSAs need special (primered) face stock in order to obtain a good die cuttability. Heavy voluminous papers increase the die cuttability also. Removable water-based acrylic PSAs also need primered papers.

Film. Water-based and solvent-based acrylic PSAs can be used without restrictions for different film materials. The same acrylic can be used for paper as well. Direct coating on nonpolar films requires a surface treatment.

Acrylic PSAs can be coated on paper-based release liners as well as on film-based release materials. Solvent-based or solventless siliconizing can be used. Direct/transfer coating provides another possibility for release level modification. Transfer coating of water-based acrylic PSAs on solventless-silicones remains the cheapest way.

Solvent-based acrylic PSAs are the most expensive adhesives. Hence:

- The use of filled solvent-based acrylic PSAs decreases the adhesive costs.
- Compounding solvent-based acrylic PSAs also decreases the adhesive cost.
- Tackifying solvent-based/water-based adhesives decreases the costs.

- Common water-based acrylic PSAs are less expensive than solvent-based ones (price ratio 2/1).
- Tackified water-based acrylic PSAs are the most cost effective materials within the range of acrylic PSAs adhesives.

Converting Costs. Converting acrylic PSAs is possible in solution or as an aqueous dispersion. Converting solvent-based acrylic PSAs is more expensive than water-based acrylic PSAs, whereas converting solvent-based rubber-resin PSAs is less expensive than acrylic PSAs. Converting low viscosity water-based acrylic PSAs costs less than coating highly viscous ones. As pointed out before, acrylic PSAs can be converted directly or by transfer, and transfer coating remains less expensive than direct coating.

REFERENCES

1. T. G. Wood, *Tackification of water based acrylic PSA,* Pressure Sensitive Tape Council, Technical Seminar, 6th May, Itasca, IL, 1987.
2. M. Sherrif, R. W. Knibbs and P. G. Langley, *J. Appl. Polymer Sci.*, **17**, 3423 (1973).
3. A. Zosel, *Colloid Polymer Sci.*, **263**, 541 (1985).
4. C. M. Chum, M. C. Ling and R. R. Vargas, (Avery Int. Co, USA), EP 1225792/18.08.87.
5. J. Johnston, *Adhesives Age*, (11) 30 (1983).
6. *Glossary of Terms used in Pressure Sensitive Tapes Industry*, PSTC, Glenview, IL, 1974.
7. ASTM, D-1878-61T; *Pressure Sensitive Tack of Adhesives*, American Society for Testing and Materials, Philadelphia.
8. H. Mizumachi, *J. Appl. Polymer Sci.*, **30**, 2675 (1985).
9. H. Barquins, *Kautschuk, Gummi, Kunststoffe*, (5) 419 (1987).
10. M. Barquins, R. Courtel and D. Maugis, *Wear*, **38**, 385 (1976).
11. K. A. Grosch, *Proc. Roy. Soc.*, **A274**, 21 (1963).
12. J. Wiedemeyer, *VDI Berichte*, **600** (3) 72 (1987).
13. G. R. Hamed and C. H. Hsieh, *J. Polymer Physics*, **21**, 1415 (1983).
14. A. N. Gent and A. J. Kinloch, *J. Polymer Sci.*, **A-2** (9) 659 (1971).
15. A. N. Gent and J. Schultz, *J. Adhesion*, **3**, 281 (1972).
16. C. H. Andrews and A. J. Kinloch, *Proc. Rov. Soc.*, **A332**, 385 (1973).
17. R. S. Chang, *Rubber Chem. and Technol*, (20) 847 (1957).
18. J. Johnston, *Adhesives Age*, (11) 30 (1983).
19. F. Egan, PSTC Technical Seminar, March 1971, Chicago.
20. S. N. Gan, D. R. Burfield and K. Sogh, *Macromolecules* (18) 2684 (1985).
21. *Adhesives Age*, (9) 19 (1986).
22. Shell, *Technical Bulletin, Polybutenes.*
23. A. Midgley, *Adhesives Age*, (9) 17 (1986).
24. D. J. Harrison, J. F. Johnson and J. F. Yates, *Polymer Eng. Sci.*, (14) 865 (1992).
25. *Adhesives Age*, (10) 24 (1977).

26. P. A. Mancinelli, New Developments in Acrylic HMPSA Technology, in *TECH 12., Advances in Pressure Sensitive Tape Technology, Technical Seminar Proceedings*, Itasca, IL, May, 1989, p. 165.
27. Polysar, *Product and Properties Index*, Arnhem, 02/1985.
28. *Adhesives Age*, (5) 32 (1987).
29. *Adhesives Age*, (9) 37 (1988).
30. E. G. Ewing and J. C. Erickson, *Tappi J.*, (6) 158 (1988).
31. Dow Corning, *Silicone PSA, Application information*, 1986.
32. *Adhesives Age*, (10) 16 (1977).
33. S. A. Parrie and P. F. Ritchie, *Adhesion*, (5) 201 (1967).
34. *Pack Report*, (10) 15 (1986).
35. J. Johnston, *Adhesives Age*, (12) 24 (1983).
36. A. Bates, *Adhesives Age*, (12) 26 (1983).
37. A. Kamagata, T. Saito and M. Toyama, *J. Adhesion*, (2) 279 (1972).
38. P. J. Counsell and R. S. Whitehouse, *Development in Adhesives* (W. C. Wake, Ed.), Vol. 1, Applied Science Publishers, London, 1977, p. 99.
39. W.R. Dougherty, 15th Munich Adhesive and Finishing Seminar, 1990, p. 70.
40. D. H. Kaelble, *Trans. Soc. Rheology*, (2) 135 (1965).
41. S. Benedek, *Adhäsion*, (5) 16 (1986).
42. T. K. Sherwood, *Ind. Eng. Chem.*, **21**, 12 (1929).
43. J. Hausmann, *Adhäsion*, (4) 21 (1985).
44. W. D. Freeston, Jr., *Coated Fabrics Technology*, Technomic Publ. Co., 1973, p. 25.
45. S. S. Hwang, *Chem. Eng. Sci.*, **34**, 181 (1979).
46. Backofen & Meier AG, Bülach, *Reverse gravure coating of emulsion PSA*, Private Communication.
47. F. Weyres, *Coating*, (4) 110 (1985).
48. H. D. Patermann, *Coating*, (6) 224 (1988).
49. E. Brada, *Papier u. Kunststoffverarb.*, (5) 170 (1986).
50. EP 0244997.
51. *Adhäsion*, (10) 399 (1969).
52. *Die Herstellung von Hafiklebstoffen*, Tl.2.2, 15d, Nov. 1979, BASF, Ludwigshafen.
53. *Adhäsion*, **9** (9) 349 (1965).
54. *Coating*, (11) 4 (1988).
55. *Coating*, (11) 380 (1986).
56. C. P. Iovine (National Starch Chem. Co., USA), EP 0212358 A2/04.03.87, p. 3.
57. H. Müller, J. Türk and W. Druschke, BASF, Ludwigshafen, EP 0.118.726/11. 02. 83.
58. J. A. Fries, *Tappi J.*, (4) 129 (1988).
59. British Petrol, *Hyvis*, Prospect, (1985).
60. *Tappi J.*, **67** (9) 104 (1983).
61. P. Caton, *European Adhesives and Sealants*, (12) 18 (1990).
62. J. Boutillier, 11th Munich Adhesive and Finishing Seminar, 1983, p. 3.
63. J. R. Wilken, *apr.*, (5) 122 (1986).
64. M. Mazurek,(Minnesota Mining and Manuf. Co., USA), EP 4.693.935/15.09.87.
65. *Adhäsion*, (10) 401 (1968).

66. *European Adhesives and Sealants*, (9) 4 (1987).
67. EP 0251672.
68. Firestone, *Technical Service Report*, 6110, August, 1986.
69. A. W. Aubrey, G. N. Welding and T. Wong, *J. Appl. Chem.*, (10) 2193 (1969).
70. E. Gronewaldt, *Coating*, (5) 144 (1969).
71. M. E. Fowler, *Polymer*, (11) 2146 (1987).
72. *Adhesives Age*, (12) 19 (1986).
73. *Die Herstellung von Haftklebstoffen*, T1-2, 2-14d, Dec. 1979, BASF, Ludwigshafen.
74. J. Wilken, *apr.*, (42) 1490 (1986).
75. M. Schlimmer, *Adhäsion*, (4) 8 (1987).
76. G. Henke, Schmelzhaftklebstoffe für non wovens, 9th Munich Adhesive and Finishing Seminar, 1985.
77. *Adhesives Age*, (6) 32 (1986).
78. *Adhäsion*, (6) 24 (1982).
79. J. Kim, K. S. Kim and Y. H. Kim, *J. Adhesion Sci. Technol.*, (3) 175 (1989).
80. J. Skeist, *Handbook of Adhesives*, 2nd Edition, Van Nostrand-Rheinhold Co., New York, 1977.
81. *Adhäsion*, (5) 202 (1986).
82. *Adhäsion*, (8) 237 (1976).
83. *Coating*, (1) 35 (1989).
84. H. Müller, J. Türk and W. Druschke, BASF, Ludwigshafen, EP 0.118.726/ 11.02.83.
85. S. G. Obran, US Pat. 4.260.659, in EP 0.202.831.
86. *Coating*, (10) 271 (1988).
87. *apr.*, (14) 340 (1987).
88. W. Druschke, *Adhäsion*, (5) 30 (1987).
89. G. Hombergsmeier, *Papier u. Kunststoffverarb.*, (11) 31 (1985).
90. L. M. Prytikin and W. M. Wacula, *Adhäsion*, (12) 44 (1983).
91. *Adhesives Age*, (7) 36 (1986).
92. A. Dobmann and H. Braun, Neues Engineering Konzept für Etikettenindustrie, 9th Munich Adhesive and Finishing Seminar, 1985.
93. EP 3344863.
94. Bevitack, Technical Bulletin.
95. *Adhesives Age*, (5) 24 (1987).
96. J. C. Fitch and A. M. Snow Jr., *Adhesives Age*, (10) 23 (1977).
97. Novamelt Adhesive Technology, Technical Bulletin (1994).
98. A. C. Makati, *Tappi J.*, (6) 147 (1988).
99. M. H. Tanashi, K. Yasuaki, M. Tetsuaki and K. Yunichi (Nichiban Co, Japan), OS/DE 3.544.868 A1/15.05.85.
100. G. L. Schneeberger, *Adhesives Age*, (4) 21 (1974).
101. J. Lin, W. Wen and B. Sun, *Adhäsion*, (12) 21 (1985).
102. *Coating*, (4) 123 (1988).
103. *Coating*, (1) 13 (1970).
104. H. Reip, *Adhesives Age*, (3) 17 (1972).

105. H. K. Porter Co., USA, US. Pat., 3.149.997, in *Adhäsion*, (2) 79 (1966).
106. *Offset Technik*, (8) 8 (1986).
107. *Offset Praxis*, (11) 98 (1986).
108. J. Merretig, *Coating*, (3) 50 (1974).
109. Kjin Co., Japanese Pat., 28.520/70, in *Coating*, (2) 38 (1972).
110. I. J. Davis (National Starch Chem. Co., USA), US Pat., 4.728.572/01.03.88.
111. *Tappi J.*, (5) 25 (1988).
112. G. Schweizer, *Das Papier*, **10A**, 1 (1985).
113. *Coating*, (6) 184 (1969).
114. *Coating*, (11) 89 (1984).
115. T. J. Bonk (Minnesota Mining and Manuf. Co., USA), EP 120.708.31/03.10.84.
116. R. Schuman and B. Josephs (Dennison Manuf. Co., USA), PCT/US86/02304/28.10.86.
117. K. Goller, *Adhäsion*, (4) 101 (1974).
118. D. Satas, *Handbook of Pressure Sensitive Technology*, Van Nostrand-Rheinhold Co., New York, 1982.
119. *Coating*, (4) 23 (1969).
120. German Pat., 2. 110. 491, in *Coating*, (12) 363 (1973).
121. U. E. Krause (Hensel Textil Co.), EP 0.238.014 A2/23.09.87.
122. *Coating*, (3) 42 (1969).
123. M. Hasegawa, US Pat., 4.460.634/17.07.84, in *Adhesives Age*, (3) 22 (87).
124. Eastman Kodak, US Pat., 359. 943, in *Adhäsion*, (5) 328 (1967).
125. Vitta Co., US Pat., 3. 598. 073, in *Adhäsion*, (5) 328 (1967).
126. Japanese Pat., 2.736/75, in US Pat., 3.691.140/18.09.72.
127. E. Pagendarm, OS/DE 3.628.784 A1/25.08.86.
128. 3M Kokkai, Tokyo Koho, Japanese Pat., 63.175.091, in *CAS*, **25** (7) 1988).
129. EP 0100146.
130. P. Tkaczuk, *Adhesives Age*, (8) 19 (1988).
131. *Adhesives Age*, (12) 43 (1981).
132. *Tappi J.*, (9) 104 (1983).
133. K. N. Nakao, *J. Adh. Soc. Japan*, (6) 6 (1970).
134. J. Andres, *apr* (16) 444 (1986).
135. US Pat., 4.563.389, in S. Krampe and C. L. Moore (Minnesota Mining Manuf. Co., USA), EP 0.202.831 A2/26.11.86.
136. A. W. Norman, *Adhesives Age*, (4) 36 (1974).
137. R. A. Fletscher, The temperature factor in compounding thermoplastic rubber based HMA, 11th Munich Adhesive and Finishing Seminar, 1986, p. 36.
138. A. Sustici and B. Below, *Adhesives Age*, (11) 17 (1991).
139. A. Bell, *Shell Bulletin*, TB, RBX/73/8/6.
140. J. A. Schlademan, *Coating*, (1) 12 (1986).
141. L. M. Schrijver, *Coating*, (3) 70 (1981).
142. J. M. Gordon, G. B. Rouse, J. H. Gibbs and W. M. Risen Jr. *J. Chem. Phys.*, (66) 4971 (1977).
143. G. Meinel, *Papier u. Kunststoffverarb.*, (10) 26 (1985).
144. H. J. Boob and H. M. Ulrich, *Kautschuk, Gummi, Kunststoffe*, (3) 209 (1984).

145. L. A. Sobieski and T. J. Tangney, *Adhesives Age*, (12) 23 (1988).
146. M. Ulrich, US Pat., 24. 906, in *Handbook of Pressure Sensitive Technology* (D. Satas, Ed.), Van Nostrand-Rheinhold Co., New York, 1982, p. 203.
147. EP 0213860 / p. 2.
148. *Coating*, (2) 43 (1987).
149. W. J. Whitsitt, *Tappi J.*, (12) 163 (1988).
150. *Tappi J.*, (5) 182 (1988).
151. L. Krutzel, *Adhesives Age*, (9) 21 (1987).
152. *Coating*, (7) 186 (1984).
153. *Adhesives Age*, (11) 27 (1983).
154. C. K. Otto, *apr.*, (16) 438 (1986).
155. J. P. Keally and R. E. Zenk (Minnesota Mining and Manuf. Co., USA), Canada Pat., 1224. 678/ 19.07.82 (USP. 399350).
156. *Adhäsion*, (2) 43 (1977).
157. R. P. Mudge (National Starch Chem. Co., USA), US Pat., 4.753.846/28.06.88.
158. *Tappi J.*, (9) 105 (1984).
159. *apr.*, (5) 228 (1987).
160. R. P. Mudge (National Starch Chem. Co., USA), EP 0225541/11.12.85.
161. J. A. Fries, *New Developments in PSA*, Tappi HM Seminar, 2-4 Juni, 1990 Toronto.
162. Minnesota Mining and Manuf. Co., USA, EP 4.693.935/18.09.87.
163. C. Donker, R. Luth and K. Van Rijn, 19th Munich Adhesive and Finishing Seminar, 1994, p. 87.
164. L. Jacob, New development of tackifiers for SBS block copolymers, 19th Munich Adhesive and Finishing Seminar, 1994, p. 87.
165. Novamelt Research GmbH, Technical bulletin.
166. Poli-Film Kunststofftechnik GmbH, Technical bulletin.
167. Novacel GmbH, FRG, Technical bulletin.
168. *Paper, Film and Foil Converter*, (3) 76 (1988).
169. Vinamul GmbH, Nacor 90, data sheet.
170. D. G. Pierson, J. J. Wilcynski, *Adhesives Age*, (8) 53 (1990).

7
Converting Properties of PSAs

Convertability is the ability of the adhesive to be used for coating and laminating, and the property of the laminate to be used for label (or tape) manufacturing. The most important factors which contribute towards good convertability of a self-adhesive laminate include the nature of the face paper, the type of the adhesive and silicone, the release force, the properties of the siliconized release liner, and the die-cutting properties of the adhesive and of the face paper. There are two different aspects of the convertability of an adhesive, namely the convertability of the fluid adhesive (solvent-based, water-based or hot-melt), and the convertability of the coated adhesive (processability); good convertability implies the suitability of the adhesive fluid to be coated without any problem, and the capability of the laminate to be processed (cutting, die-cutting, printing, labeling, etc.) without any problem.

1 CONVERTABILITY OF THE ADHESIVE

Because of the different materials used as face stock materials and the different technologies and adhesives used, the coatability of the liquid adhesive is a function of many different parameters (e.g., the physical state of the adhesive, the nature of the face stock, the coating technology, the laminate structure, the end-use properties, the cost, and the nature of the release).

1.1 Convertability of Adhesive as a Function of the Physical State

The convertability (coatability) of the adhesive is characterized by the parameters such as the wetout, the coating rheology, coating speed and versatility. All of these parameters depend on the physical state of the adhesive whether it concerns a diluted liquid system, yielding an adhesive layer through the evaporation of the carrier liquid (water or solvent) or a 100%-solid system, giving a solid adhesive layer by a physical (hot-melt) or chemical (UV or electron beam) process. Hence the coatability of diluted and undiluted systems needs to be discussed separately.

Coatability of Diluted Systems

Diluted adhesive systems are defined as solutions or dispersions of adhesives. There are aqueous or solvent-based adhesive solutions, and aqueous or solvent-based dispersions. The most important ones are organic solvent solutions of adhesives (solvent-based adhesives) and water-based dispersions of adhesives (water-based adhesives). In fact most of the solvent-based adhesives also contain (at least partially) dispersions (like rubber-resin adhesives), and most of the dispersion-based aqueous adhesives contain some dissolved materials (e.g., thickeners, tackifying polymers, etc.). In Chapter 2, the special features of the wetting out of solvent-based adhesives as well as their coating rheology were discussed. For the coating practice more information is given by the versatility of the PSAs.

Coatability of Solvent-Based Adhesives

Factors influencing the coatability of solvent-based systems include the influence of the solvent content on the coating thickeners, coating appearance, and the mechanical-hydraulic forces in the coating nip [1]. Solvent-based adhesives are mostly solutions of rubber-resin mixtures or acrylic copolymers. Their physical properties depend on both components, namely on polymer and solvent, and on the composition of the adhesive solution.

Solvent-based adhesives are generally highly viscous systems with low surface tension solvents as the carrier; they are (at least partially) directly coated onto high surface energy webs. The wetout of solvent-based PSAs is less difficult than with water-based PSAs. The coating rheology of solvent-based PSAs concerns the rheology of a true solution, thus is mainly viscosity driven and more simple than that of water-based systems. Coating versatility is the capability of an adhesive to be coated on different machines, using different coating technologies without changes in its composition or performance characteristics. Because of the high viscosity of the solutions, there are only a limited number of coating devices able to use solvent-based PSAs.

Coating speed is the most important coatability related property of solvent-based PSAs. The key issues include:

- different coating devices for different viscosities;
- base viscosity as a function of the polymerization recipe;
- thickening/diluting response of the adhesive;
- different technologies (transfer/direct) for different adhesives;
- different technologies for different end uses.

In general the most important parameters of the liquid solvent-based adhesive (solids content and viscosity) can be easily and precisely adjusted because the (mechanical) stability of the system is independent of viscosity (or solids content). On the other hand, there is no need for an exact adjustment of these properties because of the higher viscosity. Thus solvent-based PSAs are suitable to be used on different coating devices. The (possible) different nature of the used solvents only acts as a limiting factor. Therefore,

$$\text{Versatility: Solvent-based PSAs} > \text{Water-based PSAs} \tag{7.1}$$

The coating speed remains the most important converting property of solvent-based adhesives. In contrast to water-based PSAs, there is no limit (minimum speed) for the coating speed because of the unlimited wetout of these adhesives. The limits of the running speed for a given machine are dictated only by the explosion limits of the solvents. Theoretically, a high solids content and low boiling point solvent allow high drying speeds; thus an increase in solids content and the appropriate choice of adequate solvents ensure better running speeds. The newest experimental products generally contain a higher solids content. Theoretically, a 100%-solids content (at very low molecular weight) is possible; however, the most concentrated aqueous dispersions have 67–69% solids content (theoretically 75% is possible). Thus ways to achieve higher coating speeds appear to be quite different for solvent-based or water-based PSAs.

Coatability of Water-Based Systems

The coatability of water-based adhesives also is a function of the wetout, coating versatility, coating rheology, and coating speed. The wetout of water-based systems constitutes a difficult problem because of the high surface tension of the carrier (water), the low viscosity of the most important ready-to-use formulations, the low density, and the transfer technology used for coating. The use of cosolvents is limited, while higher viscosities need different coating devices. In contrast to solvent-based PSAs, an improvement of the wetout characteristics (with surface active agents) induces irreversible changes to the adhesive properties. Thus the coatability of water-based formulations is more sensitive than that of solvent-based PSAs.

For water-based PSAs, the coating rheology concerns a dispersion, not a solution, and is thus time dependent. No changes in carrier material are allowed and the dispersed polymer is of high molecular weight; coalescence yields a composite, anisotropic layer. Coating devices are very sensitive towards viscosity and thus their coating versatility is reduced; the coating speed depends on solids content as well as on dispersion characteristics. A minimum level of surface tension is required for dynamic wetting, the maximum speed for a machine is dictated by rest water content (humidity) of the dried layer. Rest humidity content is an equilibrium value which depends on the wetout additives; thus wetting out is the most important convertability parameter. The coating speed is more dependent on the coating weight for water-based PSAs than for solvent-based PSAs. In addition, the formulating freedom with coating speed remains very limited, unlike with solvent-based PSAs. The special features of the wetting out for aqueous dispersions were discussed in Chapter 2 concerning the rheology of water-based dispersions. Summarizing, the most important key issues concerning the wetout are:

- wetout as the combination of wetting/dewetting;
- dynamic versus static wetting;
- factors acting against dewetting (shear, coalescence);
- parameters influencing wetting/dewetting (surface tension, contact angle, viscosity, density);
- surface tension and parameters influencing it;
- contact angle, and the problems of measuring this parameter;
- the influence of the viscosity (versatility, wetout, coating weight).

The theoretical aspects of the coating rheology of water-based adhesives were discussed in Chapter 2.

There is a qualitative difference between the coating rheology of the liquid and that of the semisolid (dried) adhesive layer. The former is like solvent-based (solute) rheology, whereas the latter is more that of a reinforced matrix. For the first step of the coating, temporary and reversible shear sensitivity of the water-based PSAs determines their behavior on and after the metering roll. In the second stage, the diffusivity of the coated layer and its flow determine the drying speed and the aspect of the dried, coalesced adhesive film. Coating rheology of solvent-based adhesives is a function of polymer and solvent quality, whereas coating rheology of water-based adhesives is a function of the dispersion structure. On the other hand, the influence of the metering device, and of the drying "regime" is more pronounced than for of solvent-based PSAs. Slight shear rate changes may avoid "stripe" structures, air and heating adjustment may avoid adhesive blow-away, viscosity control may help adjust the smoothness of the final coating. Unlike solvent-based adhesives formulating only has a limited influence on the coating rheology of the adhesive. Key parameters include:

- adhesive build-up on the machine, mechanical stability, due to insufficient flow;
- coagulum build-up on the machine and during formulation (mechanical stability);
- coating uniformity (striped, textured structures, insufficient flow);
- stability of the viscosity (thixotropy and decrease of the viscosity, either reversible or irreversible);
- foam formation.

Coating versatility refers to the suitability of the water-based PSAs to be coated by different coating technologies. Today in Europe two coating methods for emulsion PSAs dominate their respective fields [2], namely: metering bar systems for tapes and protective films, and reverse gravure for labels. The coating technology will be discussed in more detail in Chapter 9. The mechanical stability of the water-based adhesive on the coating machine is characterized by the following parameters: coagulum build-up on the machine, grit build-up in the dispersion, stability of the viscosity, and foaming.

Several latex adhesives have a tendency for coagulum build-up on the metering roll of the reverse roll coater. Such coagulum may arise either from poor mechanical stability under shear loading or from imperfect doctoring leading to a thin coating on the roll, which dries and cannot be redispersed so that these particles eventually transfer to the applicator roll [3]. Dispersions with a high solids content and a high surface tension tend to build up dry, tacky residues. Frequently changing the rheology (diluting and thickening) leads to deposit-free water-based adhesives. Often the adhesive layer build-up on the machine is associated with unsufficient fluidity of the "almost dried" adhesive, giving a coating appearance that is textured with stripes. Diluting/thickening also helps to avoid this phenomenon.

Unlike adhesive build-up, grit (coagulum, discrete deposits) build-up is associated with insufficient mechanical stability of the dispersion (i.e., an insufficient stabilizing agent level in the formulation). Generally, tackified formulations are more sensitive; rosin-ester tackified PSAs are more sensitive than acid-resin tackified formulations (e.g., an acid-rosin with 60 ppm standard coagulum content does not influence negatively current paper coating formulations; rosin-ester dispersions must possess less than 20 ppm coagulum). Theoretically, high shear stirring tests show the grit build-up and the lack of mechanical stability; generally, most PSAs formulation withstand this test. Real data can be obtained with a two-cylinder laboratory device, or practically, a 3-day running test on the production line.

Most unthickened PSAs dispersions exhibit only slight thixotropy (i.e, changes in the viscosity after formulation that are generally irreversible). It is to be noted that 0.5–1.5 s (Ford Cup sec) are normal absolute values for observed changes in the viscosity during 8–24-hr production runs.

```
                        ┌─────────────┐
                        │    COSTS    │
                        └─────────────┘
                              ↗
                           ╱
┌───────────────┐       ╱
│ COATING SPEED │─────
└───────────────┘       ╲
                           ╲
                              ↘
                        ┌──────────────────┐
                        │  COATING WEIGHT  │
                        └──────────────────┘
```

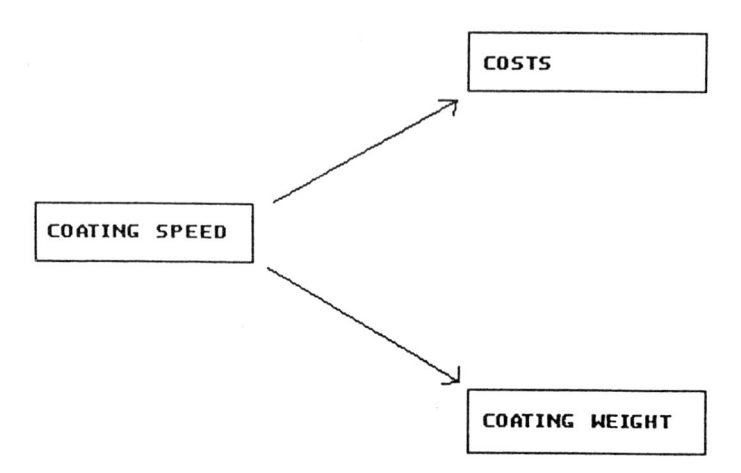

Figure 7.1   Parameters of the PSA laminate influenced by the coating speed.

Foaming does not really constitute a problem as most machines do not attain critical speeds. Most dispersions possess viscosities below 500 or above 5000 mPa·s. Also there are a variety of very efficient defoamers.

Water-based PSAs require longer drying times (and hence longer dryers) than solvent-based adhesives, and 15–150 m dryers are currently in use. The coating speed of the adhesive not only has economical importance, but ease of adjustment also influences the tolerances of the coating weight. For gravure roll coating a short contact time between web and pressure roll means a low coating weight, whereas a long contact time (low speed) leads to a high coating weight. With a given gravure size it is possible to change the coating weight within a range of approximately 10% [2].

Aqueous systems require more drying heat. The heat of evaporation of water amounts to 540 cal/g, while that of typical coating solvents, such as methyl ethyl ketone (MEK) and toluene, is 100–200 cal/g. This difference translates into longer ovens, lower line speeds, and higher utility costs [4] for water-based systems. Therefore the running speed has its economical importance, possibly leading to lower costs and higher productivity (Figure 7.1).

Originally several technical parameters were evaluated in order to improve the drying speed of the diluted systems, namely an increase of the solids content and the transfer coating technology. With transfer coating a high coating speed of the nonwater sensitive web was possible; an increase of the solids content allows (at least theoretically) a reduction of the volatiles and a higher drying speed. The speed limiting steps are the foaming, the appearance of the coated layer, and the residual water content of the dried adhesive.

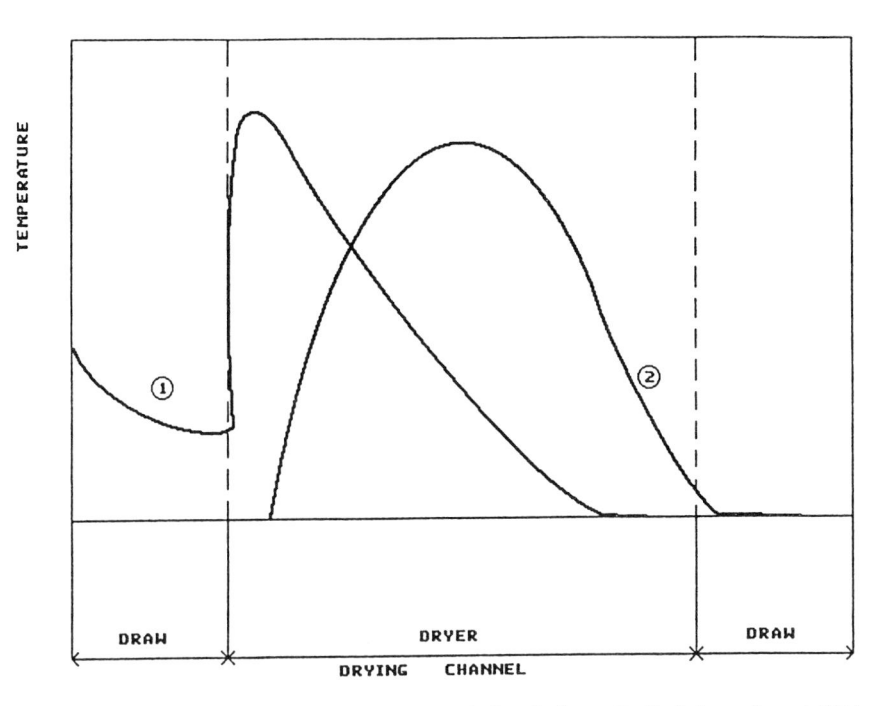

Figure 7.2 Temperature gradient required for drying of: 1) Solvent-based PSA coating; 2) water-based PSA coating.

It is evident that for a water-based PSAs exhibiting an adequate rheology the upper limit of the running speed is:

Running speed = Maximum drying speed for normal dispersions (7.2)

For high solid dispersions:

Running speed = Maximum speed leading to a smooth coated layer (7.3)

For medium viscosity dispersions:

Running speed = Maximum speed which limits foam formation (7.4)

The physical state of the PSA influences its drying properties [5]. Solvent-based and water-based PSAs display a quite different evaporation behavior (i.e., they require a quite different temperature gradient during drying) (Figure 7.2).

The coating and drying speed will depend on the proper adjustment of the drying (oven) temperature. Key aspects include the economical/technical importance of the running speed, the built-in characteristics of the water-based PSAs influencing the running speed, and the technical limits of the running speed.

Coatability of Hot-Melt PSAs

According to their physical state the coatability of hot-melt PSAs is limited. Their viscosity depends on the temperature and on their chemical composition/aging. Because of the restricted range of raw materials available for their formulation, there is a narrow temperature interval suitable for them to be coated. On the other hand, this interval is situated at relatively high viscosities. Thus special metering devices only (see Section 1.4 of Chapter 9) may be used; on the other hand, the high viscosity in the molten state allows a good wetout.

Coating Versatility. The coating versatility of hot-melt PSAs remains limited. Generally, special agents (e.g., waxes, oils) are used in the formulation (see Section 2.9 of Chapter 8) in order to extend the coatability range of the adhesive.

Coating Rheology. Hot-melt PSAs based on Kraton G are coated at a viscosity of 25,000–80,000 mPa·s (at 177°C) [6]. Acrylic hot-melt PSAs possess a viscosity range of 8000–30,000 mPa·s (350°F) [7]. As illustrated by these examples the viscosity of current hot-melt PSAs is of an order of magnitude higher than that of solvent-based PSAs, and even higher than water-based PSAs. It is not possible to change this with dilution agents or shear forces (machine).

Coating Speed. Because of the lack of carrier or solvent, hot-melt PSAs are ideal for the application of thick films onto webs at speeds of hundreds of ft/min [7] versus acrylic PSAs coated at 500 ft/min. Other data show that in comparison with water-based PSAs (200–250 m/min) hot-melt PSAs can be coated at higher speeds (300 m/min) [8] (see Chapter 9).

1.2 Convertability of Adhesive as a Function of Adhesive Properties

The adhesive properties influence the coatability of PSAs. A certain level of initial flow (tack and peel adhesion) is needed to achieve the anchorage of the adhesive onto the face stock or release liner. In some cases a primer coating is required. A pronounced influence of the tack and shear may be observed during the converting of the laminate.

1.3 Convertability of Adhesive as a Function of the Solid State Components of the Laminate

The coatability of the adhesive depends on the face stock and release liner. Wetting out, drying parameters, and running speed depend on the surface characteristics of the face stock and release liner. Machine parameters also are function of the bulk properties (mechanical characteristics) of the web and influence the coatability of the PSAs. Other bulk properties like porosity, chemical affinity, aging resistance, etc., also influence the coatability.

1.4 Convertability of Adhesive as a Function of Coating Technology

The convertability of the adhesive depends on the coating technology. For all types of PSAs and independent of their physical state, it is theoretically possible to use direct or transfer coating. Practically,, the selected coating technology (direct or transfer) supposes quite different rheological behavior. As an example, without thickening, a low viscosity conventional latex will not completely wet a silicone-coated release liner. In this case the application, convertability of the adhesive depends on the coating technology [3].

1.5 Convertability of Adhesive as a Function of End-Use Properties

Transfer coating is preferred for heat-sensitive or porous face stocks; transfer-coated PSAs have to be formulated for better wetout. Hence the convertability of the adhesive is a function of its end-use properties. On the other hand, a better anchorage (see Section 1.1 of Chapter 9) is obtained through direct coating. Certain adherents require a special adhesive smoothness or texture, which is a function of the metering device.

2 CONVERTING PROPERTIES OF THE LAMINATE

In the present context *laminating* means the production of a layered material from two or more webs using a bonding medium. In the case of PSA laminates, the face stock and release liner are connected together temporarily by means of a PSA. The PSA-coated material will be processed (converted) and used in the form of the laminate. Generally, the face stock will be printed either in-line or off-line, but after coating with a PSA; the liner also may be printed. A coated, laminated material usually starts as a continuous roll which is then cut in discrete pieces. Cut material (printed before or after cutting) will be used for labeling purposes. Labeling (separation of the liner from the adhesive and affixing the adhesive-coated label on the final substrate) can be done manually or mechanically, this process being influenced by the face material–adhesive-release properties. Hence the convertability of the laminate will be influenced by printability, cuttability, and labeling properties. All of these parameters depend on the solid state components of the pressure-sensitive laminate and on the adhesive. Therefore a detailed examination of these parameters requires a short description of the pressure-sensitive laminate, its components, and its construction.

2.1 Definition and Construction of the Pressure-Sensitive Laminate

Generally, pressure-sensitive laminates are sandwich structures where two (different or identical) solid state components (generally flexible sheets) are temporarily bonded together with the aid of a PSA. The pressure-sensitive composite is a temporary structure where one of the solid state components is acting as a temporary shield only, in order to avoid the deterioration of the PSA layer. This part, known as release liner, is in some cases (mainly for tapes) identical with the face stock material. Such a construction is possible and necessary for pressure-sensitive tapes used in continous reels. Pressure-sensitive labels are discontinuous materials where the release layer has to be built in as a separate solid state component. Therefore, it may be assumed that pressure-sensitive labels contain the following main components: a face stock material, a pressure-sensitive adhesive, and a release liner. All these components are relatively thin; paper and soft or hard plastic films are used as solid components. Soft plastic films can be bonded only by using PSAs where the use of rigid adhesives is not possible because of stress concentration.

Main Laminates

The components of the PSA laminate and its build-up depend on the end-use, on the application technology of the labels, and PSA-coated laminates.

Sheet Material. Pressure-sensitive labels are temporarily built-up composites. The end-use (application) of the label assumes peeling off the label from the release liner and the subsequent bonding of the free adhesive surface onto a substrate. This end-use called labeling is carried out by hand or automatically, with the aid of labeling guns. The main criteria concerning the feasibility of hand applied or automated labeling is the dimension of the label. It is evident that large size labels have to be considered as sheet-like materials. On the other hand, economical considerations also influence the labeling. Unicates or small series of labels have to be manufactured by technologies using sheet-like materials. Big series of relatively small and low cost labels have to be applied automatically with labeling guns. Printing, cutting, and debonding conditions for sheet and reel materials are different and therefore the quality requirements for both these label categories also differ. Sheet materials were used with priority some years ago; films as face stock material were tested for sheet-like labels first. Because of the relatively big surface and the low delaminating and labeling speed of these labels, the face stock quality and the cohesion of the adhesive play an important role for sheet-like pressure-sensitive labels.

Roll Material. Labels applied from a reel have to be tacky in order to ensure a suitable adhesion on the adherent after a very short dwell time; at the same time they have to be very slightly bonded to the release liner in order to display a low debonding (delaminating) resistance.

Laminate Build-Up. Theoretically, a PSA laminate consists of at least two solid state components (face stock and release liner) and a liquid phase (PSA) present as a continuous layer. The face stock or release liner itself may have a composite structure, the components of the laminate may show a discontinuous nature, and the laminate may possess a multilayer structure. Therefore the construction of the laminate exerts a strong influence on its properties.

Laminate Components

The pressure-sensitive layer (a liquid adhesive layer) is enclosed between two solid state sheet (forming) materials. Generally, there is only one liquid state component in the laminate, namely the pressure-sensitive adhesive layer between the face stock material and release liner. In some cases an additional primer coating is applied between the face stock material and the PSAs. Because of its reduced and limited flow properties, the primer coating may be considered (like the release coating) as a solid state component of the laminate.

The main solid state components of a pressure-sensitive label are the face stock material and the release liner. There are also sophisticated sandwich structures which possess a lot of face stock and release layers, where in some cases one or more face stock materials should also act as release liner. There are multilayer self-adhesive labels which comprise a carrier layer containing a silicone layer, an adhesive layer, a printed message, a carrier layer (e.g., polyester), a release layer (e.g., silicone), a carrier layer (e.g., polyethylene), and a printed message. These labels are useful when a printed message is to be attached to two different objects [9].

Face Stock Materials. The face stock material is the solid state component of the pressure-sensitive label with a permanent character (i.e., the adhesion of the PSAs on the face stock material—the anchorage—should be higher than its bonding to the release liner). Generally, the adhesion of the PSAs to the face stock material (i.e., anchorage) should be superior than its bonding to the substrate. The most important face stock materials are paper, cloth, plastic film, metallized film, metal foils, elastomeric foams, and nonwoven materials. Face materials include specially modified papers, as well as plastic films such as PVC, cellulose acetate, polyolefins, polyester, and polystyrene, and aluminum foil as well as metallized papers and films. Market requirements impose certain characteristics of the label material [10]. Labels for oil containers need to display durability, visual impact, and flexibility. Toiletries and cosmetics require a "no label" look, a high consumer image, and stain resistance; the electrical market needs include visibility, durability (e.g., vinyls, PET) and heat resistance. Toys require nontoxic labels with long life and visual impact. The chemical industry requires resistance to chemicals, visibility, and durability. Thus quality requirements for the label indicate the need for adequate face stock materials. Laminate manufacturers must make an adequate choice of the face stock material to be coated. In most cases this choice is limited, either imposed by the customers or, in some cases, determined by the manufacturers' own capabilities (i.e., coating machine or adhesive system).

The most important decision is the choice between a paper or film-based face stock material and in each of these domains between permanent or removable labels. Different mechanical performance characteristics are needed for tear-debonded (peeled) permanent and easily removable products. A different material stiffness is required for squeezable or "rigid" labels or for conformable face materials. For special uses preferred face stocks are those which permit transpiration of moisture; for other applications migration-resistant barrier layers are needed. On the other hand, storage or weather conditions may point towards quite different materials. Processing of the laminate (as sheet or roll material) and its printing determine the machining ability and surface properties. Economical considerations may interfere with technical requirements. The following parameters influence the choice of face stock materials: conformability, grain direction, appearance, printing method, mechanical properties, chemical and environmental resistance, and primer or barrier coating [11,12].

Edge lifting of labels can occur if a label stock is too stiff to conform to a curved surface. For foil laminates and whenever possible, the layout should be in the long-grain direction to provide maximum conformability. Many plastic containers are labeled with 60-lb high-gloss paper or foil-laminated papers selected for their attractive appearance. Special inks and printing techniques may be used because of the label end-use. Cosmetic bottles, for example, are commonly labeled with a PVC-coated foil laminate which resists alcohol [11]. The required tear and tensile strength of the face material of the

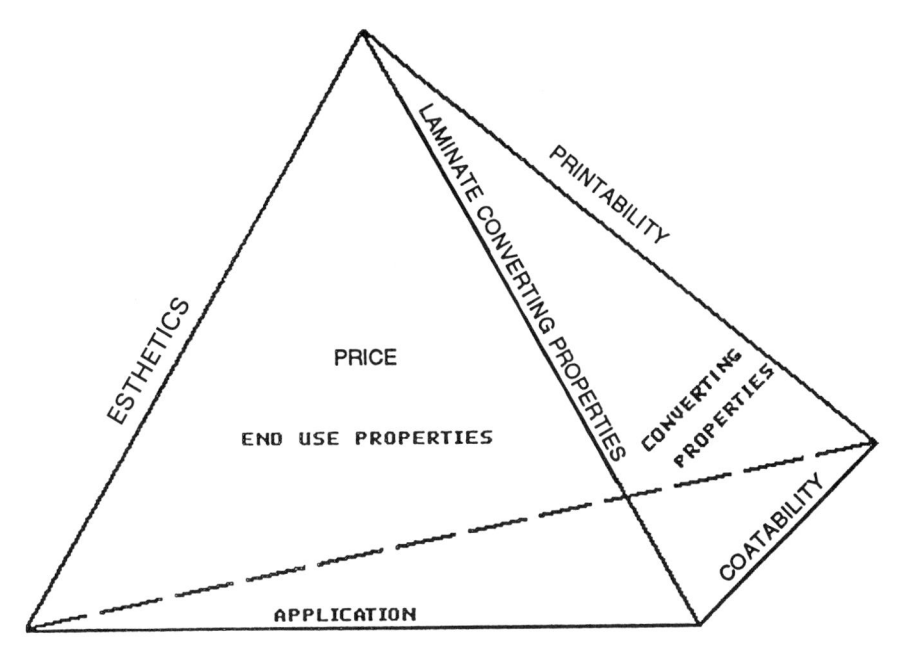

Figure 7.3 The main criteria for the choice of the face stock material.

label depend upon the label converting method and the method of applying the finished labels. Additional factors to be considered include UV light resistance and stability, cold temperature application and resistance, water immersion or high humidity resistance, and resistance to various chemicals. A primer coating which improves the anchorage of the adhesive to the label stock may also provide a barrier to prevent components in the adhesive from penetrating the label stock. Figure 7.3 summarizes the main criteria for the choice of the face stock material for PSAs laminates.

Through the choice of the face stock materials new trends may be observed [13]. Some years ago, price and productivity factors were predominant. Today environmental, esthetical, chemical resistance, flexibility, and recycling considerations appear more important. The most important face stock materials used are fibrous materials and films.

Fibrous materials are mostly paper-like products (paper and synthetic paper), textiles, and nonwoven materials. Films may be made of plastic, metal, or a combination thereof (metallized plastic) [14,15]. The most commonly employed film face stock materials in PSAs are polyester, polypropylene, polyethylene, polyvinyl chloride and, of course, reinforced paper [16]. Paper was the traditional face material for PSA labels for most of the almost fifty years of existence of the industry. This is changing with estimates for nonpaper materials running as high as 40% by 1990 [17]. There are several reasons for this trend; pulp prices have risen dramatically while petroleum prices have declined.

Changing packaging concepts such as the squeezable plastic bottles dictate the use of flexible face materials. Eye-catching labels compete more effectively. Durability of the label and retention of its appearance also is a factor, with the "last look" prior to discarding the package forming a mental image, which can influence future buying decisions [18]. Previously, the only opportunity for film-based primary labels was in situations where the package might be exposed to harsh environments or where superior graphics were necessary [19]. Actually the packaging industry was looking for striking graphic effects and chose relatively expensive plastic face stocks because of their smoother, more uniform printing surface [20].

Paper Used as Face Stock. Paper is the main face stock and release liner material actually used. Its good wettability, dimensional stability during coating, printability, and cuttability as well as its well-known technical history as an information carrier made it possible to use it from the beginning of the label industry; an ideal face stock paper possesses a high gloss, a high surface strength, a good wetout, and good anchorage for printing inks [15]. A high surface strength (tear) may be achieved with a high binder level. Gloss and wetting out are influenced by the pigments used. The most important mechanical properties of paper type as a function of its weight are listed in Table 7.1.

Noncoated papers, clay-coated papers, colored papers, paper-film laminates, and special papers also are used as face stock material [17]. The most important factors for base papers used for pressure-sensitive face materials are the required quality level, quality consistency, and efficiency [20]. Special paper grades can be selected which exhibit barrier properties (e.g., to prevent the paper surface being marred by bleedthrough of adhesive components). The pressure-sensitive label papers must be suitable for reel-to-reel conversion, reel-to-reel printing, electronic data processing (EDP) label printing, and for sheet printing. Demands on label paper from the convertor's point of view are printability, diecutting, lay-flat and matrix stripping properties. Suitable pressure-sensitive label papers must be printable in all commercial printing systems: letterpress, flexo, rotogravure, offset, silk screen printing.

Years ago, either simple, uncoated grades or cast-coated papers were used for pressure-sensitive labels. The properties of different paper qualities have been studied [21]. The sensitivity of paper towards humidity should also be taken into account (see Chapter 3) since paper is made up of both crystalline and amorphous cellulose, with the latter undergoing softening as the moisture

Table 7.1 Quality Criteria for Face Stock Paper

		Properties						
Weight (g/m^2)	Thickness (mm)	Tensile strength (N/cm)		Elongation at break (%)		Bekk smoothness (sec)		Bursting pressure (N/cm^2)
		MD	CD	MD	CD	Top	Bottom	
50	0.050	40	14	4	6	2000	200	15
60	0.050	40	25	4	6	1500	400	15
70	0.075	40	18	4	6	300	30	18
80	0.080	60	25	4	6	1500	500	25
80	0.085	45	20	4	6	200	200	18
110	0.150	100	50	20	20	—	—	—
125	0.180	100	50	20	20	—	—	—
150	0.150	80	60	7	18	—	—	—
330	0.430	250	100	4	6	50	10	—

content increases. The elastic modulus of dry paper is usually considered to be temperature dependent. The decrease in relative modulus with increasing moisture content is different for papers of different crystallinity [22]. There are very different paper qualities used as face stock, and there are many quality criteria recommended for selecting adequate face stock materials. In a first step, the paper weight is used as a simple quality index. Almost without exception, standard labels are printed in color on 85–100-g paper [23]. A high stiffness can be obtained by using a high weight or a more voluminous paper grade. The strong influence of the calendering should also be noted. Papers used as face stock material have a weight ranging from 30 to 250 g/m^2 and weigh generally about 80 g/m^2 [14,17].

Paper for data systems is primarily uncoated, engineered for strength, and has a high absorbency to capture ink from a computer terminal [24]. Coated papers constitute the larger market segment, ranging from cast-coated sheets to enamel and mat finish grades. These include thermal sensitive and other specially treated coated products. Specialty structures are another major market that includes fluorescent-coated papers, latex impregnated papers, polyethylene and paper laminations, and metallized laminates [24]. Uncoated and coated papers are used in impact printing: flexography, gravure, dot matrix, ultraviolet-cured letterpress, and offset. Both types of papers also are used in nonimpact printing, laser, inkjet, ion deposition, magnetography, and direct thermal and thermal transfer printing. In thermal transfer two contradictory properties are required: a high degree of smoothness and a surface that at the same time will absorb ink. For special uses there are different kinds of paper; humidity resistence, fat resistance, and a flame-proof character are characteristics of these papers. Laminated papers are manufactured in order to achieve chemical resistance. Metallized papers are used to get a better lay-flat; aluminium/paper laminates (7 μm aluminium, 50-g paper) also are used. For dispersion coating, paper heavier than 80 g/m^2 should be used in order to ensure enough dimensional stability [25]. On the other hand, the porosity of paper is more critical when direct coating. For labels for high speed printing applications, the absorbency and porosity of the paper should be improved [26]. The technical requirements for thermal papers were described by Park [27].

Paper types used include calendered Kraft for paper roll labels with the following advantages: good surface holdout, smoothness, dimensional stability, and an average weight of 65 g/m^2 [28]. The effect of supercalendering on paper properties is better gloss, smoothness, and ink holdout; supercalendering increases the stiffness of the paper [29]. Clay-coated paper is used for sheet labels with the following advantages: smoothness, dimensional stability, lay-flat during printing and converting, with an average weight of 85 g/m^2 [28]. The stiffness of machine-finished, machine-glossed papers is higher than that of colored clay-coated and calendered paper [30]. Polymer-coated papers for transfer coating of aqueous adhesives [28] are suitable for very low coating weights, possess a water proof layer under the silicone, and exhibit good solvent resistance, as well as good dimensional stability with a double- sided coating. Clay-coated papers make up more than 40% of the market for paper face stock materials [31]. The overlayer of coated papers may contain chemically very different components like caolin, casein, acrylics, polystyrene-butadiene latex, clay, starch, etc. [32].

Films Used as Face Stock. In addition to the traditional base materials within the range of high quality papers, films made of thermoplastic materials are becoming more and more important. The volume of PSA labels made of paper face stocks remains significantly larger than the volume of labels made with film face stocks. This is primarily due to the relative cost differences between the face stocks. Hence the benefits of film substrates for PSA labels are centered on appearance and performance, and the dominance of paper in the PSA label market is being challenged; this trend is more pronounced in Japan and Europe [33]. The advantages of face stock material other than paper include weather resistance, durability, resistance against packaged fluids, squeezability, esthetics, and transparency [34].

The following materials are used as film face stock: polyamide, polyamide/aluminium, polyvinyl chloride, oriented polypropylene, polyethylene, polyethylene terephtalate, and metallized plastics [12,35]. In Europe, 12% of the labels were film labels, on the basis of PVC, PET, polystyrene, polypropylene, and extended polyethylene [10]. Other films used as face stock materials include cellulose acetate, cellulose triacetate, cellulose acetobutyrate, and cellulose hydrate [36]. Nonpaper products have a market share of 12% by volume and 25% by value in Western Europe [37]. It was estimated that the cost of using synthetics is 4–5 times that of paper. It seems that synthetics are only used when the properties of strength, fine print acceptance, and moisture resistance are essential. For PSA labels with synthetic face materials, nature, caliper, gloss, opacity, and corona treatment are detailed in the specification. Generally, films must display good coatability and convertability in a laminate, using the same manufacturing conditions as paper-based laminates. The wetting out of the film materials is more difficult because of the smoothness, the lack of porosity, the low surface tension, and additive migration from the film. Coatability is influenced by the high deformability (elongation) and low temperature resistance of most films; convertability of the laminate is influenced by the plasticity/elasticity balance of the material.

The films most commonly used as face stock materials are PVC, polypropylene, and PET [38]. Packaging tapes are dominated by oriented polypropylene (OPP) and rigid PVC [39]; 150-μm PVC is also used for labels [40]. Calendered PVC and biaxially-oriented polyester face stocks have dominated film applications in primary labels over the last ten years. Table 7.2 lists the plastic materials used (alone or as laminate) as face stock materials for PSAs.

Generally, both soft PVC (20–50% plasticizer) and hard PVC are used as face stock materials. For tapes 25–40 μm PVC, polypropylene, and PET were used [38, 42]. PVC (polyvinyl chloride), or "vinyl" was the first major plastic to be used as face stock material. It was first used in the 1950s and remains a significant factor in the market place [33]. Its primary disadvantage results from the need to add plasticizers to achieve flexibility and stabilizers to prevent degradation. Plasticizers and stabilizers can migrate and affect the adhesive performance, so careful selection and testing are required. The problem of changes in peel value by plasticizer migration is a common phenomenon for removable PSA-coated soft PVC. The properties of hard PVC depend on the manufacturing procedure. Suspension-made PVC has superior electrical properties to emulsion-based PVC. Hard PVC possesses a good resistance against chemical oils and greases with good mechanical properties and inflammability resistance [43]. The drying temperature of soft PVC should not exceed a maximum of 45–50°C, for hard PVC 75°C. Printability depends on the nature of plasticizers used in the PVC recipe [43]. Generally, monomeric plasticized PVC is used as face stock for labels [44].

Table 7.2 Plastics Used as Face Stock Materials

Material	Primered Yes	Primered No	Commercial products Trade name	Commercial products Supplier/Converter	Alternative to	Ref.
PVC		x	—	—	—	[33]
PE	x		Compucal	Flexcon	—	[33]
PE	x		Flex	Flexcon	—	[33]
OPP	x		Kimdura	Kimberley-Clark	PVC	[33]
OPP		x	Oppalyte	Mobil	—	[33]
Polyolefin		x	Prymalin	3M	—	[33]
Polyolefin			Opticlear	Mactac	—	[33]
PE/PS	x		Polyart	BXL	—	[41]
Polyolefin		x	Maclitho	Mactac	Paper	[33]
PS		x	Opticite	Dow	Paper	[33]
PET	x		—	Decora		[33]
Cel-acetate	—	—	—	—	PVC	[33]

Lay-flat for paper requires the control of the humidity with a precision of 0.3–0.6%. Lay-flat of printed PVC film (coated with PSAs) depends on the thickness of the film, the thickness of the ink layer, the bonding forces between label and release, and the elasticity of the film and ink [45].

Film face stock materials may be coated directly or by transfer. Although new, high productivity machines use the transfer process, there remain some older special machines using the direct process. The surface tension of the web and surface tension of the water-based PSAs influence the wetout. In the most extreme case, when the surface tension of the dispersion differs greatly from the critical surface tension of the film, fish eyes and cratering will result (when the film surface tension is less than the dispersion's surface tension). Fortunately PVC displays an increased surface tension (40 dyn/cm) [46] as compared to the surface tension of the common water-based acrylic PSAs (28–38 dyn/cm). Soft and hard (plasticized) PVC exhibit different levels of critical surface tension. The critical surface tension of rigid PVC is 39 dyn/cm whereas flexible soft PVC has a critical surface tension with values of 33–38 dyn/cm [47]. The most important disadvantages of soft PVC films are the shrinkage and the flagging of the edges. Both phenomena are most apparent for extruded PVC film. Shrinkage concerns the change of the original dimensions during storage at room or higher temperatures due to chemical or physical (environmental) influences. Shrinkage is caused by the residual stresses

Converting Properties of PSAs 271

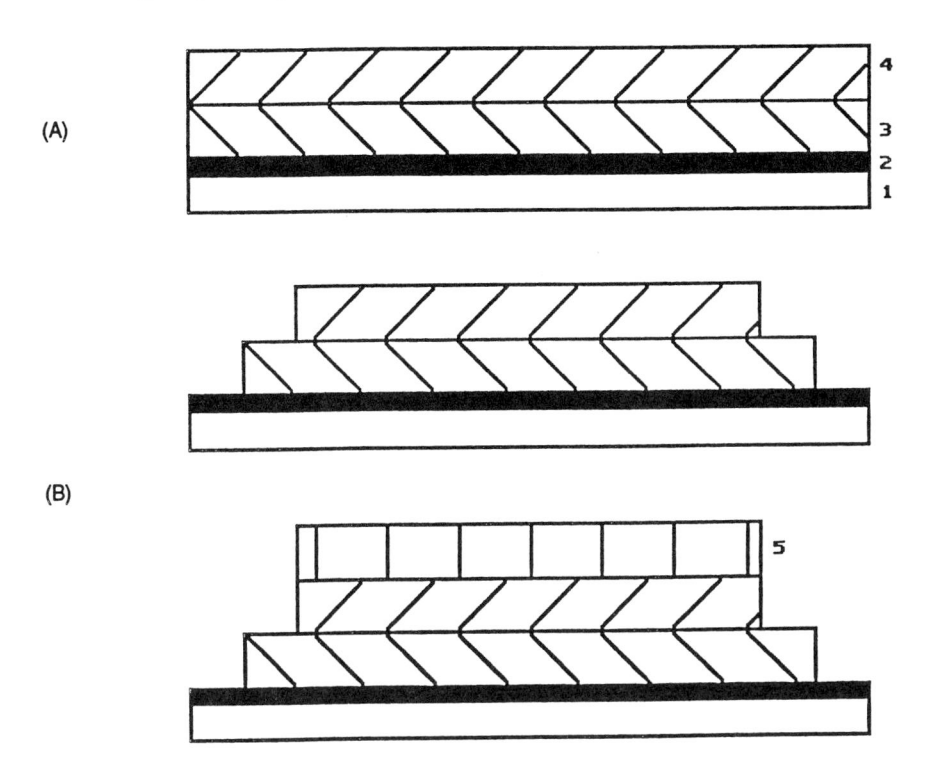

Figure 7.4 Schematical representation of the shrinkage of a PSA laminate: A) nonprinted material; B) printed material 1) release paper; 2) release layer; 3) adhesive layer; 4) face stock material; 5) printing ink.

in the material (i.e., by relaxation). In the manufacture of a PSA laminate and during PSA coating or laminating, the PVC film is stretched and strained; the strained film will be laminated with the dimensionally more stable paper-based release liners. During storage relaxation occurs and the strained material returns to its original length (i.e., it shrinks) (Figure 7.4).

The shrinkage of soft PVC does not depend on the tension of the film. It is caused by the laminate manufacture. Another part of the shrinkage is introduced during the manufacture of the film. This part is more pronounced for extruded films. Usually during the manufacturing of the base film, heat is applied to convert dry powder blends or granules into a molten state for calendering or extrusion into a plastic film [48]. Whilst still warm the unsupported film is often subject to tension which can cause stretching in the longitudional direction. This manifests itself as shrinkage (recovery of strain) when the product is reheated to the temperature at which the tension was

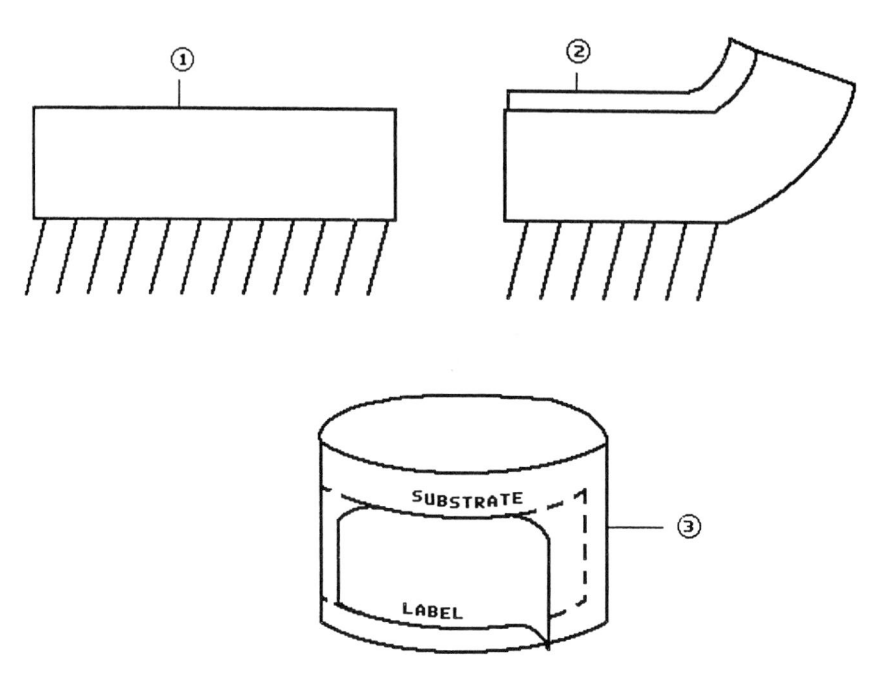

Figure 7.5 Flagging after printing of a PSA film. 1) nonprinted; 2) printed; 3) applied.

applied. The laminator must act to eliminate this hazard with film made under controlled conditions. Additionally, product shrinkage can be induced by residual solvents after printing, or through plasticizer migration out of the film into the adhesive or substrate.

Flagging or wing-up of the labels on curved surfaces due to insufficient adhesion of the PSAs to the substrate is different from flagging after coating (i.e., the loss of the lay-flat). This phenomenon is due to the chemical influence of the adhesive or printing inks [45]. Wing-up depends on the following factors: the adhesive/printing ink composition, the thickness of the film, the adhesive properties of the PSAs, the thickness of the adhesive/ink film, the differences in the modulus of coating/coated material, and the coating width (full or limited). Figure 7.5 illustrates flagging after printing of a PSA film.

Another disadvantage of PVC used as face stock material is the requirement for special coating machines to avoid tensile forces during coating and laminating [49]. Plasticizer migration from the PVC may not only influence the face stock shrinkage and thickness, but also the adhesive properties. Peel adhesion values change when removable PSAs are used on plasticized face stock [50]. An additional disadvantage of the use of PVC as face stock is its sensitivity towards adhesive rheology. This is a general phenomenon also characteristic for other film face stock materials. On the other hand, this phenomenon is induced by and related to the adhesive. Practically, the poor coating rheology typical of conventional latex adhesives manifests itself in several ways: first, there is a tendency in roll applications to form ridges parallel to the direction of coating. Ridges adversely effect adhesive performance by reducing the effective area of contact with the secondary substrate, but more importantly they substantially detract from the appearance of a lamination when a clear film is used (see Section 2.9 of Chapter 8). PVC has a temperature resistance of about 70°C, which is low compared to cellulose acetate (120°C). Processing of PVC needs special machines because of its high elongation. Peel values on plasticized PVC may vary [51]. Stabilizers of PVC (e.g., Pb, Sn) cause oxidation of the tackifier resin and thus loss of the tack. Thickness variations of vinyl labels often cause problems in the applications [52]. Generally, thickness variations of ±10% are observed during the manufacture of PVC film. Other data show that ±8% profile tolerances are common [53].

The environment and product resistant properties of film labels are significant advantages [54]. These advantages are obvious in high humidity environments and they also provide a significant marketing advantage. The polar surface of PVC provides good printability (i.e., PVC displays good processability and printability). PVC films used as face stock materials may be cured via electron beam [55]. They may be coated with silicones using electron beam-curing. The back side of the film may be coated via screen printing with a PSAs; thus a "monoweb" material may be manufactured.

In the range of plastic films PVC displays the best balance between coating, converting, and end-use properties (e.g., good wettability, good anchorage, high drying temperature resistance, good printability, good weatherability, and nonflammability). Its replacement with polyolefins is due mainly to environmental considerations. There are some differences between the properties of clear and opaque PVC:

- Soft, clear PVC contains more plasticizer; consequently, shrinkage values are higher for clear PVC than for opaque film.
- Soft, clear PVC is more flexible than opaque film, thus bonding wet adhesion (BWA) values are better.
- Soft, clear PVC is more "plastic" than opaque film, hence cutting properties are poorer than those for opaque film.
- Soft, clear PVC has lower surface tension values, therefore wetout is better on opaque film.

Table 7.3 Performance Characteristics of Different Plastic Films Used as Face Stock Materials

Criteria	Performance characteristics of plastic films				
	LDPE	HDPE	OPP	Hard PVC	PET
Stiffness	Low	Medium	Good	Very good	Very good
Die cuttability	Very low	Medium	Good	Very good	Very good
Thermal resistance	Very low	Low	Good	Good	Very good
Ductility	High	Medium	Low	Low	Low
Tear resistance	High	Medium	Very low	Low	Very low
Coatability	Low	Low	Low	Very good	Medium

The range of polyolefins used as face stock materials is shown in Table 7.3, with each kind of plastic film displaying its special advantages; in parallel to PVC other PSA film materials were developed [56].

The most dynamic segment of other face stock materials is that of the polyolefins, mainly polyethylene and polypropylene. The properties of face stock materials used for tapes were summarized by Meinel [57]. A detailed description of the polyolefin films used as packaging and face stock material is given by Placzek [58–60] and their printability is discussed by Verseau [36]. Polyethylene face stocks offer excellent chemical, solvent, moisture, and shrinkage resistance. Because of its flexibility, polyethylene was used successfully as squeezable label material on plastic packages [19]; there are different kinds of polyethylene, the most important materials being LDPE and HDPE. LDPE is clear and HDPE has a slight white color (opacity) [61]. HDPE is generally used for labels dispensed with labeling guns; it is used for plastic bottle labeling (oils, chemicals, cosmetics). LDPE possesses a good reversible deformability and is adequate for squeeze bottles. Polyethylene is sensitive towards solutions of surface active agents and environmental stress cracking [43]. It is temperature sensitive, therefore the drying temperature should be less than 50°C during printing. Polyethylene is used as face stock material alone or as a composite (coextrudate or colaminate). Polyethylene-coated paper (cardboard) or nonwoven coated paper are used for special labels. The use of polyethylene as release liner will be discussed later (see Section 1.5 of Chapter 9). It should be mentioned that top-coated polyethylene may be used also.

Polypropylene films offer high clarity and good resistance properties. Polypropylene labels offer potential as lower cost alternatives to polyester labels in numerous applications [19]. Polypropylene is stiffer, cleaner, and has better temperature resistance than polyethylene. Special polypropylenes may be subjected to radiation sterilization. The drying temperature when printing on polypropylene may attain 60°C [43], oriented polypropylene may withstand 110°C. Oriented polypropylene face stock materials are manufactured with a resulting thickness range of 19–60 µm [62]. Generally, OPP face stock materials for labels display the following benefits [63]: strength and durability, versatility, choice of finishes, printability, cost effectiveness, compatibility of polypropylene label and container, environmental acceptability, a noncrosslinked character and excellent die cutting properties. There are matte and glossy, transparent, white, and pearlized, corona-treated (on both sides) films, with a gauge of 55–80 µm [63].

A range of oriented polypropylene (OPP) labeling films for in-mould and self-adhesive applications has been introduced. These films offer a choice of finishes. Printable with all usual methods, they are expected to be suitable for end-uses ranging from toiletries to food, car care, and gardening products [64]. Mobil is manufacturing clear and opaque OPP films as face stock material (Labe-Lyte®), with a gauge of 19–60 µm, used in pressure-sensitive label applications. They are said to be adequate for label converters using different printing, cutting techniques, to ensure sensitive roll stock laminates of superior resistance to tearing and moisture, and breakage reduction during stripping [65]. There are different polypropylene film qualities used as face stock material, each with its advantages. A comparison of the main properties of polyethylene and cast and oriented polypropylene used as face stock material is given in Table 7.4.

Table 7.4 CPP/OPP/PE, Used as Face Stock Materials: Comparison of the Main Properties

	Face stock material		
Properties	CPP	OPP	PE
Thickness	+ +	– –	+ +
Profile	– +	+ +	– –
Transparency	+ –	+ +	– –
Shrinkage	+ –	+ +	– –
Storage dimensional stability	+ –	+ +	– –
Strain resistance	+ –	+ +	– –
Price	+ +	– –	+ +

Table 7.5 The Choice of the Film Face Stock Material as a Function of the End-Use of the Laminate

End-use requirements	Performance characteristics, film material			
	PE	PS	PP	Soft PVC
Chemical resistance	Good	Medium	Good	Low
Moisture resistance	Very good	Very good	Very good	Medium
Shrinkage resistance	Good	Very good	Medium-low	Low
Flexibility	Good	Low	Medium-low	Very good
Clarity	Low	Very good	Very good	Low

Polypropylene films manufactured via the chill roll procedure show less thickness variations than blown film (Table 7.4). A surface treatment on blown polypropylene films lasts longer than on cast polypropylene films [36]. Taking into account its mechanical and thermal characteristics (stiffness and temperature resistance) as well as the optical ones, polypropylene seems more appropriate as replacement for PVC than polyethylene. A detailed discussion of PVC versus polyolefins as face stock material was given by Hammerschmidt [66]. For large surface labels cast PVC should be selected; a lower price alternative is calendered PVC. Polyester displays better temperature resistance, cellulose acetate better clarity. In any case, the choice of an adequate material also is influenced by its stiffness (E modulus) (Table 7.5). In the same range of materials, the required thickness may influence the formulation. As an example, for a thickness of up to 60 μm hard PVC is used; above 80–120 μm soft PVC is suggested.

Low Surface Energy Face Stock Materials. Materials with low surface energy such as polyethylene and polypropylene are difficult to be coated. Wetting out on these materials is problematic and they exhibit a low bond strength (anchorage) unless the surface is pretreated to improve wettability and adhesion. The surface tension of water is 72 dyn/cm, while a typical solvent has a surface tension of 25–30 dyn/cm. With many substrates having surface free energies in the range of 30–40 dyn/cm wetting proves to be problematic. In the case of water-based PSA technology, the surface energy of the web must be raised 7–10 units above the dyne level of the coating [67]. Polyethylene for tapes and labels has to be treated [68]. Numerous pretreatment methods in order to improve the wettability and adhesion on the surface of nonpolar face stock materials are known [69]. Generally, they improve the polarity of the material through physical or chemical effects [70].

The most common classical treatment methods are chemical treatment [71], flame (thermal) treatment [72,73], and corona treatment [74,75]. Chemicals or flames produce a dehydrogenation/oxidation of the surface. No universal theory exists for corona treatment, but the number of CO groups increases as a function of the used electrical energy (power) [76]; for example, about 2–3 $\times\ 10^4$ carboxyl groups per cm^2 are formed for an energy level of 500 J/m^2.

Corona treatment works by setting up an electrical field around the film to be treated. This field ionizes the air molecules directly in its vicinity. These ionized air molecules, which are usually in the form of ozone and/or nitrous oxides, actually oxidize the surface of the film. The untreated surface is composed entirely of carbon/hydrogen repeating units. After the corona treatment, the carbon/hydrogen units are oxidized to include hydroxyls, carboxylic acid, and occasionally amine, or nitrous groups. The chemical change of the surface accounts for the increase of the critical surface tension. The effect of this oxidation only penetrates 5–10 Å beneath the surface of the material; this may not yield a permanent change of the film surface tension as plastic films are in a constant state of change and untreated polymers can in time migrate to the surface and cover the treated polymers. Generally, different frequencies between 10–80 kHz are used; higher frequencies yield a better treatment. The final surface tension depends on the material treated. Polypropylene is more difficult to be treated than polyethylene.

A disadvantage of the corona treatment is its short duration. The surface tension of a corona-treated blown film decreases from 56 to 36 dyn/cm; this effect is due to the migration of slip additives [77].

Untreated polyethylene and other polyolefins are difficult face materials for acrylic adhesives, and a large amount of effort has gone into the search for a technique which will easily and reliably increase anchorage without a significant change in the bulk properties. The most extensively studied techniques for enhancing adhesive bonding are treatment with helium gas plasma, oxygen gas plasma, or chromic acid. These and other surface modification procedures suffer from the same common shortcoming as corona treatment: the short life time of the treatment [78]. Corona treatment is possible for different face stock materials, including polymers [79–84], aluminum [85], or other similar materials [86,87]. During corona treatment (in a 12–20 kV, 20 kHz electrical field) chain scission and oxidation occur. Energy rich electrons and ions break carbon–carbon (CC) and carbon–hydrogen (CH) levels (3.7 and 4.3 eV, respectively). The distance of the electrodes from the web influences the efficiency of the processes; the best results are obtained with a gap of about 1 mm [88]. Although corona treatment is a complicated subject, one can conclude from current data that corona treatment generates polar sites and then increases the polar surface component of the free energy [89]. The possible functional groups that may be formed are shown in Table 7.6 [90].

Table 7.6 The Influence of the Corona Treatment on the Polyolefinic Surface. Possible Polar Functional Groups Formed [91, 92]

Oxygen	Nitrogen	Halogen
R–OH	R–N–H	Cl–R–C–H
R–C(=O)OH		Cl–R–C–Cl
R₂C=O	R–C(=O)–NH–CH	
R–C=C– with R–O, O–R	R–C(NO$_2$)–	Cl–R–CH$_2$–C–H

The surface tension of polypropylene amounts to 28 dyn/cm; corona treatment raises the surface tension up to 50 dyn/cm [88]. For printing and adhesion the surface tension should be 40 dyn/cm [93]. HDPE is more difficult to treat than LDPE. After a day the concentration of oxygen in the top (corona-treated) layer decreases by 13% [88]; after a year this decrease is 50%. Metallized plastic films lose their treatment more rapidly. Generally, the loss of the treatment effect (i.e., the decrease of the surface tension as a function of the time) depends on the material nature (formulation) and environmental conditions. Table 7.7 shows the decrease of the surface tension of corona treated polyolefin films as a function of the time. Due to the slight degree of crosslinking in the modified surface, the mobility of the modifying chemical groups is high in the case of polymer materials, and therefore a decrease of the pretreatment effect occurs during storage.

Table 7.7 Shelflife of the Corona Treatment. Changes in the Surface Tension During Storage of a Blown LDPE Film

Storage time (weeks)	Surface tension (dyn/cm)
0	49
1	47
2	46
3	46
4	46
5	50
6	49
7	49
8	48
9	49
10	48
11	50
12	50

More stable surface energies can be obtained with semiorganic or organic layers, only a few nanometers thick, which may be deposited by plasma polymerization. Compared to conventional "wet" processes, the plasma process offers new facilities for adhesion pretreatment [94], namely fats, oils, and waxes can be removed without residues; polymer surfaces can be roughened gently and without the use of agressive chemicals baths; the wettability of nonpolar plastics can be improved by gently oxidizing the surface; undefined plastics can be coated with a thin primer film. The films are treated for several seconds or minutes in a "cold" plasma generated in a common device, applying a high voltage to the process gas. Special advantages of the plasma pretreatment are: easy control and automation, difficult geometries can be treated, and nontoxic gases generally are used in small quantities [94]. Adhesive anchorage may be improved by grafting of the polyolefin film. Monomers suitable for graft polymerization onto the polyolefin film to promote the anchorage of normally tacky acrylic PSAs to the film include acrylic acid, methacrylic acid (and esters thereof), acrylamide, methacrylamide, sterically nonhindered tertiary alkyl acrylates and methacryl amides, secondary alkyl acrylamides and methacryl amides having three or less carbon atoms in the alkyl group, and N-vinyl pyrrolidone. Crosslinking agents may be added to

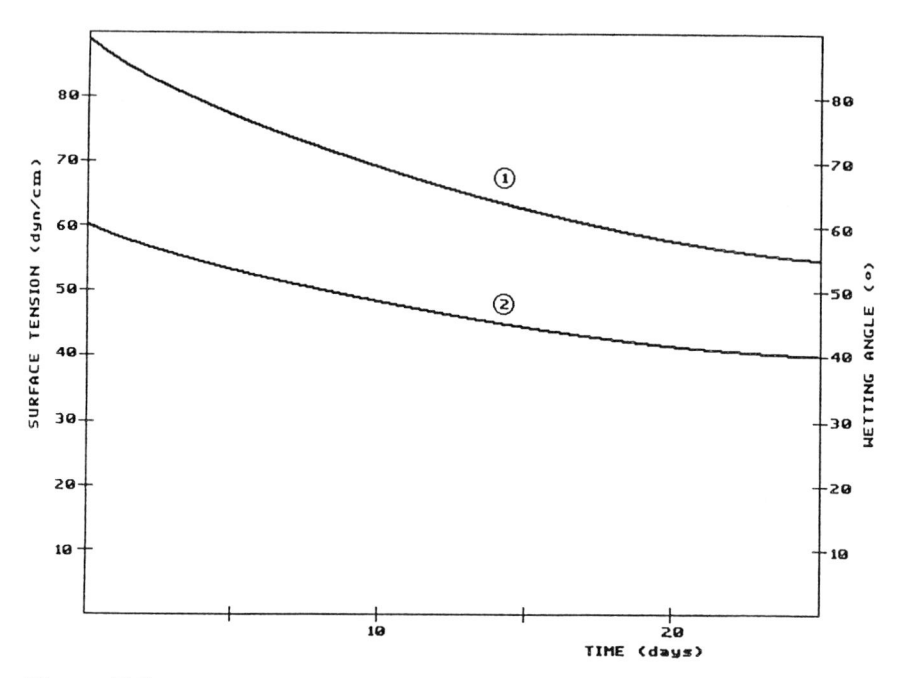

Figure 7.6 The change of the wetting angle 1) and surface energy 2) after plasma treatment.

enhance the resistance of the product [78]. For plasma treatment a high frequency electrical field, at 0.2–2 mbar is applied [95,96]. Corona treatment is oxidation, without monomers, and without crosslinking reactions; totally differently the plasma treatment occurs as a sum of oxidation, reduction, functionalizing, crosslinking, and deposition, with the aid of monomers. Figure 7.6 illustrates the change of the surface energy and wetting angle after plasma treatment. Figure 7.7 illustrates the change of the wetting angle after plasma and corona treatment in comparison to chemical treatment [97].

Plasma treatment yields homogeneous and stable surface tension values superior to flame treatment [70]. Another possibility is sulfonation; this procedure makes it possible to increase the surface tension from 21–23 dyn/cm to 41 dyn/cm. Technical data about low pressure plasma technology are given by Liebel [98]. Irradiation of polyolefin substrates, such as with an electron beam, to improve the adhesion of various coating also has been used [78, 99–104].

The flame treatment (flame process of Dr Kreidl) mainly is used on hollow packaging items [105,106]. Gas flame treatment is known as the Kritshever

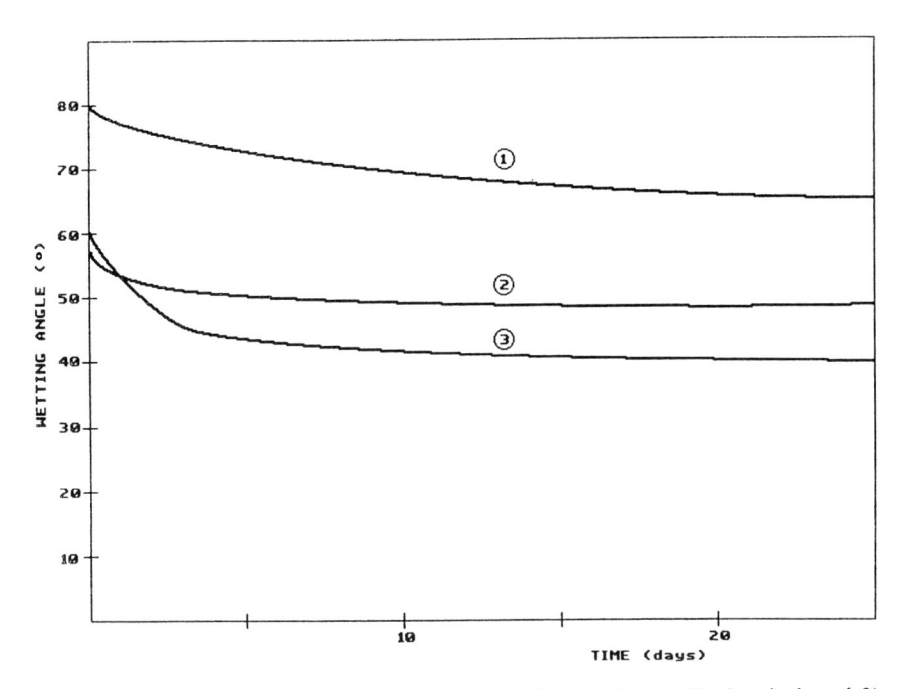

Figure 7.7 The change of the wetting angle after: 1) plasma, 2) chemical, and 3) corona treatment.

process [107]. Polyethylene for tapes also may be chemically pretreated [68, 108]. Recipes for chemical treatment of polyethylene films used as face stock material have been described [68,109]. Milker and Koch [110] give a detailed description of the chemical treatment of face stock materials. The Lohmann/ Ahlbrand procedure works with a fluor/inert gas mixture containing a 5–10% fluor volume. For polyolefins, the surface tensions values have attained 50 dyn/cm (Table 7.8) using about 2.5 kg fluor/100,000 m^2 web [110].

The improvements in storage stability (shelf life) through the treatment with fluor is better than for other procedures, generally lasting longer than eight months. As an example, the immediate value of the surface tension for polypropylene is higher than 54 dyn/cm; after 2 months it is 50 and it is 45–50 after eight months [110].

Polyolefins contain slip agents, antistatic agents, low molecular weight oligomers, and waxes. Erucaamide or slip agent is commonly used in polyethylene films to reduce the film's coefficient of friction. A slip agent is an effective modifier of the polyethylene film because of its low solubility in the polyethylene matrix. This low solubility forces the slip agent to migrate

Table 7.8 Surface Treatment with Fluor: Comparison of the Surface Tension Values Versus Corona-Treated and Untreated Polyolefin Films

Face stock material	Surface tension, dyn/cm		
	Without treatment	Corona-treated	Fluor treated [110]
LDPE	30–32	35–44	> 54
HDPE	32	32–36	> 54
BOPP	32	35–38	> 54

Table 7.9 Comparison of the Versatility of PVC and PP as Face Stock Materials

	Face stock material, application field				
	Tapes		Labels		
Criteria of evaluation	PVC	PP	PVC	PP	Ref.
Machining costs	—	Higher	—	—	[112]
Primer for printing	—	Yes	—	Yes/no	[39, 66, 112]
Release on the printed surface	—	Yes	No	No	[66, 112]
Low noise	—	Yes	No	No	[39, 66]

Table 7.10 The Main Performance Characteristics of PP and PE as Label Face Stock Material

	Performance characteristics			
	PE		PP (nonoriented)	
Requirements	LDPE	HDPE	Homopolymer	Copolymer
Stiffness	Low-medium	Medium-high	High	Medium-low
Flexibility	High	Medium-low	Low	Medium
Dimensional stability	Medium	High	Low	Low
Cuttability	Medium	High	Low	Low
Printability	Low	Very low	Medium	Medium
Temperature resistance	Low	Medium	High	Low-medium
Clarity	Low	Very low	High	Medium
Weatherability	Low	Medium	High	Medium

to the film surface, causing the usually tacky film surface to become slippery and easily handled by man and machine [111]. The migration of slip agents does not allow the anchorage of the adhesive, thus polyethylene-based face stock material should contain little or no amount of slip agent.

Table 7.9 shows a comparison of the versatility of PVC and polypropylene as face stock material for labels and tapes. The polarity and stiffness of PVC allows its use in both application fields. The opportunities of polypropylene as face stock material are superior for tapes. Table 7.10 displays the main characteristics of polyethylene and polypropylene used as face stock material.

There are opportunities for other plastics in the development of security, computer, battery, and other specialty labels. Several different film materials like cellulose derivatives, polyester, polystyrene, etc., were evaluated.

The following cellulose derivatives are used as face stock material: cellulose acetate, cellulose triacetate, cellulose acetobutyrate, and cellulose hydrate. Cellulose acetate possesses good dimensional stability, printability, and oil resistance, and it is waterproof, but sensitive towards solvents. Because of its tendency to be electrostatically loaded it remains difficult to process [36,43]. Based on a 50-µm cellulose acetate film which was modified to produce a brittle film with a low tear strength, the film can be used as an overlaminate seal on labels which are actually impossible to remove without cleavage [113]. Tamper-proof, fragile films guarantee that the printed label will simply be destroyed if it is attempted to remove it. Cellulose hydrate (Zellglas) is generally plasticized and overcoated in order to improve its printability or chemical resistance [43]. As plasticizers cyclohexyl phtalate, as laquers nitrocellulose, maleinate resins, sulphonamide resins, or dammar are used. Waxes for reducing the water sensitivity also are used; these may make the printing difficult. MS-quality (moisture proof and sealable) films need a higher drying temperature than normal Zellglas (P).

Polystyrene and rubber-modified polystyrene films are considered semi-squeezable because of their stiffness [19]. These films are mostly used as front and back labels on plastic containers. Polystyrene does not show enough stability against fats and oils but possesses a good resistance against water, alkaline and acid materials. Polystyrene is sensitive towards high temperature and environmental stress cracking [43]. Polystyrene films have to be corona charged to a level of 55 dyn/cm, which is maintained for a year; good dimensional stability, lower thickness, and good dispensing characteristics are obtained.

The properties of polyester (PET) films have been described [114,115]. Polyester exhibits very good mechanical properties and chemical resistance, but is difficult to print [43]. PET labels are used on polycarbonate (PC) bottles. Squeezable bottles need squeezable labels. Hot-stamp PVC- or PET-based films are manufactured (36–60 µm); metallized PVC-, PET-, and OPP-based films also are used for UV protection purposes. The coated and uncoated PET are used as face stock material [116]. For untreated polyester the adhesive should have improved humidity resistance, good film clarity before and after humid aging, no adhesive transfer failure mode, and a good cohesive strength [117].

Polyamide, with a good abrasion resistance, resistance against scratching, mechanical resistance, good resistance against oils, very good printability, also may be used as face stock material [118]. Polycarbonate may be used for touch sensitive electrical connections as a 250-µm PSA-coated film.

Polymer films used as face stock material show different water sensitivity. Polyethylene, polypropylene, and PVC possess a water absorbtion of about 0.1% after immersion for 24 hr; PET absorbs more (0.3–0.5%) [119]. Laminates of white, opaque polypropylene films with PET are suitable for soft drink labels. The film has a very low coefficient of friction of film to metal, which allows trouble free high speed application; the film shrinks 4.5% in the machine direction [120].

Comparison Between Paper and Plastics. The nature of the face stock will be determined by the intended end-use. Typically it will be made of paper, polymeric films, foils, or similar materials. Usually, it will be a face material (mass per unit area) of at least 50 g/m^2, although for paper it will typically be at least 70 g/m^2 as lower weight materials will make it difficult to provide the necessary strength to be peeled off without tearing. Face materials may include other flexible and rigid (solid) natural and synthetic materials and their combinations, such as plastics, elastomers, solid metals and foils, ceramics, wood, cardboard, and leather; essentially, they may be of any form including film, solid articles, woven and nonwoven textile materials, etc. Representative porous materials include paper and nonwoven fabrics, which may be composed of cured pulp, rayon, polyester, acetate fibers, cotton fibers, and blends thereof. The thickness of the film materials is determined by the end-use of the adhesive-coated product, for polyolefins in the range of about 0.025–5.0 mm. The most used film sheet materials, PVC and polyolefins, both contain other micro- or macromolecular additives. The polyolefin sheet material may contain additives such as carbon black, calcium carbonate, silica, titanium dioxide, crosslinking agents, dispersants, and extrusion aids; PVC contains plasticizers and stabilizers. Both paper and plastic-based face stock material may have a multilayer structure like top-coated paper, co-extruded or laminated films. The most important difference between paper and plastic film face stock materials lies in their porosity, polarity, and mechanical and chemical/thermal resistance.

Paper is porous; its porosity allows the penetration of the adhesive and improves its anchorage or the anchorage of top-coated layers. Therefore it is

Table 7.11 The Influence of the Humidity on the Modulus of Paper at 65°C

Moisture content (%)	Specific elastic modulus (kNm/g)
4	9.1
6	8.2
8	6.5
10	5.5
12	4.5
14	4.0
16	3.3

easier to manufacture removable labels with paper as a face stock material. On the other hand, the penetration of the adhesive may cause bleedthrough of the adhesive (staining) or of the water (from water-based adhesives) changing the dimensional and mechanical characteristics of the paper. The porosity of the paper and its polarity due to its cellulosic nature allow a good anchorage of adhesives of quite different chemical nature. The most important mechanical characteristics of the paper are its stiffness and its low creep compared to plastic films. The modulus of elasticity for paper is about 2929–3750 mPa·s, its creep rate is 4.42–1.07 Pa^{-1} [121]. On the other hand, the modulus of the main polyolefins is about 200–1400 mPa·s [122], which is much lower. The modulus of paper is strongly influenced by its water content. Through full saturation paper losses 95% of its mechanical stability (Table 7.11).

On the other hand, the modulus of plastic films may be improved through the choice of the chemical basis. The toughness properties of polymers increase with increasing molecular weight. The modulus of plastics also may be improved using fibers. Another possibility is to use oriented plastic films, where mechanical characteristics are superior, but depend on the degree of orientation and direction (Table 7.12).

Table 7.12 The Influence of the Orientation of Plastic Films on Their Modulus

Material	E-Modulus (N/mm^2)			
	PVC	PET	MOPP	BOPP
Value	4000	4500	200	2000

Another way to improve the stiffness is to build a sandwich structure of plastic films as coextruded or adhesive bonded laminates. Top coating improves the stiffness also. Quite different film qualities are manufactured via blown extrusion and casting. The chill roll process is generally superior, with better transparency, dimensional tolerances, and dimensional stability, but remains more expensive [66]. The low creep resistance of plastics (cold flow) caues high and partially uncontrolled elongation, and deformation of the web during film extrusion or coating; therefore curling and shrinkage may occur. These defects may be at least partially avoided using laminates or films with a special profile (i.e., a higher thickness at the "edge" of the film) [123]. The low stiffness of the films makes their high speed application (with labeling guns) difficult.

Advances in label dispensing have also influenced materials used for high speed applications. Conventional dispensers for paper-based PSAs do not require special application systems because of the inherent stiffness of the paper. With today's conformable and squeezable PSA film products sophisticated application systems are equipped with conventional dispensing heads [19]. As seen in Table 7.13, the stiffness, the dimensional stability and the die-cutting properties of paper as face stock material are generally superior to those of different films [124].

On the other hand, plastic films bring a new level of transparency and humidity resistance. Generally, the choice of an adequate plastic material depends on the end-use of the label. Cast PVC is used for large labels, calendered PVC for cheaper ones [66]; PET is used for heat resistant labels, cellulose acetate for clear ones. HDPE is recommended for dimensionally stable items, pearlescent polypropylene for toiletries, extended polyethylene in sleeving applications, etc. With regard to their stiffness the different face stock materials may be listed in the following ranking [125]:

Table 7.13 Stiffness, Dimensional Stability and Die-Cutting Properties: Common Films Versus Paper

Material	Stiffness	Dimensional stability	Die-cutting properties
LDPE	Low	Fair	Low
HDPE	Fair	Good	Fair
OPP	Good	Good	Good
H-PVC	Very good	Fair	Good
PET	Very good	Very good	Very good
Paper	Very good	Fair	Very good

Converting Properties of PSAs

Table 7.14 Application and Mechanism of the Primer Coating

Primer effect as a function of the application method of the PSA	
Direct coating	Transfer coating
1. Improves the anchorage of PSA: — no smoothening effect required — increase the chemical affinity of the contact surface	Improves the anchorage of PSA: — smoothens the paper face stock surface — increases the chemical affinity of the contact surface
2. Strengthens the face stock: — reinforcement of the fiber structure — strengthening of the surface layer	Strengthens the face stock: — reinforcement of the fiber structure — strengthening of the surface layer
3. Absorbs stresses	Absorbs stresses
4. Stiffens the face stock material; changes the peel and cuttability	Stiffens the face stock material; changes the peel and cuttability
5. Limits adhesive bleedthrough	Less important
6. Influences drying	No importance

$$\text{Stiffness: OPP} > \text{PVC} > \text{Paper} \geq \text{Polystyrene} \geq \text{Polyethylene} \quad (7.5)$$

As far as their cost effectiveness is concerned, there is a quite different ranking (which may differ as future trends in plastic prices vary):

$$\text{Cost effectiveness: Paper} > \text{OPP} \geq \text{Polyethylene} > \text{PVC} \geq \text{Polystyrene}$$

$$(7.6)$$

As alluded to earlier porosity of the face stock material allows penetration of the adhesive. Smoothness and the nonpolar character of the face stock surface impart slight anchorage of the adhesive. In the case of film-based face stock materials, plasticizers or slip additives may migrate to the surface of the film. Emulsions or suspensions make polymers exude surfactants on the film surface.

Fibrous materials may suffer fiber tear during rapid debonding. In all these cases a top coat (primer) on the face stock surface is required. Table 7.14 summarizes the application and mechanisms of the primer coating.

Release Liners. The release liner does not play any role in the bonding of the label, but it influences the coating, converting, and labeling properties. Like the face stock material, the release liner is a sheet-like, thin, solid state component, and is usually paper or film based. Its release, dehesive nature is given by a special polymer coating (see Chapter 5). The most important factors affecting the release liner include the nature of the adhesive (chemical type, thickness, modulus, diluents), the nature of the silicone coating (chemical composition, coating weight, film continuity, degree of cure, crosslink density), the face material variations (roughness, porosity), the laminate characteristics (paper age, laminate age, thickness and modulus, adhesive coating method), and the matrix stripping operation (speed, angle, physical dimensions) [126].

The most widely used material for release liner remains paper [15]. More than 50% of the liner material is based on calendered paper, more than 20% on top-coated paper and only 2% on films. The liner is the "packaging material" for the adhesive, the carrier material for die-cutting, and the transport material for the labels. The role of the paper as transport (end-use) material for the label is replaced by plastic films in some specific application fields such as medical labels, where polyethylene with adequate stiffness and good die-cuttability is used [56]. Another application field concerns metallic labels or PSA-coated materials, where the quite different toughness of the metal and plastic substrate brings about low peel adhesion or where the plastic-based release liner displays a better weatherability.

Release papers are specially coated materials, normally made from 30–60 g/m^2 high density papers (glassine types) which have a special coating that repels the solidified PSAs. A silicone coating is generally used, but release papers can carry a wax or another type of nonsticky coating. The nature of the paper influences in a decisive manner the dehesive properties of silicone liners. Low profile tolerances, chemical inertness, high barrier properties, good wetting out, smoothness, no yellowing, and high dimensional stability are necessary [127]. The kind of pore formation and surface structure of the base paper exert a determining influence on the consumption of silicone and the degree of film forming during the coating process. By means of various measuring techniques for paper it is now possible to record the influence of the base paper on the silicone coating. These include the measurement of rigidity, porosity, and above all, penetration. The presence of migrating species in a release coating can be disastrous to adhesive performance. Very small quantities of silicone fluids such as the ones included in antifoam agents can detackify rubber-based adhesives.

Release liners must fulfill a variety of demands. They must offer the correct physical properties for the envisaged end-use, as well as allowing the convertor to use existing equipment and chosen silicone systems to obtain the required release properties. Glassines, treated krafts, polyolefin-coated papers, and pigment-coated grades are available. A typical base paper comprises clay, cellulose pulp fibers for strength and short fibers for smoothness and uniformity. In order to achieve a closed silicone film, with less than 1 g/m^2 of silicone, a very smooth and even surface is necessary. The pigment coating must be free of pits and flaws if subsequent release problems are to be avoided.

For several years PVC films have been used as release liner for rubber and silicone PSAs. Better results were obtained using stiffer or siliconized films. The main films used as liners for siliconizing include polyester, OPP, LDPE, HDPE, and polystyrene. Films used as release liner make labeling easier (photosensitive regulation), faster (the ratio stiffness/thickness is better for films than paper), and different colors may be produced; there is no humidity influence on the film, no rehumidification after drying necessary. Film release liners are smooth and impart smoothness to the PSAs. No blocking through die-cutting occurs. As they can be molten, recycling as polymer is possible.

2.2 Printability of the Laminate

The printing methods used for labels and the most important problems occuring during printing PSAs materials are summarized by Fust [128,129]. There are close correlations between printability and other technical requirements of the material used for the production of labels [130]. If a large metallized high gloss surface such as gold, silver or copper is required, this can be best achieved with thermoplastic foils like PVC, PET, or OPP. While PVC films can be easily printed, OPP foils may require considerable development efforts in order to fulfill the special requirements made on the quality of the printed surface. Furthermore, the mechanical resistance which plays a significant role for the printing, die-cutting or matrix stripping, and for the application of PSA labels, must be included into these basic considerations. The caliper of the substrate also is important.

There are some differences between the printability of polyethylene and polypropylene [36]. Polyethylene is printable by gravure, with a minimum film thickness of 0.025 mm and polypropylene at 0.012 mm. Polypropylene is superior to polyethylene concerning the resistance towards fatty materials, has better gas barrier properties and higher elasticity. On the other hand, the anchorage of printing inks on polypropylene improves only after 12–24 hr. For paper the printability depends on the lay-flat properties. Lay-flat depends on waving which is a function of the flexibility; the latter depends on the paper weight. Paper with a 40–80 g/m^2 base weight shows waving with short wavelength, heavy papers (100 g/m^2) show waving with longer wavelengths [131].

The use of screen printing for PSA laminates was addressed by Perner [132]; screens with 21–43 mesh are used. Flex printing uses up to 48 gravure lines (Raster). Gravure printing uses 60 line numbers; offset printing also is carried out [129]. The most explosive growth area is label stock for nonimpact printing applications [24]. Direct thermal printing is the largest segment, followed by toner-based printing (laser, ion deposition, magnetography) etc. The nonimpact techniques of direct thermal and thermal transfer printing are the most reliable and cost effective means of bar coding.

Printability depends mainly on the quality of the face material. For laminates the nature of the face is a function of the adhesive also; adhesive/face interactions change the face quality. Generally, the following phenomena influence the printability, namely the lay-flat (flatness) of the laminate (curl control), the dimensional stability of the laminate (and uniform caliper), and the surface quality of the laminate (gloss, smoothness, opacity).

The lay-flat of the laminate generally depends on the face/adhesive equilibrium, the moisture content, and is especially important for paper-based labels. The dimensional stability of the laminate is a function of the face material and its environment/adhesive resistance. For paper labels it depends more on the laminate/moisture interaction; for film labels it is a function of the liner/adhesive interaction, or ink/film face stock/adhesive interaction.

Temperature sensitive cold flow of the adhesive (edge bleeding or oozing) influences the printability (especially for heat generated laser printing, or xerox printing). The surface quality of the laminate can be influenced through adhesive migration in the face, or migration of the face components in the adhesive layer (soft PVC). For paper labels adhesive penetration changes the visual appearance of the face; special additives from the adhesive (e.g., antioxidants) can make the anchorage of the ink difficult. In general the mechanical stability of the laminate has its importance also. Labels are now printed at speeds of 175–200 m/min and higher [24].

Lay-Flat

Pressure-sensitive adhesive laminates may lose their lay-flat properties (i.e., flagging, curling, wing-up of the laminate occurs). The curling of PSA films depends on the thickness and on the stiffness (elasticity) of the film [133]. For paper or other fiber like materials the construction of the web (fiber direction) and fiber sensitivity towards water may have an influence on the lay-flat properties [134]. On the other hand, shrinkage of the paper forming fibers as a function of the chemical composition, respectively manufacturing technology, determines the wet elongation of the fibers. For multilayer papers, wet elongation, elasticity, and thickness of the paper are the most important factors influencing the lay-flat [134]. Wet elongation also depends on the humidity of the medium, thus generally in industrial practice webs (especially paper) are rehumidified in order to avoid curling due to changes in the water content [8]. In general, sheet or roll materials have different lay-flat problems. The "roller effect" (concavity) is mainly known for paper rolls; the "tunnel effect" (convexity) is displayed by sheets (or films).

Eliminating preheating and reducing the humidity content of the release liner (to 4.2–4.5%) may improve lay-flat. These phenomena may be explained on the basis of the theoretical knowledge concerning the paper structure. By water absorbtion or disorbtion, changes in the volume of cellulosic fibers occur [135], they swell or shrink (20% changes in the fiber diameter). These changes occur across the original machine direction: at 30% relative humidity (RH) the water content of the paper increases to 4%; at 50% RH the humidity of the paper reaches 8%, at 70% RH a 9% absolute value of the humidity is observed. Practically, lay-flat may be improved using special release liners, with higher thickness, and with controlled humidity (corresponding to the equilibrium content at 55% RH and 20°C). "Hygrorest" [136] is such a product. Single cut labels after curing do not lie flat in the delivery boxes. Differences in moisture disorbtion of face papers cause edge wrinkling upon conversion [137]; lay-flat depends on the wet elongation of the paper [138]. For a multilayer construction the deformation depends on the tension σ; this tension is a function of the modulus E, elongation ε, and wet elongation e, or

$$\sigma = E(\varepsilon - e) \tag{7.7}$$

Taking into account the distance between medium and bottom layer (Z), the elongation is given by the following equation:

$$\varepsilon = \varepsilon^o + Z \cdot k \tag{7.8}$$

where k is the inverse of the radius of curvature.

Dimensional Stability

Dimensional stability is exemplified by shrinkage. Shrinkage of soft PVC films is affected by the stability of the film and its anchorage (if mounted) on the substrate (Figures 7.8 and 7.9). Prestretched (e.g., calendered or cast)

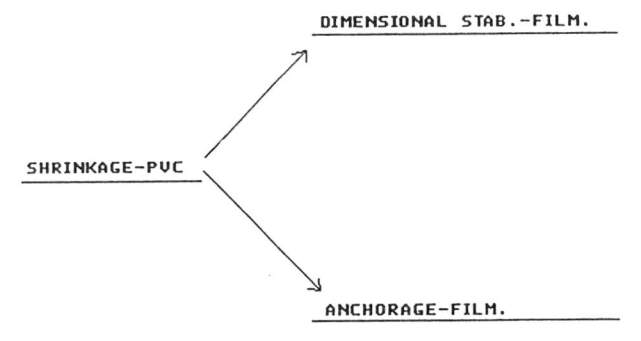

Figure 7.8 Parameters influencing the shrinkage of a PVC film.

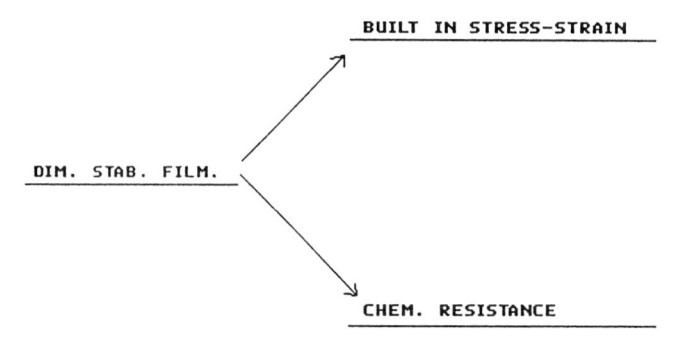

```
                    BUILT IN STRESS-STRAIN
                   ↗
                  /
DIM. STAB. FILM. 
                  \
                   ↘
                    CHEM. RESISTANCE
```

Figure 7.9 Parameters influencing the dimensional stability of a PVC film.

plastic films tend to shrink if subjected to elevated temperatures. Mounted films shrink back less than unmounted ones, because of the fixing effect of the adhesive. In the case of labels, cohesive strength prevents shrinkage and subsequent dirt accumulation on the exposed adhesive. Cohesive strength depends on the built-in cohesive strength of the adhesive and on the time-temperature stability of this one (Figure 7.10).

Cellulose acetate used in tapes exhibits a shrinkage of 0.6% (24 hr at 71°C) [139]. Thermofixed OPP is shrink resistant up to 110°C [140]. On the other hand, humidity changes can cause a shrinkage of 1% in cellulose films. Drechsler [141] demonstrated the influence of the surface layer on the base film, and on the shrinkage of the base film. He states that a 30-μm biaxially oriented polypropylene (BOPP) film shrinks at 70°C. A polymer lacquer on both sides increases the temperature resistance to 75–78°C; a coating of acrylic PSAs increases the shrinkage resistance to 75–80°C. The shrinkage of

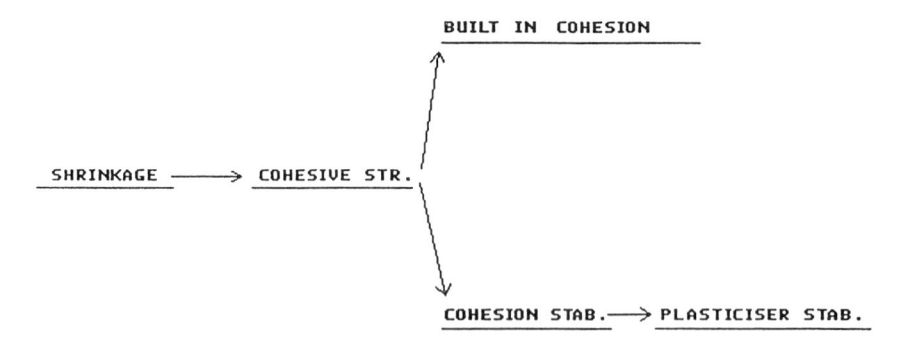

```
                              BUILT IN COHESION
                             ↑
                            /
                           /
SHRINKAGE ──→ COHESIVE STR.
                           \
                            \
                             ↓
                              COHESION STAB. ──→ PLASTICISER STAB.
```

Figure 7.10 Dependence of the shrinkage on the cohesive strength of the adhesive and factors influencing it.

biaxially oriented polypropylene (8% MD, 4% TD) is more pronounced than for oriented polyester (3% MD, 2% TD). Paper also undergoes shrinkage. Drying of the paper during laminating causes a shrinkage of 0.8–1% TD (corresponding to a decrease of the relative humidity from 50 to 30%). Dimensional changes of the label paper produced by environmental humidity may occur during printing, or cutting too [142]. The test methods to quantify the shrinkage will be discussed in Section 3.1 of Chapter 10.

Surface Quality of Labels

The surface appearance of the label is influenced by the adhesive (i.e., its migration, penetration, bleeding, and staining). Migration may also produce coating and/or printing defects [143]. Penetration is a function of the face stock material, of the adhesive and of the manufacturing parameters, and depends on the porosity of paper, its humidity content, and pH [144]. Dispersion-based PSAs display a higher tendency to migration than solvent-based ones. Therefore migration tests are carried out for water-based PSAs at a lower temperature (i.e., 60°C), than for solvent-based adhesives (i.e., 71°C) [145]. The activation temperature of the migration depends on the formulation recipe [146]. For tackified PSAs the activation temperature of the migration decreases with increasing resin content. There are two separate kinds of staining: pressure-induced (roll products) and heat induced (see Chapters 2, 8, and 10). The converting properties of the laminate also are influenced by its aging. The change of the laminate properties due to aging is discussed in Chapters 8 and 10.

Cuttability of the Pressure-Sensitive Laminate

There are only a few data available in the literature about the cuttability of self-adhesive laminates. Die-cutting properties of the PSA label depend on the strength and type (failure) of the face paper (matrix removal), the type of adhesive and silicone, the adhesive-silicone release force, the die-cutting properties of the adhesive, and of the face paper, the properties of the siliconized paper, and the cutting process parameters. The influence of cutting rate and temperature was studied by Matschke [147]. The manufacturing parameters of the film and its thickness also influence the cuttability. The adhesive affects cuttability via its recipe and aging. The converting (i.e., unwinding and slitting of tapes) is strongly influenced by the machine parameters. For tapes the unwinding rate and the temperature of the film influence the adhesion. The unwinding (peeling) force f is a function of the adhesion, modulus, and thickness (coating weight) of the PSA, and the thickness, modulus, and width of the tape [147]. The modulus of elasticity of the adhesive depends on the speed of unwinding and temperature. A similar behavior is observed when cutting/diecutting PSA label materials. Different cutting angles are necessary for cutting different labels: gummed paper (22°), plastics (19–24°), soft PVC (22°), and polyethylene (25°) [148]. On the other hand, low coating weights provide "clean" converting conditions (no oozing) [149]. For good cuttability the PSAs should exhibit moderate cohesion; too high values of the cohesion results in "pull out," while too little strength allows smearing [150]. The cuttability of paper can be improved using lubricants [151].

Cuttability of the laminate is a cold-flow related characteristic; hence there are two distinct aspects of this property: static and dynamic cuttability. Static cuttability is the time-dependent behavior of the cut material, this one displaying more or less cold flow of the adhesive, and smearing of the side of the roll/sheet material. It is very important to remember that low level forces (due to the material's own weight or to the winding tension) may cause cold flow of the adhesive. On the other hand, dynamic cuttability related to the cold flow of the adhesive during cutting includes two different aspects: cuttability and die-cuttability. Cuttability is necessary for the initial cutting (dimensioning) of the roll/sheet materials, and it means cutting through the whole laminate. Die-cuttability refers to cutting-out the labels, while leaving the carrier release sheet intact. Adhesive cold flow influences in the same manner both cuttability and die-cuttability; the influence of the liner and of the nature of the release may be different.

Cuttability of Sheet Materials—General Aspects

Paper label products have to display cohesion and heat resistance needed for sheet stock subjected to flying knife (cutting of the rolls in TD) and guillotine blade converting (cutting of the sheets) (Figure 7.11). Cutting, guillotining, and die-cutting are mechancial operations carried out at high frequency. The less viscous and more rigid the response of the polymer the cleaner the process tends to be. If viscous flow within the polymer is significant during the

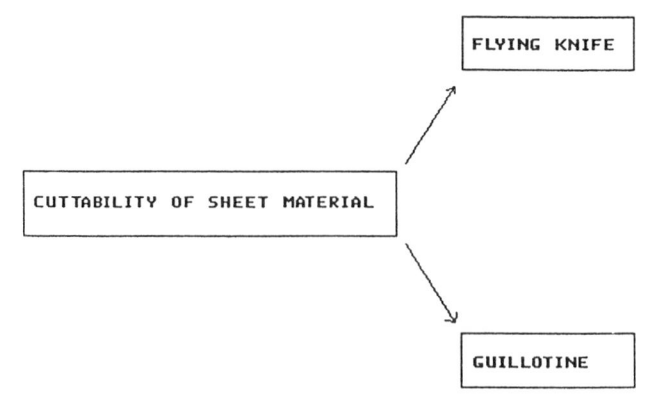

Figure 7.11 The main cutting methods influencing cuttability.

converting generation, poor die-cutting (characterized by unsatisfactory matrix stripping) or poor guillotining (knife fouling) can result. Formulating approaches which increase the high frequency modulus will enhance the overall convertability.

Evaluation of Cuttability. There are no laboratory methods allowing the evaluation of cuttability properties. Different laboratory tests only help to optimize an adhesive formulation, its final performance characteristics being dependent on the face stock, liner, and their interaction with the adhesive. Final conclusions about cuttability can only be drawn on the basis of real cutting tests, made on an industrial guillotine. Generally, every converter has established specific evaluation methods for laminates. As general criteria the smearing of the knife (front-side and back side) and the smearing of the cut material should be considered. The level of the adhesive deposits on the knife, the smearing of the cut strips by the adhesive layer, and blocking of the cut sheets (no moving capability) characterize the cuttability of the laminate. Table 7.15 summarizes the cuttability results for different laminates illustrating how to quantify the different aspects of the cuttability. Steps of the cutting process (cutting through, guillotining) and the phenomena associated with the cutting were shown in Figure 2.16.

The following phenomena can be observed during the cutting operation:

- smearing of the knife;
- smearing of the cut material (sheets and strips);
- blocking of the cut material, and
- adhesion of the cut material to the knife.

Blocking and pulling up of the cut material are secondary phenomena, produced by the smearing of the cutting and cut surfaces. In order to obtain a good cuttability one requires no smearing of the surfaces and low adhesion of the smeared surface. There is a need for no adhesive flow out (AF), absorption of the smeared adhesive by the cut surfaces (AA), and the lack of adhesion of the smeared adhesive (AC). Hence,

$$\text{Cuttability} = C = f\left(\frac{AA}{AF, AC}\right) \tag{7.9}$$

Adhesive Flow Out. Important parameters include smearing of the edges (oozing), edge flow, face bleed, and gumming. Adhesive flow out is a function of the cold flow or creep of the adhesive. Creep of the pure (bulky) adhesive is different from that of the adhesive in the laminate and AF depends on the adhesive and on the laminate:

$$AF = f \text{ (Adhesive, Laminate)} \tag{7.10}$$

Table 7.15 Cuttability for Different Laminates (Evaluated Using an Internal Cuttability Index)

Laminate characteristics				Cutting characteristics							
				Aspect of the cut stripes				Aspect of the cut surface			
				Comments		Index		Comments		Index	
		Number of cuts		Number of cuts		Number of cuts		Number of cuts	
Adhesive type	Coating weight (g/m²)	Paper weight (g/m²)	Paper weight (g/m²)	10	35	10	35	10	35	10	35
WB AC	24	80	100	DM	DM	1	1	sac	PM, ac	2	4
WB AC	24	80	100	DM	DM	1	1	sac, PM	ac, NM	4	5
WB AC	20	80	85	DM	DM	1	1	sac, PM	ac, PM	3	4.5
WB AC/CSBR	26	80	85	DM	sac, PM	1	2	ac, PM	ac, NM	5	5
WB AC/CSBR	28	80 primered	85	DM	DM	1	1	DM	DM	1	1
SB AC	19	70	88	DM	DM	1	1	DM	DM	1	2
WB AC	20	78	95	DM	DM	1	1	DM	sac, PM	1	3
Crosslinked HM	22	79	100	DM	DM	1	1	DM	DM	1	1

Converting Properties of PSAs

	Apect of the cutting knife					Aspect of cut 4-cm stripes		
	Ahesive build-up on the knife							
	Front side, number of cuts: 35		Back side, number of cuts: 35			Number of cuts: 35		
Adhesive type	Width (mm)	Index	Width (mm)	Index		Comments	Index	Overall index
WB AC	2	3	30	4		Blocked	1	16
WB AC	2	3	30	4		Pulled up	5	23
WB AC	2	3	25	3.5		Blocked	1	17
WB AC/CSBR	2	3	40	5		Blocked	1	22
WB AC/CSBR	2	3.5	25	3.5		Blocked	1	12
SB AC	0	1	20	2		Blocked	1	9
WB AC	0.5	2	30	4		Blocked	1	13
Crosslinked HM	0	1	0.5	1		Blocked	1	7

The lower the overall index, the better the cuttability.
WB AC, water-based acrylic; CSBR, carboxylated styrene-butadiene rubber; SB AC, solvent-based acrylic; HM, hot-melt; DM, dry and removable; PM, partially removable; NM, not removable; sac, slight adhesive coated; ac, adhesive coated

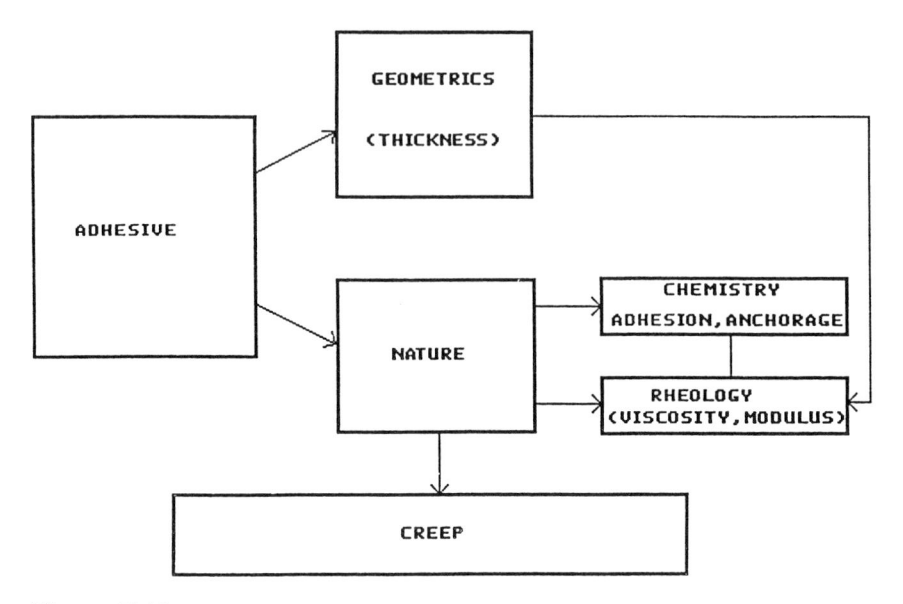

Figure 7.12 The influence of the adhesive on the creep.

The adhesive flow-out is a function of the nature and amount of the adhesive, the laminate components (nature and dimensions), and laminate structure (see Chapter 2). On the other hand, creep also depends on the cutting conditions (nature and level of the mechanical stress); the rate and distribution of the stress as well as the working temperature influence the cold flow. Finally, AF is a function of the adhesive, the laminate, and working conditions:

$$AF = f \text{ (Adhesive, Laminate, Working conditions)} \qquad (7.11)$$

Like thermoplastics PSAs are viscous fluids exhibiting cold flow. A PSA in a laminate is a viscous fluid, flowing under hydrodynamically limited conditions (i.e., like a thin fluid layer between parallel surfaces). The flow velocity depends on its distance from the top surface of the fluid layer (i.e., from the coating weight):

$$AF = f \text{ (Adhesive nature, Adhesive amount)} \qquad (7.12)$$

The adhesive nature influences the creep through the viscoelastic properties and the creep from the laminate through its interaction with the laminate components, which impart a resistance against the flow (Figure 7.12).

The fluidity of the PSAs depends on its viscoelastic properties characterized by its viscosity and modulus and the dependence of these properties on

Converting Properties of PSAs

the stress rate and temperature (see Chapter 2). The cohesion of the adhesive and the cohesion/adhesion balance characteristize the viscous behavior. The modulus and its dependence on the temperature must be known in order to make a first approximation about the fluidity of the adhesive. For a better understanding the frequency dependence (temperature dependence) of the modulus (storage and loss modulus) must be known. The value of the modulus (viscosity) and parameters influencing it at ambient temperature, and the temperature/speed sensitivity of the modulus (viscosity) must be discussed separately. The viscoelastic properties of the adhesive depend on the viscoelastic properties of the bulk adhesive, and on those of the composite adhesive layer, or

$$\text{Viscoelastic properties} = f\,(\text{Bulk adhesive, Composite material}) \quad (7.13)$$

The viscoelastic properties of the bulk adhesive are different from those of the composite material. Generally, the viscoelastic properties of composite solvent-based adhesives are superior to those of the bulk material. The viscoelastic properties of water-based (composite) PSAs are inferior to those of the bulk adhesive. This behavior can be explained by the different nature of the additives. The viscoelastic properties of the bulk material depend on its chemical nature, molecular weight, MWD, and structure (crosslinked, branched, or linear) (see Chapter 2); hence,

$$\text{Viscoelastic properties of the bulk adhesive} =$$
$$f(\text{Nature, Molecular weight, MWD, Structure}) \quad (7.14)$$

The tackified adhesive should also be considered as the bulk adhesive although tackification (with plasticizer or resin) decreases the cohesion level. Generally, low cold flow (good cuttability) implies more elasticity and a less viscous character of the viscoelastic fluid (i.e., higher cohesion). Knowing the inversely proportional relation between cohesion and adhesion, in a first approximation, good cuttability means high cohesion and low adhesion:

$$\text{Cuttability} = f\left(\frac{\text{Cohesion}}{\text{Adhesion}}\right) \quad (7.15)$$

The chemical/physical structure of the adhesive determines its viscoelastic characteristics and the dependence of the viscoelastic behavior on the temperature/stress rate. During cutting, the temperature of the laminate increases and consequently the viscosity of the material decreases. Thus the value of viscosity at higher temperature (cohesion at higher temperature) is very important for cuttability purposes. The most simple method of measuring the temperature dependence of the viscosity is the test of the shear at elevated temperatures, and the test of the hot shear (HS) at different temperatures gives

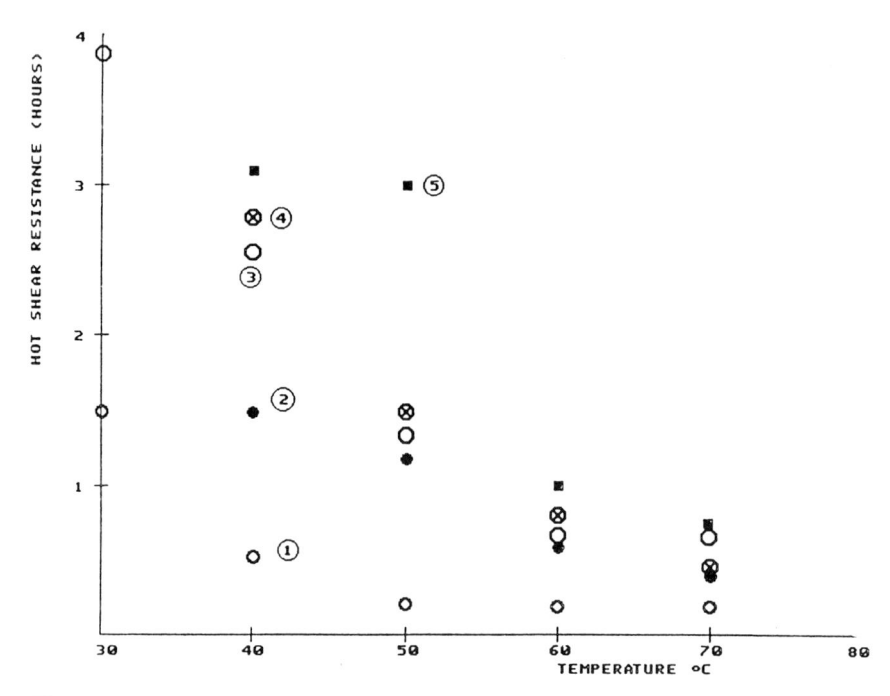

Figure 7.13 The dependence of the hot shear for different tackified formulations. 1), 2), 3), and 4) are different water-based acrylic PSA formulations.

the hot shear gradient (HSG) (see Chapters 6 and 10). Figure 7.13 illustrates the dependence of the hot shear on the temperature for different tackified formulations.

Different formulations display a different temperature sensitivity of the cohesion (Figure 7.13). On the other hand, cutting tests carried out with different formulations demonstrate that formulations with a limited temperature sensitivity of the shear display better cuttability. Room temperature shear values are more important for the cohesion/adhesion balance, influencing mainly the peel on polyethylene.

Cuttability and Adhesive Properties of Resin. The adhesive properties of the resin influence the cuttability of the laminate. Unfortunately, there is no theoretical or empirical equation clarifying the dependence of the cuttability of the laminate on the adhesive properties (shear, peel, and tack) of the adhesive;

$$\text{Cuttability} = f \text{ (Adhesive properties)} \tag{7.16}$$

Converting Properties of PSAs

Table 7.16 Dependence of the Hot Shear and Hot Shear Gradient on the Formulation. The Influence of the Resin Nature for Water-Based Formulations

Tackifier			Hot shear values, min (°C)				
Code	Nature	Supplier	30	40	50	60	70
CF 52	Acid rosin	A&W	90	35	12	10	11
DEG G	Rosin ester	DRT	390	170	72	45	21
MBG 64	Rosin ester	Hercules	330	90	72	38	23
CF 8/21	Rosin ester	A&W	> 1500	192	160	55	26
E-9241	HC-resin	Exxon	230	144	90	45	40

Acrylic = a 40/60 (wet/wet) weight blend of V205/80D (BASF).

The best documented interdependence is the correlation of cuttability and shear. Unfortunately the best known shear values are room temperature shear values, and there are only a few hot shear data available; no hot shear gradient data are known from the literature. Cuttability improvement is a function of the shear increase (order of magnitude) and of the hot shear gradient (HSG). Thus, the cuttability improvement CI depends on the shear as follows:

$$CI = f(\log \Delta HS) \left(\frac{1}{HS_{n+i} - HS_n} \right) \qquad (7.17)$$

where HS_n denotes the hot shear at a given temperature, HS_{n+i} is the hot shear at another temperature with ΔT at least 10°C, ΔHS is the improvement of the $HS = HS_i - HS_o$, where HS_o is the hot shear of a known product with an average cuttability. Table 7.16 illustrates the dependence of the hot shear and HSG on the formulation (resin nature).

Although changes in the room temperature shear and slight changes in hot shear are possible, no cuttability improvement can be observed (except for CF 52). On the other hand, the study of the HSG and the dependence of the cuttability (and peel) on HSG denotes the necessity of a minimum HSG for good cuttability (Figure 7.14). It is evident that slight changes in the formulation (using resins with a higher melting point) do not necessarily impart an improved cuttability.

For shear tests the modulus of the adhesive is important also. The contact area A between the adhesive and the face stock/release liner may be estimated according [85] as a function of the modulus E, the roughness of the surface $1/\beta$, the limited contact area A_l and the density ρ of the adhesive:

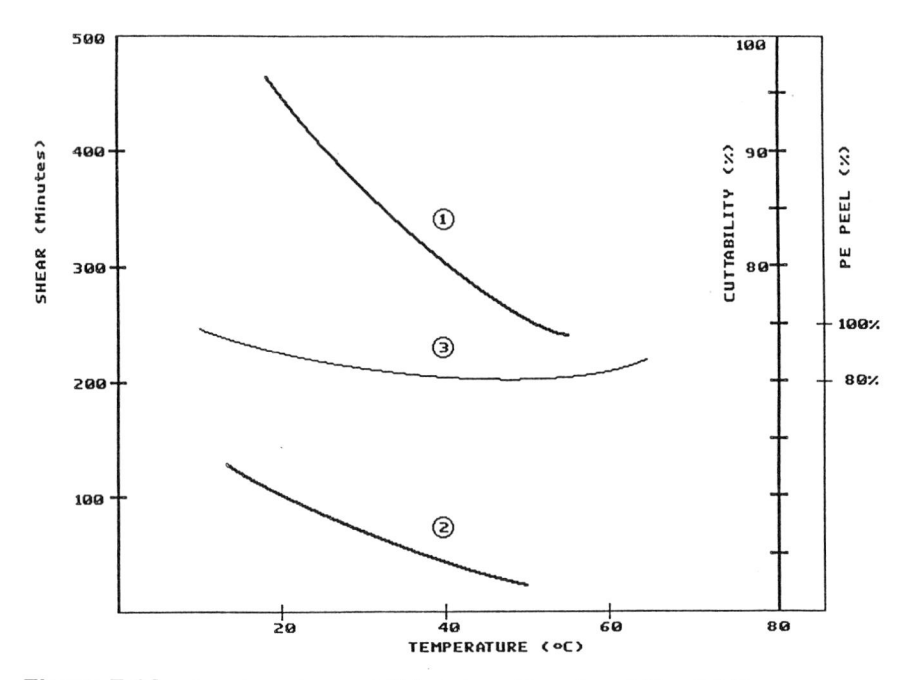

Figure 7.14 Interdependence of 1) hot shear, 2) peel, and 3) cuttability.

$$A = A_I \left(1 - e^{(-\beta \rho)/E}\right) \tag{7.18}$$

The viscous flow, creep, or cold flow of the adhesive may be characterized by its cohesion (see Chapters 2 and 3). Smearing by cutting will be described by the elongation viscosity and relaxation of the fluid adhesive. The coefficient of viscous traction $\eta_o(t)$ as a ratio of tensile stress $\sigma(t)$ and elongation rate $\dot{\varepsilon}$ is [152]:

$$\eta_o(t) = \frac{\sigma(t)}{\dot{\varepsilon}} \tag{7.19}$$

On the other hand, the section $A(t)$ during elongation also depends on the elongation rate:

$$A(t) = A_o \cdot e^{-\dot{\varepsilon}t} \tag{7.20}$$

The coefficient of viscous traction (elongation viscosity) includes an ideal elastic and a viscoelastic component both dependent on the modulus E. The viscoelastic part depends on the relaxation time of the material. If elongation occurs rapidly, the tension is given by the following equation:

$$\sigma(t) = [E(t) + \psi(t)]\varepsilon_o \qquad (7.21)$$

where $E(t)$ is the tension relaxation modulus (a ratio of tension and elongation) and $\psi(t)$ is the relaxation function:

$$E(t) = \frac{\sigma(t)}{\varepsilon_o} - \psi(t) \qquad (7.22)$$

$\psi(t)$ is approximated for rubber according to the following equation:

$$\psi(t) \sim f(1.0) t^{-n} \qquad (7.23)$$

For tests carried out at a constant rate:

$$\sigma(t) = K(t - t_o) \qquad (7.24)$$

where K is a constant and t is the time. The relaxation modulus for natural rubber is not zero at temperatures as high as 100°C (i.e., rubber alone does not behave like a fluid). Tackified rubber compositions (i.e., PSAs) display a quite different behavior. These considerations show that the modulus and generally the complex time/temperature dependent rheology of PSAs influence the cold flow (shear/cuttability behavior) (i.e., it is difficult to find a simple relation between shear resistance and cuttability).

Concerning the peel dependence on the cuttability, a 10–30% cuttability improvement needs a 1000% peel decrease and/or a 50–100% tack decrease (Figures 7.15 and 7.16). Changes in the tack are common, but changes in the peel are rarely possible (e.g., sheet/roll material). On the other hand, changes in the hot shear of an order of magnitude are generally possible only through crosslinking. In conclusion the improvement of the cuttability of a given formulation from the point of view of the bulk adhesive remains limited. Actually cuttability improvement by means of the composite structure of the adhesive (laminate) seems to be more important.

Influence of Adhesive Nature on Creep in the Laminate. Creep in the laminate differs from the creep of the bulk material because of the mechanical/chemical interaction between adhesive and solid laminate components. For this mechanical/chemical anchorage a minimum level of fluidity and tack is necessary. That is the reason why hard low-tack formulations display poorer cuttability (showing fiber tear) than low-tackified medium-tacky formulations. Thus, in a first approximation the creep in the laminate is inversely proportional to the anchorage:

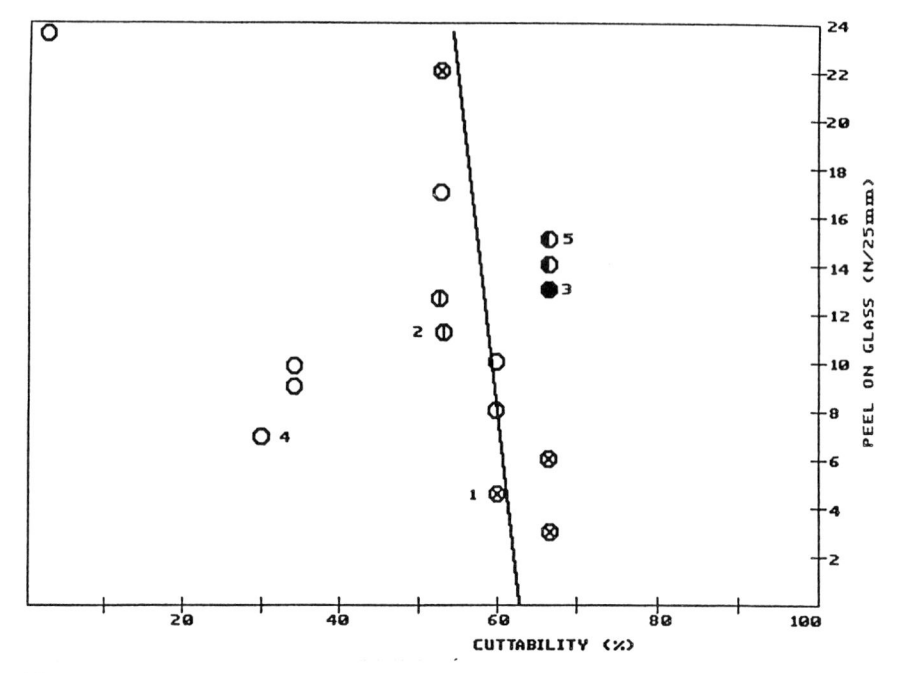

Figure 7.15 Interdependence peel/cuttability. The dependence of the guillotine cuttability on the peel of PSA film laminates that have a different peel level: 1), 2), 3), 4), and 5) are different water-based PSA formulations.

$$\text{Creep} = f\left(\frac{1}{\text{Anchorage}}\right) \tag{7.25}$$

$$\text{Cuttability} = f\left(\frac{1}{\text{Creep}}\right) = f(\text{Anchorage}) = f(\text{Adhesion}) \tag{7.26}$$

Creep also depends on the cutting process conditions. The amplitude of the deformation increases as a function of the nature of the applied stresses as follows: 1, 1.3, and 1.5 for compressive, shear, and tensile forces respectively [153]. Thus laminates with elastomers have to be designed to resist tensile stresses. On the other hand, it is to be taken into account that the temperature increase of the elastomer through hysteresis increases with the frequency and the square of the amplitude of the deformation

Fillers increase and softeners decrease the wall slip behavior of polymer blends [85]. For rubber slippage through a high pressure capillary viscometer,

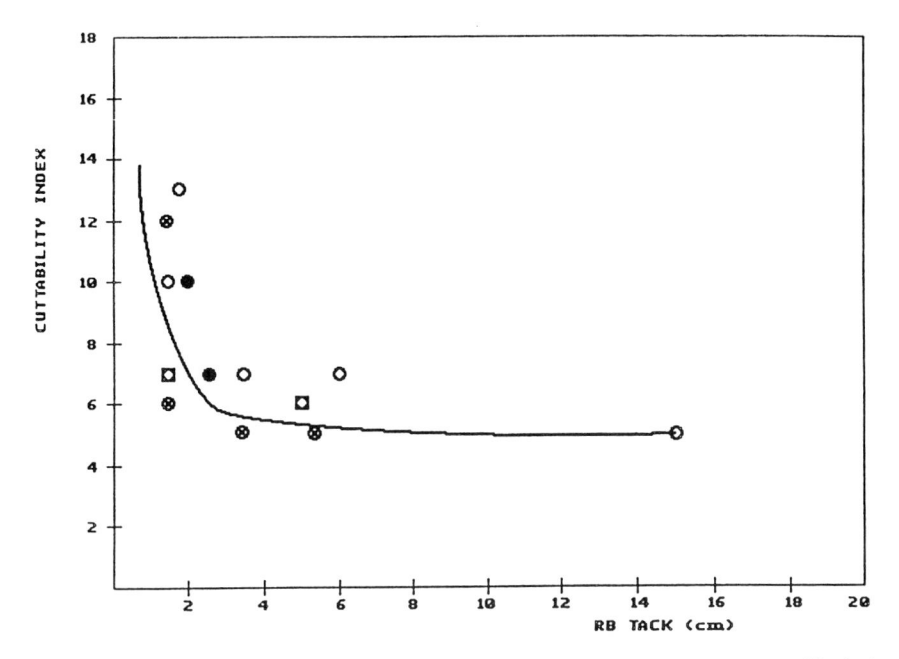

Figure 7.16 The dependence of the cuttability on the tack. Cuttability (guillotine) as a function on the rolling ball tack for different PSA formulations.

the flow volume possesses three different components, a wall slip component V_g, a shear component V_s, and a middle layer component V_k, so that:

$$V_g = \pi \cdot R^2 \cdot w_g(\tau_w) \tag{7.27}$$

$$V_s = \frac{\pi R^3}{\tau_w^3} \int_{t_o}^{t_w} \frac{\tau^3}{\eta(\tau)} d\tau \tag{7.28}$$

$$V_k = \pi \cdot r(\tau_w)^2 \cdot w(r) \tag{7.29}$$

where w_i is the flow rate, η is the viscosity, τ is the stress, R is the radius of the capillary, and r is the radius of the middle (shear and slip free) layer [85]. The flow volume generally depends on the stress applied, on the geometry of the layer and (for the middle layer only) on the viscosity. Therefore it can be suggested that the stress distribution in the layer (i.e., anchorage of the adhesive and stiffness of the solid state components) and geometry of the layer (coating weight) are more important for cold flow than the viscosity of the adhesive (see Section 3.2 of Chapter 2). In a similar manner, plasticizers and temperature promote wall slip, while fillers decrease it. The same interdependence may be supposed for the cold flow (bleeding) of the adhesive.

Dependence of Cold Flow on Adhesive Coating Weight. Cold flow of the adhesive in a laminate is hindered by marginal interaction of the adhesive with the "walls" of the laminate. Because of the local character of this interaction the middle layer displays free flow. It is evident that the width of the free flowing layer depends on the coating weight and on the width of the anchored layer. The anchored layer depends on the coating technology and the nature of the solid laminate components; thus the free flowing layer depends on the coating weight and coating technology.

$$\text{Cold flow} = f(\text{Free flowing layer})$$
$$= f(\text{Coating weight, Coating technology}) \quad (7.30)$$

In a first approximation the cuttability depends on the coating weight as follows:

$$\text{Cuttability} = f\left(\frac{\text{Coating technology}}{\text{Coating weight}}\right) \quad (7.31)$$

Dependence of Cold Flow on the Laminate. Cold flow in the laminate depends on the width of the free flowing region, this one being a function of the anchorage. Anchorage of the adhesive depends on the face material (i.e., the anchorage in the laminate depends on the adhesive bonding on the face and liner). On the other hand, the uniformity of the forces acting on the laminate, their distribution, the compression stress character in the first cutting step are determined by the rigidity of the face. Paper face materials are hygroscopic and there is a continuous water transfer between adhesive and face, thereby influencing the composite character of the adhesive. The solid components of the laminate influence the cold flow in the laminate through the anchorage of the adhesive, the nature of the stress on the adhesive, and the change in composite structure.

The anchorage of the adhesive on the face material (F) and liner (L) depends on the smoothness and chemical character of both. As to the smoothness, it is well known that papers of different quality are used for face materials and release liners. The chemical affinity of the surface may be changed using a dry coating (primer). The primer is used mainly for removable coatings (with a softer formulation) and improves the cuttability (see Chapter 2). Figure 7.17 summarizes the factors influencing the anchorage.

The water content of the paper influences the anchorage. "Wet" anchorage is weaker than dry anchorage; thus:

Converting Properties of PSAs

```
                    SMOOTHNESS OF THE SOLID COMPONENTS
                   ↗
                  /
                 /
                /        HUMIDITY
  ANCHORAGE    /
               \
                \              ↓
                 \      CHEMICAL AFFINITY OF
                  ↘     THE SOLID COMPONENTS
                        + ↑
                        PRIMER
```

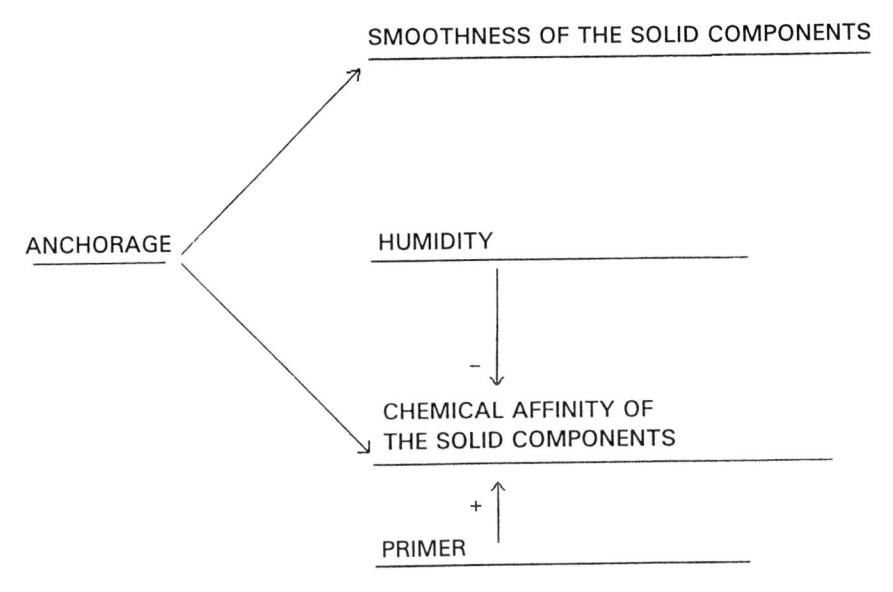

Figure 7.17 Parameters influencing the anchorage of a PSA on the face stock material.

$$\text{Cold flow} = f\left(\frac{\text{Smoothness of F\&L}}{\text{Chemical affinity of F\&L}}\right) \quad (7.32)$$

$$\text{Cuttability} = f\left(\frac{\text{Chemical affinity of F\&L}}{\text{Smoothness of F\&L}}\right) \quad (7.33)$$

$$\text{Cuttability} = f\left(\frac{\text{Chemical affinity, Primer of F\&L}}{\text{Smoothness, Humidity of F\&L}}\right) \quad (7.34)$$

The chemical affinity of the face/release liner also depends on the adhesive nature. Different adhesives display different adhesion (release force) on different liners (Figure 7.18). The release peel force influences the composite behavior of the laminate (cuttability) and the labeling properties. A high release force increases the yield stress of the liner resulting in unsatisfactory matrix stripping (die cuttability) possibly causing liner breakage.

As incompressible liquids PSAs support compression forces. In the first step of the cutting operation the force acts perpendicularly on the whole laminate (or laminate assembly). This situation changes only by the local stress concentration and weakness of the solid laminate components, giving rise to the flexure of the face stock and shearing of the adhesive. The stiffness of the solid laminate components influences the stress distribution and the

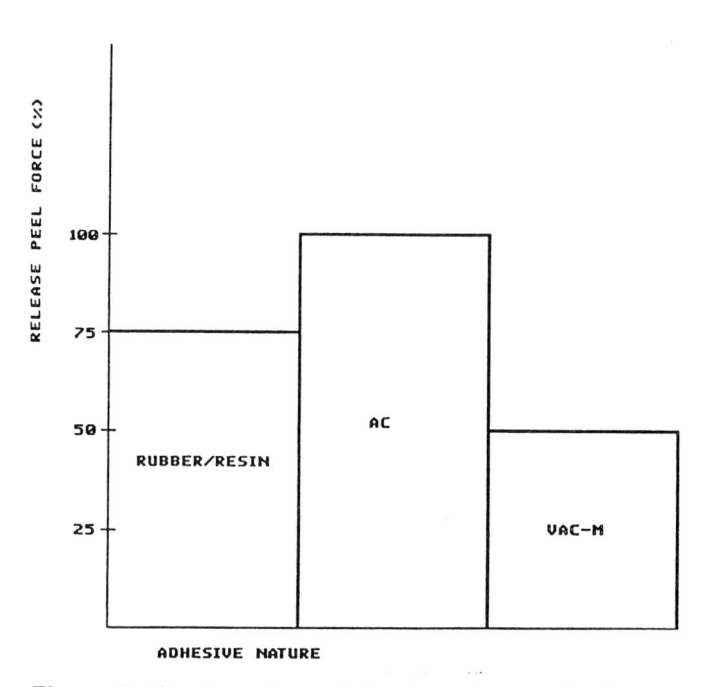

Figure 7.18 Dependence of the release force on the chemical nature of the PSA.

creep of the adhesive. Generally, the stiffness of the paper depends on its composite structure and its humidity content (Figure 7.19); both cold flow and cuttability depend on the composite structure and humidity:

$$\text{Cold flow} = f\left(\frac{\text{Humidity of F\&L}}{\text{Composite structure of F\&L}}\right) \tag{7.35}$$

$$\text{Cuttability} = f\left(\frac{\text{Composite structure of F\&L}}{\text{Humidity of F\&L}}\right) \tag{7.36}$$

Another parameter influencing the cuttability is the stiffness of the laminate components. The flexural resistance of a laminate depends on the siliconizing, on the adhesive nature and state (crosslinking, humidity), on the laminate age, on the coating technology (direct/transfer), on the adhesive amount (coating weight) and on the face stock/release nature. Generally, PSAs lower the flexural resistance (stiffness) of paper. A systematic study of these parameters identifies the following factors influencing the stiffness of the label: the materials (nature of face stock, release and adhesive), the thickness of the components (face stock, release liner thickness and coating weight), and the

Converting Properties of PSAs 309

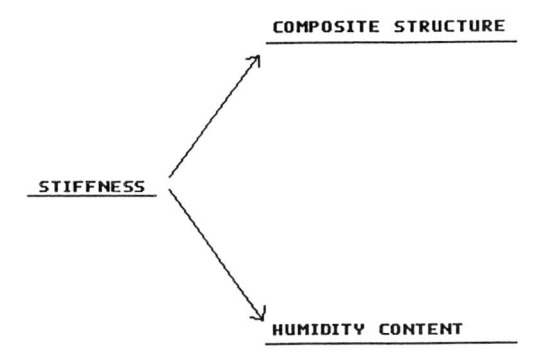

Figure 7.19 Factors influencing the stiffness of a PSA laminate.

construction of the solid components. In this section, the influence of these parameters on the stiffness of the label, and on the properties (adhesive and end-use) of the PSA label will be illustrated. As can be seen from Table 7.17, film laminates (plastic face stock/paper release liner or filmic face stock/filmic release liner) are softer than paper-based ones. The number of layers and their build-up into sandwich constructions also influences the stiffness too (Table 7.18). The nature and softness of the adhesive also can change the stiffness of the label (Table 7.19) (see Chapter 2). Siliconizing (one side or both) modifies the stiffness of the labels, eventually increasing it (Table 7.20). The stiffness of the label also is a function of the coating weight (Table 7.21). Engraved (embossed) face stock gives (as supposed) better stiffness (Table 7.22). A primer coating (with a soft primer) changes the stiffness only slightly, but improves the peel (Table 7.23). The humidity modifies the modulus and stiffness of paper (Table 7.24).

The coefficient of friction of the paper (slip in the cutting machine) increases with the humidity in the air [154].

The influence of the face stock material on the cuttability was discussed in Chapter 2. Here the dimensions and the mechanical characteristics (mainly the modulus) of the face stock and their influence on the cuttability are discussed. A high stiffness of paper face stock or liner can be obtained by using a higher weight or a more voluminous paper grade (thickness). The finish or type of fiber used also affect the stiffness; sulphite pulps are generally stiffer and the difference in stiffness between papers made from sulphite and sulphate pulps is about 20%. Mechanical strength and to a lesser degree smoothness also are affected. However, this approach also reduces ink hold out (or water hold out), thereby resulting in a higher stiffness (at least theoretically). Generally, the thickness of the laminate is an index of the stiffness

Table 7.17 Stiffness of Film Labels Versus Paper Labels with Different PSA Coatings

Adhesive code	Face stock material		Stiffness (mN/25 mm)
	Paper	Film	
01		x	650
01	x		814
02		x	682
02	x		1017
03		x	632
03	x		961
04		x	148
04	x		331
05		x	648
05	x		833

Table 7.18 The Influence of the Number of Layers and Layer Build-Up on the Stiffness of the PSA Laminate

	Stiffness (mN/m)			
	Number of layers/build-up			
	2	3	4	5
Coating weight (g/m^2)	FS-A	FS-A-FS	FS-A-FS-A	FS-A-FS-A-FS
12.0	13	117	249	99
17.0	23	32	223	122
20.0	22	106	193	103
30.5	22.5	157	210	158
35.5	18	171	227	121
48.5	26	149	154	176

FS, Face stock; A, Adhesive

Table 7.19 The Influence of the Nature and Softness of the Adhesive on the Stiffness of the Label

Adhesive nature					Stiffness
WB	SB	AC	RR	CSBR	(mN/10 mm)
X		X			698
X				X	480
X		X		X	515
	X		X		483
X		X			723

Table 7.20 The Influence of the Siliconizing on the Stiffness of the Paper

Number of silicone coating layers on the paper	Stiffness (mN/10 mm)
0	78.0
1	83.7
2	169.6

Table 7.21 The Influence of the Coating Weight on the Stiffness of the Laminate

Laminate code	Coating weight (g/m^2)	Laminate stiffness (mN/10 mm)
1	12.0	127
	22.0	122
	51.0	146
2	14.5	145
	22.0	107
	43.0	113

Table 7.22 Dependence of the Stiffness on the Embossing of the Face Stock Material

Face stock quality	Stiffness (mN/m)		
	Face stock	Release liner	PSA laminate
Normal	77	40	208
Embossed	155	40	470

Table 7.23 The Influence of the Primer on the Stiffness of the Laminate; the Influence of the Dwell Time on the Stiffness

Dwell time (days)	Stiffness of the laminate and laminate components (mN/cm)				
	Face stock	Adhesive	Liner	Laminate	Laminate with primer
0	162	96	74	332	346
3	162	96	74	334	367
4	162	101	74	337	385

Table 7.24 The Influence of Solvents on the Stiffness of Face Stock Paper; Immersion of Paper in Solvents

Solvent	Stiffness of the paper (mN/10 mm)
None	166
Water	22
Toluene	153
Plasticizer	161
Oil	150

Converting Properties of PSAs 313

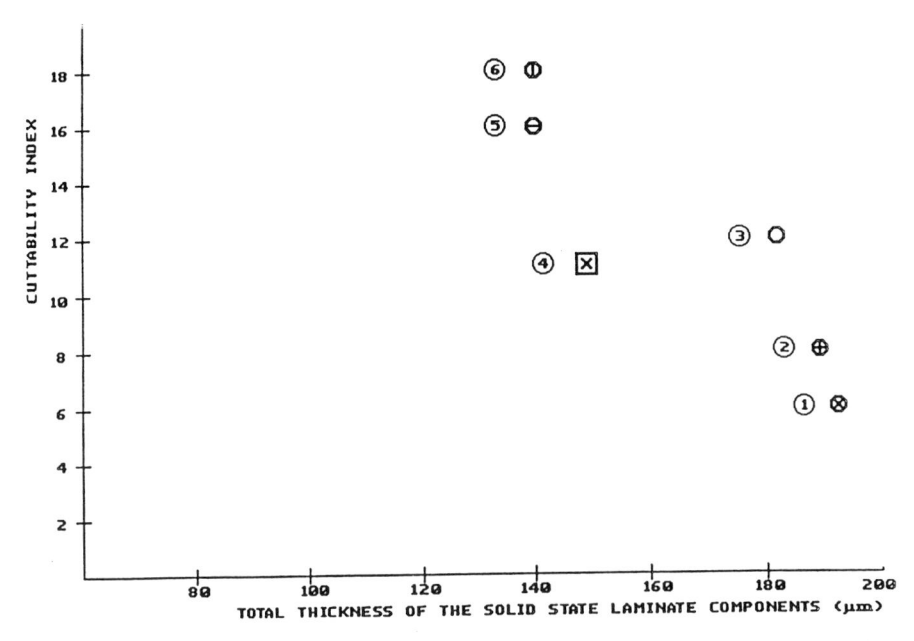

Figure 7.20 The influence of the laminate thickness on the cuttability. 1) through 6) are different PSA laminates.

of the laminate; laminates with a higher thickness exhibit a better cuttability (Figure 7.20).

Generally, the stiffness of the liner is tested in order to evaluate its machinability [155]. The stiffness of the liner depends on its specific weight; the used specific weight (stiffness) also is a function of the face material. For paper-based liners 55–100 g/m^2 material, for film coatings 120–180 g/m^2 papers are proposed [155]. The specific weight of the liner and its flexural resistance (stiffness) MD and CD should be carefully controlled. In order to evaluate the main parameters influencing the flexural resistance of the label, the stiffness of a label can be estimated as follows. The stiffness S is a function of the modulus E and thickness h of the material:

$$S = f(E, h) \tag{7.37}$$

For a sandwich structure (label) the stiffness is the sum of the components, where the final modulus and thickness are the sum of the components:

$$E = \sum E_i \tag{7.38}$$

$$h = \sum h_i \tag{7.39}$$

where

$$h_{sandwich} = h_{facestock} + h_{liner} + h_{adhesive} \tag{7.40}$$

Assuming that

$$h_{liner} = 1.5 h_{facestock}$$
$$h_{adhesive} = 0.5 h_{facestock} \tag{7.41}$$

then

$$h_{sandwich} \approx 3 h_{facestock} \tag{7.42}$$

According to the dependence of the moment of inertia on the thickness, the stiffness depends on h following the equation:

$$S = f(h^3) \tag{7.43}$$

Thus the increase of S with thickness changes for a label comparatively to face stock material is given by:

$$\Delta S = f(3h^3) \tag{7.44}$$

On the other hand, there is a decrease of the laminate modulus caused by the adhesive; in a first approximation:

$$E_s = \frac{E_{fs}}{3} + \frac{E_{fs}}{3} + \frac{0.1 E_{fs}}{3} = \frac{2.1 E_{fs}}{3} = 0.7 E_{fs} \tag{7.45}$$

where E_s is the modulus of the laminate and E_{fs} is the modulus of the face stock. Thus the change of the stiffness for a label, as compared to the pure face stock material, will be given in a first evaluation, by the following equation:

$$\Delta S = f[3h^3, 0.7E] \tag{7.46}$$

The stiffness of the label is higher than that of the face stock material and the most important factor influencing it, is the thickness of the components. An exponential dependence of 2.5 was found for cardboard stiffness as a function of its thickness [156]; 4–5% thickness tolerances lead to a ±15% change in stiffness.

Influence of Other Parameters on Cuttability. Improved crossdirection tensile strength allows faster die cutting and stripping during converting [157]. Laminates with a different humidity content display different cuttability. The residual water content of water-based adhesives is less than 5%, and as the equilibrium water content of the paper exceeds 5% the paper acts as a continuous water reservoir for the adhesive; therefore conditioning of the laminate before cutting is necessary.

$$\text{Cold flow} = f(\text{Humidity of F\&L}) \tag{7.47}$$

$$\text{Cuttability} = f\left(\frac{1}{\text{Humidity of F\&L}}\right) \tag{7.48}$$

There are two kinds of cold flow conditions: cold flow of PSAs under high speed and high forces (cutting, die-cutting, slitting), and cold flow of PSAs under low speed and low forces (storage). High speed stresses act like normal stresses at low temperature. Low speed stresses act at normal temperatures, thus their action is continuous and cannot be avoided; they are characteristic for both sheet and roll materials. Cuttability depends on the laminate and the cutting machine. "Engraved" knifes lead to better results than normal ones. Static cold flow depends mainly on the storage conditions (temperature and humidity); for roll materials static cold flow is also a function of the winding tension.

The scope of the formulation is to design formulations exhibiting no adhesive flow (oozing). The fluid character of the PSAs is necessary for its bonding, thus it is not possible to completely avoid the flow of the adhesive. Taking into account a certain level of adhesive flow, there is a need to minimize its tack, in order to avoid smearing of the cutting and cut surface, and the displacement of the cut sheets. Smearing of the cut and cutting surface, and the instantaneous displacement of the cut strips are related to the elevated tack level of the adhesive. Thus, in a first approximation:

$$\text{Displacement} = f(\text{Smearing}) = f(\text{Tack}) \tag{7.49}$$

Smearing and displacement are also functions of the peel, but in this case one can assume that the instantaneous peel is not greater than the instantaneous tack (the real debonding force is zero), thus the phenomenon is regulated by tack;

$$\text{Cuttability} = f\left(\frac{1}{\text{Tack}}\right) \tag{7.50}$$

On this basis the known (and potential) use of a tack decreasing agent (silicones) for high tack solvent-based or hot-melt PSAs can be explained (see Chapter 8).

Smearing of the surface and displacement of the strips are tack related, thus they are coating weight-dependent phenomena. The coating weight—the adhesive amount deposited as a consequence of the flow out and transfer to the knife—depends on the total surface absorbtivity of the paper. Thick, voluminous papers contain a smaller (relative) surface of adhesive and thus display less tack:

$$\text{Displacement} = f(\text{Smearing}) = f\left(\frac{\text{Coating weight}}{\text{Surface}}\right) \quad (7.51)$$

$$\text{Cuttability} = f\left(\frac{1}{\text{Smearing}}\right) = f(\text{Front surface}) \quad (7.52)$$

Thick, voluminous release liners should be used for water-based coatings on soft PVC. The influence of the kind of the liner (stiffness) on the cuttability of the laminate was confirmed in the field.

Cuttability Differences Between Paper/Film Laminates. Generally, paper laminates are based on tackified formulations, with a high tack and low cohesion. Tack values of untackified film coatings are generally 0.1–0.5 from those of tackified paper laminates. On the other hand, cohesion values of the nontackified film coatings are 5–10 times those of the tackified ones. Thus flow out of the film coatings is less prevalent than cold flow from paper laminates; on the other hand, adhesion of the smeared surface is low. The most important problem for film coatings is the absorbtion of the smeared adhesive. In order to quantify the parameters influencing the cuttability of the film/paper laminates, one should differentiate between paper and film laminates. On the basis of practical experience the following coefficient can be associated with the parameters used in the former equation 7.9:

$$C_{\text{film}} = f\left(\frac{1AA}{0.5AC, 0.8AF}\right) \quad (7.53)$$

$$C_{\text{paper}} = f\left(\frac{0.5AA}{0.8AC, 1AF}\right) \quad (7.54)$$

where AA is the absorbtion of the smeared adhesives on the cut surfaces, AC is the lack of adhesion of the smeared adhesive, and AF is adhesive flow out.

Quantitative Evaluation of Cuttability. On the basis of the above equations the following interdependence between the parameters influencing the cuttability C can be defined:

$$C = f\left(\frac{A_{coh}, A_{anc}}{A_t, A_{dim}}\right), \left(\frac{F_{dim}, F_q}{F_{hum}}\right), \left(\frac{R_{dim}, R_q}{R_{hum}}\right) \quad (7.55)$$

where:

Converting Properties of PSAs

A_{coh} = cohesion of the adhesive
A_{anc} = anchorage of the adhesive
A_t = tack of the adhesive
A_{dim} = dimensions of the adhesive layer
F_{dim} = dimensions of the face material
F_q = quality of the face material
F_{hum} = humidity content of the face stock
R_{dim} = dimensions of the release liner
R_q = quality of the release liner
R_{hum} = humidity content of the release liner

The above relation becomes:

$$C = f\left(\frac{(A_{coh}, A_{ctt})(A_t)}{(A_t)(A_{ct}, C_t)}\right), \left(\frac{(F_{th})(F_{sm}, F_{st}, F_{af})}{F_{hum}}\right), \left(\frac{(R_{th})(R_{sm}, R_{st}, R_{af})}{R_{hum}}\right) \quad (7.56)$$

where:
A_{ctt} = time/temperature dependence of the adhesion/cohesion
A_{ct} = adhesive coating weight
C_t = coating technology
F_{af} = face material affinity toward the adhesive
F_{sm} = face material smoothness
F_{st} = face material stiffness
F_{th} = face material thickness
R_{af} = release liner affinity toward the adhesive
R_{sm} = release liner smoothness
R_{st} = release liner stiffness
R_{th} = release liner thickness

The following coefficients should be associated with the key parameters:

$$C = f\left(\frac{(1A_{coh}, 2A_{ctt})(0.2A_t)}{(0.7A_t)(1.2CW)(A_{ct}, 0.3C_t)}\right), \left(\frac{(3F_{th})(0.2F_{sm})(2F_{st})(0.4F_{af})}{10F_{hum}}\right), \left(\frac{(4R_{th})(0.1R_{sm})(2R_{st})(0.1R_f)}{10R_{hum}}\right)$$

(7.57)

$$C = f\left[\left(\frac{3.2A}{2.2A}\right), \left(\frac{5.6F}{10.0F}\right), \left(\frac{6.2R}{10.0R}\right)\right] \quad (7.58)$$

$$C = f\left[\left(\frac{21.3\%A}{14.6\%A}\right), \left(\frac{37.3\%F}{66.7\%F}\right), \left(\frac{41.3\%R}{66.7\%R}\right)\right] \quad (7.59)$$

$$C = f(35.9\%A)(104.0\%F)(108.0\%R) \quad (7.60)$$

Calculating the relative influence of the adhesive A, face material F, and release liner R on the cuttability one finds

$$C = f[(15.5\%A)(42.0\%F)(43.6\%R)] \qquad (7.61)$$

Summarizing, the cuttability is more a function of the laminate structure and coating technology than of the adhesive.

REFERENCES

1. P. Herzog, 15th Munich Adhesives and Finishing Seminar, 1990, p. 3.
2. Bachofen & Meier AG, Bülach, *Reverse gravure coating of emulsion PSA*.
3. C. P. Iovine, S. J. Jer and F. Paulp (National Starch Chem. Co., USA), EP 0.212.358/04.03.87, p. 3.
4. D. J. Zimmer and J. S. Murphy, *Tappi J.*, (12) 123 (1989).
5. C. Massa, *Coating*, (7) 239 (1989).
6. *Coating*, (11) 316 (1984).
7. *Tappi J.*, (9) 104 (1983).
8. *Coating*, (4) 123 (1988).
9. U. P. Seidl (Schneider Etiketten u. Selbstklebetechnik GmbH und Co.), OS/DE 3.625.904/ 04.02.88; in *CAS*, **16**, (3) (1988), 109, 23972b.
10. M. Bateson, *Labels and Labelling International*, (6) 36 (1986).
11. A. W. Norman, *Adhesives Age*, (4) 36 (1974).
12. R. Jordan, *Adhäsion*, (1/2) 17 (1972).
13. *Verpackungsberater*, (5) 108 (1980).
14. H. Müller, J. Türk and W. Druschke, BASF, Ludwigshafen, EP 0.118.726/ 11.02.83.
15. J. Paris, *Papier u. Kunststoffverarb.*, (9) 53 (1988).
16. W. C. Perkins, *Radiation Curing*, (8) 8 (1980).
17. R. Hinterwaldner, *Coating*, (4) 86 (1984).
18. M. Larson, *Packaging*, (11) 10 (1986).
19. J. M. Casey, *Tappi J.*, (6) 151 (1988).
20. T. Marti and R. Bidie, European Tape and Label Conference, Exxon, Brussel, April 1989, p. 171.
21. *Coating*, (2) 58 (1988).
22. L. Salmen, *Tappi J.*, 7 (12) 190 (1988).
23. A. Ryan (Avon Overseas Ltd.), private communication.
24. J. Young, *Tappi J.*, (5) 78 (1988).
25. *Coating*, (9) 264 (1984).
26. L. Placzek, *Coating*, (11) 411 (1987).
27. E. Park, *Paper Technology*, (8) 14 (1989).
28. I. Kesola, European Tape and Label Conference, Exxon, Brussel, April 1989, p. 9.
29. P. E. Patt, *Tappi J.*, (12) 97 (1988).
30. H. Senn, 12th Munich Adhesive and Finishing Seminar, 1987, p. 93.
31. *Druckprint*, (10) 32 (1987).
32. G. Jayme and G. Traser, *Das Papier*, (10A) 694 (1969).

33. J. A. Fries, *Tappi J.*, (4) 129 (1988).
34. R. Hinterwaldner, *Coating*, (4) 86 (1984).
35. H. P. Ast, *Seifen, Öle, Fette-Wachse*, (9) 289 (1985).
36. J. Verseau, *Coating*, (7) 189 (1971).
37. P. Thorne, *Finat News*, (3) 150 (1988).
38. W. Graebe, *Papier u. Kunststoffverarb.*, (2) 49 (1985).
39. M. Bowtell, *Adhesives Age*, (12) 37 (1986).
40. *Der Siebdruck*, (5) 21 (1988).
41. R. Wilken, *Finat News*, (1) 53 (1988).
42. *apr.*, (14) 340 (1987).
43. K. Taubert, *Adhäsion*, (10) 377 (1970).
44. *Die Herstellung von Haftklebstoffen*, T1.2.2, 15d, Nov. 1979, BASF, Ludwigshafen.
45. *Der Siebdruck*, (1) 14 (1988).
46. A. Zosel, *Colloid Polymer Sci.*, **263**, 541 (1985).
47. J. A. Fries, *New Developments in PSA*, p. 27.
48. R. Lowman, *Finat News*, (3) 5 (1987).
49. *Adhäsion*, (1/2) 20 (1987).
50. *Adhesives Age*, (7) 12 (1986).
51. *Adhesives Age*, (7) 36 (1986).
52. H. K. Porter Co., USA, US Pat., 3.149.997; in *Adhäsion*, (2) 79 (1966).
53. *Coating*, (1) 20 (1984).
54. *Coating*, (1) 89 (1984).
55. P. Holl, *Verpackungs-Rundschau*, (3) 214 (1986).
56. T. Zepplichal, *Adhäsion*, (9) 19 (1984).
57. G. Meinel, *Papier u. Kunststoffverarb.*, (10) 26 (1985).
58. *Adhäsion*, (1) 11 (1974).
59. *Adhäsion*, (8) 296 (1984).
60. L. Placzek, *Coating*, (6) 140 (1980).
61. *Kunststoffverpackung*, (2) 7 (1988).
62. *Topics*, Courtaulds Films, (1) autumn, 1991.
63. Mobil, OPP Art., (5) 1 (1991).
64. *Converting Today*, (3) 5 (1981).
65. *Converting Today*, (11) 7 (1991).
66. P. Hammerschmidt, *apr.*, (4) 190 (1986).
67. D. A. Markgraf, *Converting and Packaging*, (3) 18 (1986).
68. E. Djagarowa, W. Rainow and W. L. Dimitrow, *Plaste u. Kaut.*, (2) 100 (1970); in *Adhäsion*, (12) 363 (1970).
69. *Adhäsion*, (5) 27 (1982).
70. K. Armbruster and M. Osterhold, *Kunststoffe*, (11) 1241 (1990).
71. W. Brockmann, *Adhäsion*, (12) 38 (1978).
72. W. H. Kreidl, *Kunststoffe*, (49) 71 (1959).
73. H. Büchel, *Adhäsion*, (10) 506 (1966).
74. G. Kühne, *Neue Verpackung*, (34) 434 (1981).
75. J. Hansmann, *Adhäsion*, (12) 136 (1979).

76. E. Prinz, *Coating*, (12) 269 (1979).
77. R. Van Linden, *Kunststoffe*, (69) 71 (1979).
78. T. J. Bonk and S. J. Thomas, (Minnesota Mining and Manuf. Co., USA), EP 120.708.31/ 03.10.84.
79. Eckhard Prinz, *Coating*, (1) 20 (1978).
80. G. Höfling and H. Breu, *Adhäsion*, (6) 252 (1966).
81. *Coating*, (9) 119 (1969).
82. *Coating*, (9) 296 (1969).
83. E. Prinz, *Coating*, (2) 56 (1978).
84. P. B. Sherman, *Papier u. Kunststoffverarb.*, (11) 62 (1986).
85. C. Jepsen and N. Räbiger, *Kautschuk, Gummi, Kunststoffe*, (4) 342 (1988).
86. *Converting Magazine*, (3) 54 (1986).
87. E. Prinz, *Coating*, (10) 374 (1988).
88. G. Menges, W. Michaeli, R. Ludwig and K. Scholl, *Kunststoffe*, (1) 1245 (1990).
89. R. M. Podhajni, *Converting and Packaging*, (3) 21 (1986).
90. D. Briggs, C. R. Kendall, A. R. Blythe and A. B. Wooton, *Polymer*, (24) 47 (1983).
91. B. Märtens, *Coating*, (6) 187 (1992).
92. Softal, *Report*, (102) 2 (1994).
93. E. H. Prinz and K. H. Meyer, *Papier u. Kunststoffverarb.*, 10/1984, in /505/88.
94. H. Grünewald, Non-Poluting Plasma Pretreatment for Better Adhesion, 15th Munich Adhesives and Finishing Seminar, 1990.
95. F. A. Sliemens, *Papier u. Kunststoffverarb.*, (4) 11 (1988).
96. W. Eisby, *apr.*, (829) 794 (1988).
97. Paul Swaraj, *Papier u. Kunststoffverarbeiter*, (1) 52 (1988).
98. G. Liebel, *Vorbehandlung von Kunststoff Oberflächen mit Niederdruck Plasma*, Fachtagung, Kunststoffoberflächen des Praxis Forums, Berlin, Bad-Nauheim, 1977.
99. Heger, US Pat., 4.041.192 in ref. 78, p. 3.
100. H. Fydelor, US Pat., 4.148.839 in ref. 78, p. 3.
101. Magat, US Pat., 3.252.880 in ref. 78, p. 3.
102. Garnett, US Pat., 4.179.401 in ref. 78, p. 3.
103. S. Yamakawa and F. Yamamoto, *J. Appl. Polymer Sci.*, **25** 31 (1980).
104. S. Shkolnik and D. Behar, *J. Appl. Polymer Sci.*, **27** 2189 (1982).
105. *Adhäsion*, (12) 506 (1966).
106. *Adhäsion*, (10) 6 (1983).
107. D. K. Perino, *Paper, Film and Foil Conv.*, (11) 85 (1969).
108. *Coating*, (6) 174 (1972).
109. L. Devine and M. Bodnar, *Adhesives Age*, (5) 35 (1969).
110. R. Milker and Arthur Koch, *Coating*, (1) 8 (1988).
111. F. T. Kitchel, *Tappi J.*, (2) 156 (1988).
112. *Papier u. Kunststoffverarb.*, (10) 2 (1985).
113. *Converting Today*, (11) 9 (1981).
114. A. M. Slaff, *Coating*, (7) 198 (1973).
115. J. Patschorke, *Adhäsion*, (3) 37 (1970).
116. R. Hummel, *Adhäsion*, (1/2).

117. *Adhesives Age*, (11) 40 (1988).
118. F. Weyres, *Coating*, (4) 110 (1985).
119. L. Placzek, *Coating*, (8) 199 (1980).
120. *Paper, Film and Foil Conv.*, (9) 28 (1989).
121. H. E. Krämer, *apr.*, (31) 843 (1988).
122. U. Eisele, *Kautschuk, Gummi, Kunststoffe*, (6) 539 (1987).
123. A. Masaaki, S. Sadaji, N. Hiroshi and O. Tatsuhiko (Nitto Electric Ind. Co. Ltd.) Japanese Pat. 63.142.086/14.06.88; in *CAS*, **25** (6) (1988).
124. *Adhäsion*, (5) 76 (1984).
125. *Neue Verpackung*, (5) 176 (1980).
126. D. Satas, *Adhesives Age*, (8) 28 (1988).
127. *Der Polygraf*, (5) 368 (1986).
128. K. Fust, *Coating*, (2) 65 (1988).
129. *Pack Report*, (11) 47 (1983).
130. R. Hummel, 12th Munich Adhesive and Finishing Seminar, 1987, p. 58.
131. *Druckwelt*, (13) 34 (198).
132. G. Perner, *Coating*, (5) 150 (1992).
133. *Der Siebdruck*, (1) 14 (1988).
134. M. Schlimmer, *Adhäsion*, (4) 8 (1987).
135. A. Zimmermann, *Verpackungs-Rundschau*, (3) 298 (1972).
136. *Paper, Film and Foil Conv.*, (10) 39 (1971).
137. *Adhäsion*, (1/2) 27 (1987).
138. O. Fellers, L. Salmen and M. Htun, *apr.*, (35) 1124 (1986).
139. B. Wright, *Adhesives Age*, (12) 25 (1971).
140. *Coating*, (3) 64 (1974).
141. C. W. Drechsler, *Coating*, (1) 10 (1985).
142. H. Salmen and M. Htun, *apr.*, (35) 28 (1986).
143. E. Böhmer, *Norsk Skogindustrie*, (8) 258 (1968).
144. E. Brada, *Papier u. Kunststoffverarb.*, (5) 170 (1986).
145. A. W. Dobmann and H. Braun, Neues Engineeringkonzept für Etikettenindustrie, 10th Munich Adhesive and Finishing Seminar, 1985.
146. *apr.*, (37) 1046 (1988).
147. K. Matschke, Hoechst Folien, 5. Klebebandforum, Frankfurt/M, Nov. 1990, p. 141.
148. *Druck-print*, (11) 33 (1986).
149. *European Adhesives and Sealants*, (9) 46 (1987).
150. J. A. Fries, *Tappi J.*, (4) 129 (1988).
151. Westwald Co., US Pat., 2.140.710; in *Coating*, (7) 184 (1974).
152. F. Grajewski, A. Limper and G. Schwarter, *Kautschuk, Gummi, Kunststoffe*, (12) 1188 (1986).
153. T. Timm, *Kautschuk, Gummi, Kunststoffe*, (1) 15 (1986).
154. *apr.*, (42) 1152 (1988).
155. H. Senn, *Papier u. Kunststoffverarb.*, (4) 470 (1986).
156. G. Renz, *apr.*, (24/25) 860 (1986).
157. S. W. Medina and F. W. Distefano, *Adhesives Age*, (2) 18 (1989).

8
Manufacture of Pressure-Sensitive Adhesives

The manufacture of PSAs includes the manufacture of the raw materials as well as their formulation. PSAs include solutions, dispersions, or hot-melt adhesives. Water-based PSAs can be used in certain cases as such without any additional formulation. Solvent-based acrylic PSAs also may be used unformulated. The origin of the PSA manufacture is based on the blending and dissolution of different raw materials, none of which exhibits a PSA character by itself. Therefore the role of the formulator is especially important in the development of PSAs.

1 MANUFACTURE OF PSA RAW MATERIALS

Manufacturing the raw materials refers to the manufacture of the components of the PSA recipe. There are rubberlike (elastic) and liquidlike (viscous) PSA components (see Chapter 5). Rubber-resin PSAs are mixtures of both components. Synthetic PSA raw materials (e.g., acrylics, vinyl acetate, ethylene, and maleinate copolymers) possess viscoelastic properties obtained through synthesis. In this case the manufacture provides a viscoelastic material in the desired physical state (solid state, solution, or dispersion) through the synthesis.

1.1 Natural Raw Materials

Natural raw materials with rubberlike properties and natural resins or plasticizers may be used in order to formulate an adequate recipe (see Chapter 5). Of course these natural raw materials are not really manufactured, but modified chemically or physically. Physical modification allows the use of these materials in a suitable state (i.e., in solution or dispersion).

Rubber

Natural rubber is used as a solution, and manufacture of the PSAs implies dissolution of the rubber. In fact a slight mechano-chemical destruction of the rubber occurs in parallel with the dissolution process. At least theoretically a precalendering (i.e., a controlled depolymerization of the natural rubber) is possible, but not economically viable. There are only a few special uses (like protective films) were mastication of the rubber must be carried out. Masticated rubber was used for tapes [1]. Pressure-sensitive adhesives exhibiting low tack, for bonding onto itself, also are formulated on the basis of masticated natural rubber [2].

The high molecular weight portion of natural rubber is insoluble in solvents and therefore natural rubbers must be milled to a Mooney viscosity of 53–75 or below to obtain solubility at 10–30% solids in solvent [3,4]. As shown in Figure 8.1 and Table 8.1 unmilled natural rubber displays a broad range of molecular weight (based on viscosity measurements) as a function of its

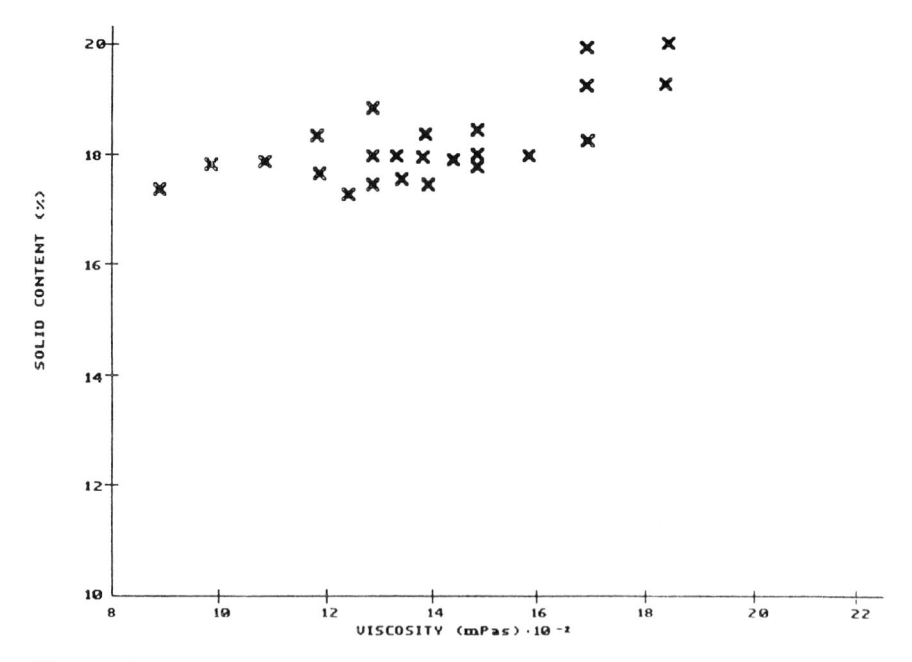

Figure 8.1 Variation of the solution viscosity values for a given solid content, for rubber/resin PSAs on the basis of milled rubber.

Table 8.1 Dispersion of the Viscosity Values for Unmilled Rubber-Based PSA

Batch	Brookfield viscosity (mPa·s)
01	11,000
02	20,000
03	17,000
04	19,000
05	14,000
06	12,000
07	16,000
08	10,000
09	18,000
10	13,000

A 1/1 (wet weight/wet weight) rubber/resin blend was dissolved and diluted to the same solids content.

natural product character; milling reduces the viscosity (Figure 8.1) and the variation of the viscosity values (Table 8.1). Milling reduces the molecular weight and thus the peel adhesion (see Chapters 5 and 6).

Resins

Resins are used in solution, in dispersions, or as liquid resins. Manufacturing the raw materials means a technical process in order to get the resin in the desired end-use form. Unfortunately, natural resins possess a special chemical structure which is oxidation and polymerization sensitive; therefore natural resins have to be modified chemically in order to be stabilized. Generally, disproportionation and hydrogenation improve the aging stability. Hydrogenation also increases resin compatibility [5]. Dimerization and esterification lead to high melting point resins (see Chapter 5).

1.2 Synthetic Raw Materials

Synthetic raw materials for PSAs are mostly viscoelastic materials which need no or only a low level of viscous formulating additives; hence manufacturing refers to the synthesis and formulation of a PSA adhesive. On the other hand, the development of the polymer chemistry in the last decade made it possible to obtain synthetic viscous components (tackifier resins or plasticizers). Recent progress in the macromolecular chemistry led to the synthesis of special (stereoregulated or crosslinkable) elastomers. Today synthetic raw materials for PSAs are available as elastomers and viscous components.

Synthetic Elastomers and Polymers

Manufacturing synthetic rubber (as a raw material for PSAs) implies the synthesis of special rubbers (elastomers) and special viscoelastic materials. The rubbers possess a special chemical composition or structure. In the class of new elastomers with a special structure, the stereoregulated diene-based tactic elastomers (used mainly in rubber-resin solvent-based formulations) and the crosslinkable, block copolymers (used mainly as raw materials for hot-melt PSAs) should be mentioned. In this range of materials the halogen containing elastomers should be noted also. Among the rubberlike raw materials with a viscoelastic character there is a broad range of chemically quite different polymers, like the hydrocarbon-based carboxylated styrene-butadiene rubber (CSBR; used mostly in aqueous dispersions), the acrylic-based homo- and copolymers (used for solvent-based, water-based, or hot-melt PSAs), the silicone elastomers (used mainly as solutions), and polyurethanes (see Chapter 5).

Synthetic Resins

In Chapter 5 a detailed description of the resins used for PSA manufacture was given and the chemical nature of synthetic resins discussed. Details about the use, the level, and recipe of synthetic resin-based PSAs are given in Chapter 6. Synthetic resins are mostly based on hydrocarbons, but derivatives of heterocyclical compounds also are used. Like natural resins, these compounds also can be used in liquid, solvent, or dispersed state.

Some of the synthesic resins are prepared by polymerizing special monomers. Such monomers are usually derived from the so-called C_9 and C_5 cuts in the fractionation of crude oil or similar materials. Such synthetic hydrocarbon tackifiers generally have ring-and-ball softening point temperatures between 10°C and about 100°C. The polycarboxylic acid ester tackifier resins are polymerized from one or more monomers such as acrylic acid which may be substituted with alkyl or alkoxyl radicals having one to four carbon atoms or with alkyl or alkanol esters of such acids in which the alkyl or alkanol moiety contains from one to about six carbon atoms. The most important raw materials are C_4–C_5 diolefins, styrene, α-methyl styrene, vinyl toluene, dicyclopentadiene, α and β-pinene, d-limone, isoprene, and piperylene. The polymers may be classified as aliphatic, aromatic, alkyl aromatic, monomer-based, hydrogenated, functionalized hydrocarbon resins, or terpene resins. In comparison with natural resins, hydrocarbon resins display only a limited compatibility. Aliphatic resins are suggested for aliphatic polymers or block copolymers where the aliphatic midblock has to be tackified, like natural rubber, S-B-S or S-I-S block copolymers. Aromatic resins are used for aromatic or polar polymers, like acrylics, styrene butadiene rubber, or ethylene vinyl acetate copolymers. The aging stability of the resin is very important. Resins displaying good aging stability include aliphatic hydrocarbon resins, aromatic hydrocarbon resins, dicyclopentadiene resins, and pure monomer-based resins.

Additives

The additives used in PSA formulations are included in order to modify the adhesive, converting, or end-use properties. Most of them produce changes in the rheology of the adhesive; others serve to modify the liquid state properties, the coatability of the adhesive, or the convertability of the PSA laminate. There is a special class of additives used in order to improve the chemical stability of the adhesive. Other additives may be selected in order to improve the antimicrobial resistance of the adhesive. The adhesive raw materials (i.e., base elastomers and tackifiers) are processed during the manufacture of PSAs together with other additives (i.e., they are formulated).

2 FORMULATING PSAs

Formulating means changing the adhesive composition, by addition of different macro- or micromolecular compounds, having elastomeric, or viscous, surface-active, or chemically stabilizing properties, in order to prepare recipes for PSAs; it is advisable to add antimicrobial agents to the adhesive. The moderate pH and presence of polyvinyl alcohol makes latexes sensitive to bacteria.

PSAs should have a T_g of about −15 to −5°C. Blending a high T_g, hard resin with a lower T_g elastomer provides a desirable formulating latitude [6]. On the other hand, PSA coating and converting, and end-use of the PSA label require other special performance characteristics. These can be achieved only through formulating. The scope of the formulation of PSAs is to produce tailored adhesives for each one of the intended end-uses. Generally, the coating technology of the adhesive, the converting technology of the adhesive, the end-use properties, and economical considerations influence the formulation of PSAs.

The existing coating and converting machines, as well as environmental considerations, determine the choice of an adhesive. End-use properties of the label require special face stock materials and/or PSAs with certain adhesive properties. Economical considerations also limit the choice of the laminate components and formulating components. Generally, PSA formulations contain polymers (elastomers), tackifiers (resins or plasticizers), fillers, stabilizers, and carrier agents (solvent, water) [7]. This constitutes a base formulation, where all the components (except the carrier agent) contribute to the adhesive properties.

Table 8.2 Tackifier Level Required to Improve the Tack

Formulating components			Chemical composition, parts by weight (wet)														
Code	Supplier	Nature	1	2	3	4	5	6	7	8	9	10	11	12	13	14
3703	Polysar	CSBR	100	80	33	—	33	—	43	40	—	—	—	—	—	—
CF52	A&W	AC rosin	—	20	23	—	23	—	19	26	—	—	—	—	—	—
80D	BASF	AC	—	—	33	100	43	—	—	34	—	—	—	—	—	—
V205	BASF	AC	—	—	—	—	—	100	38	—	—	—	—	—	—	—
Exoryl 2001	Exxon	AC	—	—	—	—	—	—	—	—	100	—	—	—	—	—
SE-351	A&W	Ester rosin	—	—	—	—	—	—	—	—	—	100	—	20	—	—
5650	Hoechst	AC	—	—	—	—	—	—	—	—	—	20	100	100	—	20
Nacor-90	National	EVA copolymer	—	—	—	—	—	—	—	—	—	—	—	—	100	100
Rolling ball tack (cm)			15.0	3.0	1.5	7.0	2.5	6.0	1.8	2.6	5.9	4.5	3.0	2.8	11.0	8.0

2.1 Adhesive Properties

There are many reasons for formulating. The first one is the need to improve the adhesive properties. Rubberlike base materials used for PSAs do not possess sufficient intrinsic tack and peel adhesion. Most viscoelastic materials with built-in tack through the synthesis do not possess enough peel adhesion to nonpolar surfaces, and soft, tacky PSA raw materials exhibit too low a cohesion level. Therefore the whole range of the adhesive properties has to be achieved through formulation.

Adhesion/Cohesion Balance

A proper balance between high tack, peel adhesion, and high cohesion is necessary in most cases. It should be mentioned that formulating for high tack and peel (discussed generally in terms of tackification), using viscous, formulating agents (tackifiers and plasticizers) decreases the shear resistance. A high cohesion level is given by the rubberlike component of the adhesive or by partial crosslinking.

Tack. The main goal of the formulation is to improve the tack and the peel adhesion of PSAs. Generally, this is achieved by the use of plasticizers and tackifiers. Tackifying in order to obtain an improved tack is easier. As can be seen from Table 8.2, a tackifier level of 10–20% is enough for tack improvement of acrylic-based PSAs. A higher tackifier level is imposed mainly for the improvement of the peel adhesion, not of the tack. This statement appears very important for formulations of PSAs coated on soft PVC, where the adhesive should contain a minimum resin level in order to avoid resin/plasticizer interaction.

Trends for the improvement of the tack also are observed. New, wider converting machines (more than 2 m wide) are built, giving rise to a higher variation of the coating weight (12–26 g/m^2). New face stock and substrate materials, based on polyethylene-polypropylene copolymers, are introduced with poorer anchorage and peel values. New PSA raw materials with lower tackifying response (EVAc, CSBR) are now used. Hence, new film coatings with better tack properties are necessary. More removable, and semipermanent labels (with lower coating weights) are needed. All these changes require adhesives with improved tack properties (e.g., more tackifier or more efficient tackifiers).

Peel. High peel adhesion requires a certain tack level for bonding and a certain cohesion for debonding. Tack increases continuously upon adding soft, viscous components to the formulation. The dependence of the peel on the ratio of elastic/viscous components is more complex, going through a maxi-

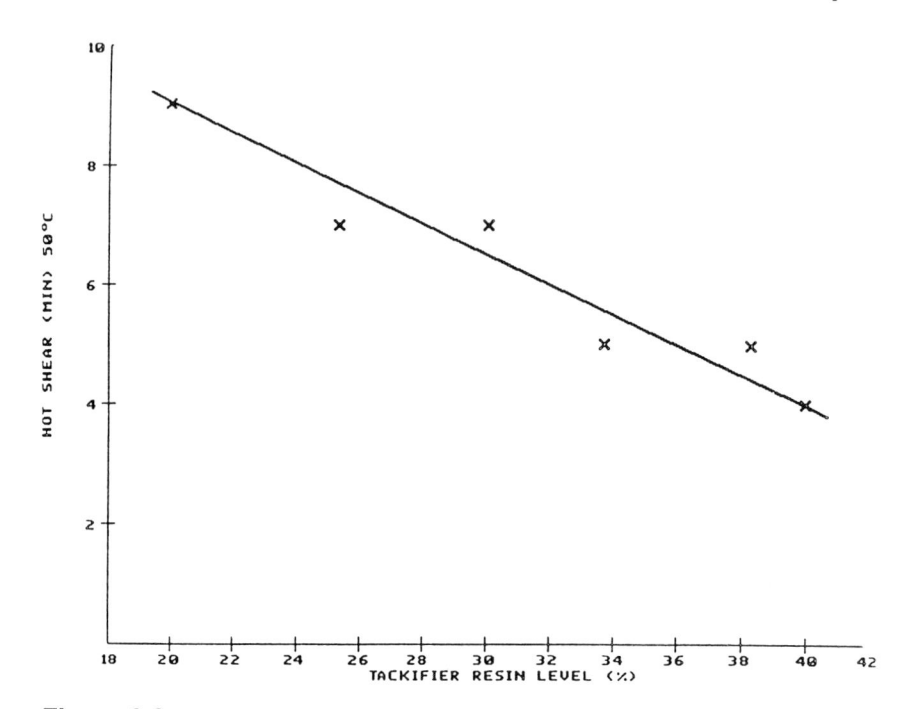

Figure 8.2 The influence of the tackifier level on the shear resistance. Hot shear (50°C) dependence on the tackifier resin concentration.

mum as a function of the level of the soft component. The same ambiguous influence of the crosslinking on the peel should be noted. There are adhesive formulations with crosslinking agents leading to low peel adhesion (i.e., removable PSAs), but in some cases crosslinking leads to high peel adhesion (see Section 1.2 of Chapter 6).

Shear. The most important means to influence the cohesion are tackification and crosslinking. Commonly-used tackifier levels cause a decrease in shear strength. The loss of the shear resistance due to the tackification process is a function of the elastomer and tackifier nature and their ratio. Figure 8.2 indicates that a tackifier level above 20% decreases the shear strength of acrylic PSAs.

It is desireable to obtain as much tack as possible without losing a significant amount of internal strength (cohesion). The better balance of adhesion and cohesion properties is difficult to obtain through polymerization only. The addition of hard monomers to increase the cohesive strength generally results in a decrease of the tack [8].

2.2 Formulating Opportunities

It is possible to modify the adhesive properties of a PSA by changing the base rubberlike polymer (i.e., through blending different elastomers). For rubber-resin formulations this implies the use of different qualities of natural rubber, or blends of natural rubber with synthetic elastomers (e.g., stereoregulated ones). For synthetic PSAs mixtures of acrylics, or acrylics with EVAc or CSBR can be used. Although generally used, this formulating freedom is less well known or applied than blending the rubberlike component with low molecular, viscous materials (mainly resins) known as tackification.

2.3 Tackification

Tackifying or tackification refers to the increase of the tack and peel adhesion through formulation: as formulating agents low molecular weight, soft materials such as resins and plasticizers or special elastomers can be used.

Tackification with Resins. The T_g and modulus of acrylic PSAs can be adjusted by polymerizing with different ratios of acrylic monomers. Resin addition offers an additional possibility for this modification [3]. Tackifying resins have found widespread use in the PSAs industry since its inception. There are practical and theoretical reasons for their use and there exist technical and commercial reasons which make the use of tackifiers necessary.

It is a fact that the tack and peel adhesion (especially the peel on nonpolar, untreated surfaces) of PSAs (independent of their chemical basis, converting technology, and end-uses) do not meet the practical requirements. Therefore, there is a need for formulating (compounding) the raw adhesives with chemicals providing a better tack/peel level. Traditionally these materials were employed in rubber-based adhesives, where their use is often necessary in order to achieve a desirable balance of properties. Acrylic PSAs can be designed in such a way as to be suitable for some applications without the need for modification. The addition of tackifiers to latex acrylic systems can, however, widen the performance window available to these materials, and thus can be a useful tool to complement the existing methods of performance modification. Several methods of performance modification do exist: the best known formulation methods include the compounding with other PSAs and/or adding plasticizers. Unfortunately these methods only have a limited use. The formulation of PSAs with other PSAs does not always result in the desired tack/peel balance. Compounding with macromolecular plasticizers (like polybutenes) may give rise to compatibility and cold flow problems. Addition of plasticizers changes the rheology of the adhesive, involving the decrease of the peel and staining resistance. Thus the most important method of tackification remains formulating PSAs with resins. The successful incorporation of

Table 8.3 Glass transition temperature and molecular weight of common tackifier resins and PSAs

Formulating component

		Tackifier resin					Base polymers		
Code		Supplier	Nature	MW	T_g (°C)	Code	Supplier	Nature	T_g (°C)
Regalrez	1018	Hercules	Hydrogenated HC-resin	407	−22	80 D	BASF	AC	−53
	1033			493	−7	V205	BASF	AC	−40
	1065			723	17	102L	BASF	AC	−58
	1078			819	29	A120	BASF	AC	−50
Piccolyte	590	Hercules	Terpene resin	—	32	S304	BASF	SAC	−22
Piccotex	75	Hercules	Aromatic resin	—	29	DH130	Hoechst	EVAc	−5
	9191/92	Exxon	C_5 aliphatic HC	—	46	DH137	Hoechst	EVAc	−25
	323		Modified C_5 HC	—	−12	AK600	Condea	AC VAc	−24
	9251		Modified C_5 HC	—	12	AK610	Condea	SAC	−12
	339		Cyclic hydrogenated HC	—	29	XZ87543	DOW	CSBR	−43
	9271		Modified C_5	—	25	3703	Polysar	CSBR	−54
	348		C_5/C_9 HC	—	−26	3958	Polysar	CSBR	−31

The glass transition temperature T_g was measured by DSC.

Manufacture of Pressure-Sensitive Adhesives

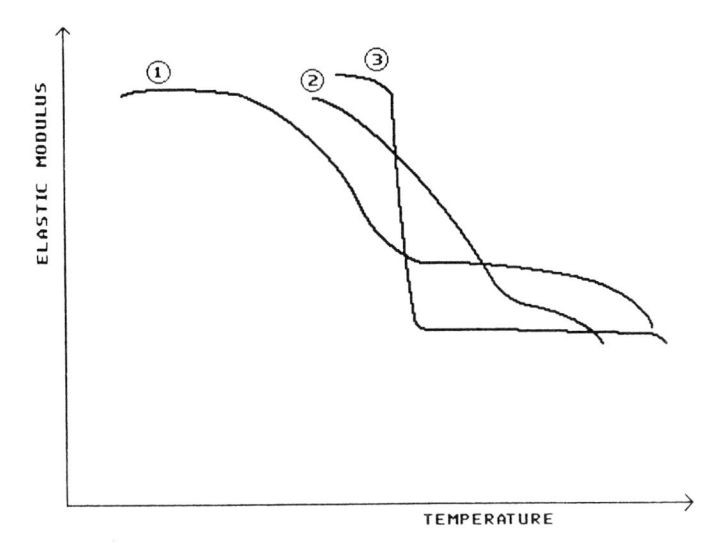

Figure 8.3 The effect of the tackifier on the modulus. The different behaviour of the modulus for 1) a pure and 2) tackified formulation versus 3) ideal comportment. Modulus change with increasing temperature/decreasing strain rate.

tackifier resins can lead to significant cost savings. Therefore cost reduction also implies the use of tackifiers.

The resin functions as a solid solvent for the polymer. Generally, resin tackifiers are macromolecular compounds with lower molecular weight and modulus and a higher T_g than current PSAs (see Chapter 3). Table 8.3 lists some data concerning the T_g and molecular weight of common tackifier resins and acrylic PSAs.

Compounding PSAs with the tackifier resin lowers the modulus and increases the T_g of the blend. The minimum modulus depends on the resin/rubber solubility. On the other hand, a high resin loading increases the modulus also. A lowered modulus must promote bond formation (creep compliance), a higher T_g must make bond rupture difficult (debonding resistance). It can be seen from Figure 8.3 that tackification lowers the rubbery plateau region. Suitable formulations must display a longer and lower rubbery plateau region, but a rapid (almost vertical) transition (with increasing strain rate) in the glassy region.

In order to meet these requirements, a good tackifier must have a low molecular weight, a high T_g, and exhibit good compatibility (see Chapter 3). In order to evaluate the tack of formulated PSAs, the following examination criteria will be considered, namely the ease of tackifying, the tackifier level and the obtained tack. The suitability of PSAs to be tackified (ease of tackification) should be examined with respect to the chemical nature and the physical state of the base polymer and of the tackifier. Elastomers with different chemical bases used as main PSA components have different tackifying responses (i.e., their compounding at a same level of tackifier results in different levels of improvement of tack and peel, and a decrease of the shear). As discussed in Chapter 6, acrylic PSAs display a special, high tackifying ability compared with EVAc or CSBR. Acrylic hot-melt PSAs display an excellent compatibility with the same tackifiers and plasticizers as S-I-S block copolymers [9]. Acrylic PSAs display good compatibility with common tackifiers (plasticizers or tackifier resins). They are compatible with tackifier resins of a different chemical basis (e.g., rosin acid, rosin ester, hydrocarbon, urea-formaldehyde resins). CSBR show better compatibility with hydrocarbon resins; EVAc possesses only a limited compatibility with all kinds of resins. There is a general compatibility between low tack/high peel and high tack/low cohesion acrylic PSAs, or between acrylic and other PSAs (EVAc or CSBR). Compounding of acrylic PSAs with EVAc yields high shear; compounding with CSBR leads to high tack. The data of Tables 8.4 and 8.5 demonstrate the adjustment possibilities of the peel adhesion based on PSA compounding.

The use of chloroprene latex as shear modifier is well known. Neoprene 102 latex is a noncrystallizing low gel dispersion; its low gel content allows good quick stick and a good holding power when properly formulated [10]. Tack improvement can be achieved using soft, tacky acrylic dispersions or polyvinyl ether derivatives [11].

It is difficult to predict the resin level required to achieve a certain tack or peel value because of the dependence on the nature of the tackified elastomer and on the nature of the tackifying resin. The nature of the face stock/adherent should be taken into account as well. A level of up to 30–40% (by weight) tackifying resin is used in formulated acrylic PSAs [12–14]. Generally, acrylic PSAs require a lower tackifier loading level than CSBR- and EVAc-based PSAs. Acrylic PSAs need 30–40% tackifier in order to ensure a high level of polyethylene peel. The use of more than about 25 wt% of tackifier in the total adhesive is common, and more than 50% will lead to the adhesive being nonpeelable.

The modulus increases at higher resin loadings. A modulus increase will usually begin in the range of 40–60% resin loading [13]. Hence this is the upper limit of the tackifier amount. On the other hand, it was mentioned before that more than 20% tackifier resin leads to lower shear strength. The minimum/maximum amount of the tackifier depends on the acrylic nature (chemical composition, molecular weight), the nature of the tackifier, and on the converting and end-use properties (face stock, edge bleeding or oozing, migration, die cutting, etc.). The tackifier level is a function of the desired

Table 8.4 Peel Modification Through Tackification of AC PSAs with Other AC PSAs

	Formulating components							Adhesive properties		
	1			2			Coating weight (g/m^2)	Peel (N/25 mm)		Rolling ball tack (cm)
Code	Supplier	Basis	Level (pbw, wet)	Code	Supplier	Basis	Level (pbw, wet)		Glass	PE	
FC88	UCB	AC	100	PC80	UCB	AC	—	18	8.0	4.0	3.5
FC88	UCB	AC	—	PC80	UCB	AC	100	19	6.0	5.0	2.5
FC88	UCB	AC	80	PC80	UCB	AC	20	30	—	4.0	3.5
FC88	UCB	AC	70	2313	Nobel	AC	30	23	11.0	5.0	4.5
FC88	UCB	AC	50	2313	Nobel	AC	50	24	8.0	3.0	7.5
FC88	UCB	AC	50	880D	BASF	AC	50	21	13.0	8.0	7.0
FC88	UCB	AC	70	880D	BASF	AC	30	21	10.0	5.0	6.5
V205	BASF	AC	100	80D	BASF	AC	—	19	—	4.0	5.0
V205	BASF	AC	80	80D	BASF	AC	20	20	16.0PT	9.0	10.0

Note: The samples 1 through 7 are film coatings; the samples 8 and 9 are paper coatings. pbw = parts by weight; PT = paper tear

Table 8.5 Adjustment of the Adhesive Properties Via PSA Compounding

Formulating components			Chemical composition, parts by weight (wet)											
Code	Supplier	Base	1	2	3	4	5	6	7	8	9	10	11	12
V205	BASF	AC	100	—	—	30	40	—	—	—	—	—	—	—
1360	Hoechst	EVAc	—	100	—	30	30	50	70	50	25	—	—	64
FC88	UCB	AC	—	—	100	30	20	—	—	—	—	—	—	—
913	UCB	AC	—	—	—	—	—	50	—	—	—	—	—	—
1370	Hoechst	EVAc	—	—	—	—	—	—	30	—	—	—	—	—
85D	BASF	AC	—	—	—	—	—	—	—	50	100	100	—	16
3703	Polysar	CSBR	—	—	—	—	—	—	—	—	—	—	100	20
Adhesive properties														
Peel (N/25 mm) from glass			—	4.0	8.0	10.0	10.0	—	—	11.0	12.0	12.0	10.0	10.0 PT
		PE	4.0	2.0	4.0	5.6	5.0	0.2	5.0	—	—	8.0	—	—
RB tack (cm)			6.0	6.0	3.5	3.5	3.5	7.0	3.0	—	—	—	—	8.0
Coating weight (g/m^2)			19	18	18	24	24	21	23	20	21	20	18	—

Note: Samples 1 through 7 are paper coatings; samples 8 through 12 are film labels.

adhesion/cohesion balance, and of the end-use properties of the laminate. Theoretically, it also is a function of the unformulated PSA and of the tackifier nature (compatibility). Thus for natural rubber, a two-phase structure for natural rubber-resin mixtures was proposed [15]. More resin and low molecular rubber have migrated to the surface. Only a 12% resin rich phase was identified optically [16]. In some cases of tackified PSAs, if the ratio of the energy loss modulus and energy storage modulus (tan δ) is examined by dynamic mechanical analysis (DMA), another peak appears. This extra peak may be related to a resin-rich phase; there could be some phase separation between the resin and the rubber so that the rubbery matrix exists as a two-phase material. When tackifying block copolymers the compatibility of the additives with both phases must be taken into account [17]. Aromatic resins (coumarone-indene) associated with the pressure-sensitive domains give nontacky materials. Aromatic associating resins influence the tensile strength less than resins associating with the elastomer. There are compatibilizers like resin acids.

In hot-melt PSAs the aromatic resins are compatible with the polystyrene domains, and increase (or decrease) the stiffness depending on the melting point (> 90°C, or < 70°C). For better shear and peel coumarone-indene resins (melting point = 110°C) or α-methyl styrene resins (melting point = 145°C) should be used [17]. The aliphatic resins are compatible with the rubber segments (see Chapters 2 and 5). Resins of choice for CSBR-based adhesive systems are the rosin acid and rosin ester resins and some of the terpene phenolics. The hydrocarbon resins as a general rule are not as compatible with CSBR polymers as the rosin acids and rosin esters. For nitrile and neoprene polymers the best choice remains the phenolics and modified phenolic resins [18]. Some manufacturers of tackifier dispersions provide data about the compatibility of the tackifier with PSA dispersions of different chemical basis. Compatibility is examined according to the optical appearance (clear, hazy, opaque, separation of components) of the film [19]. Tackifiers may be included in either minor or major amounts. The adhesives may contain very small amounts of tackifier, to increase the tack of the total composition, or they may contain up to 150 parts (by weight) or more of tackifier per 100 parts of one or more of the polymers. For the same amount of resin addition, a natural rubber-based system behaves quite differently from one based on S-B-S block copolymers. For the same resin loading the two systems would lie in different areas of the window of application, due to the difference in the loss tan δ peak temperature of the polymers. For S-B-S multiblock copolymers the midblock loss tan δ peak temperature of the polymer is much lower than that of natural rubber and CSBR copolymers. At room temperature the storage modulus value is greater by two decades than that of natural rubber; thus the concentration of compatible resin in the styrene-butadiene system is much greater (see Chapters 2 and 3). This cannot be predicted by the Fox equation because of the varying compatibility of a resin with the midblock and the endblock of a block copolymer. A rheology modifier must be added to reduce the storage modulus value, but not significantly affect the loss tan δ peak temperature; this modifier is usually an oil or a liquid resin [20].

As an example for natural rubber-based PSAs maximum tack is obtained for 62 parts resin/100 parts rubber [19]. Maximum quick stick is obtained at 87% resin and optimum peel adhesion is obtained at 100/100 tackifier level [21]. Tackifying natural rubber with a hydrocarbon resin gives maximum tack at 50–65% resin, maximum quick stick for 85% resin and maximum peel at a 100/100 resin/rubber ratio [21]. For polybutadiene 60–150 parts resin per 100 parts rubber should be added [22]. For S-I-S maximum tack is given at 50–87% resin and optimum quick stick at 75–125% resin [19]; a loading level of 150/100 is required for maximum peel. Rolling ball (RB) tack and quick stick show quite different maximum levels as a function of the resin concentration [23]. Maximum tack is achieved at lower resin concentrations (55–70 parts resin/100 parts of S-I-S rubber), quick stick has a maxima at 95–100 parts of resin/100 parts of rubber. S-I-S rubber displays a maximum rolling ball tack at 40–100 parts hydrocarbon resin/rubber ratio and maximum peel adhesion is achieved at a 100/100 ratio [21]. As can be seen from the above examples the tackifier level depends on the elastomer nature. It should be stressed that a maximum peel level implies more tackifier than a maximum tack level. Tack for acrylic hot-melt PSAs goes through a maximum at 45–60 parts tackifier with properties also dependent on the tackifier composition [24]. In experiments with ethylene/maleinate copolymers, a 30/100 resin/elastomer ratio was used [25]. In tackifying studies of acrylic latices using rosin derivatives 0–70 parts resin/100 parts latex were used [3].

In a series of experiments testing electron beam-cured PSAs, the resins and plasticizers were screened at a simple 50/50 polymer/resin ratio [26]. Generally, a lower resin/elastomer screening ratio is used for acrylic PSAs (solvent-based or water-based), for water-based PSAs a 30–40% tackifier level is suggested. Concerning the evaluation of different tackifiers for rubber-resin formulations the rubber quality should also be specified. As an example crêpe with a Mooney viscosity of 55 was used for the screening of tall oil resins [27].

The peel of tackified CSBR adhesives increases with the tackifier content and attains a maximum at a certain tackifier content [28]. A similar behavior may be observed for PSAs on other chemical basis. Generally, shear values decrease with increasing tackifier level, but in some special cases (e.g., acrylic hot-melt PSAs) the cohesive strength increases with the tackifier level. Table 8.6 illustrates the variation of the shear as a function of the adhesive/tackifier nature and level.

Table 8.6 Shear Dependence on the Adhesive/Tackifier Nature/Level

Formulating components

PSA		Tackifier		Chemical composition, parts by weight (wet)								
Code	Supplier	Nature	Code	Supplier	1	2	3	4	5	6	7	8	9
3703	Polysar	CSBR	—	—	69	52	37	36	43	38	24	—	—
3958	Polysar	CSBR	—	—	—	—	—	—	—	—	10	37	—
6157	Polysar	AC	—	—	—	—	—	—	—	—	—	—	50
80 D	BASF	AC	—	—	—	—	34	16	39	33	32	32	—
V 205	BASF	AC	—	—	31	48	32	48	49	38	33	—	50
—	—	—	CF52	A&W	—	—	—	—	—	—	—	33	—
—	—	—	CF301	A&W	—	368	29	270	136	50	—	—	16
Hot shear (70°C)		min			22	—	—	—	—	—	> 22	> 24	—
		hr											

Table 8.7 Tackifying with a Low Resin Loading

Tackified PSA dispersion		Adhesive characteristics	
		Peel at RT (N/25 mm)	Shear at 50°C (min)
1090	(UCB)	20.0 PT	100
5650	(Hoechst)	16.5	70
2001	(Exxon)	4.5	60
85 D	(BASF)	16.8	30
80 D	(BASF)	6.5	90
20094	(Syntho)	0.2	120

A blend of 100/20 PSA/Tackifier (wet/wet) was used. Tackifier: SE 351 (Akzo-Nobel); Face stock: paper; RT, room temperature.

The relatively high tackifier level required for high peel adhesion is explained by its influence on the modulus (modulus increase begins in the range of 40–60%). The resin level required for PSAs is generally higher than that for "classical" adhesives. A difference between the current state and future trends concerning the tackifier loading has to be made. Today the best overall balance of performance properties is obtained using moderate (e.g., about 30%) levels of high softening point tackifiers. Practically, concentrations between 30 and 40% of resins with a 50–70°C softening temperature are used. As an example for an acrylic a 30 parts (dry) tackifier level of acid rosin is suggested. Unfortunately the shear level of these formulations is not sufficient to give good die cuttability with normal (not voluminous or multilayer) face stock and release paper. There are however trends concerning the improvement of the cohesion.

In the future the first possibility to improve the shear strength of the tackified dispersion is a decreased tackifier loading level. For this purpose one needs high tack dispersions which require only a minor amount of tackifier in order to obtain a high tack/peel level. Table 8.7 illustrates the range of some recent developments in this field; it can be seen, that special dispersions need no more than 20% tackifier. The second possibility is to use special tackifier-friendly acrylic-based PSAs which can absorb more than 40–50% tackifier, without the loss of the shear. Table 8.8 illustrates some new developments in this field. Although both categories of products possess higher shear, the practical shear/die cutting tests do not lead to superior results compared to current products.

Table 8.8 Tackification of Tackifier-Friendly Acrylic PSAs

Formulating components				Adhesive characteristics		
AC-PSA		WB tackifier resin		RB tack (cm)	180° peel (N/25 mm)	RT shear (min)
Code	Level	Code	Level			
80 D	72					
V 205	38	324	40	10	12	30
V 205	70	8/21	30	15	13	30
80 D	60	8/21	40	13	9	200
2395	60	TX	60	8	13	460
2395	60	124T	60	8	9	600
2395	60	301CF	60	9	12	34
1304	60	SE80CF	40	14	16	90
620	60	SE80CF	40	9	15	14

AC, acrylic; WB, water-based; RB, rolling ball; RT, room temperature

The obtained tack level depends on the base elastomer (chemical and macromolecular characteristics), on its physical state, on the tackifying agent (chemical and other characteristics), and on the tackifier level and technology. Generally, the obtained tack level should be examined for the same kind of PSAs. A detailed analysis of the obtained tack level is given in Chapter 6 (Adhesives Characteristics), including a comparative evaluation of the adhesive performance characteristics for PSAs on a different chemical basis.

Choice of Tackifier. To modify properties many variations on the starting formulation are possible [10], namely the ratio of hard to soft resin, a change of the softening point of the hard or soft resin, and change in the pH of the compound. Generally, tackifiers are employed to impart or control one or more of the following properties in PSAs (i.e., tack, peel strength, cohesive strength, staining, migration or adhesive bleedthrough, bleeding or oozing, stringing or legging, and aging characteristics). Their use also influences the most important coating characteristics (e.g., the rheology, foam generation, substrate wetout, and mechanical stability). Generally, the most important criteria for the choice of tackifiers include the nature of the base PSA (to be tackified), the balance of the desired adhesive, converting and end-use properties, the tackifying technology, and economical considerations.

Compatibility of Resin. Compatibility is a general requirement for optimum tackification [29] (see Chapters 2 and 3). The resin functions as a solid solvent for a portion of the polymer, and the resin and the polymer must display similar solubility characteristics if the resin is to act as a tackifier (mutual compatibility). Several papers were published on the effect of the resin on PSAs properties and on the compatibility of resins with elastomers, especially in combination with block copolymers [30,31]. Hydrogenated hydrocarbon resins (e.g., Regalrez, Hercules) may be selected according to their cloud points [32]. Cloud point measurements reveal compatibility with aliphatics and incompatibility with aromatics. As discussed in Chapter 2, for a first test of compatibility the T_g is used. For hot-melt PSAs the position of the T_g shows whether the resin is compatible with the end- and midblocks of the polymer. Polystyrene displays a limited compatibility with natural rubber only [33]; it is adequate for butadiene styrene copolymers. Polyvinyl cyclohexane possesses good solubility in natural rubber, but it cannot be dissolved in butadiene styrene copolymers. Poly-*t*-butylstyrene may be used in natural rubber- and butadiene styrene-based PSAs. Traditionally most hot-melt PSAs systems are based on S-I-S block copolymers. They are usually easier to tackify than S-B-S materials. Their better compatibility with tackifying resins and greater elasticity have ensured S-I-S block copolymers a larger part of the hot-melt PSA market. In recent years more attention was paid to the tackification of S-B-S-based block copolymers. These polymeric materials are cheaper than S-I-S copolymers, but nevertheless offer interesting performance characteristics. A clear color and good aging resistance for the tackifier resins are needed [34]. S-B-S does not exhibit enough elasticity, flexibility and tackifier compatibility [35]. Tackifiers for these rubbers must display good compatibility, a light color, good aging stability, and a satisfactory cohesion/tack balance. The most important resins for tackifying S-B-S block copolymers are Escorez ECR-366, ECR-368 and ECR-369 (Exxon) [35], where ECR-366 is a modified C_5 aliphatic hydrocarbon resin.

Tackification supposes mutual solubility [30,36–38] and is achieved if on the mutual solubility diagram of the elastomer and tackifier the solubility domain of the elastomer is enclosed within the domain of the resin [39]. This does not happen for hydrocarbon resins and rubber; however, as seen earlier, hydrocarbon resins are adequate tackifiers for natural rubber. For classical hot-melt adhesives the compatibility of the waxes depends on their MWD and melting point, and influences the bleeding [35].

Compatibility also depends on the molecular weight and MWD; a narrow MWD indicates better compatibility. As an example Regalrez resins have a MWD characterized by a ratio M_n/M_w = 1.1–1.4 [32]. Block copolymer acrylic-based PSA display very good compatibility with hydrogenated rosin esters and C_5/C_9 hydrocarbon-based resins [24]. Generally, compatibility is a function of the melting point of the resin and its composition. Regalrez as an example shows good compatibility with acrylic-based PSAs up to a ring-and-ball softening point of 78°C. Thus, on the basis of mutual compatibility, there are general recommendations concerning the use of certain tackifiers for different classes of latexes [4]:

- For natural rubber latexes, petroleum-based alifatics, petroleum aromatics terpenes, and rosin esters are proposed.
- For CSBR, rosin esters, petroleum-based aromatics, low molecular weight, pure monomer aromatics, α-pinene, and terpenes are suggested.
- For acrylic PSAs, rosin esters, pure monomer aromatics, low molecular weight polystyrene, and copolymers of α-methyl-styrene-vinyl toluene are proposed.

Hydrocarbon-based resins display lower overall compatibility than natural products. Aromatic resins are proposed for acrylic-, CSBR-, and EVAc-based PSAs [40]. For example Alresen PT 214 is proposed by Hoechst to be used as tackifier for Mowilith LDM 1360 EVAc. Because the main criterion for the choice of a tackifier is its compatibility with a PSA, it is possible to cover the needs of different PSAs with a limited range of tackifiers. For example [20]. DRT (France) is supplying a disproportionated rosin, a terpene phenol resin, and a stabilized rosin ester for chemically very different PSAs [20].

There is a fundamental difference with regard to the choice of tackifiers for PSAs and other adhesives. Bonding of PSAs is (mainly) a physical process. No chemical interaction is needed between tackifier and/or elastomer and substrate. Thus the chemical structure of the resin is less important than its compatibility and its influence on the adhesion/cohesion balance. The molecular weight should be less than 1000 and the resin should have a narrow MWD (e.g., Snowtack CF 301, from Tenneco) [29].

Electron beam-crosslinkable S-I-S rubber develop a crosslinked network with some resins (i.e., higher gel content) at a lower electron beam dose than with other electron beam-cured PSAs. The average gel content for the adhesives using the saturated resin was 0.7% at 3 Mrad, while the average for those using the unsaturated resin was only 0.15%. Rolling ball tack values of PSAs prepared with the saturated resins stayed at about the same level after irradiation; rolling ball tack of PSAs using unsaturated resins increased from 4.7 to 6 cm [26]. While there is a major difference between the adhesives based on saturated hydrocarbon resins and those based on unsaturated resins, there also are differences within these classes. For the adhesives with unsaturated resins these differences can be large. Therefore the selection of tackifier resins is of prime importance in obtaining the maximum amount of polymer crosslinking with a minimum of electron beam radiation. For electron beam-cured hot-melt PSAs based on Kraton D 1320 X (a branched styrene-butadiene copolymer with functionalized end groups) Nitzl used hydrocarbon resins of the Escorez series, mainly the E-1310, E-1401, E-5300, and E-5380 types [14].

Hydrocarbon resins together with terpene-phenolic resins also are recommended for polyvinyl acetate dispersions. A narrow MWD imparts better compatibility also [41]. For rubber-resin PSAs β-pinene derivatives, for acrylic PSAs hydrocarbon-based resins were suggested as tackifiers at the early stages of PSA manufacturing [42]. Hydrocarbon resins (like Zonarez B115) are recommended for natural rubber-based, or S-I-S-based PSAs (e.g., Zonarez Z115). In both cases maximum tack is achieved at resin/rubber ratios exceeding 50/100; maximum peel is obtained at resin/rubber ratios above 1.

The major factors in determining the type of resin to use with a particular latex are its composition (chemical structure), and molecular weight. Resins of appropriate structure are compatible, but only to some limited concentration. Acrylic PSAs show the best tackifying ability, with a broad compatibility with most tackifying resins, independently from their composition. Styrene-butadiene rubber, for example, with a higher styrene content does not respond well to resin addition. EVAc shows lower compatibility with tackifiers than acrylic- or CSBR-based PSAs. One can conclude that the types of tackifiers used with acrylate copolymers and styrene-butadiene rubbers are generally useful with the new EVAc-based polymers as well. Somewhat lower softening points than normal may be desirable depending upon the particular EVAc. Suitable tackifiers for EVAc dispersions are polyvinylether, colophonium resins, low molecular resins, and plasticizers. As an example a level of 30% polyvinylether for 70% MV 704 dispersion (Hoechst) is recommended. As indicated earlier hydrogenated hydrocarbon resins display better compatibility towards EVAc with low vinyl acetate content (< 40%). A narrow MWD also implies a better compatibility [43].

Balance of Adhesive Properties, as a Function of the Resin Nature and Softening Point. The addition of resin lowers the modulus of the rubber and increases the T_g [44,45]; accordingly a suitable tackifying resin should produce these changes. The lowering effect of the modulus and the increase of the T_g are dependent on the resin's own characteristics (E and T_g) and its compatibility. In a simplified approximation for resin/elastomer systems with good compatibility, the most important parameter is the softening domain (melting point) of the resin (also related to E and T_g). In general, soft resins impart aggressive grab and quick stick, while harder resins help retain good cohesive strength and creep resistance; resins with a low softening point impart tack, but give poor cohesive strength, while those with a high softening point give good adhesive strength but poor tack. For example, glycerine esters are tackier than pentaerithrite esters. On the other hand, compatibility is a

Manufacture of Pressure-Sensitive Adhesives

function of the molecular weight and softening point of the resin. A higher softening point resin is less miscible with the base polymer [46]. Typical rosins and hydrogenated rosin ester tackifiers have ring-and-ball softening temperatures of about 25°C to about 115°C, while preferred tackifiers have softening temperatures of about 50°C to about 110°C. Table 8.9 illustrates the dependence of the adhesion/cohesion balance on the characteristics (softening temperature and chemical nature) of the tackifying resin.

High softening point resins and ester resins impart less tack (and peel) than low softening point acid ones. The lower acid level increases the sensitivity of tack and shear to molecular weight. In the tackification of natural rubber with hydrogenated rosin esters, a low softening point resin (75°C) gives maximum tack at higher resin concentration (110 parts resin/100 parts rubber) than a high softening point resin (95°C), which displays a maximum tack at 60 parts resin/100 parts rubber ratio [47]. Evidently, the absolute value of the final tack obtained with the low softening point resin is higher. In a manner quite different from tackifying S-I-S with hydrocarbon resins, in some cases a maximum value of the peel adhesion was observed as a function of the resin concentration, which increases continuously with the resin concentration when tackifying natural rubber with resin. Resins with a softening point of about 50°C impart tack but give poor cohesive strength, while those above about 70°C give good cohesive strength but poor tack [10]. To modify the properties the following variations on the starting formulation are possible, namely the ratio change of the hard to soft resin or a softening point change.

The addition of higher T_g polymers and/or the addition of higher softening point tackifiers, enhances the convertability and provides specific converting properties [48]. Increased thermal or pressure sensitivity can be obtained by increasing the level of tackifier and/or changing to a softer, more heat sensitive tackifier. This may cause a reduction of the high temperature bond strength. The tackiness of a high acrylate copolymer can be increased by using a lower molecular weight (lower softening temperature) tackifier [49]. Tables 8.10 and 8.11 illustrate the influence of the low softening temperature resins on the tack and shear. It can be seen that chemically different resins (polar melamine derivatives, nonpolar hydrocarbons) cause a similar shear decrease. On the other hand, it should be noted that acid and ester rosins with the same softening point interval, impart different adhesion/cohesion levels (Table 8.12).

The influence of the chemical nature and melting point of the tackifier resin should always be examined with respect to the required adhesive property. For properties where bonding affinity is more important a change of the chemical nature is preferred. For mechanical properties of the bulk polymer (inside the adhesive layer) the change of the melting point should be considered preferentially. As an example, different tackifiers (stabilized rosin acids,

(text continues on p. 350)

Table 8.9 The Influence of the Tackifier Resin Nature and Softening Point Range on the Adhesive Characteristics of WB AC PSAs

		Formulating components					Acrylic PSA		Adhesive characteristics			
		Tackifier resin dispersion										
	Code	Supplier	Softening point (°C)	Chemical basis	Level (wet)		Code	Level (wet)	Coating weight (g/m²)	Rolling ball tack (cm)	180° peel (from PE, N/25 mm)	Shear at 50°C (min)
1	CF 52	A&W	55–65	Rosin acid	21.4		V 205 / 80 D	35.7 / 21.4	21	13	25 PT	102
	H5786	Hercules	95–105	Rosin ester	21.4		V 205 / 80 D	21.4 / 35.7	16	> 30	14 PT	98
2	H5786	Hercules	95–105	Rosin ester	42.8		V 205 / 80 D	35.7 / 21.4	21	5	20 PT	250
3	MB698	Hercules	Liquid	Rosin ester	7.2		V 205 / 80 D	35.7 / 21.4				
	4576	Akzo	95–105	Rosin ester	35.7							
4	MB6152	Hercules	55–65	Rosin ester	42.8		V 205 / 80 D	35.7 / 21.4	16	2	20 PT	37
5	511	DRT	55–65	AC-rosin	48.4		V 205 / 80 D	19.3 / 32.2	19	4	25 PT	55
6	DEG	DRT	35–45	Rosin ester	48.4		V 205 / 80 D	35.7 / 21.4	20	2.5	23 PT	40
7	DEG 80	DRT DRT	35–45 60–80	Rosin ester Rosin ester	14.3 28.6		V 205 / 80 D	35.7 / 21.4	21	5	20 PT	47
8	9251	Exxon	—	HC	48.4		V 205 / 80 D	35.7 / 21.4	18	1.5	25 PT	15
9	9251 9271	Exxon Exxon	— —	HC HC	33.4 15		V 205 / 80 D	35.7 / 21.4	19	2.5	20 PT	25
10	OU80 OU30	Oulu Oulu	— Liquid	— —	33.4 15.0		V 205 / 80 D	35.7 / 21.4	20	7	17 PT	20
11	CF 52	A&W	55–65	AC-rosin	48.4		V 205 / 80 D	35.7 / 21.4	18	2.5	18	5

Manufacture of Pressure-Sensitive Adhesives

Table 8.10 The Influence of Low Softening Temperature Resins on the Adhesive Characteristics of PSAs

Product / Code	Formulating components					Adhesive characteristics			
	Tackifier resin			Level (parts, wet)	Coating weight (g/m²)	Rolling ball tack (cm)	180° Peel (N/25 mm)	Shear at 50°C (min)
	Supplier	Softening point range (°C)	Basis					
Polyetheneimine	BASF	RT	Polyetheneimine	20	17.0	3.5	12.0	15.0
Hercolyn D	Hercules	RT	HC/ester	20	21.0	4.0	16.0	15.0
CF 301	A&W	60–65	Acid rosin	100	19.0	6.0	10.0	200.0
Cymel	Cyanamid	RT	Melamine	20	20.0	3.0	10.0	20.0
CF 52	A&W	55–65	Acid rosin	100	18.0	2.4	14.0	100.0

Base PSA: CSBR dispersion (100 parts).
RT, room temperature; HC, hydrocarbon.

Table 8.11 The Influence of High Cohesion AC Components and Liquid Tackifier Resins on the Adhesive Characteristics

	Formulating components							Adhesive characteristics				
	Tackifier resin			Base Acrylic PSA			Coating weight (g/m²)	Rolling ball tack (cm)	Peel (N/25 mm)	Shear at 50°C (min)	Remarks
Code	Supplier	Softening point range (°C)	Level (parts, wet)	Code	Supplier	Level (parts, wet)					
CF 52	A&W	55–65	33	V 205	BASF	27	20	3.0	19	7	
				80 D	BASF	20					
CF 52	A&W	55–65	33	V 205	BASF	47	19	3.0	6.0	10	
				80 D	BASF	20					←— Shear increases
				FC88	UCB	5					Tack increases —→
				20094	Syntho	5					
CF 52	A&W	55–65	33	V 205	BASF	47	20	6.5	19	18	
				80 D	BASF	20					
				FC88	UCB	10					
				20094	Syntho	10					
CF 52	A&W	55–65	33	V 205	BASF	47	20	6.5	15 PT	20	
				80 D	BASF	20					
				20094	Syntho	10					
CF 52	A&W	55–65	33	V 205	BASF	47	20	9.0	14 PT	20	
				80 D	BASF	20					
				20094	Syntho	20					
CF 52	A&W	55–65	33	V 205	BASF	47	22	1.4	14 PT	19	
Hercolyn D	Hercules	RT	10	80 D	BASF	20					

RT, room temperature

Table 8.12 Influence of the Resin Nature on the Adhesive Characteristics

Formulation components							Adhesive characteristics			
Tackifier resin				Acrylic PSA						
Rosin acid		Rosin ester					Coating weight (g/m²)	Rolling ball tack (cm)	180° peel (N/25 mm)	Shear at 50°C (min)
Code	Supplier	Code	Supplier	Code	Supplier	Formulation (parts, wet)				
CF 52	A&W	53251	A&W	V 205	BASF	35.7	21	5.0	20 PT	25
				80 D	BASF	21.4				

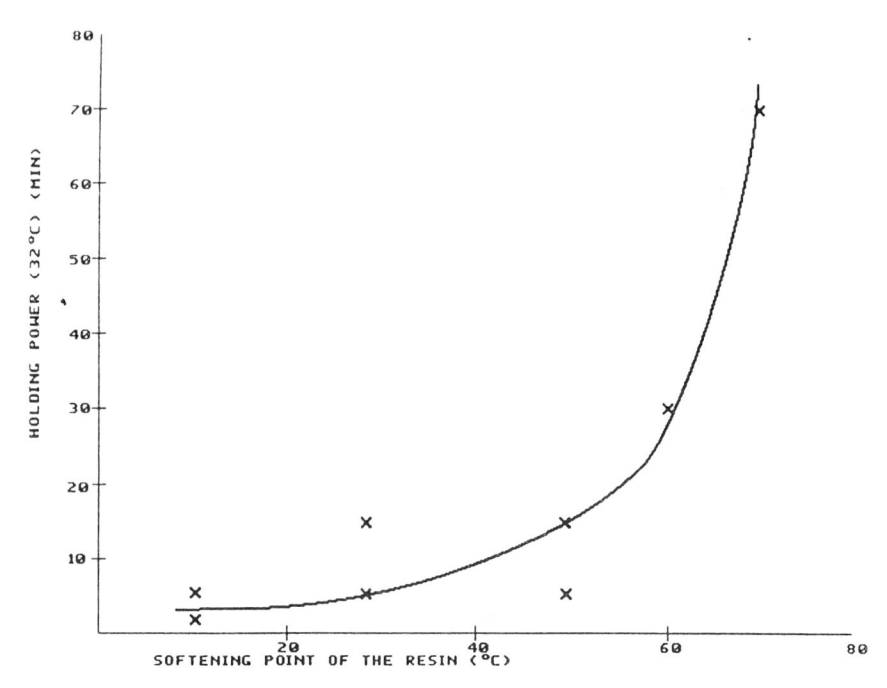

Figure 8.4 The influence of the softening point of the tackifier resin on the holding power of a PSA formulation. The increase of the holding power with the increase of the softening point of the resin.

disproportionated rosin acids, hydrocarbon resins, rosin esters, hydrogenated rosin esters) were tested for CSBR [50]. The best performance characteristics concerning the peel were obtained with a stabilized rosin acids (softening point 52°C). The best shear resistance is given by a high melting point hydrocarbon resin (65°C) and the best loop tack is obtained when using stabilized rosin acids. The best polytack is given for disproportionated rosin acids. The complex dependence of the adhesion/cohesion balance for a neoprene latex on the melting point and chemical nature was described [10].

The dependence of the shear (SAFT) of a rubber/hydrogenated resin formulation is illustrated in Figure 8.4. As can be seen from the figure the shear resistance increases with increasing softening point of the resin. In order to improve simultaneously the chemical nature and melting point dependent properties, mixtures of resins with different melting points and on different chemical basis (liquid resins) are suggested. Resin mixtures with different melting points (1 to 1 blends of resins with melting points of 32°C and 64°C, and 26°C and 82°C) were suggested for alkylene-maleinate copolymers [25]. A detailed discussion of the choice of tackifier depending on its nature and softening point will be given later (see Section 2.8).

Influence of Physical State of the Tackifier. Both solvent-based and water-based tackifying resins can be added to water-based acrylic PSAs. Generally, it is not possible to add molten tackifiers to water-based acrylic PSAs, but there are some special halogenated copolymer dispersions with good thermal resistance. Relatively low melting point resins (95°C) may be added to aqueous adhesive dispersions (polyvinyl ethers) in a molten state. Some liquid resins can be added as such (without emulsifying) to compound a latex (e.g., Cymel 301 in neoprene latex) using high speed agitation; because of the good mechanical and chemical stability of the latex, there is no coagulation problem. The use of liquid resins as tackifiers also depends on the properties of the face stock material. Ordinary low cost papers without an overlayer or special thermal papers are more sensitive to adhesive bleedthrough. The use of liquid resins or a high surface-active agent concentration in the tackifier [10] must therefore be avoided.

Different acrylic latexes were tackified with polybutene and polybutene emulsions [51]. From comparative testing in both systems it is anticipated that the tack and adhesion results are similar, independently of the system used. On the other hand, the shear strength for carrier-free resin tackified acrylic PSAs is higher than with resin emulsions [13]. In the formulating practice, the use of tackifying resin dispersions is more widespread because of technological advantages [52]. In the early stages of PSA formulation, rubber-resin PSAs were tackified with resin solutions. The first tackified acrylic formulations also used tackifier solutions. Later tackifier dispersions in combination with water-based PSAs were developed. Solvents may be used as an additive (10–20%) in latexes [53,54]. In a similar manner, the solids content of latexes may be improved with rubber solutions (20%) [53].

Tackification of Water-Based PSAs. The same tackifying resins used for solvent-based PSAs also are envisaged for water-based PSAs. Theoretically, there are three different ways for the additon of tackifier resins into water-based systems (i.e., as resin solutions, resin dispersions, and as a bulky, molten material). From a technical point of view the most convenient method is the use of resin dispersions. From the point of view of the adhesive properties, solution or molten resin feed-in seems to be better because there are no water-sensitive additives in the resin. Practically, resin solutions only have a limited use (for PSA coating onto film); molten resin may be added in heat-resistant dispersions (like chloroprene) only. Resin-tackified water-based PSAs acquire a composite character with limited compatibility of the emulsion borne components; this tackifying effect is less effective than in the case of solvent-based systems. On the other hand, resin fluidity in the adhesive layer is enhanced by its composite structure, thus strike-through (penetration) is more prevalent in water-based systems. Water-soluble chemicals from the paper or film also may negatively influence the aging properties of the resin. Because of the composite structure of the resin particles, the heat resistance of the resin dispersions may differ from that of the bulk material, influencing the storage and converting properties of the laminate. In summary, the most important factors influencing the choice of tackifiers for water-based systems are the limited number of resins available as an emulsion, and the limited allowable loading level of liquid low cohesion resins, because of the decrease of converting properties and heat resistance, and increase of the penetration (adhesive bleedthrough), and the limited number of resins with good aging stability.

A detailed study of the tackification of water-based dispersions with rosin-type tackifier dispersions with a broad range of T_g and softening point (-16°C to +45°C and 25–90°C, respectively) and molecular weight between 720–1150 was done by Wood [55]. He stated that suitable tackifier resins have to display low molecular weight, high T_g, and good compatibility. There are some features common to the modification of rubber-resin-based and water-based acrylic PSAs. The addition of certain resins reduces rolling ball tack in direct proportion to the softening point of the resin and the concentration employed, except for the 50% level of the 90°C softening point resin [55]. On the other hand, the best overall balance of performance properties is obtained using moderate (30%) levels of high softening point tackifiers. A similar influence of T_g, molecular weight, and MWD on the tackification of hot-melt PSAs (S-I-S triblock copolymers) was demonstrated by Tse and McElrath [56]. They showed that if narrow MWD resins are blended with S-I-S, the position of loss tan δ on the temperature scale corresponds to the molecular weight of the resin. This is because the resin T_g is controlled by its molecular weight (see Chapters 2 and 3). Film coatings require high transparency and no yellowing during end-use. For this purpose hydrocarbon resins are preferred.

Tackifiers and Converting Properties. Good converting properties of the aqueous dispersion (coating properties) and of the laminate (processing) are essential (i.e., an adequate rheology, low foam generation, good substrate wetout, excellent mechanical stability, and high converting speed or drying). Aqueous dispersions exhibit non-Newtonian shear thinning characteristics and variations in the viscosity recovery. This difference may produce an undesirable "corduroy" effect which is termed ribbing or striation. Acid resin-tackified acrylic-based PSAs show better flow properties. Generally, tackified dispersions display a less adequate rheology.

Tackified aqeous dispersions contain more surface-active agents than untackified ones; hence they tend to generate more foam than pure PSAs. The

Table 8.13 Wetting Ability of Different Aqueous Tackifier Dispersions

	Water-based tackifier		
Code	Supplier	Wetting agent (%)	Wetting ability
MBG-152	Hercules	3	Very low
DEG/G	DRT	1	Average
MBG-64	Hercules	2.5	Low
50 D	Oulu	3	Very low
EH-4576	Akzo	4	Very low
E-9251	Exxon	1	Good
E-9271	Exxon	1	Good
CF 52	A&W	1	Good
SE 351	A&W	1.5	Low
CF 8121	A&W	2	Low

The wetting ability was tested on solventless silicone coated release paper at a viscosity of 400 mPa.s; Lumithen IRA (BASF) was added as wetting agent.

wetout behavior of tackified water-based PSAs is generally a function of the base acrylic (CSBR or EVAc) and of the tackifying resin. For the same base, PSAs using different tackifiers and sulfosuccinates as wetting agents, the wetting out on the siliconized release paper depends on the pH; at higher acidity one needs less wetting agent for an adequate wetout. Generally, the surface tension and the wetting out of tackifier dispersions from different suppliers vary quite a bit (as a function of the experience and knowledge in dispersing technology). Table 8.13 illustrates the wetting characteristics of different tackifier dispersions. Wetting ability was evaluated by the amount of added surface agent necessary for good wetout.

An additional deficiency of tackified water-based adhesives is the tendency for coagulum to build up on the metering roll when conventional emulsions are applied by a reverse roll coater. The use of higher solids emulsions increases opportunities for particle-particle interaction, particularly when the emulsions are subject to mechanical stress. If there is an insufficient protective layer on the particle surface to repel other particles, the polymer particles will fuse, eventually yielding a significant mass of gel particles. Such coagulum may arise either from poor mechanical stability under shear conditions, or from imperfect doctoring leading to a thin coating on the roll, which dries and cannot be redispersed, so that these particles eventually transfer to the application roll and disturb the appearance of the coating. Generally, acid

Table 8.14 Drying Speed of Pure and Tackified Water-Based PSA

Time (min)	Weight loss of the wet coating (%)	
	Pure PSA	Tackified PSA
0.0	0.0	0.0
1.5	0.8	0.9
2.0	1.4	1.5
3.0	2.3	2.7
4	3.3	3.8
6	5.4	5.6
8	7.4	7.2
9	7.7	8.0
10	8.5	8.7
12	9.9	10.0

Evaluation of a 100 μ coating on Mylar of a V 205/80 D/CF 52 blend at room temperature.

rosin tackifiers show better mechanical stability than ester-based ones. In order to increase the mechanical stability of the tackified dispersion, it is proposed for ester-based tackifiers to neutralize the acrylic mixture, and to add the tackifier dispersion in a second step. Laboratory tests show a slight difference between the drying speed of the tackified and pure unformulated acrylic-based PSAs (Table 8.14).

Industrial experience generally indicates a higher converting (coating) speed for tackified water-based acrylic-based PSAs (a 5–20% increase of speed). A possible explanation of this increase is based on the dependence of the coalescence on the surface tension. Coalescence is proportional to surface tension, thus tackifier dispersions with a higher surface tension display higher coalescence and allow higher coating speeds (see Chapter 3).

Die cutting is the most important feature of the laminate influenced by the tackifier resin. Tackifying resins influence the cohesion of the adhesive, and thus its shear strength and its die cuttability. It is to be expected that high softening temperature resins impart better shear and die cuttability than the low softening point ones. On the other hand, laminate shear and die cuttability depend on the hygroscopy of the components also. Tackifiers from different suppliers exhibit different die-cuttability properties (Figure 8.5). A comparison

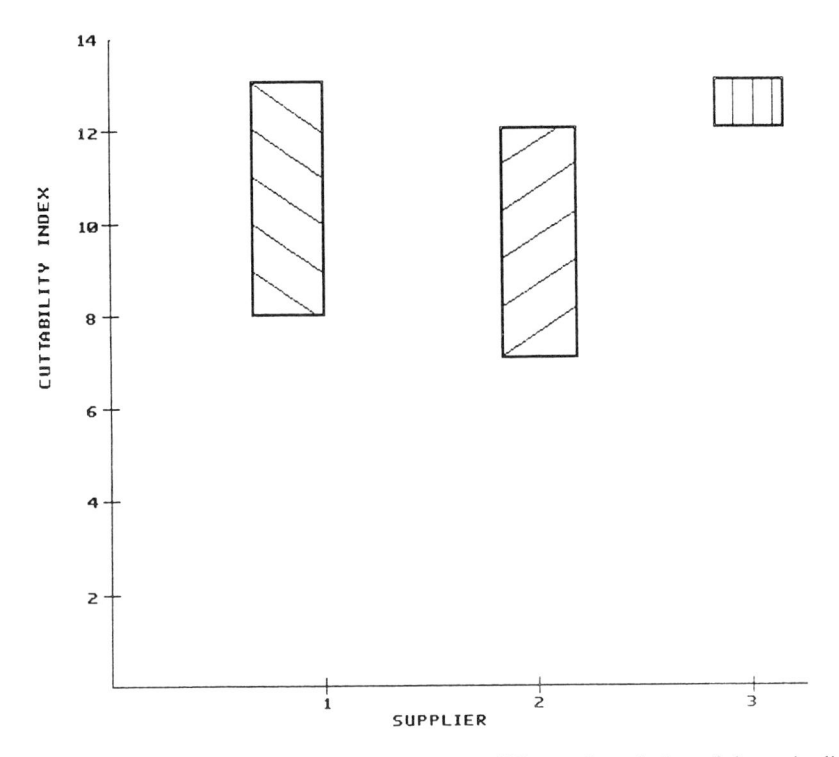

Manufacture of Pressure-Sensitive Adhesives 355

Figure 8.5 The influence of the supplier (different formulation of the resin dispersion) on the cuttability of the PSA paper laminate. 1), 2), and 3) are formulations based on the same resin from different tackifier resin dispersion suppliers.

of the die cuttability of different tackifiers (tackified adhesives and laminates) indicates that ester-based tackifiers are superior to acid-based ones.

Tackifier and End-Use Properties. As discussed in Chapter 5, the choice of the base elastomer and in a similar manner the choice of the tackifier influence the end-use properties of the PSA laminate and the label; quite different end-use properties can be achieved. As an example, for clear face stock material the adhesive color (resin color) is very important. Lightly colored hydrocarbon tackifiers (liquid resin tackifiers) were suggested for EVAc- and S-B-S-based adhesive systems [57]. Generally, water-white- and UV-stable water-based PSA formulations can be obtained on the basis of hydrogenated hydrocarbon resins [58].

Japanese PSA-based tape production started with natural rubber-resin or polyterpene/natural rubber combinations [59]. The resin and polyterpene were replaced by C_5 and C_5/C_9 petroleum resins due to their cost/performance balance and grade diversification for meeting the needs of the quality-oriented Japanese PSA tape market. The main reason for the use of hydrogenated water-white resins in PSAs seems to be their excellent weathering performance rather than their color and transparency characteristics. Phenolformaldehyde resins give better plasticizer resistance than rosin esters [60]. Special resins and resin formulations are suggested for repositionable and removable PSAs. As an example an aqueous dispersion of a methyl ester or glycerol ester of hydrogenated rosin is recommended [61] (see Chapter 6). The tackifying results demonstrate the necessity for the use of soft, essentially aliphatic resins, since the more aromatic resins exhibit a high degree of initial tack, which increases upon aging, thereby no longer retaining the removable character of the formulation [52]. The nonreactive resins impart a tack increasing with time to synthetic rubbers, and the reactive resins function as curing agents for synthetic rubbers [63]. For removable aqueous natural rubber-based PSAs, preferred classes of tackifiers are anionic aqueous dispersions of hydrogenated rosin esters and anionic aqueous dispersions of hydrogenated rosins [61].

The resin stability influences the end-use properties of the PSAs also. Abietic acid contains a conjugated double bond and will react with oxygen. This will increase the softening point of the resins and the subsequent tack of the adhesive will decrease with time. Crystalline tackifiers are more stable than amorphous ones. Amorphous tackifiers provide higher tack at room temperature, but they display lower tack at lower temperatures [64].

Regarding the use of tackifiers for S-I-S/S-B-S-based formulations, while the temperature has little effect on the properties of PSA laminates produced on the closed die coater, it clearly has a considerable influence on those of the roller-coated products. At the same time, tapes coated with the die coater are superior than those produced with the roll coater [65]. Therefore the aging stability of the resin is influencing the formulation and end-use properties of the PSA laminate. The tackifier resin type also influences the staining of CSBR-based PSA formulations. Rosin acid-based materials are outperforming hydrocarbon resins for example, as do higher softening point resins within any type [50].

Special Features. It should be stressed that it remains very difficult to carry out an elastomer-independent evaluation of the tackifier performance characteristics. Solvent-based PSA formulations use natural rubber (crêpe or smoked sheets), styrene butadiene copolymers (S-B-S and S-I-S block copolymers) or elastomers, as well as acrylic PSAs. Rubber as elastomer is used for hot-melt PSAs almost only as block copolymer, or for water-based PSAs almost only as CSBR.

Therefore the differences in the chemical nature, structure of the rubber-based elastomers cause differences in the behavior of the same resin upon tackification. Additional effects arise from the dispersed state of the components. Styrene-butadiene rubbers may have a linear, branched, or crosslinked structure. Their properties depend on molecular weight, T_g, colloid factors, and stabilizer systems. Block copolymers commonly have a linear structure although some radial copolymers have now become available. Generally, rosins and its derivatives are more closely associated with emulsion CSBR latexes [66]. With CSBR the latex with the higher butadiene content accepts more resin and the ultimate tack is nearly double that of the reduced butadiene system. For this system, if a 80°C SP resin is used, the optimum tack occurs at a slightly lower resin content than for a 52°C SP resin [66]. On the other hand, CSBR generally has a higher modulus and requires more resin. As discussed earlier in hot-melt PSA formulations (based on S-I-S), the molecular weight of the resin is more important (hydrocarbon versus rosin). The types of tackifiers used with styrene butadiene rubbers are generally suitable for EVAc-based polymers as well; a somewhat lower softening point than normal may be desirable.

Polyterpene Resins. Polyterpene resins offer an adhesive formulation the lightest color ranges of natural tackifiers in addition to the advantages of good thermal stability and color retention. They are good choices for "water white" formulations. Although hydrogenated hydrocarbon resins are frequently selected for these applications, polyterpenes can often meet the color appearance of a hydrocarbon with better heat, aging, and tackifying properties [67].

Polyterpenes are available in a wide softening point range, from 10°C to 135°C. They exhibit good compatibility with backbone polymers, especially those with a high ethylene content (EVAc); they also display good compatibility with S-I-S and S-B-S block copolymers. Liquid polyterpenes act not only as viscosity diluents (hot-melt PSAs) like waxes, but also as cotackifier. Aromatic modified polyterpenes offer improved tackifying capabilities with high molecular weight polymers (S-B-S, S-I-S) when compared with hydrocarbon, rosin esters, and modified polyterpenes. Terpene phenol resins offer the best thermal stability of any adhesive tackifier [67].

2.4 Rosin-Based Tackifiers

Rosin-based tackifiers are actually the most important class of tackifiers for PSAs. Their success is based on the experience made with them in other fields than PSAs and on the relatively broad range of data issued from tackifying solvent-based rubber-resin and solvent-based acrylic PSAs. In order to evaluate their advantages and disadvantages from a technical point of view, their contribution to the adhesion/cohesion balance, and their influence on the converting, processing, and end-use properties will be examined. On the other hand, the compounding technology influences their technical suitability as well. The adhesive properties of PSAs tackified with rosins are characterized by the adhesion/cohesion balance and the time/environment stability. First, the most important adhesive properties (e.g., tack, peel, and shear) will be reviewed using acrylic PSAs as base materials. The main types of rosin resins used include acid rosins and rosin esters.

Acid Rosins

The first group are thermoplastic acidic resins, derived from pine trees, resulting in a mixture of organic acids (abietic acid and its derivatives). These products are modified by disproportionation, polymerization, hydrogenation, and esterification. For a tackifier to impart the desired performance to an adhesive, it is necessary that the tackifier exhibits at least partial compatibility with the polymer. Typically, decreasing the polarity or increasing the molecular weight of a tackifier will decrease its compatibility. In general, the adhesive properties of acid rosin-tackified PSAs are superior to those of the ester-based ones [48]. On the other hand, rosin esters display a higher water resistance than acid rosins [68]. This feature appears important when tackifying PSAs for film coating and also concerning the shear stability of dispersions with a high resin loading. The holding power of an adhesive is basically a viscosity driven effect. Samples with a high degree of entanglement (higher viscosity) would be expected to exhibit a higher resistance to shear than a sample with few entanglements (low viscosity). It is assumed that a tackifier loosens the entanglement network and therefore increases compliance in the entanglement structure. Increases of T_g are due to segmental friction. This stands in contrast to a plasticizer, the addition of which causes the T_g to decrease due to a loosened entanglement network and decreased segmental friction. For the improvement of the shear of acrylic PSAs, it appears preferable to have a broad molecular weight distribution resin including some very high molecular weight species, so as to obtain a high viscosity [69]. One could expect that a molecular weight increase (and a high melting point) imparts a similar shear improvement when tackifying with rosin esters; high melting point, high molecular weight ester resins yield better shear strength. This hypothesis is only partially confirmed for the properties of the bulk adhesive; molecular weight and softening point were only secondary factors in controlling shear/adhesion [48]; there are some parameters which override the effect that softening point and molecular weight have upon shear. The tackifier's specific formulation (i.e., surfactant type and level and other additives) appears to have the largest impact on shear (Figure 8.6).

The evaluation of the tackifying effect of rosin acid versus rosin ester is more important with respect to the tack/peel balance than the shear. The tackification of different acrylic PSAs and other PSAs with various resins

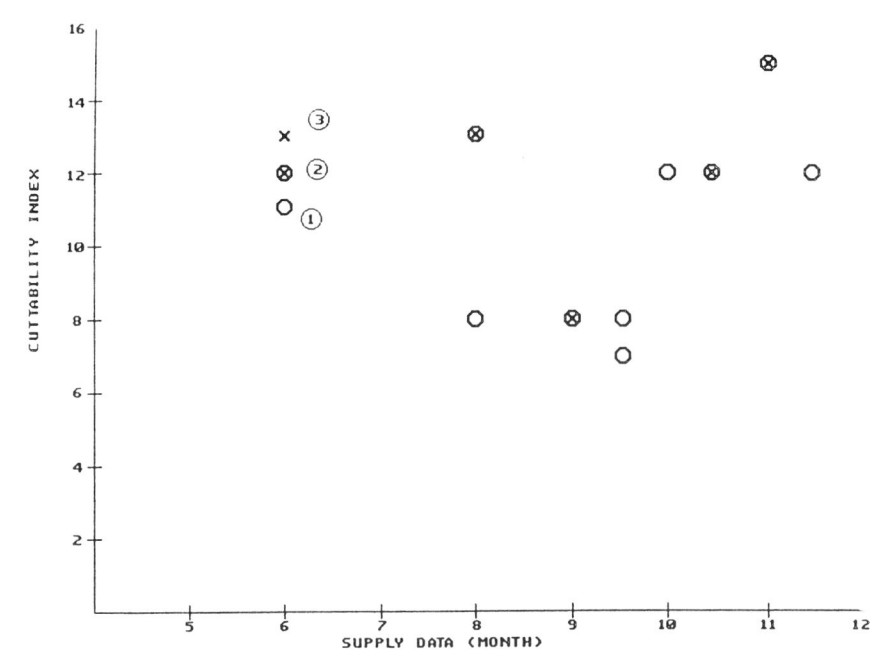

Manufacture of Pressure-Sensitive Adhesives 359

Figure 8.6 Cuttability dependence on the formulation of the tackifier resin dispersion. Variation of the cuttability values (cuttability index) as a function of the formulation tolerances. The change of the cuttability for a water-based acrylic PSA-coated laminate, where the same tackifier resin has been used as dispersion from different suppliers (1, 2, 3) during an extended period of time.

from several suppliers will be covered next. The most important tackifier dispersion commercialized since 1985 was the Snowtack CF52 (Albright & Wilson); its main competitor is CF-301 (an acid resin with a narrower MWD that imparts better shear and water resistance, but lower polyethylene-peel). Originally intended to be used at a level of about 20% (on a dry basis) both resin dispersions were used in formulations with 30–40% tackifier. Polymers with a different chemical natures require different amounts of tackifier resin; additions of tackifier causes the decrease of the modulus and an increase of the T_g. The latter appears more pronounced for acrylic VAc than CSBR in comparison to acrylics alone. At high resin levels the modulus and shear strength also are increasing; the resin concentration at which this change occurs is higher in blends with acrylic dispersions than in mixtures with vinyl-acrylic dispersions [70]. Consequently acrylic PSAs exhibit a better tackifying ability. The molecular weight of the polymers also influences their tackifying

Table 8.15 Tackification of Different AC PSAs with the Same Acid Rosin-Based Tackifier

	Base PSA					Adhesive characteristics		
Code	Supplier	Nature	Chemical basis	Characteristics	Level (parts, wet weight)	Rolling ball tack (cm)	180° peel (N/25 mm)	Shear at 80°C (min)
V 205	BASF	AC	EHA	Soft AC	67	2.5	18 PT	5
R3375	Harco	AC	EHA	Soft AC	67	2.0	18 PT	3
FP2313	Nobel	AC	EHA	Soft AC	67	1.0	18 PT	1
PC 80	UCB	AC	EHA	Soft AC	67	2.0	17 PT	2
80 D	BASF	AC	EHA-AN	Medium AC	67	6.0	17	60
85 D	BASF	AC	EHA-AN	Medium AC	67	4.0	19	70
2338	Synthomer	AC	EHA-AN	Medium AC	67	6.0	16	60
FC88	UCB	AC	EHA-AN	Medium AC	67	7.0	14	40
552	Exxon	AC	EHA	Hard AC	67	8.0	15	50
20088	Synthomer	AC	EHA-EA	Hard AC	67	12.0	12	90
20094	Synthomer	AC	EHA-MMA	Hard AC	67	>20	10	200
V 205	BASF	AC	EHA	Soft AC	45	2.5	22 PT	5
80 D	BASF	AC	EHA-AN	Medium AC	22			
R3375	Harco	AC	EHA	Soft AC	45	3	20 PT	4
80 D	BASF	AC	EHA-AN	Medium AC	22			
PC 80	UCB	AC	EHA	Soft AC	45	3.5	18 PT	4
FC 88	UCB	AC	EHA-AN	Medium AC	22			
3703	Polysar	CSBR	—	Soft CSBR	67	1.5	16	100

Tackifier: CF 52 (A&W) at a level of 33 parts (wet/wet). All formulations tested as paper labels.
AC, acrylic; EHA, ethyl hexyl acetate; MMA, methyl methacrylate; CSBR, carboxylated styrene butadiene rubber; PT, paper tear.

ability; acrylic PSAs show excellent tackification with acid rosins, but the range of the obtained adhesion/cohesion values is dispersion related (Table 8.15).

Table 8.15 shows that the same tackifier acts differently on PSAs with a different adhesion/cohesion balance. A medium (33%) resin concentration generally provides good polyethylene peel adhesion and excellent tack values for soft and medium PSAs. High molecular weight, highly cohesive acrylic PSAs do not show enough tack and polyethylene peel; these compounds cannot be tackified sufficiently without the presence of another PSA. Measurements aimed at comparing the effect of CF 52 and Dermulsene 511 do not reveal any differences concerning the adhesion/cohesion balance, converting and processing properties of the finished product using a V205/80D blend as reference.

Table 8.16 summarizes the adhesion properties of tackified acrylic PSAs, formulated with rosin-based tackifier resins from the same supplier. The agressivity of the formulations decreases from 1 through 5 (i.e., from acid to ester), but acid rosin tackifiers also may yield different results (1, 3, 5). Here, tackifier 3 is less agressive (lower tack and peel, although an acid rosin) but it brings higher shear, better water resistance, and better aging properties. The better processability of tackifier 2 is probably due to its lower water sensitivity.

Acrylic PSAs display the best tackifying response with rosins due to their general compatibility. Acid rosins are superior, providing a more agressive tack than rosin esters. Thus it can be expected that rosin esters would exhibit poorer adhesive characteristics with other PSAs as well. Comparative tests of CSBR tackification with acid rosins and rosin esters show only a slight tack decrease, but more polyethylene peel decrease for the same resin level. The aging resistance of the acid rosin does not meet the future, high requirements. Although the resin's affinity towards crystallinity, and the level, nature, and technology of the antioxidants/formulating affects the aging resistance, the less agressive but more stable resin esters should be used for demanding applications. Ester rosins tackify base polymers in a less pronounced way than acid ones. Generally, the tack and polyethylene peel of the rosin ester-tackified acrylic PSAs is lower than that of the acid-based ones, and due to the migration (of the low molecular weight components) paper coatings with these resins display a higher dispersion of adhesive property values. Butyl acrylate-based formulations are less shear sensitive and therefore may contain (at least theoretically) high (up to 50 wt%) tackifier loading levels. This statement points toward the use of ester rosins at a higher level, reaching the same tack level as obtained with rosin acids. Unfortunately, although shear performance was improved, this formulation approach does not yield a better adhesion/cohesion balance and processing properties are not superior (Table 8.17).

Acid rosin-based dispersions possess a higher surface tension, a lower solids content, and consequently poorer wetout properties than pure PSAs; therefore formulated PSAs need additional wetting agents. The amount of

Table 8.16 Tackifying the Same Base PSA with Different Rosins from the Same Supplier

	Water-based tackifier		Properties				
				Adhesive characteristics			
			Rolling ball tack (cm)	Peel (N/25 mm)		Shear at 50°C (min)	Cuttability
	Code	Chemical basis		Glass	PE		
1	CF 52	Acid	2.0–2.5	21 PT	15 PT	5–10	Poor
2	CF 51	Ester	2.5–5.0	22 PT	10 PT	8–15	Fair
3	CF 301	Acid	4.0–5.0	20 PT	9–10	35	Good
4	SE 8/21	Ester	5.5	20 PT	16 PT	50	Fair
5	CF SP	Acid	8.5–10.0	18 PT	15	100	Fair

Base AC PSA: blend of V 205/80 D. Tackifiers supplied by A&W.
PE, polyethylene; PT, paper tear.

Table 8.17 Tackifying Acrylic PSAs with a Different Chemical Basis (from Various Suppliers) with Different Rosins

Nr	Code	Supplier	Base						Formulation (parts wet weight)							
				1	2	3	4	5	6	7	8	9	10	11	12	13	14
1	V 205	BASF	EHA	100	38	39	50	50	50	50	50	—	—	—	—	—	—
2	80 D	BASF	EHA-AN	—	22	32	10	10	10	10	10	—	—	—	—	—	—
3	CF 52	A&W	Acid rosin	—	40	29	40	—	—	—	39	—	—	—	—	—	—
4	CF 50X	A&W		—	—	—	—	40	—	—	—	—	—	—	—	—	—
5	CF 51 (SE351)	A&W	Rosin ester	—	—	—	—	—	40	—	—	—	—	—	—	—	—
6	SE 62	A&W	Rosin ester	—	—	—	—	—	—	40	—	—	—	—	—	—	—
7	Bacote 20	—	Crosslinked	—	—	—	—	—	—	—	1	—	—	—	—	—	—
8	E-2395	Rohm & Haas	BuAc	—	—	—	—	—	—	—	—	60	60	—	—	—	—
9	3385	BASF	BuAc	—	—	—	—	—	—	—	—	—	—	70	—	—	40
10	SE80CF	A&W	Rosin ester	—	—	—	—	—	—	—	—	40	40	30	40	—	—
11	XZ95043	Dow	BuAc	—	—	—	—	—	—	—	—	—	—	—	60	—	—
12	XZ95044	Dow	BuAc	—	—	—	—	—	—	—	—	—	—	—	—	60	—
13	Revacryl 491	Harco	EHA	—	—	—	—	—	—	—	—	—	—	—	—	—	—
PE Peel	N/25 mm	FTM 1		—	13	13	6.9	6.8	6.7	8.2	7.0	9.3	7.6	6.7	8	5	9
CB Peel	N/25 mm																
							18	11	14	16	13	8	7	6.0	7.2	6	7.2
Loop SS	N				11	15	22	21	27	26	23	26	18	24	23	12	21
Loop PE	N				5	5	15	12	14	10	12	16	13	13	10	9.4	5
Shear	hr	RT		18–50	—	—	11	28	22	38	27.6	66	54	60	30	78	42
	min	50°C					16	114	42	168	40	12	—	—	—	—	—

Face stock: PET (Mylar)

added wetting agent depends on the product quality and formulator skill. Because of the acid pH of the mixture, there are no stability or thickening problems with sulfosuccinates. The order of blending of the components is arbitrary. However, changes in drying conditions could give rise to wet coatings, telescoping, or poor die-cutting properties. Acid rosins do not possess enough aging stability. Thus, partial or total loss of the adhesion (end-use property) may be observed.

Rosin Esters

Rosin esters are newer products in the development of PSAs. Their dispersing technology requires other dispersing agents and level than with acid rosins. Rosin ester dispersions possess inferior converting properties, such as a poorer wetout, a lower mechanical (stirring, shear) stability of the liquid dispersions, and a lower pH stability. The shear of the bulk PSAs tackified with high melting point, high molecular weight rosin esters is theoretically superior to that of acid-tackified formulations. For a dispersion the shear depends upon the recipe of the adhesive/resin and on the humidity balance of the laminate. Therefore die-cutting properties of the ester-tackified formulations are not better than those of the acid-based ones. Die cutting depends on the anchorage of the adhesive; consequently it can be expected that adhesive formulations with good anchorage display better die-cutting properties. The anchorage on the face stock and substrate depends on the acidity of the resin. Thanks to their high acid number certain resins (e.g., Oulumer 70 and 75E) have very good adhesion to cellulosic materials. However, their observed die-cutting properties are not superior because of their higher water sensitivity. In this case the highest improvement of the cuttability may be observed by lowering the tack. Generally, rosin esters impart lower tack and polyethylene peel than acid rosins. Many formulations with rosin ester tackifiers are binary mixtures with a low softening temperature component. Migration of this component causes loss of coating weight and lowering of the tack peel values in time. Hence, the end-use properties of the rosin ester-based formulations are inferior to those of rosin acid-based ones. Acid rosins may be blended in without regulating pH. Direct compounding of rosin esters with common acrylic PSAs gives rise to mixtures with a higher mechanical sensitivity. Therefore some suppliers of rosin esters propose a two-step compounding process. In the first step the formulation of the acrylic PSAs occurs; next, after increasing the pH of the acrylic mixture to 7 or 8, the tackifier is added. Adjusting the pH may cause thickening of the whole mixture and change its rheology.

Summarizing, rosin esters were introduced as replacement of acid rosins, due to the concerns regarding the loss of adhesive properties with time. Their unique advantage is their supposed better aging stability. Rosin esters require a higher loading level for the same level of adhesion. Because of their binary formulation, rosin esters cause migration and give rise to higher dispersion of the peel-tack values. Rosin esters possess a lower mechanical stability and display poorer wetout. Compounding rosin esters requires adjustment of the pH (associated with thickening), thus it is a more complex operation than with acid resins. Rosin esters do not provide any better processing or end-use properties, and are more expensive than acid resins. Practically, the raw materials base for rosin remains limited as they are derived from natural products. Resources of tall oil resins remain limited also. Because of the classical nature of the raw materials and of the old processing technology, there are no possibilities to reduce manufacturing costs. Thus one can assume that the price level of rosins will continue to increase.

2.5 Hydrocarbon-Based Tackifiers

In order to determine the technical merits and shortcomings of hydrocarbon-based tackifiers, their adhesive, converting, and processing properties need to be examined comparatively to their main competitors (i.e., the rosins). Hydrocarbon resins possess a limited compatibility, thus it can be expected that from the point of view of the dry adhesive, they would show better or similar tackifying properties only for certain PSAs (e.g., acrylic PSAs). However, difficulties may arise when tackifying high molecular weight BuAc. However, it appears possible to design better converting or processing properties of the laminate, depending on the dispersing skill of the resin manufacturer. Consequently, hydrocarbon resins are not yet universal tackifiers, but for the most important PSAs (e.g., acrylic- and CSBR-based PSAs) they can provide technical advantages for the dry material as well as for the aqueous dispersions.

Adhesive Properties of Hydrocarbon-Resin Tackified Formulations

Some years ago a trend was observed in the converting industry, namely a switch from rosin acids to rosin esters, caused by the low aging stability of the acid rosin-based formulations. Acid rosins are the best tackifiers, providing a good adhesion/cohesion balance for all the commonly used PSAs. Hydrocarbon resins are generally inferior tackifiers comparatively to rosin-based ones. This holds true for the instantaneous adhesive properties of the acid rosins. Aged formulations with acid rosins, including emulsions, show poorer characteristics than hydrocarbon resin-based ones. Rosin ester tackifiers generally display inferior adhesive characteristics to acid rosins; thus, opportunities exist for hydrocarbon resins mainly in comparison to rosin esters. Hydrocarbon-based tackifiers exhibit an adhesion/cohesion balance similar to acid rosins. Slightly better shear values are possible, but they are not high enough to improve the processing properties. For CSBR high levels of hydrocarbon resins are necessary (like acid rosins) in order to achieve sufficient polyethylene peel (Table 8.18).

As the tackifier loading increases from 25 to 30%, the failure mode changes from adhesive failure to paper tear in both the 180° peel and loop tack tests. However, if the tackifier level is increased above 35% the shear strength drops dramatically. Results indicate that for rosin acid-tackified acrylic PSAs, the optimum tackifier range is approximately 30–35 wt% [48]. Shear depends mainly on the molecular weight of the elastomer and the resin level [13]. As the resin level is increased, the shear failure time generally decreases exponentially. This behavior is generally valid (i.e., the resin level has more influence on the shear than the nature of the resin). It can be expected that an equal resin loading (independent of the resin nature) would cause the same shear decrease. Consequently, at a level of 30–35%, hydrocarbon resins should give the same tack and cohesion as acid rosins. However, experimental results indicate a better tack of the hydrocarbon resin-tackified dispersion at the same resin loading level (Table 8.19).

Acrylic-hydrocarbon resin mixtures exhibit slightly better tack values than acid rosins at the same shear and peel level. Like rosin acids hydrocarbon resins provide a more pronounced tack improvement than an increase in polyethylene peel. Practically, the tackifying response in combination with common acrylic PSAs may be considered similar to that of acid rosins.

Aging Properties of Hydrocarbon Resins

Acid rosin tackified acrylic formulations could suffer a partial or total loss of the adhesive properties due to chemical changes in the resin. Under environmental influence abietic acid undergoes chemical transformations giving rise to hydroperoxides. Dimerization with metal salts (rosinates) also is possible. This phenomenon is responsible for the increase of molecular weight, and for the loss of the tackifying effect. Aging due to acid groups should be avoided when using rosin esters or hydrocarbon resins. Theoretically, hydrocarbon resins possess superior aging properties, but this hypothesis needs to be confirmed in practice.

Converting Properties

Wetout, mechanical stability, and foam resistance of hydrocarbon resin dispersions are equivalent to those of rosin acids and superior to those of rosin esters. Generally, the additional wetting agent level needed for good wetting reaches 1.0–1.5%, comparatively to the 2.0–3.5% necessary for rosin esters.

Processing Properties

At the same adhesion/cohesion balance die-cutting properties depend mainly upon the humidity balance of the laminate. This also is a function of the

Table 8.18 Tackification of CSBR Dispersions with Acid Rosin/Hydrocarbon-Based Tackifiers

	Formulating components			Chemical composition					
Nr	Code	Supplier	Chemical basis	1	2	3	4	5	6
1	CF 52	A&W	Acid rosin	50	—	43	33	—	20
2	E-9251	Exxon	Hydrocarbon resin	—	50	—	—	33	—
3	3703	Polysar	CSBR	50	50	57	66	66	80
Adhesive performance characteristics	Peel (N/25 mm)		on glass	19 PT	17 PT	17 PT	15 PT	20 PT	14 PT
			on PE	14 PT	15 PT	16 PT	11	15	10
	Rolling ball tack (cm)			1.5	2	1.5	2.5	2.5	3

Table 8.19 Tackifying Acrylic PSA with Hydrocarbon Resins and Rosins

Nr	Component	Supplier	1	2	3	4	5	6	7	8	9	10	11
1	Acronal 85D	BASF	21.5	25.0	29.0	83.0	77.0	74.0	83.0	66.0	83.0	40.0	21.4
2	Acronal V205	BASF	35.5	34.0	32.0	—	—	—	—	—	—	40.0	35.7
3	Snowtack CF52	A&W	43.0	41.0	39.0	27.0	23.0	26.0	27.0	—	27.0	20.0	42.8
4	E-9251	Exxon	—	—	—	—	—	—	27.0	38.0	—	—	—

Properties

	1	2	3	4	5	6	7	8	9	10	11
Coating weight (gsm)	24	25	25	25	26	26	19	20	20	—	18
Peel (N/25 mm)	20 PT	20 PT	20 PT	19	9	14	14	15	15	17	18 PT
Rolling ball tack (cm)	7.5	2.5	4.5	2.0	2.5	1.7	2.4	2.5	—	3.0	7.5
Shear											
Room temperature (hr)	40	40	30	—	—	—	45	29	40	—	—
50°C (min)	16	23	20	—	—	—	—	—	—	—	—

Table 8.20 Hot Shear and Cuttability Performance for Acrylic Water-Based PSAs Tackified with Rosin and Hydrocarbon Resins

	Hot shear values (min) Water-based tackifier				
Hot shear test temperature (°C)	CF 52 A&W	DEG/G DRT	MBG 64 Hercules	E-9251 Exxon	8/21 A&W
30	90	402	330	264	1440
40	100	162	90	102	192
50	12	72	72	90	192
60	10	45	38	46	51
70	11	22	23	40	26
Cuttability	Fair	Good	Poor	Fair	Fair

Adhesive: Blend of Acronal V 205/80 D with 40% resin (wet weight) was used; face stock: paper.

hygroscopy of the adhesive layer; the hygroscopy of the adhesive layer depends on the nature and level of the wetting agent. Hydrocarbon resin dispersions with less surface-active agents display lower hygroscopy and better cuttability. The few experimental measurements do not confirm this hypothesis. The dependence of the shear on the drying time (i.e., humidity content) of the layer was demonstrated. Table 8.20 lists the hot shear gradient (HSG) and cuttability results of different formulations. Because of the exponential dependence of the hot shear on the temperature, an increase of the cohesion of some orders of magnitude is required for a better shear/die cuttability. Unfortunately, hydrocarbon resins do not impart this shear increase.

Compounding hydrocarbon resins with acrylic- or CSBR-based PSAs does not cause problems with the pH level of PSAs. Unfortunately, the mechanical stability properties of some hydrocarbon resin dispersions are not sufficient. From the point of view of the instantaneous adhesive properties, there are no advantages to the use of hydrocarbon resins. Their adhesion/cohesion level is similar to the level reached by acid rosins, but without a similar general applicability. However, hydrocarbon resins age better than acid rosins. In a first approximation the user has to choose between rosin acids with insufficient aging stability but general suitability, and hydrocarbon resins with good aging stability but limited usability. The development of hydrocarbon resin dispersions was slow. Therefore, customers have replaced rosin acids with rosin esters which display a better aging stability but carry a higher price tag and exhibit lower tack (see Figure 8.7). Hydrocarbon resins are superior to rosin esters, concerning the adhesion/cohesion balance and the price level.

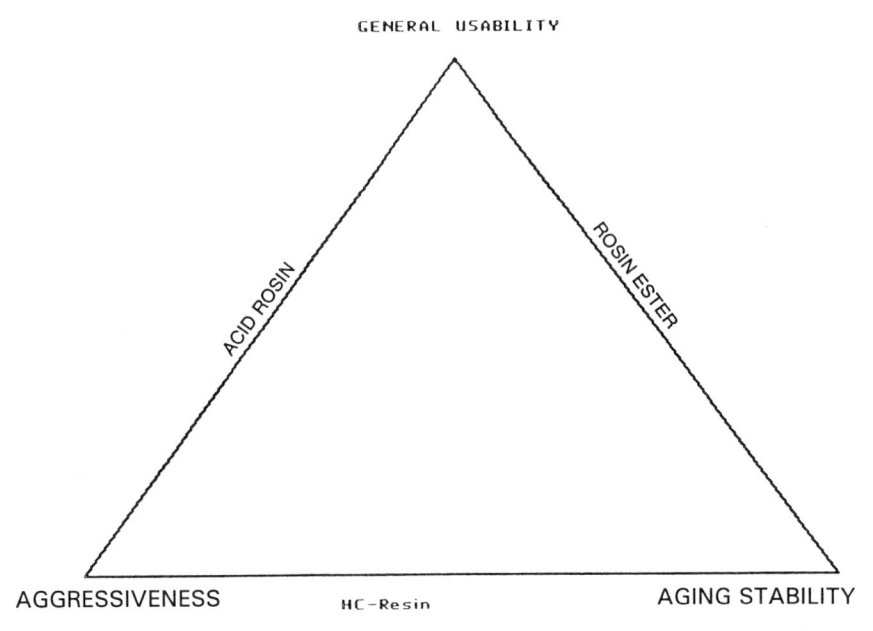

```
                    GENERAL USABILITY
                          /\
                         /  \
                        /    \
                   ACID/      \ROSIN
                   ROSIN       ESTER
                      /        \
                     /          \
                    /_____\
AGGRESSIVENESS      HC-Resin        AGING STABILITY
```

Figure 8.7 General applicability, agressivity (tack) and aging stability of different tackifier resins for PSA formulation purposes.

Raw Material Basis

Raw materials for hydrocarbon resins are naphta derivatives; their supply depends on the oil processing industry. Price fluctuations for the crude oil are possible, but there is no danger of a general raw material shortage, as the refining/cracking capacity is more than satisfactory. Hydrocarbon-based solid resins are manufactured in many different plants. The forecast concerning a general shortage of C_4 fraction from the steam crackers has not been proven [71]. Steam-cracked naphta leads to aromatic, polycyclic, C_5-modified hydrocarbons, and C_5 hydrocarbon resins. Hydrocarbon-based tackifiers do not perform better than acid rosins, but are superior to rosin esters. The large raw material basis, the low price level of the petrochemicals (compared to natural products), the simple manufacturing technology, and the competition between suppliers ensures a lower price level for hydrocarbon resins. Table 8.21 summarizes the main technical and commercial advantages/disadvantages of hydrocarbon resins versus rosin resins. Consequently, hydrocarbon resins should be used in the following applications:

Table 8.21 Comparison of Rosin and HC Resins: Technical and Commercial Strengths and Weaknesses

Technical			
Strengths		Weaknesses	
Rosin	Hydrocarbon	Rosin	Hydrocarbon
Experience in use	—	—	Limited experience
General usability	—	—	Restricted usability
General availability	—	—	—
Broad range of grades	—	—	Limited range of grades
Good adhesive properties	Good adhesive properties	—	—
Fair converting properties	Fair converting properties	—	—
—	Good aging resistance	Poor aging resistance	—
—	Colorless	Yellowing	—
Commerical			
—	Unlimited raw material basis	Limited raw material basis	—
—	Unlimited technology	Limited technology	—
—	Lower price	Higher price	—

- permanent label (paper) adhesives, as hybrid tackifiers together with rosin esters;
- permanent label (paper) adhesives, as hybrid tackifiers together with low tack rosin acids;
- permanent label (paper) adhesives, as replacement of rosin esters, providing a similar adhesion/cohesion balance like acid rosins, but better aging stability;
- permanent label (paper) adhesives, as a less expensive replacement of acid rosins (but not generally applicable);

- permanent label (film) adhesives for clear coatings onto PVC or other films;
- removable label (paper) adhesives, but only for formulations containing a nontacky low peel natural latex and CSBR.

From a technical point of view rosins actually perform better than hydrocarbon resins. However, there are two major disadvantages to the rosin tackifiers, namely the low level of adhesive properties of rosin esters, the low aging stability of acid resins and the inadequate coating/processing properties of the aqueous dispersion (related to its water sensitivity). Formulations containing some hydrocarbon resins and rosin esters offer an adhesion/cohesion balance like acid rosins. Consequently, there is a real opportunity for a partial replacement of rosins in current recipes. Today about 60% of converters are using rosins. An amount of 10–20% of the rosin ester could be replaced by hydrocarbon resins (i.e., a maximum 12% of the current resin consumption). About 70% of labels are paper laminates with a maximum of 40% tackifier resin or 28% of the whole PSAs production is tackifier based. Assuming the 12% hydrocarbon resin level, an actual use of about 4% hydrocarbon resin (related to the whole amount of raw materials) seems possible.

Plasticizing

Tackification may be carried out using resins or plasticizers. The tackifying mechanisms and the influence of the tackifying agent on the T_g is quite different for plasticizers and tackifiers. The addition of resins to an adhesive increases the T_g of the base polymer, the addition of plasticizers decreases it [72] (see Chapters 2 and 3). Tackifier resins may increase the tack, without an important decrease of the shear resistance; plasticizers increase the tack, but lower the cohesion. It remains difficult to make a sharp distinction between tackifiers (resins) and plasticizers (solvents) as a result of their chemical basis. Actually both low molecular weight liquid resins and relatively high molecular weight plasticizer liquids (solvents) are known. The use of liquid tackifier resins induces the same rheological changes as with plasticizers. As an example, tackifying styrene butadiene copolymers is more difficult than tackifying S-B-S ones. A solvent-based copolymer with a plateau modulus of 55×10^7 dyn/cm^2 needs a much lower T_g tackifying resin in order to decrease the modulus than an S-B-S copolymer (plateau modulus 4.3×10^6); here the tackifier should act like a plasticizer [73].

The influence of the chemical nature, structure, and molecular weight of solvents used as plasticizers was discussed [74]. Generally, plasticizers are reactive liquid components that interact with PSAs. Adhesives based on polyisobutylene (PIB), EVAc, polyvinyl ether (PVE), and thermoplastic elastomers (S-I-S, S-B-S) are very sensitive towards compounds like dioctylphthalate

[75]. Aromatic plasticizers like dioctyl phthalate (DOP) associate with the polystyrene domains and act like aromatic oils (i.e., soften the polystyrene domains). Therefore, a barrier layer must be present between PVC and Cariflex TR-1000 (Shell). Including plasticizer into polybutyl acrylate (PBuAc) dispersions brings good removability after long time storage, but produces migration, low cohesion, and adhesive transfer [76]. Plasticizers increase tack but lower cohesion. Common plasticizers are low molecular (micromolecular) compounds; other macromolecular products also may be used. As an example Parapol is used as tackifier/plasticizer. It is a viscous, liquid copolymer of n-butene and isobutylene. In hot-melt PSAs it increases tack and also is applied to obtain softness and low temperature flexibility, and to control cohesive strength [77]. For many years polybutenes were successfully used as tackifiers/plasticizers in hot-melt and solvent adhesive systems. Later the polybutene dispersions were incorporated in water-based PSAs. Acrylic latexes were used because of their increased compatibility with polybutenes. The successful incorporation of polybutenes (at 20–30% addition levels based on the final "dry" film) would result in significant cost savings. In hot-melt PSAs polybutenes plasticize and provide tackiness by associating with the isoprene midblock, which has negligible inherent tack. Polybutene contributes to the agressive tack and quick stick properties necessary for adhesive applications [78]. Generally, polybutenes can be incorporated into acrylic latex systems by either of two methods, either by addition to the acrylic polymer latex, where it acts as an external plasticizer, or by addition to the acrylic monomers, prior to emulsion polymerization, where it acts as an internal plasticizer.

Classical plasticizers are used on a large scale in PSA applications. For removable PSAs, recommended compounds include: 2–diethyl hexyl esters of ftalic acid and 2–diethyl hexyl adipate, esters of sebacic and aceloinic acids with 6–10 alcanols, alkyl esters of succinic, glutamic, adipic acid with epoxydized soybean oil, phenol alkyl sulphonic acid ester, and propylene glycolalkylphenyl ether, di-2–ethyl hexyl ester of thiodipropionic acid, acetyltributhylcitrate, and glycoldibenzoate; phtalic acid esters like di-n-butyl phtalate, diisobutyl phtalate, di-2–ethylhexyl phtalate, and polypropylene glycolmethylol ether [74,76]. The choice of the plasticizer depends on the chemical basis of PSAs, application technology, and end-use requirements. The compatibility of different adhesive raw materials with plasticizers was discussed in a detailed manner [79]. Plasticizers used may be solid (sulfonamide) or liquid (phtalates, phosphates, glycolates, etc.) [80]. The plasticizer nature depends on the adhesive nature. As an example, for S-B-S butyl-benzyl phtalate was suggested [40]. For electron beam-cured hot-melt PSAs Parapol 350 (Exxon) and Flexon 1076 (Shell) are used as plasticizers. In this case formulation ingredients like plasticizers have to be inert when subjected to irradiation. To obtain optimum crosslinking with a minimum of radiation, the selection of suitable resins and plasticizers is necessary (e.g., PSAs based on multiarmed S-I-S block copolymers crosslink much faster with saturated than with unsaturated resins and plasticizers). On the other hand, the rheology changes due to the plasticizer have a quite different scope in permanent and removable adhesives. In permanent adhesives they improve the tack, in removable ones, they have to make the adhesive softer in order to absorb the peel energy. Solvents and plasticizers increase the viscosity of PSAs dispersions [81]. The sensitivity of different plasticizers toward antimicrobial agents also is quite different. Their environmental relevance, toxicological findings and suitability for medical products are other aspects to be taken into account at the time of their selection [82]. A common level of 20% (wet/wet) is proposed for water-based permanent adhesives. Compounding water-based acrylic PSAs with more than 10% plasticizer requires special skills (e.g., plasticizing of Acronal 120).

The most important additives influencing the adhesive properties are the tackifiers (resins and plasticizers). Detackifying agents are known too (see Section 2.8). Tackifiers change tack and peel in parallel. Other ingredients like crosslinking agents have a similar effect (see Chapter 6). Their choice will be discussed in Section 2.8. There are many formulating additives which are used for other purposes, but also influence the adhesive properties; such ingredients include fillers, antioxidants, surfactants, etc.

2.6 Coating Properties

In order to wet the face stock or release liner material during coating PSAs must display a good wetout ability. They must also resist shear forces during coating (metering device) and thermal shocks during drying. Formulating the adhesives in order to optimize the coating properties intends to improve these characteristics. Wetting out depends on the receiver surface. It is evident that direct coating on paper or film is less difficult than indirect coating via a release liner, because of the fairly different surface tensions of these materials. On the other hand, direct coating on paper is easier (concerning the wetout) than direct coating onto film as the nature of the face stock film influences the wetting ability. Polar surfaces (PVC, PC, PET) are easier to coat than polyolefins. The nature of the silicone (solventless or solvent-based) influences the wetting behavior as well. Depending on these factors, different wetting agents at different levels can be added to the formulation. When direct coating bleedthrough (migration) should be avoided, a high viscosity and a high solids content is needed.

Drying when direct coating is limited by the temperature resistance of the face stock material. Formulating is limited by the chemical resistance of the face stock material. Different coating devices also require different coating viscosities. The drying oven capacity implies a lower or higher solids content. These few examples illustrate how coating properties of PSAs affect its formulation. Next, the influence of the coating technology on the formulation will be discussed.

Influence of Coating Technology on Formulation

The most important feature of the coating technology is its direct or transfer character. In direct coating the adhesive is deposited on the face stock material, mostly as a low viscous liquid. In transfer coating PSAs are temporarily coated on a siliconized web, the adhesive transfer onto the face stock material is carried out later after the drying operation (i.e., the coating properties of the liquid adhesive have to be designed for the release liner). Therefore quite different formulations are required for direct and indirect or transfer coating. In some special cases (Monoweb) the PSA coated on the face stock material is protected by the back side of the face stock material. Such applications are encountered mainly for protective films or tapes where a release layer is coated on the back side of the face stock material or the chemical composition of the face stock material allows the peel off of the adhesive.

The quite different requirements for the formulation of PSAs as a function of the coating technology are illustrated by the design of water-based adhesives. In this case for direct coating, neither liquid resins nor peel modifiers should be used, and only a low level of crosslinking agents; highly viscous, thickened dispersions are preferred. When necessary, surfactants with good water resistance should be used. For indirect or transfer coating, as a function of the liner nature, varying levels of wetting agents are needed. Defoamers may be added also, but only at a low level. A higher mechanical stability of the dispersion is required. The formulation for direct/transfer coating influences the release force from the release liner also. Aqueous dispersions damage the siliconized release paper, and therefore the release force between a transfer-coated adhesive and the release liner is higher. An adhesive required to be coated directly onto a web via Meyer bar application, requires a different set of intermediate properties (e.g., wetout, total solids, viscosity) compared to one designed for transfer coating on a reverse roll coater via a silicone release liner.

Influence of Face Stock Material on Formulation

Wetting out on paper is easier than on films and a higher surface tension can be tolerated; for the same reasons a lower viscosity may be used. Unfortunately, low viscosity dispersions or solutions migrate easier through porous paper. Wetting out on polymer films is more difficult. Here the polarity and the roughness of the film influence the coatability. As an example, white or clear PVC displays a different wetout ability and needs different surface-active agent levels. Polyolefins are difficult to be coated as their slip agent content influences the wettability. The porosity of paper and the nature of the coating method influence the coating weight as well as the adhesive/converting/end-use properties.

Influence of Coating Machine on the Additive Choice

Shear on the coating machine depends on its construction and functional parameters, as does foaming. Shear contributes to good wetting; foaming can be avoided with the use of defoamers, therefore it works against wetout. Too high shear on the metering device or in pumps produces coagulum; hence poststabilizing with surface-active agents can be necessary. Insufficient thermal (drying) capacity requires the improvement of the volatility of the adhesive carrier (solvents or water).

A given metering device can be used for a certain viscosity range. Engraved cylinder rolls with different line screen use different viscosities and solids contents. Different flow properties are required for different types of metering devices (e.g., Meyer bar or gravure cylinder) and even the same formulated adhesive coated on different coating devices may lead to quite different coating weights and adhesive properties (Table 8.22).

Therefore different versions of the flow and adhesive properties are needed as a function of the coating machine. Adjusting the adhesive's viscosity (via formulation) for any machine is important for all kinds of adhesives. For water-based PSAs, dissolving nonionic alkylphenoloxyethylates may be accompanied by an increase of the viscosity. Therefore warm water should be used; later it can be diluted with cold water [83]. In some cases (certain sulfosuccinates) the improved wettability is given by the increase of the viscosity (and decrease of the surface tension). In other gravure coating

Table 8.22 The Influence of the Coating Technology on the Adhesive Properties

Theoretical coating weight (g/m^2)	90° Peel (N/25 mm)	
	Coating device	
	Mayer bar	Gravure roll
0.7	0.37	0.18
0.8	0.42	0.27
0.9	0.48	0.31
1.0	0.60	0.40

applications dilution of the raw dispersion and subsequent rethickening is used in order to avoid gravure-induced structures (i.e., optical defects in the coating).

The choice of the processing oil has the greatest impact on hot-melt PSAs performance. The greater solvency power of naftenic oil produces lower adhesive viscosities and shear adhesion failure temperatures, while increasing the tack. Peel values strongly decrease with the use of fully saturated mineral oils [84]. The flow properties of rubber block-copolymer hot-melt PSAs are improved by polyethylene-based polymers. Modification with low molecular weight polyethylenes results in significantly better flow properties, allowing for less energy intensive mixing and more uniform deposition of the molten adhesive [85].

Polyacrylate rubbers with relatively low molecular weight can be dissolved directly in a polar solvent without milling. However, milling the elastomer would further help in lowering the viscosity, this being required for certain coating devices. On the other hand, in general, milled elastomers possess less cohesive strength (i.e., require less tackifiers).

Precautions must be taken to avoid degradation of hot-melt PSAs by oxidative attack during the application of the adhesive. During coating the thin film of adhesive is spread on the web at elevated temperatures. In roller coaters where the adhesive is transferred from the heated roller, the adhesive is exposed to air, while in die coaters the adhesive is extruded onto the web in a closed system and air exposure is less critical. Therefore when formulating the type of the coater should be considered [86]. The nature of the adhesive influences the drying process [87]; different adhesives require different driers (length of drying oven, temperature, air volume, temperature gradient). Conversely, for a given oven, the adhesive formulation should be adjusted in order to obtain an optimum drying velocity.

2.7 Converting Properties

The processing properties of the laminate will be discussed separately. In this section only the most important of them will be covered; they are printability and cuttability. These are influenced by properties like migration, lay-flat, and reel stability.

Migration

The influence of the face stock material on migration was covered in Chapter 2. It can be difficult to store label stock for any length of time due to the aging/bleeding problems. The adhesive bleedthrough often interferes with the quality of the art work and printing. Reels often bleed at the bottom and dry out at the top (gravity bleeding). There are a lot of difficulties with high gloss labels bleeding through. Generally, migration leads to the deterioration of the face stock, deterioration of the adhesive properties by lowering the coating weight, phase separation in the adhesive, and enrichment of one of the components [88]. Phase separation is caused by the different melting points of the components and different molecular weight distributions. Acrylic PSAs are sensitive to this phenomenon, especially when used on face materials which promote migration of oils into the adhesive, thereby lowering the cohesive strength [89]. Generally, the following factors influence the migration (penetration, bleedthrough, staining) of the adhesive in the face stock: the characteristics of the adhesive and face stock, and the machine conditions. The influence of the face stock on the migration was discussed earlier (see Section 3.2 of Chapter 2). Next the influence of the adhesive characteristics on the migration will be covered.

Influence of Adhesive Characteristics. The chemical composition of the adhesive, the formulation of the adhesive, the tackifier, and the additives, their chemical and macromolecular characteristics all influence the migration. Migration starts with an almost sudden temperature increase which depends only on the formulation of the adhesive, but not on the paper properties [88]. The depth of the migration is given by the apparent density of the paper. The phenomenon of penetration was well studied for paper manufacturing [90], where ink penetration depends on the concentration, the viscosity, the water retention properties, the temperature, and the chemical composition. In a similar manner the same parameters influence the migration of PSAs. Water retention and viscosity of water-based PSAs depend on the surface-active agent content and its nature. In the particular case where the adhesive is to be used to form laminates and at least one of the surfaces is a printed surface, the presence of any residual surfactant can lead to discoloration or bleeding [91]. Staining is a common characteristic with water-soluble adhesives [92].

There are two separate conditions causing staining: pressure (reel-wound laminates) and heat (storage). For CSBR, in both cases the level of staining is related to the butadiene content and molecular weight distribution of the base polymer, the tackifier resin type, and its softening point and emulsification system [50]. Generally, staining increases with increasing butadiene content, but decreases at higher gel contents, although not proportionally. The tackifier resin type has a definite effect; rosin acid tackifiers outperform hydrocarbon resins for example, as do higher softening point resins within any type.

Molecular Weight of the Tackifier Resin. Hot-melt PSAs usually contain S-I-S or S-B-S rubber block copolymers, tackifiers, and hydrocarbon-based oil. The use of low softening point tackifiers and/or a high concentration of an oil may be necessary to achieve higher tack, better low temperature properties, reduced melt viscosity, removability, and lower the cost [85]. However, upon aging the much lower molecular weight components can migrate from

Manufacture of Pressure-Sensitive Adhesives 379

the hot-melt PSAs. Plasticizers migrate from flexible PVC film and upon entering a PSA cause loss of adhesion due to a lowering of the PSA's cohesive strength. Because they are based on relatively low molecular weight components, hot-melt PSAs are more adversely affected by migration of plasticizers. Generally, low molecular fractions penetrate more [88]. Measuring the molecular weight distribution in the area where migration occurred, it was found that a high molecular weight fraction has the same concentration in the middle and at the extremities (i.e., center and outer range), but the low molecular weight portion has a higher concentration at the extremities. Some agreement could be found between the starting temperature of migration and the softening point temperature, as measured by the ring-and-ball method [41]. According to the theoretical model, the self-diffusion rate R_d is inversely proportional to the molecular weight [93]:

$$R_d = f\left(\frac{1}{MW^2}\right) \tag{8.1}$$

For rubber the gel/sol ratio influences both cohesive strength and migration. For those skilled in the art of formulation the dependence of the migration on the molecular weight is well known from the practice of thickeners. Thus different grades of polyvinyl pyrrolidone (PVP) with different molecular weights cause different levels of migration. As an example PVP-K-15 (low degree of polymerization, DP) gives migration, where PVP-K-90 (high DP) does not lead to any penetration in paper. When Hyvis 10 polybutene is used, the polybutene exudes from the adhesive and soaks through the backing release paper; therefore only the high molecular weight Hyvis grades (i.e., 200 and above) should be used [51].

Machine Influence. The coated release liner may be laminated to the face stock, in a pressure nip between a rubber roll and a steel roll. This technique accomplishes the transfer of the adhesive mass to the face stock with a minimum of penetration [94].

The incorporation of 1.5% amorphous copolyethylene (A-CPE) in a S-B-S copolymer-based hot-melt PSA prevents paper face stock bleedthrough upon aging at 158°F for 2 weeks. The ability of these low molecular weight polyethylene and polyethylene copolymers to gel organic liquids is responsible for the reduced migration of low molecular weight tackifiers and oils [85]. Another possibility to avoid bleedthrough is the use of barrier coatings. Resin compatibility also influences migration; colloidal silica reduce the bleeding as well.

Cuttability

In Chapters 2 (Rheology) and 7 (Converting Properties) the parameters influencing the cuttability were outlined. The adhesive nature (solvent-based, hot-

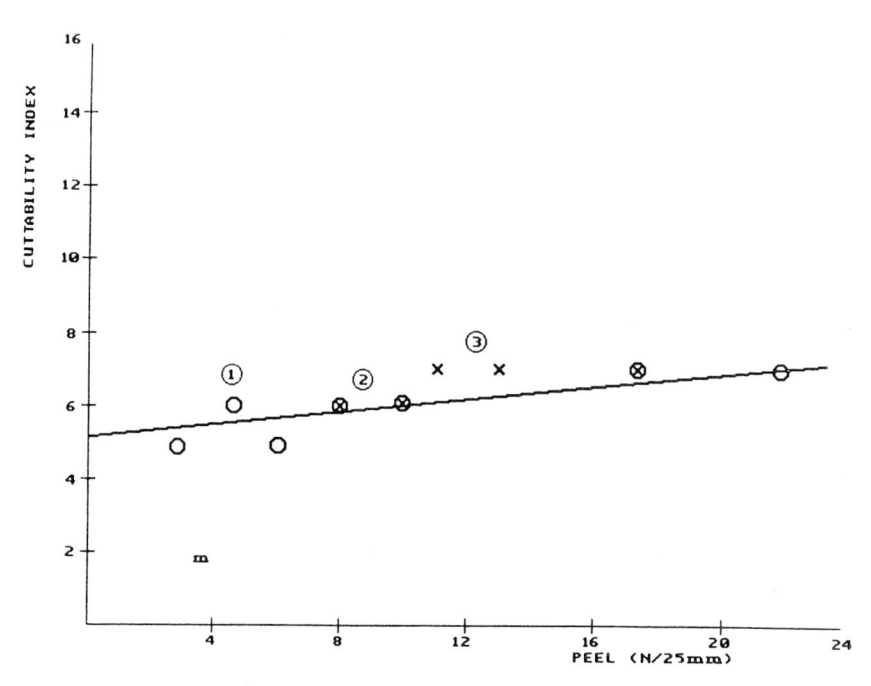

Figure 8.8 Cuttability of film laminates as a function of the peel. 1), 2), and 3) are different water-based acrylic PSAs.

melt, or water-based) influences the cuttability, but within the same class of adhesives the adhesion/cohesion balance also plays an important role. Hot-melt adhesives display good overall adhesive performance on a wide range of substrates including polyolefins, and show high resistance to humid conditions. In the United States, hot-melt technology represents about 34% of the label stock production. In contrast, the convertability of hot-melt adhesives, which had been questioned at the early stages of development of hot-melt technology, has limited the potential in Europe [95]. Poor cuttability of hot-melt PSAs limits their use for labels. The formulator has the option to use other PSAs when good cuttability is required. Later on formulations were modified in order to improve the cuttability of hot-melt PSAs. Thus silicones (polysiloxane additives with a molecular weight of up to almost 10,000) are mixed with tackified synthetic rubber-based PSA compositions to reduce edge ooze or cold flow upon cutting sheets coated with such components. Either non-reactive or reactive polysiloxanes can be utilized [96]. Figures 8.8 and 8.9 illustrate the die-cuttability of film laminates, as a function of the peel for

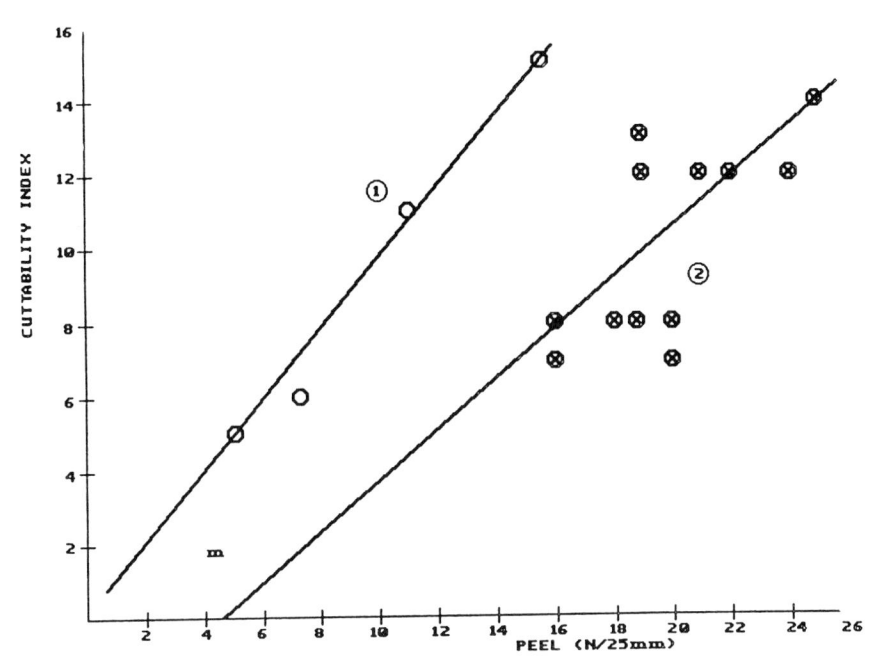

Figure 8.9 The influence of the formulation on the cuttability. Cuttability of a paper laminate as a function of the peel. 1) SB formulation; 2) water-based formulation.

water-based and solvent-based formulations. The formulator should use quite different recipes for a given peel or cuttability target (Figures 8.8 and 8.9).

2.8 End-Use Properties

Technical and commercial factors influence the choice and formulation of PSAs. The most important technical parameters include the end-use, coating, and converting properties of the adhesive. In this section, the influence of the end-use properties of the adhesive on its formulation will be examined. The end-use properties of the labels differ according to the face stock material used (paper or film) or to the character (permanent or temporary) of the adhesive bond. Therefore it is recommended to examine the principles of formulating separately, in light of these criteria.

There is a wide variety of quality requirements corresponding to different end-uses. In some cases different performance characteristics such as clarity, temperature resistance, removability, water resistance and solubility, etc. are required. Generally, both of the main adhesive components affect these performance characteristics (see Chapter 5), but in many cases formulating additives play a decisive role. As an example, acrylic hot-melt PSAs are of great interest for clear film applications and other applications where UV stability is critical. On the other hand, hydrocarbon resin-based tackifiers also are recommended for such uses. Hydrogenated pure monomer resins have a better color stability [41]. On the other hand, a crystalline rosin is thermally more stable than an amorphous one [97]. An improvement of the UV stability may be achieved with stabilizing agents.

Medium softening point tackifiers provide balanced adhesive properties, including strong specific adhesion to polyethylene and other polyolefins. For deep freeze labels a hydrogenated, water-white liquid resin was proposed, with good tack at temperatures down to $-30°C$ [98]. Lower softening point tackifiers help formulations to improve low temperature adhesive properties for frozen food packaging, cold temperature tapes, and labels [67].

The resin level has its influence also. For the curved panel lifting test (carried out at $150°C$) one should use more than 10% tackifier; above 40%, the label is less likely to be stripped off cleanly. Water-based ethylene copolymers possess a better resistance towards plasticizers migrating from PVC [99]. A comparison of acrylic PSAs with CSBR and vinyl acetate/ethylene copolymers show that acrylic PSAs possess an adequate plasticizer resistance, aging stability, and heat resistance; CSBR exhibits better adhesion to polyolefins and EVAc displays good plasticizer resistance, aging stability, heat resistance, adhesion to polyolefins, and an adequate rheology [100].

Plasticizer resistance may be improved by the appropriate choice of the formulating additives. Phenolformaldehyde resins provide better plasticizer stability than rosin esters [60]. The addition of rubber dispersions may improve the plasticizer stability also [101]. Test data indicate that the incorporation of 1–5% acid-polyethylene derivatives into hot-melt PSAs contributes significantly to obtaining a stable bond to plasticized PVC [85]. The choice of an adequate base elastomer with plasticizer or low tackifier level leads to removable PSA formulations. On the other hand, the proper use of cross-linking agents, plasticizers, or slip agents can also ensure removability. High temperature resistance may be obtained with highly cohesive base elastomers containing self-crosslinking units, with high melting point tackifiers or crosslinking agents and fillers. A high temperature resistance may be given by small glass microbubbles [102]. On the other hand, incorporating glass microbubbles also may impart a good removability [103].

For many different end uses a general problem remains the adjustment of the release force (i.e., formulating for release and labeling ease), which depends on the interaction between the adhesive and the release liner. This interaction depends on the adhesive's nature, the release liner nature, as well as on coating and converting conditions. Adhesives with a different chemical basis show different release levels from the same release material. On the other hand, solvent-based, water-based, or solventless silicones display different release forces with the same adhesive. A chemical interaction between water-based acrylic PSAs and solventless silicone is possible also. The release force also is a function of the age of the laminate.

Permanent and Removable Labels

Pressure-sensitive adhesives for labeling applications are usually classified as either removable or permanent. The removable adhesive, once applied to a substrate, must be able to be removed after a residence time on the substrate and at the various temperatures the substrate may be subjected to [62]. High tack, high peel adhesion and in most cases high cohesive strength PSAs are used for permanent labels. Good anchorage on the face stock and on the substrate is needed. Affinity towards nonpolar surfaces is necessary. Low peel, medium tack, and medium cohesion are needed for removable PSAs. No adhesive build-up and adhesive break is allowed. Evidently these very different requirements naturally lead to quite different formulations. Adhesive break and cohesive failure is characteristic for common hot-melt S-I-S-based PSAs. Therefore common hot-melt PSAs are not adequate for removable labels [34]. In Chapter 2, the parameters influencing the removability and the raw material basis were examined in detail. Here, the special features for the formulation of removable PSAs are covered.

The nature and level of the tackifier finally depend on the end-use requirements of the laminate. Removable or permanent use, and film or paper coating require different tackifiers and perhaps a different tackifying technology as well. Cost factors (the price level of the final label) also influence the choice of the tackifier. The properties which contribute to peelability are principally limited tack, a low build-up factor, and a relatively soft adhesive. These properties can be achieved conventionally by omitting or only using a low level of tackifier, including tack deadeners (such as waxes), and by including plasticizers. Some removable formulations include natural rubber and/or CSBR latexes; hence they need hydrocarbon tackifiers. On the other hand, soft removable adhesives tend to migrate; therefore they do not need liquid tackifiers. When using natural rubber-based formulations, the only possibility is to change the molecular weight of the base elastomer (i.e., to use milled or unmilled rubber) in order to modify the cohesion and the peel adhesion (i.e., removability) of the adhesive. In this case the choice and level of tackifiers used is more important. Another possibility exists in the use of plasticizers or crosslinking agents. Experimental results demonstrate the necessity for the use of soft, essentially aliphatic resins, since the more aromatic resins exhibit a high degree of initial tack which increases on aging, so that they are no longer removable [62]. Glycerol esters of highly hydrogenated rosins also may be used [104]. Common liquid plasticizers include polybutene/isobutylene derivatives. Polybutene and polyisobutylene display a quite different storage and aging behavior; polybutene suffers upon aging and crosslinking, and becomes hard. Polyisobutylene depolymerizes during aging and becomes soft and tacky. The natural rubber latex component also imparts a releasability characteristic to the adhesive, such that the adhesive-coated face material after it was pressed onto a contact surface, can be easily removed by simply manually lifting the face material from the contact surface. The natural rubber latex component in combination with the tackifying resin has the additional property that little adhesive residue is left behind on the contact surface when the adhesive-coated material is removed. A level of about 30% natural rubber latex in a vinyl acetate maleinate dispersion yields a removable PSAs, leaving no residues behind upon debonding.

Styrene multiblock butadiene/styrene copolymers (43% styrene) were proposed as removable hot-melt PSAs [84]. The suitability of a block copolymer as a removable PSA is heavily dependent on other ingredients used in the adhesive formulation, such as tackifiers and processing aids; for hot-melt PSAs, S-I-S-type rubbers are preferred since S-EB-S shows excessive build-up of the peel adhesion. An S-B-S rubber (30% styrene/70% butadiene) in addition to exhibiting some build-up of the peel upon aging, leaves a substantial amount of residue on the substrate [62]. Hence removable hot-melt PSAs should be formulated on the basis of S-I-S block copolymers, low softening point aliphatic resins, and salts of fatty acids. The resins have a softening point below 30°C, as determined by the ASTM E-28 ring-and-ball method [62]. Commercially available low softening point aliphatic resins recommended for removable hot-melt PSAs include Regalrez 1018 (Hercules), Escorez 1401 and ECR-327 (Exxon), Wingtack 10 (Goodyear), and Zonarez 25 (Arizona/Bergvik). These resins are used in removable hot-melt PSAs in amounts of 20–50 wt% [62]. Removable hot-melt PSAs may contain up to about 25 wt% of plasticizing or extending oil in order to provide wetting and/or viscosity control; such components include paraffinic and naphtenic oils [62].

The use of liquid components is characteristic for the first removable water-based formulations as well, where plasticizers (up to 20 wt%) [11,105] were suggested in the recipe. Such a high plasticizer content, however, increases the danger of bleeding; polyisobutylene (PIB) may be used as a liquid softening component. The lack of age hardening has made butyl and PIB the first choice as the elastomer base for hot-melt removable label PSAs. Latex PIB dispersions were suggested for water-based PSA formulations. However, the staining is more pronounced for removable PSA labels with a liquid component in the formulation.

The removability of acrylic-based PSAs is improved by the addition of small amounts of organofunctional silanes (e.g., initial peel strength of 0.057 kg/cm after 20 min at 22°C, and a final peel strength of 0.065 kg/cm after 7 days at 22°C compared with 0.097 and 0.389 respectively for a similar adhesive without silane) [106]. Modified, crosslinking methyl polysiloxanes also may be used [107]. Polysiloxane grafted copolymer PSAs give repositionable labels [96]. These copolymers have pendant polysiloxane grafts which cause the exposed surface to initially have a lower degree of adhesion.

Fillers also change the peel adhesion value as contact hindrance caused by filler particles reduces the peel. Therefore in some cases inert fillers are used in removable formulations. Different mathematical formulas were proposed by different authors to describe the modulus increase by fillers [108]. Fillers reduce wetting and flow performance characteristics of hot-melt PSAs, but may improve their cold flow [109]. Satas [110] indicated that large amounts of nonreactive pigments, such as clay and calcium carbonate, can be tolerated without noticeable effect on the peel. Zinc oxide and in some cases colloid silica are active pigments and much smaller amounts can be tolerated. The properties of fillers used for PSAs have been described [111–115]. Aqueous PSA dispersions and polymer solutions can contain more than 2%, often more than 5%, and even more than 10% fillers, colorants, and/or extenders [49]. Fillers to be used in acrylic PSAs are discussed in [116]. BaO, $CaCO_3$, and caolin do not decrease tack, but ZnO and TiO_2 do. Addition of zinc oxide in formulations with carboxyl functional groups dramatically reduces the pressure-sensitive character of the adhesive [117]. Normally zinc oxide is used in PSAs for special tapes, improving the cohesion and reducing the cold flow [118]. In carboxylated neoprene latex zinc oxide fulfills a dual function: it acts as an acid acceptor to neutralize hydrogen chloride which is formed, and it also crosslinks with the carboxyl groups on the polymer, thereby increasing the cohesive strength. The fillers act as contact regulators, pH regulators, crosslinking agents, and stiffeners according to their and the adhesive's composition; all these effects influence the peel. Crosslinking is another way to achieve removability (i.e., through the choice of a self-crosslinking PSA, or an external crosslinking agent).

This use of reactive fillers and the use of isocyanate as a crosslinking agent for rubber-based PSAs is well known from the adhesive practice; reactive resins may be used also. Polyacrylic esters may be crosslinked by isocyanates [76,119]. Crosslinking also may be carried out by more sophisticated, built-in multifunctional monomers like N-methylolacrylamide [76]. Polyacrylamide was used as crosslinking agent for quite different polymers [61]. Other functional polar monomers with carboxyl, amide, or hydroxyl groups may be built-in [120]. Zinc oxide and zirconium ammonium carbonate can be used as reactive metal derivatives [91,121]. It is possible to obtain a more sophisticated structure, where a crosslinked polymer is dispersed like an inert filler in a PSA matrix [122].

Contact hindrance (i.e., the decrease of the adhesive contact surface) is a general method to improve removability. It can be achieved by coating only a minor portion of the substrate [61], coating it discontinuously [124], coating it with a structure of adhesive/nonadhesive strips [125], or with a mixture of adhesive/nonadhesive material [126]. In the latter case the formulator has to design the adhesive blend, in the former cases (discontinuous structures) only its rheology. As a special case of formulation the use of raw adhesives having a built-in emulsifier (surface-active groups) can be mentioned. Small amounts of vinyl-unsaturated emulsifier monomers like sodium-sulfoethylmethacrylate or salts of styrene sulfonate may be used [127,128]. For repositionable, re-adhering labels (like "Post It") and for protective films, both technical possibilities (i.e., coating of structured adhesive layers and the formulation "without" external surface-active agent) are used.

Another field in the formulation for removable adhesives concerns the design of primer coatings. Peelable pressure-sensitive adhesive-coated laminates are comprised of a face material with a primer coating (which includes a contact cement) and the PSA contains the contact cement as well [14]. The use of a contact cement as a primer and in the adhesive enhances cohesion of the adhesive coating and anchorage to the face material which in turn improves peelability. Generally for a contact cement, a harder elastomer may be used [14,129] (e.g., CSBR for natural rubber), a harder (high melting point) tackifying resin, or the same adhesive in a more crosslinked status. The primer coating (with a quite different or the same adhesive composition) and the PSAs may contain quite different additives (for the coating). The primer coating is applied to label grade paper and dried to form the first layer. The peelable PSA is coated onto a paper release liner with a cured silicone release layer, and than dried. The primer coating and the adhesive-coated liner are subsequently laminated together [14].

Polyisocyanate and polyurethane derivatives used as formulating additives for removable compositions may play quite different roles. In general, they are used as crosslinking agents, but it is possible to apply them as primers where they will also act as crosslinking agents. In this case the isocyanate polymerizes first with reactive groups in the face material to create a strong chemical bond in addition to conventional physical adhesion. The remaining isocyanate reacts with moisture and with the PSAs. The best results are achieved using solvent-based polyisocyanates. Dispersion-based polyisocyanates penetrate and break through (migration); solventless ones do not penetrate enough and increase the stiffness of the face material too much.

Paper and Film Labels

General Purpose Permanent Paper Laminates. Permanent paper laminates require tackified formulations where the tackifier level depends on the sheet/roll nature of the laminate. Dispersion-based tackifiers should be used and, as a function of the base PSA and face, rosin esters (acrylic), hydrocarbon (acrylic, CSBR) or rosin acid resins (acrylic, CSBR) should be selected. For light release (solventless liner) a high wetting agent level should be used, mainly on sulfosuccinate basis; the level of defoamers should be limited. The release force depends on the direct or transfer coating method used. For agressive paper labels (roll material, for labeling machines) high tack, high peel adhesion, and low controlled release is required. For sheet material less tack, but high peel, cohesion, and excellent die cutting are required. Time/temperature stability of the adhesive properties also is required. Tackified acrylic adhesives meet those requirements; both solvent-based and water-based PSAs can be used. Tackified rubber-resin solvent-based adhesives are adequate for roll/sheet material also; tackified EVAc-based PSAs are more suited for sheet laminates.

Removable Labels. The guidelines for the formulation of removable PSAs were outlined earlier. The main requirements can be summarized as follows: low peel, good removability, low migration, no bleeding, and high tack. Crosslinked solvent-based acrylic PSAs and special primered (or primerless) water-based acrylic PSAs meet these requirements; solvent-based rubber-resin PSAs and water-based PSAs meet those same requirements for other self-adhesive products (e.g., protective films, etc.). Nontackified or slightly tackified formulations should be used; plasticizers as tackifier are preferred. Crosslinking agents and peel modifiers should be used, with low levels of wetting agents (sulfosuccinates are not preferred). Direct coating appears to be the preferred coating method.

Permanent Film Laminates. These films are engineered to be durable, conformable, dimensionally stable, and able to withstand severe weather and handling conditions. They can be processed by screen printing, roll coating, steel die cutting, thermal die-cutting, and premasking. High transparency, high water resistivity, and dimensional stability are required. Roll and sheet materials are coated with solvent-based and water-based acrylic PSAs. The market segment for high polyethylene adhesion is partially covered by ethylene copolymers. Mostly nontackified formulations are used; the level of dispersion additives should be limited. Sulfosuccinates are not preferred; wetting agent free, thickened formulations (if possible) are suggested.

Removable Film Laminates Low peel, good removability, high tack, good transparency, dimensional stability and water resistivity constitute the performance requirements, which are met by solvent-based acrylic and water-based PSAs. Plasticized formulations are possible. A low level of dispersion additives should be used (direct coating is proposed); crosslinking agents may be included in the formulations.

PVC-Based Laminates. A special case of adhesive formulations concerns coatings onto PVC. In order to understand the principles of the formulation of PSAs for PVC coatings, the requirements for PVC labels will be studied first (except for plasticizer migration). The properties of adhesives for coating onto PVC include the adhesive performance characteristics (Table 8.23), water resistance, shrinkage or dimensional stability, optical appearance, aging, processability, and FDA or BGA approval. It is apparent that different peel, tack, and shear values are required for sheet or roll material applications.

Table 8.23 Adhesive Properties of a PSA-Coated PVC Label Material

		Adhesive properties			
				Values, application	
Property	Method	Substrate	Comments	Sheets	Roll
Peel adhesion	180°	Glass	N/25 mm		
			Immediate	7	10
			20 min	8	12
			1 hr	8.5	14
			24 hr	10	16–20
		Stainless steel	Immediate	6	—
			20 min	—	—
			1 hr	—	—
			24 hr	—	—
		PE film	Immediate	4	5
			20 min	5	6
			1 hr	(5)	—
			24 hr	6	8–12
Tack	Rolling ball		cm	2.5–3.5	—
	Loop tack	Glass	N/cm	6–10	10
		PE	N/cm	4	6
Shear	Hot shear	Stainless steel	min	60	20

In Table 8.24 the other characteristics of PSAs used for PVC (e.g., water resistance, shrinkage, and optical appearance) are summarized. In order to clarify the importance of the given values, a more detailed discussion of these properties follows. First the water resistance is examined. Water resistance encompasses the stability of the adhesive and end-use properties after immersion in water. For certain practical uses, the labels (label sheets) will be immersed in water in order to achieve a partial and temporary loss of the tack for a better repositioning. Clear, transparent labels should not suffer whitening under these conditions and the loss of tack should persist for a short time only in order to achieve rapid adhesion onto the final substrate. Consequently one should distinguish between label sheets (hand applied, voluminous film coatings), and labels (roll or sheet material, applied by high speed labeling machines, labeling guns, or by hand). For label sheets affixed mostly on polar surfaces water-whitening (loss of transparency) and wet adhesion are the most critical characteristics (Figure 8.10).

For labels affixed at high speed, mostly on polyolefin bottles, wet adhesion and squeeze bottle anchorage are the most important properties related to water resistance. There is no correlation between water-whitening and the other water-resistance related properties, but there is a correlation between wet anchorage and wet adhesion on glass or plastics (after immersion in water or diluted solutions of detergents, over a period of time). Wet anchorage is the adhesion of the PSAs onto the face stock (film) (Figures 8.11 and 8.12).

Hence, wet adhesion includes two separate aspects. First, one observes the adhesion of the wet PSAs coating on the substrate, and fast (or slow) increase of the adhesive forces between substrate and coating, resulting from the drying of the coating (over a short period of time/minutes). Testing occurs by trying to move the label parallel to the surface. Next, the adhesion of the PSA coating on the substrate is measured, the substrate with the affixed label being subjected to immersion in water (or water and surface-active agent) for hours, or days. Testing occurs by peeling the label from the substrate surface.

Consequently, wet adhesion on the substrate tested through "push-away" is a Bonding Wet Adhesion (BWA) test. Wet adhesion tested through peeling off is a debonding wet adhesion (DWA) test (Figure 8.13).

Formulating for Special Uses

Aging. The synthetic and natural polymers utilized in the adhesives industry are frequently susceptible to degradation and this degradation is markedly accelerated by interaction with oxygen at elevated temperatures as well as with light. On the other hand, warranty times of at least 2–4 yr are required for quality labels [130]. There are several practical data showing the influence of the formulation on the aging of PSAs. Thus, for example, U.S. Patent No

Table 8.24 End-Use Properties of PSA-Coated PVC Label Material

	Water resistance		
Property	Method	Values	Units
Water whitening	Subjectively	> 10–17	sec
Loss of transparency	Colorimetry, after 7 min	< 3–8	%
Wet adhesion on glass	Subjective push away after: 1 min 5 min 10 min 30 min 60 min		
Squeeze bottle wet anchorage	Subjectively after 24 hr	No wings, lifting, high peel adhesion, subjectively	

	Shrinkage		
Properties			
Position	Method	Values	Units
Unmounted	7 days, 60°C clear PVC white PVC	MD 0.9 CD 0.6 MD 0.5 CD 0.4	% %
Mounted	7 days, 60°C clear PVC white PVC	MD 0.5 CD 0.4 MD 0.2–0.3 CD 0.15	% %

	Optical appearance	
Property	Method	
Coagulum		No grit
Optical appearance	Subjectively	—
Optical clarity	Colorimetry	

MD, machine direction; CD, cross direction

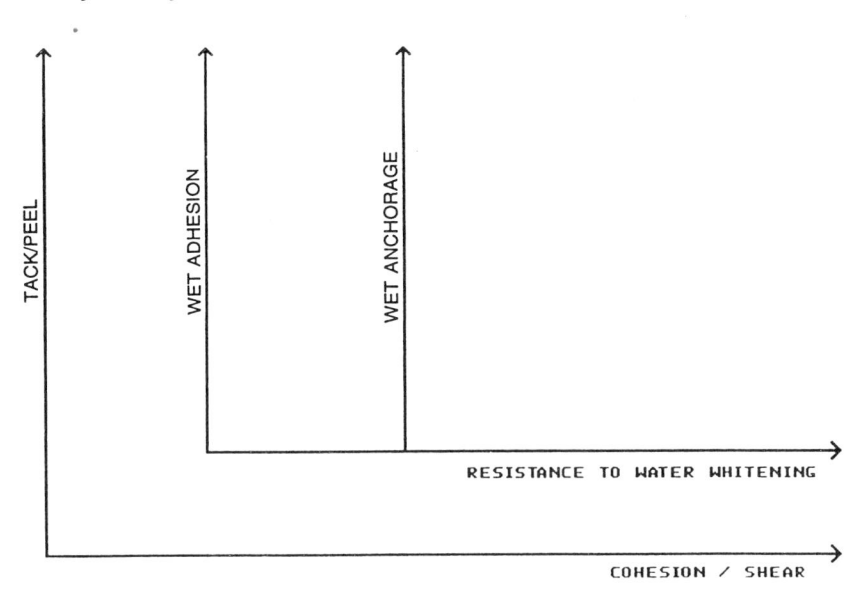

Manufacture of Pressure-Sensitive Adhesives 391

Figure 8.10 Water resistance of label sheets. Interdependence of the water resistance performance with the adhesive characteristics.

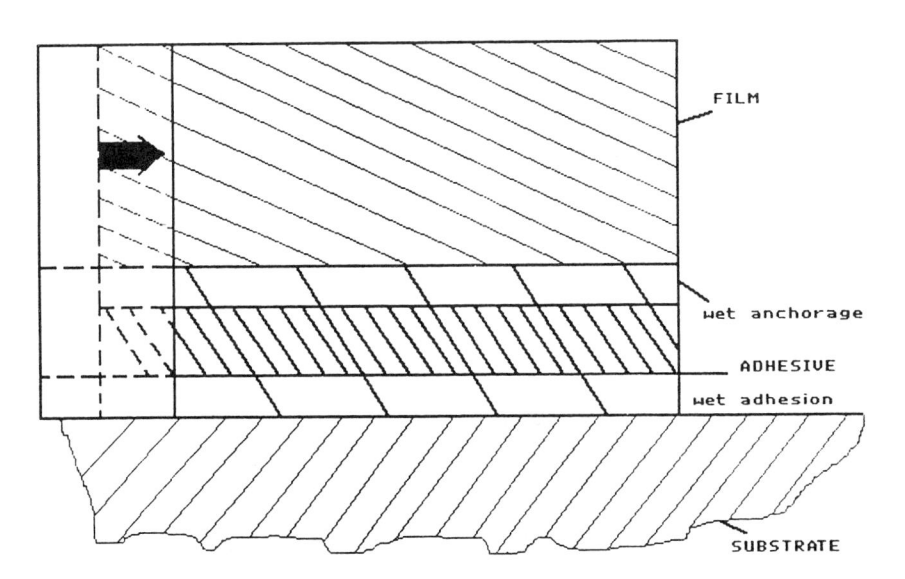

Figure 8.11 Water resistance of PSA labels. Wet anchorage and wet adhesion, where and how they work.

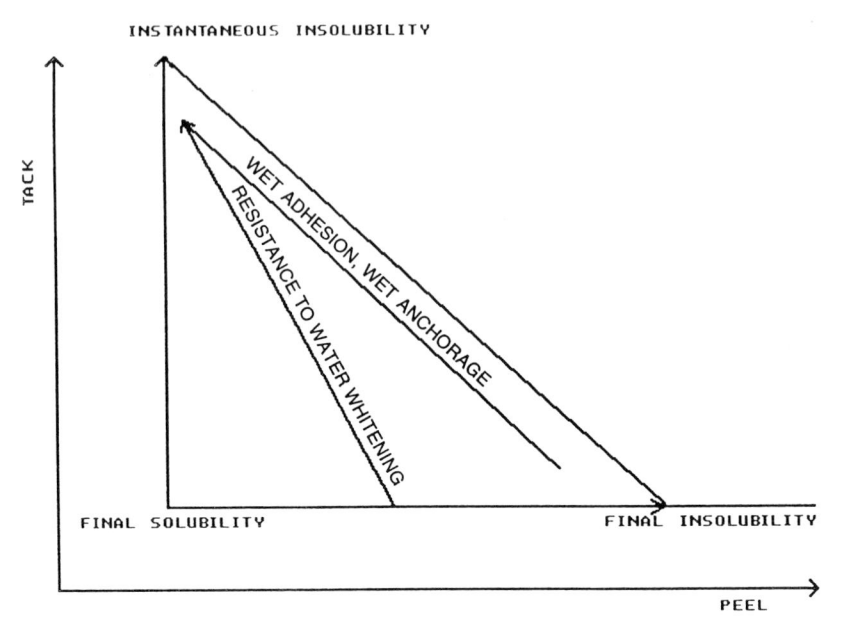

Figure 8.12 Water resistance of PSA labels. Interdependence of the water resistance parameters, peel, and tack-related parameters.

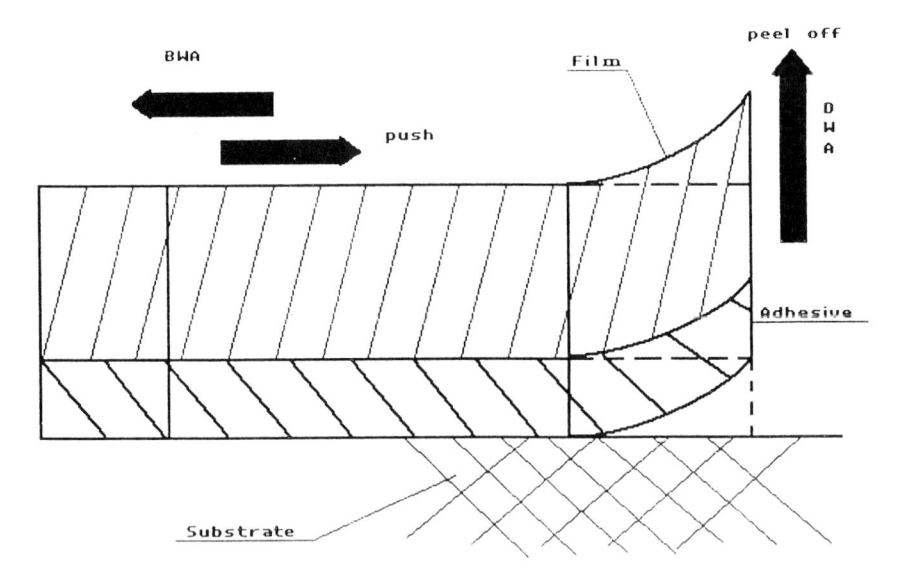

Figure 8.13 Water resistance of PSA labels. Test of the wet adhesion.

4,418,120 states: "It has long been recognized that adhesives that consist essentially of a copolymer of alkyl acrylate and a minor proportion of copolymerizable monomer such as acrylic acid, do not require a tackifying resin and are able to resist aging; thus, such adhesives have advantages over the earlier and more traditional rubber-resin adhesives" [131].

In some cases (film coating) water-white UV-stable PSA formulations are needed [32]. For special uses, like cereal inserts, water-based PSAs with low edge ooze and adhesion bleed that comply with U.S. Food and Drug Administration (FDA) regulation 175.105 are required. Such formulations need special base elastomers and surface-active agents. Any adhesive composition must include an antioxidant to inhibit oxidation of the tackifying agent and consequent loss of tackiness as the adhesive composition ages [8]. The saturated hydrocarbon-based long-chain structure of acrylic PSAs provides good chemical stability towards aging through oxygen and light. Hence acrylic PSAs can be used without protecting agents for opaque as well as for transparent face stock. Rubber-based PSAs are inadequate for transparent webs. Thus there is a general trend to give several years warranty for these laminates.

Tackified formulations show less stability upon aging. Tackified, water-based acrylic PSAs can interact with special face materials (e.g., thermal paper) giving rise to the partial or total loss of the adhesive properties. Tackified film laminates may show discoloration. The most important problem is the frequent interaction between metal ions from the paper (face stock) and acid rosin tackifiers causing the loss of tack and peel adhesion. This phenomenon influenced the formulators to change the resin base from acid- to ester-based ones, although the latter show less tack and peel adhesion. In this context hydrocarbon-based tackifiers can compete with the less agressive rosin esters. As can be seen from the DSC data in Figure 8.14 differences in T_g for fresh and aged material are very pronounced (−35.5°C versus 1.5°C, respectively). This is probably related to the polymerization of the resin. The exact nature of the phenomenon is still unknown.

Abietic acid (in rosins) contains a conjugated double bond and will react with oxygen. This will increase the softening point of the resin, and the subsequent tack of the adhesive will decrease with time. Development work was carried out to improve the aging stability of hydrocarbon-based elastomers and tackifiers. The unsaturation that exists in the diene midblocks is responsible for the aging of S-B-S. The experience with PVA-stabilized carboxylated neoprene latexes indicates that they are more environmentally stable than conventional neoprene [132]. On the other hand, the structure and properties of S-B-S block copolymers are nearly ideal for PSA formulations since they combine high cohesive strength and good aging characteristics. The importance of the compatibility between PSA and tackifier was demonstrated

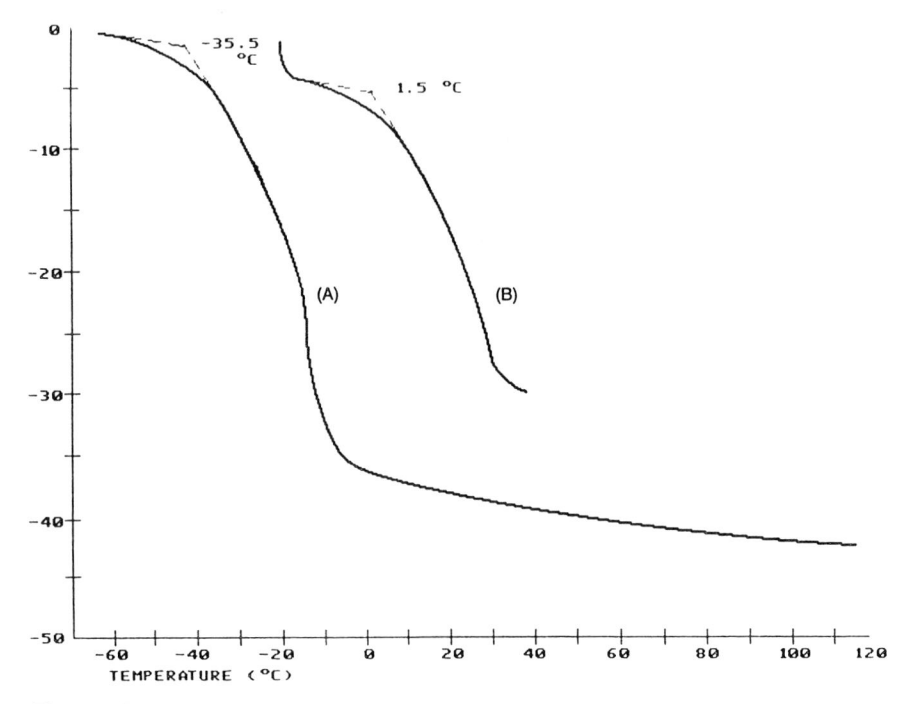

Figure 8.14 The increase of the T_g during aging for a PSA formulation using an insufficiently stabilised acid rosin as tackifier. The shift of the DSC peak to positive values for the aged adhesive (B).

earlier. Compatibility influences the time/temperature stability of the end-use properties; it also is influenced by the chemical nature of the resin. For example, unsaturation of the resin will be eliminated by disproportionation and hydrogenation. Hydrogenation brings more compatibility and aging resistance. Practically, the choice of an adequate base elastomer and tackifier, together with the use of antioxidants and UV-stabilizing agents, impart a good aging stability.

There are many components with a well-known chemical composition that are used as antioxidants for adhesives. The most important are the hindered phenols [8], ortho-substituted or 2,5–disubstituted hindered phenols, in which each substituent group is a branched hydrocarbon radical, with 3–20 carbon atoms (e.g., tertiary butyl or tertiary vinyl phenol). Other hindered phenols include para-substituted phenols where the substituent group is OH, R being methyl, alkyl, etc. Among the sulphur containing organometal salts, the nickel derivatives of dibutyl ditiocarbamate are used. Hindered phenols or sulphur-containing organometal salts are used for crosslinked acrylic PSAs, dibutyl-dithiocarbamate may be used for thermoplastic rubber-based hot-melt PSAs. Oxidation sensitivity depends on the formulation, on the tackifier level, and nature of the PSA. Thus at low resin content (for S-I-S-based PSAs) there is little or no tack retention when exposed to air; however, the tack retention is increased if the resin level is increased to 60%. UV-stabilizers are 2,4-dihydroxydibenzofenones, substituted hydroxyphenyl benzoates, octylphenyl salycilates, and resorcin-monobenzoles. Stabilizers used for block copolymers are sterically hindered phenols, thiotriazines, trimethyldihydroxychinolin polymers, alkylated bisphenol; for PSAs the best would be Zn-dibutyldithiocarbamate.

Hot-melt adhesive compositions generally contain 0.2–2% antioxidant. Among the preferred stabilizers or antioxidants utilized are high molecular weight hindered phenols and multifunctional phenols such as sulphur- and phosphorus-containing phenols. Hindered phenols may be characterized as phenolic compounds, which also contain sterically bulky radicals in close proximity to the phenolic hydroxyl group [133]. The performance of these antioxidants may be further enhanced by utilizing them in conjunction with known synergists such as thiodipropionate, esters, and phosphites [62]. For removable hot-melt PSAs that are based on styrene/butadiene multiblock copolymers, 0.3% Irganox 565 (Ciba Geigy) and Polygard (Uniroyal) were proposed [84]. Furthermore it was shown that after 7 days at 70°C an S-I-S rubber loses 99.5% of its tensile strength, with antioxidant of only 10%.

A preferred antioxidant for natural rubber-based formulations is available under the trade name Heveatex B 407A antioxidant, which is a ball-milled dispersion of 4,4 butylidene bis powder in an aqueous base [61]. A butylated reaction product of paracresol and dicyclopentadiene and a liquid carrier such as ditridecylthiodipropionate (Wingstay L by Goodyear) or an aqueous dispersion containing zinc dibutyldithiocarbamate were proposed as nonstaining, nondiscoloring antioxidants [61]. Oxidation of neoprene yields hydrogen chloride, which in the presence of moisture is converted to hydrochloric acid. Neoprene may be protected by nondiscoloring, nonstaining antioxidants of the hindered bisphenol class (like Antioxidant 2246 or Wingstay). A 50% dispersion of Wingstay L under the tradename Bostex 24 may be used for aqueous PSA dispersions. For these 0–3% w/w DDA-EM 50%, KSM-EM 33%, or Vulkanox TD-EM (Bayer) may be used also. Suitable antioxidants for polyvinylethers include Ionox 330 (Shell), Irganox 2010 (Ciba Geigy), or ZKF (Bayer).

The aging of S-B-S and S-I-S rubbers is usually noticeable by the loss of tackiness and cohesive strength, and it is reasonable that some of these changes may be related to the incompatibility of the two-phase system. However, it is more likely that the unsaturation that exists in the diene midblocks is responsible for this instability [134]. Generally, for S-I-S-based hot-melt

PSAs, amine- and phenol derivatives are used as chain stabilizing agents (antioxidants). Phosphites and thioesters are used as killers for peroxides. Oxybenzochinones, oxibenzotriazoles, and nickel chelates are used as UV stabilizers [135].

As can be seen from these examples, there are many antioxidants available with different structures, formulations, and trade names. Their choice is made on the basis of their working mechanisms, their dispersability, and on the basis of economical considerations also. The most important criteria in the choice of antioxidants and UV stabilizers is, beside their efficiency, their inert character towards the face stock and the printing materials. Nonstaining compounds are suggested only. Their selection is made mostly on the basis of practical tests (see Chapter 10).

Water Resistance. Wash off labels need good water solubility, labels for soft drink containers need time-dependent water solubility, whereas deep-freeze labels and labels for outdoor use and for textiles need no water solubility (i.e., the water resistance should be adjusted as a function of the intended end-use). Water resistance is a function of the nature of the base polymer, the particle size and particle size distribution, the surface-active agent nature and content, and the coating and converting conditions. Satisfactory results can only be obtained with crosslinked products. One component products with built in crosslinking capabilities are required. A high degree of crosslinking, however, cannot be achieved for PSAs without the loss of the pressure-sensitive performance characteristics.

On the other hand, high tack/peel adhesion are very important for both debonding resistance and bonding velocity under humid circumstances. Therefore the formulation of water-resistant or water-soluble PSAs is always associated with the loss of the main adhesive properties (i.e., water-resistant/soluble PSAs do not perform as well as common PSAs for permanent applications). The final performance of the PSAs depends on the skill to use an adequate ratio of solubilizer (being less tacky) either built-in (base polymer) or added (wetting agent), tackifier (danger of coalescence) needed for wet tack, and chemical base. The most important test criteria and characteristics are discussed in Chapter 10.

Thermal Resistance. Formulating for thermal resistance requires the use of temperature-resistant elastomers and/or tackifiers. Chloroprene latexes display good thermal resistance. Self-crosslinking, acrylonitrile containing latexes were suggested also. Reactive, high melting point resins and/or crosslinking agents also may be used.

Deep Freeze PSAs. Special demands are placed on the chemical composition and physical properties of low temperature PSAs (i.e., adhesives intended for use at relatively low temperatures). Often PSAs which have adequate cohesive and adhesive strength at low temperatures are so gummy at ambient conditions that they complicate adhesive handling at ambient temperatures and the manufacture of adhesive-containing articles. Such gumminess also causes creep and bleedthrough on labels and other PSA-coated face materials [49].

Theoretically, there are two different formulating approaches in order to achieve good deep freeze properties. The first one is the choice of very soft elastomers (acrylic PSAs) with no need for tackifying resins. The second consist in the use of low molecular weight liquid tackifiers. Both approaches can be worked out more easily using solvent-based adhesives; better results are obtained without tackifiers. For deep freeze adhesives, the application temperature is 25–100°F, the service temperature ranges from –50 to –150°F [136]. PSAs should have a lower T_g than its application temperature. Such soft PSAs often give rise to edge oozing, bleeding and converting problems. For rubber-based formulations, measurement of peel adhesion and polytack at ± 5°C and –20°C provides a reasonable indication of the final performance, both for low temperature (i.e., chill) and for deep freeze applications. As the butadiene level increases beyond 80% the suitability for deep freeze applications improves. Gel content appears to have a diminishing influence as temperature decreases [50]. Wet tack formulations are similar to those for low temperature applications [137].

2.9 Influence of Adhesive Technology

The state of the adhesive (liquid or molten) limits the use of tackifiers or other formulating additives. Generally, a good mixing of the components supposes a liquid state (solution, dispersion, or melt). Solvent-based adhesives are tackified with resin solutions, but water-based ones also have been successfully tackified with resin solutions. Latex-resin dispersion mixtures were developed. The use of solid (molten) resins is limited by the thermal stability of the water-based PSAs. Chloroprene dispersions are sufficiently stable to be tackified with molten resins. Molten resin (melting point < 90°C) may be fed into aqueous dispersions of vinyl ether and acrylic PSAs also [137].

Solvent-Based Adhesives

As the carrier solvent may influence the stability, the adhesive properties, and the coating properties of the solvent-based adhesives, it appears necessary to separate the formulating of the uncoated and coated adhesive.

In order to obtain adequate coating properties the solution polymer molecular weight and concentration must be balanced [138]. High molecular weight elastomers give viscous solutions. Therefore, the solids content of solvent-based formulations is generally less than 25%. Technical problems arise when coating highly viscous solutions and drying low solids formulations. The main components of the adhesive formulation have to be dissolved before or during blending. Mastication of natural rubber may facilitate the dissolution process. The most commonly used solvents include ethyl acetate, n-hexane or n-heptane, toluene, special boiling point petrol, acetone, and isopropyl acetate. Their efficiency, drying speed, and economical and environmental properties are the main criteria for this choice. Their recovery possibilities have also to be taken into account [139–141]. The overall performance of solvent-based formulations in comparison with hot-melt and water-based PSAs will be discussed in Section 2.11.

The most important coating characteristic of solvent-based formulations remains the viscosity which depends on the base elastomers and on the solvent used. For special cases other important parameters were identified. The viscosity of PSA solutions may be reduced using dispersed rubber (CSBR, neoprene, or polybutadiene) together with dissolved rubber (nitrile rubber, NC; butyl rubber; polyisobutylene; ethylene butydiene rubber, EBR; polyisoprene) in an isoparaffinic solvent [142]. Viscosity adjustment for solvent-based rubber-resin adhesives is achieved with toluene/hexane, where toluene is a solvent for polystyrene blocks in S-B-S and hexane is a solvent for the elastomeric domain. Other solvents (i.e., heptane, octane, petrol) are similar to n-hexane; cyclohexane and methyl isobutyl ketone (MIBK) are like toluene [17]. The viscosity of rubber-resin solutions decreases with the increase of the resin content [138]. Pressure-sensitive adhesives using the less photochemically active ethyl acetate instead of toluene or xylene are recommended for clear film labels.

Generally, the manufacture of solvent-based PSAs (i.e., manufacturing the adhesive solutions) is a process of dissolving the main adhesive components (rubber and resin) and adding other solid or liquid components. This is a simple procedure, where dissolving the elastomer is taking more time and needs, in some cases, special dissolvers (homogenization and heat transfer) and a programmed feed-in of the materials.

Hot-Melt Adhesives

Hot-melt adhesives are 100%-solid materials which do not contain nor require any solvents. They are solid materials at room temperature, but upon application of heat melt to a liquid form (fluid state) in which form they are applied to a web. Upon cooling the adhesive regains its solid form, and gains its cohesive strength. In this regard hot-melt adhesives differ from other types of adhesives which achieve the solid state through evaporation or removal of carrier liquids (solvents, water) or by polymerization. Hot-melt PSAs are carrier-free systems where formulating is necessary in order to improve the adhesive, coating, and end-use properties, mostly by changing the viscoelastic properties of the pure, bulky components. Normally no coating additives are used.

Hot-melt adhesives were developed in the 1940s. In the 1960s block copolymers for hot-melt were introduced [33]. This group of adhesives has developed particularly rapidly since the early 1970s because of their nonpolluting character thanks to the absence of solvents and the low energy input required to manufacture them. In addition thicker adhesive layers of up to 100 g/m^2 can be applied by hot-melt coating instead of using solvent-based PSAs [143,144]. Much attention is given to the proper selection of the polymer backbone component of hot-melt PSAs. While the backbone polymer determines cohesive strength, flexibility, viscosity and heat resistance for an adhesive system, the tackifier determines the color of the hot-melt PSA, as well as its wettability and thermal stability. Tackifiers also determine the entire system functionality, greatly influencing overall performance [67]. The suitability of a block copolymer for removable PSAs is heavily dependent on other ingredients used in the adhesive formulation, such as tackifying resins and processing oils [67] (see Chapters 5 and 6).

A detailed description of the characteristics of the raw materials for hot-melt adhesives was given by Fricke [145], King [146], and Jordan [147]. The cohesion determining components are the base polymers [144]; for hot-melt PSAs particularly, these are the most important S-B-S and S-I-S block copolymers and polyisobutylene, but also the rubberlike polymers such as natural rubber, CSBR rubber, acrylonitrile copolymers, polychloroprene, etc. To increase tackiness, different tackifiers are added as adhesion determining components. The resins used are either natural resins like colophonium or synthetic resins such as coumarone resins, hydrocarbon resins of different grades and others, some of which are thermoplastic and have softening points of about 70–120°C. Additional adhesion promoting substances are plasticizers, which are used (particularly for permanent adhesive formulations) with rubber-type base polymers. In order to obtain exactly defined mechanical, physical, and chemical properties in the adhesive layer, fillers are added. The addition of stabilizers protects the polymers from degradation during processing and from damage caused by heat and oxidation.

S-I-S provides a long open time for hot-melt formulations, while S-B-S is preferred for better cohesion. New triblock copolymers PS-PEB-PS contain a soft segment of PEB in a polystyrene matrix (30% PEB and 70% PS), whereas S-EB-S copolymers display better compatibility with paraffinic oils [148]. Generally, a hot-melt PSA recipe consists of a thermoplastic rubber, an endblock compatible resin, a midblock compatible resin, oil, and an antioxidant. With the choice of the resin for hot-melt PSAs, one has to separate the influence of endblock and midblock compatible resins. The main functions of midblock resins comprise the tack, improving the adhesion and their function as processing aid. New EVAc segment polymers were developed for hot-melt PSAs [58]. These products display better environmental stability. Their recipes do not contain waxes because of their mutual incompatibility. These formulations have a viscosity of 9,000–56,000 mPa·s (at 150°C).

Tackification of Hot-Melt PSAs. To gain useful PSAs properties, thermoplastic rubber, which has almost no intrinsic adhesive character, is compounded with tackifying resins and sometimes plasticizers. Those are low molecular weight materials, often unsaturated and with an aliphatic character, such as rosin esters, polyterpene resins, and C_5 hydrocarbon resins. These resins tend to concentrate in the continuous elastomeric phase of the thermoplastic rubber, thereby swelling and softening the rubber. Hot-melt PSAs based on styrene/butadiene multipolymers were designed using 25% rubber, 34–44% resin, and 30% processing oil [84]. The addition of high melting point polystyrene associating resins improves heat resistance of hot-melt PSAs based on multiblock butadiene/styrene elastomers with no or slight decrease in tack [67]. Plasticizer oils should lower the modulus, and improve the tack and deep freeze properties of hot-melt PSAs. They also lower the melt viscosity. Ideally, a plasticizer oil is not soluble in the polystyrene domains, is fully soluble in the rubber segments, has low volatility, low specific gravity, and high storage stability [147]. Plasticizer oils contain less than 2% aromatic fractions. Soft resins and polybutylene may be used as plasticizers also. By associating with the isoprene midblock (which has negligible inherent tack), polystyrene contributes to the agressive tack and quick stick properties, and helps retaining shear strength.

A screening recipe (formulation) for a hot-melt PSA designed on the basis of the literature should contain [149–151]: 100 parts S-I-S, 100 parts aliphatic hydrocarbon resin, and 5 parts antioxidant. In order to improve the tack and to lower the viscosity, a plasticizer oil should be added to the recipe: 100 parts S-I-S, 100 parts resin, 40 parts plasticizer oil, and 5 parts antioxidant. In order to reinforce the polystyrene domains, a high melting point coumarone-indene resin should be added to the recipe as well: 100 parts S-I-S, 100 parts resin, 40 parts plasticizer oil, 60 parts high melting point resin, and 5 parts antioxidant.

With a polybutene used as plasticizer and a high melting point linear homopolymer of α-methyl styrene as reinforcing resin, the formulation becomes [152]: 100 parts thermoplastic elastomer, 150 parts tackifier, 2 parts antioxidant, 125 parts polybutene, and 40 parts reinforcing resin.

It remains difficult to formulate removable hot-melt PSAs. Current removable adhesives supplied for label stock are acrylic latexes and solvent-based formulations. Both of these materials have high molecular weight polymers that reduce flow on a surface to prevent build-up of adhesion. In contrast, hot-melt PSAs are based on materials having lower molecular weight components that make reduced flow or wetting on a surface very difficult. Certain formulations containing a S-I-S block copolymer, a low softening point highly aliphatic resin, and a metal salt of a fatty acid exhibit a good balance of properties and are used as removable PSAs [62].

Stabilizers in Hot-Melt PSAs. Adhesive formulations based on unsaturated synthetic elastomers and tackifiers are highly sensitive to oxidation. These adhesives thus require the addition of stabilizers to protect them from oxidative degradation during preparation, storage, and end-use. While the use of a nitrogen blanket can offer some protection during mixing, further significant protection against oxidative degradation of hot-melt PSAs can be obtained with stabilizers. A hot-melt PSA formulation may contain several components, including a base rubber, tackifier resin, and oil. Oil is the most stable of the viscous components used in the adhesion formulation [153]. The presence of stabilizers in the rubber and in the tackifier can significantly improve the performance of the materials over the unstabilized formulation. Antioxidants like hydroxycinamate derivatives and hydroxybenzylbenzene derivatives were suggested.

A blend of rubber with tackifier resin (without oil) responds well to the stabilizers present in the rubber as well as in the tackifier resin. However, the stability of the adhesive formulation is significantly lower than that of its components [153]. Thermoplastic rubbers used in hot-melt PSA formulations like the other ingredients in the adhesive formulation, must be stabilized against oxidation attack. Phenol derivatives as antioxidants are used for hot-melt PSAs [34]; zinc dibutyldithiocarbamate also may be used [65]. Antioxidants for hot-melt PSAs are Tenol BHT (Eastman), Antioxidant 80 (ICI), Santonox R (Monsanto), or dilaurylthiopropionate and n-butyl-hydroxy toluene. Other derivatives like substituted phenols, phenotriazines, and carbamide or orthotoluylbiguanidine are suggested [154]. For electron beam-cured hot-melt PSAs (e.g., Kraton D 1320 X) Irganox 1076 is used as antioxidant [155].

Viscosity of Hot-Melt PSAs. A hot-melt adhesive must possess a sufficiently low viscosity to enable flow at melt temperatures. Traditionally the adhesive property that suffers the most in this instance is shear. Curing of the hot-melt adhesives greatly improves shear properties. While thermal cure or crosslinking is feasible, radiation curing by electron beam or UV is preferred.

Hot-melt PSAs possess viscosities of 20,000–100,000 mPa·s [156], respectively 50,000–250,000 mPa·s [157]. Hot-melt PSAs exhibit a viscosity of 11,000–16,000 mPa·s at 350°F [158]. Low viscosity hot-melt acrylic PSAs possess viscosities of less than 50,000 mPa·s at 178°C [24]. Viscosities for hot-melt PSAs based on styrene/butadiene midblock copolymers are situated between 4000 and 17,800 mPa·s at 250°F and 600–1100 mPa·s at 400°F. For high quality hot-melt PSAs a low processing viscosity at 150–175°C and high peel strength at room temperature are required [159].

Hot-melt PSAs based on thermoplastic rubbers use temperature and not solvents to reduce the application viscosity; the addition of plasticizers such as oils or low molecular weight resins is necessary to lower the viscosity of the blend to practical levels. Polybutenes also have been used successfully as tackifiers/plasticizers in hot-melt PSAs. The choice of an adequate resin as tackifier for hot-melt also may influence the viscosity. A noncrystallizing rosin as low viscosity tackifier in hot-melt adhesives (providing it is stabilized against crystallization and oxidation) offers good adhesive performance characteristics [157].

Single Component Hot-Melt PSAs. The proper selection of comonomers and synthesis additives has led to the development of olefinic copolymers that are PSAs when unformulated [160]. Because these hot-melt PSAs are pure olefin copolymers they offer several advantages over typical styrene block copolymer/oil/tackifier blends. Those include the lack of oils which can bleed into face materials, less possibility of skin sensitivity problems (which is generally associated with the tackifiers in conventional hot-melt PSAs), a good thermal stability, and a low color and odor. Other advantages are the ability to be chemically modified by reaction with α-olefin polymers, the ability to be blended with most olefinic compatible adhesive ingredients for general applications, and a lower density than conventional rubber-based hot-melt PSAs, resulting in increased durability.

Recent developments in acrylic polymer technology have led to a new family of phase-separated acrylic block copolymers. The thermoplastic elastomer characteristics exhibited by these block copolymers were utilized for hot-melt PSAs applications [24]. Acrylic hot-melt PSAs possess excellent thermal, oxidative, and photooxidation stability relative to the competitive hot-melt PSAs and solvent-based products. A range of modifiers (rosin, terpene, phenols, hydrogenated rosin esters, etc.) also may be used [161].

Manufacture of Hot-Melt PSAs. The entire operation of compounding hot-melt PSAs from the raw materials to the coating of the label or tape is divided into two heat-treatment stages: compounding and coating. In order to limit the excessive thermal history and oxidation of the materials there are three possibilities: reducing the residence time during compounding, running the coating in-line with the compounding operation, or keeping all polymers from the compounding stage to the coating line in a closed system or under inert atmosphere to avoid oxygen contact at the lowest possible temperature [144,162,163].

With the sigma blade mixer, which is the preferred mixer type for small to medium sized plants, a semicontinuous operation is possible by using the mixer with a discharge screw and a process control system. For medium sized to large batch sizes, however, compounding on an intermeshing corotating twin-screw mixer is the compounding method which leads to the best quality products. Some of those plants have been in operation for several years and are capable of compounding at rates of over 1000 kg/hr. This technology is similar to that used for compounding plastics, but the feeding is more complicated on account of the many components required. The requirements for compounding can be summarized as follows [144]:

- mastication or plastification of the polymers through shearing and heat without degradation through oxidation;
- melting of the tackifiers and homogeneous incorporation in the polymer matrix;
- homogeneous incorporation of fillers with good dispersion of the filler particles;
- homogeneous incorporation of plasticizers, waxes, antioxidants, and stabilizers;
- facilities for creating vacuum to ensure freedom from pores and remove remaining moisture from the fillers;
- constant temperature of the finished adhesive compound, so that the required viscosities can be adjusted precisely and reproducibly.

In the mixing cycle the overriding principle is that of obtaining a completely homogeneous mix in the shortest possible time at the lowest possible temperature (i.e., with a minimum thermal loading history). For continuous mixers the thermal history remains very short due to the short residence time, and no evidence of polymer degradation is usually observed. When high shear batch mixers are used in open contact with air, degradation occurs rapidly, but this can be minimized by nitrogen blanketing. Precautions must also be taken to avoid degradation by oxidative attack during the application of the hot-melt PSAs. In roller melt mill coaters when the adhesive is transferred from a heated roller the adhesive is exposed to air, while in die coaters the adhesive is extruded onto the substrate in a closed system and air exposure remains much more limited.

During the manufacture of hot-melt PSAs a part of the resin and the antioxidant are fed in first, then the block copolymer, and last the rest of the resin. Mixing in Z-blade mixers needs 30–35 min. This method is slower than using a durable (twin) screw extruder. The most important parts of a fully continuous production line for compounding adhesives are shown in Figure 8.15 [144].

Elastomers (either pellets or base product), additives, and some of the resins are premixed and gravimetrically fed into the premixer. When formulating with fillers, these also are fed into the premixer. The basic polymers

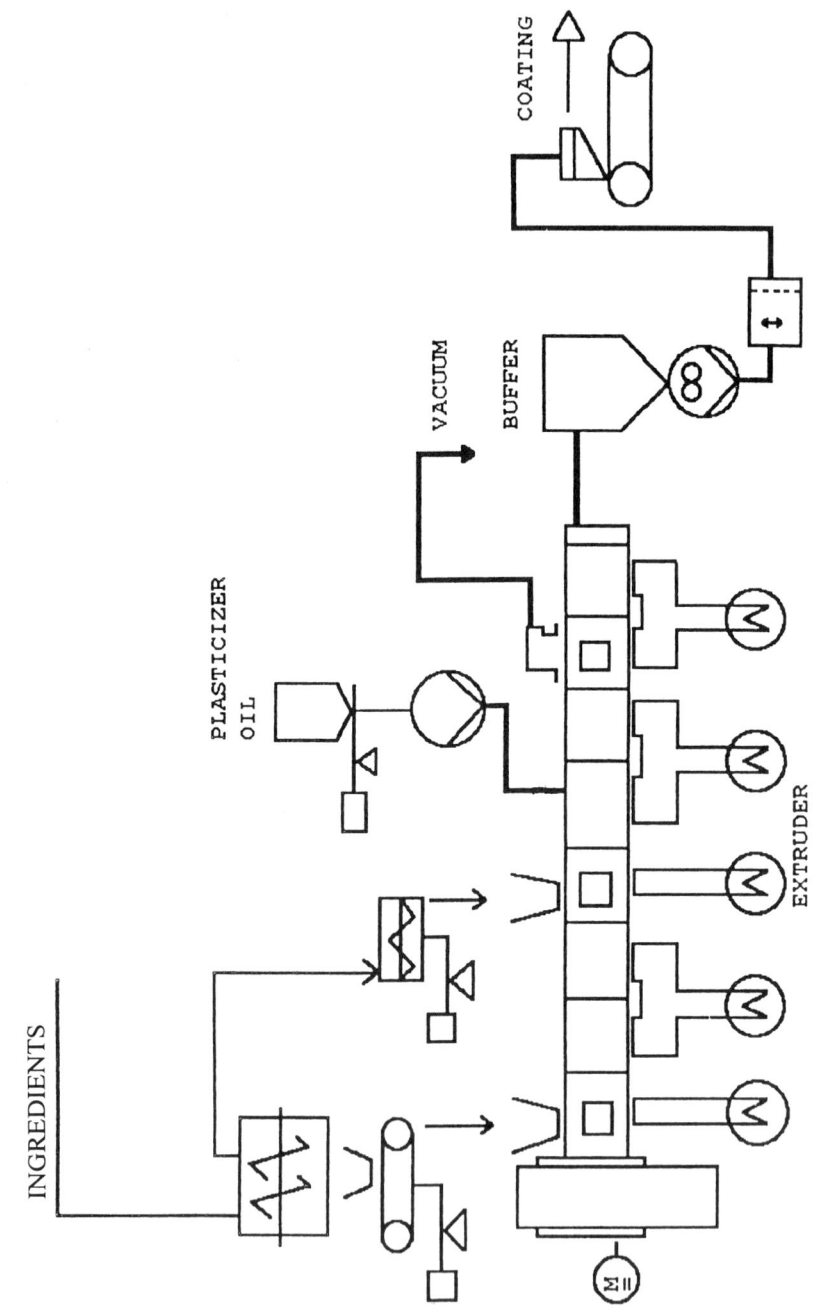

Figure 8.15 Production line for continuous compounding of a hot-melt PSA.

are masticated or, if they are thermoplastics, plasticized by special kneading elements and mixed with some of the resin. The second feed part for the rest of the resin is in another premixer and venting takes place there also. In the subsequent mixing zone the resin is melted and incorporated together with the liquid components and has to be well homogenized. The entire compound is very efficiently degassed. After a residence time of only 3–5 min the adhesive compound is ready to be transferred to the buffer vessel. From there filtering and coating are carried out via a gear pump. The main component of the compounding plant is a corotating twin-screw compounder. The screw profiles scrape each other as well as the S-shaped barrel base. This brings about self-cleaning of the screw profile, avoids dead corners, and results in a relatively narrow residence time distribution. The screw profile is characterized by a diameter ratio described by Friedrich [144]. At typical shear rates of $1000-2000^{-1}$s, the screw is in the range of high shear mixers.

Although in principle all heatable mixers can be used to make hot-melt formulations, the extent varies to which degradation of the base polymers and polymer resin components occurs [144]. The thermal history of the product, which is the result of thermal stress and its duration, is to be kept as short as possible. This can be done by using suitable continuous compounding lines, if possible in on-line operation with the adhesive coating equipment. This is a good solution to the problem for medium and large sized batches. For small and medium sized batches, however, contact between the melt and oxygen should be avoided and the possibilities of using batch mixers in a semicontinuous process with the coating line should be exploited. Generally, high degrees of degradation with open mixing systems and low degrees of degradation with closed mixing systems are observed. The possibility to carry out the entire mixing operation in a sigma blade semicontinuous mixer (medium shear mixer with a shear rate of about $100-200^{-1}$s) was investigated [144]. The mixing machine with sigma-shaped blades remains the classical mixing system for the production of hot-melt PSAs [164]; it is being used for mixing volumes of 0.5 l to several cubic meters. Batch times last from 45 min up to approximately 2 hr, depending on the size of the mixing chamber.

Hot-melt PSAs do not meet all performance requirements [143]. The melt viscosity of hot-melt PSAs may change during storage of the molten hot-melt PSAs [185]. The major benefits of hot-melt PSAs are agressive tack and peel adhesion, bonding to rough surfaces, good water resistance, and economics. Their limitations are heat resistance, aging, cohesion and plasticizer resistance. Since hot-melt PSAs are 100%-solid adhesives it is easy to apply heavy coat weights at a reasonable speed. Viscosity can be modified within limits by changing the coating temperature. No drying is required, only cooling of the adhesive. Therefore coating equipment for hot-melt PSAs requires a relatively low investment. In die-cutting and slitting operations the adhesive must be at room temperature since otherwise the risk of stringing may appear. Hot-melt PSAs need a much lower energy input than solvent-based or water-based adhesives [166], namely only 0.05 kWh/kg to 0.15 kWh/kg. The energy consumption for hot-melt coating is about 5–10% of the energy consumption for solvent-based or water-based coating (i.e., 0.0109 kWh/m^2). Furthermore hot-melt PSAs coating machines are running at 200–300 m/min, compared to 100 m/min for solvent-based or water-based coating [167]. The thermoplastic character of adhesives based on thermoplastic rubbers is a definite advantage in that it makes it possible to apply the adhesive as a solvent-free hot-melt. However, there are two major potential disadvantages: the limited maximum service temperature at which the adhesive will perform satisfactorily and the limited resistance to attack by common solvents and to PVC plasticizer migration. Conventional heat-initiated crosslinking systems as used in solvent-applied adhesives are not suitable for hot-melt processing as premature gel formation can be expected [155].

Therefore, for hot-melt PSAs radiation curing by exposure to electron beams or ultraviolet (UV) light immediately after application would be a more practical approach. Initial irradiation trials on linear S-I-S block copolymers showed that high radiation dosages were required and tack was lost, especially when polyacrylate curing agents were included. Extensive research has led to the development of new radiation crosslinkable polymers. These development polymers have a multibranched S-I-S structure (about 10% styrene content). The polymer does not contain functionalized groups and crosslinking takes place in the isoprene phase; at the same time the low polystyrene level facilitates good tack even after cure. Other work also is underway which includes the effect of modifying the endblock and the introduction of functionalized groups grafted on the rubber midblock. Radiation-cured PSAs were discussed by Hinterwaldner [168]. UV radiation produces IR radiation, therefore a thermal treatment in parallel to UV light absorption occurs. On the other hand, from both electron beam-curing methods—linear cathode and scanner method—only the scanner method ensures a low level of the dosage. Generally, the electron beam-curing generates higher crosslinking than UV.

In radiation-curing applications liquid PSAs or hot-melt PSAs are used. Liquid PSAs are blends of acrylic monomers and oligomers. The monomers are crosslinked like reactive diluents. Hot-melt PSAs are more important because the oligomers used for these have higher molecular weight and are monomer free. They may be coated at lower temperatures than normal hot-melt PSAs. A special type of electron beam-cured hot-melt PSAs are based on elastomers like thermoplastic elastomers (TPE; styrene-like copolymers). UV crosslinkable acrylic hot-melt PSAs were described in a detailed manner [168].

Water-Based Adhesives

The formulation of water-based PSAs requires special attention because of the water sensitivity and the water incompatibility of some formulating agents, and because of the fact that some additives to a water-based formulation are left in the coated adhesive and thereby modify the chemical composition of the bulk adhesive. Such additives (used mainly in order to ensure the fluid, dispersed character of the adhesive) primarily include the surface-active agents, surfactants, and defoamers. Generally, there are two categories of additives used in water-based systems: additives used for the improvement of the adhesive system and those used for the improvement of the dispersion system. Additives aimed at improving the adhesive system influence the adhesion/ converting and end-use properties; additives used for the improvement of the emulsion system influence mostly the converting properties. The most important additives used for improvement of the adhesive system comprise tackifiers, antioxidants, crosslinking agents, and peel modifiers. Additives used for improvement of the emulsion system include surface-active agents, like wetting agents, stabilizing agents, solubilizing agents, defoamers, and viscosity modifiers. Special additives which improve the water resistance also are used. One should note that additives used for the improvement of the adhesive properties of the water-based systems are similar to and chemically identical to additives used for solvent-based systems. However, their physical state may differ.

Improvement of Adhesive Properties of Water-Based Adhesives. The most important features of tackifiers were discussed in Section 2.2. The same tackifier used for solvent-based systems also can be proposed for water-based PSAs. Both micromolecular tackifiers (plasticizers) and macromolecular tackifiers (tackifier resins and other polymers) are employed. Suitable water-based tackifiers display superior adhesive performance properties after aging, peel adhesion and tack without major loss of shear, exhibit resistance to humidity, and specific adhesion to polyolefin films.

Resins preferably are added as aqueous dispersions in order to produce solvent-free adhesives. However, the use of resins as solutions in toluene and white spirit is equally possible and, in some cases, liquid resins or molten resins may be added also. There are many suppliers of tackifier dispersions with a different resin base. A detailed description of the problems concerning the use of water-based resin dispersions as tackifier was given by Dunkley [52] and Piaseczinsky [170]. Jordan [171] pointed out that it also is possible to tackify (using mostly hydrocarbon-based resin dispersions) natural latex, but the most important application of water-based resin dispersions remains the tackification of acrylic PSA dispersions. The most important theoretical and practical problem of tackification of acrylic PSAs, CSBR and EVAc dispersions was discussed in Section 2.3 in a detailed manner. Recent trends in water-based acrylic PSAs and the effects of tackification on properties and performance were described by Wood [46] and Di Stefano [172]. A quite different tackifying response is suggested in recent developments of CSBR and EVAc dispersions; CSBR emulsions typically require 50–60% tackifier to develop optimum properties. New ethylene multipolymers are less cohesive and accept less tackifier than acrylic PSAs.

Tackifying with aqueous dispersions differs from solvent-based tackification process. Because of the composition of the dispersed systems, nonadhesive (i.e., technological) additives may play an important role. As seen when tackifying solvent-based systems, the softening point and molecular weight of the resin significantly influences the performance characteristics of the adhesive. For water-based systems there are some parameters which override the effect of softening point and molecular weight of the base polymer. The tackifier supplier's particular formulation (surfactant type and level and other components) appears to have the largest impact on shear [172]. The effects of resin incompatibility on the viscoelastic, morphological, and performance properties of tackified acrylic adhesives are qualitatively very similar to those observed for rubber-based solvent-based adhesives [46]. Predicting the adhesive performance characteristics on the basis of the rheological characteristics of the raw materials remains difficult for water-based systems. Factors like pH, and surfactant nature and level play an important role; some resins are acid, other are esters (neutral). Some base dispersions are acid (CSBR), others have to be alkaline (Neoprene). On the other hand, some surfactants (like sulphosuccinates) are stable in acid or neutral medium only; some thickeners work in alkaline medium only. The surfactant in a tackifier dispersion may be incompatible with the polymer, causing opacity, even though the tackifier itself is miscible with the polymer [172]. In this situation it is possible to provide some guidelines for formulations with a particular tackifier and different base elastomer dispersions [173]. Blending acrylic dispersions requires a test of compatibility, the adjustment of the pH, and after addition of solvent, plasticizers or fillers, etc., a storage time [174]. The compounding of latex has been discussed in a detailed manner [175,176].

Tackifiers. A wide range of technically equivalent (i.e., yielding the same adhesive properties) tackifier dispersions are available. Differences exist concerning their dispersion properties (machinability), logistical and economical benefits. Most suppliers provide technical brochures containing data about the chemical composition (resin type), rheological behavior of the bulk resin (softening point and T_g of the base resin), physical characteristics of the base resin (e.g., color), and applications. The adhesive properties with different base elastomers (acrylic PSAs, CSBR, natural rubber, etc.) also are evaluated and physical properties and the stability of resin emulsions also are described. In Chapter 5 the various chemical bases of tackifiers were discussed in a detailed manner. The selection of a supplier also is determined by logistical and economical considerations. The main criteria are the existence of a convenient production site, the range of tackifiers on the market (e.g., pine/tall oil/hydrocarbon-based resins), the availability of raw materials, knowledge in tackification and emulsification, development of new resin types and improvement of resins (over the last 2 years), number of successful grades, knowledge of PSAs, estimated innovative capacity, sufficient production capability, supplier flexibility, technical assistance in converting, and last but not least price.

Plasticizers. The plasticizers used for solvent-based PSAs also are used, but at a lower level, in water-based PSA formulations. The use of plasticizers is more difficult in the case of water-based systems because of the solvent role of the liquid plasticizer, influencing the rheology and wetting properties of the aqeous system. Water-based partially soluble plasticizers may increase the viscosity. Migration of the plasticizer (strike-through) is more dangerous with water-based systems; therefore the plasticizer level used for water-based systems remains lower than the one generally accepted for solvent-based ones. The technology of addition is very important. In making up the coating mix for a peelable PSA using a hard latex with a plasticizer, there are advantages to make up the mix by first mixing the emulsion or latex with the plasticizer, which is usually supplied as a liquid. This incorporates the plasticizer in the hard (reinforcing dispersion) before adding the main adhesive polymer. This procedure appears to reduce any tendency of the plasticizer to cause undesirable migration in the finished product [14]. Polybutenes are first emulsified to form an oil-in-water emulsion, using either an anionic or nonionic surfactant. An alternative method of addition is to use a solution of the polybutene (e.g., 50% solution in toluene). However, while this may be convenient, it can only be used in plants where solvents are already being used and recovered. The polybutene emulsion or solution in solvent is blended with the acrylic latex to give the desired level using a high speed, low shear mixer [51].

Suitable antioxidants do not influence the printing properties of the laminate. Generally, antioxidants used for PSAs should be tested (before evaluation of their antioxidative effect) for their interaction with printing inks. In general, the same antioxidants used for solvent-based PSAs also are adequate for water-based PSAs. In this case they are supplied as aqueous dispersions (Bayer) or may be dissolved in alcohol (AS-14 Bayer). Table 8.25 lists some common protective agents used for PSAs.

Peel modifiers. Tackifiers (plasticizers and resins) act as peel modifiers. There are other additives with a tackifying effect that exert a more pronounced effect on the peel, like polybutenes (e.g., Hyvis). These additives added to

Table 8.25 Common Protective Agents Used for PSAs

		Application field		Protection against	
Product code	Supplier	Water-based PSA	Solvent-based PSA	UV light	Heat, oxygen, aging
Vulkanox	Bayer	—	x	—	x
BKF	Bayer	—	x	—	x
NKF	Bayer	—	x	—	x
SKF	Bayer	—	x	—	x
AS13	Bayer	x	x	x	—
Naugard 445	Lehmann	—	x	—	x

the acrylic latex yield a more pronounced peel increase than their aqueous dispersions. Thus, not only their amount, but their form (state) appears important. The compatibility of macromolecular additives should also be taken into account. Polybutenes show incompatibility with acrylonitrile (AN) copolymers, causing a marginal decrease in tack. Another class of peel modifiers are the silicone derivatives (e.g., Goldschmidt). These are liquid compounds added in 0.5–2.0% wet weight to the aqeous dispersion in order to reduce the peel. Unfortunately they only act at room temperature. High temperature aging of removable formulations on the basis of these additives yields no reduction of the peel adhesion.

Crosslinking agents. Crosslinking agents interact with functional groups from the bulk polymer. Crosslinkers in the form of water-soluble agents only have a restricted interaction with the bulk polymer (from the dispersion particles) but an unlimited interaction with the aqueous medium (dispersion additives) and the moisture-sensitive liner. The use of postadded crosslinking agents for water-based systems should be limited. On the other hand, some organic crosslinking agents are micromolecular, with a delayed crosslinking action up to the drying step, thus having enough time to give a penetration in the uncrosslinked state (e.g., Bayer, aqueous dispersions of PUR oligomers). External crosslinking agents include polyaziridines, epoxies, aminoplasts (such as melamine formaldehyde resins), and metal salts or oxides such as zinc oxide or zirconium ammonium carbonate, etc.; N,N'-bis 1,2-propenylisophthalimide can be used also [177]. Amine formaldehyde condensates (0.8–10%) may be used together with substituted trihalomethyltriazine as additional crosslinking agent.

The most important class of crosslinking agents used are the polyfunctional aziridines. Polyaziridines are suggested for use with Nacor 523 and other crosslinkable water-based PSAs. The most well-known commercial product used is Neocryl-CX100. The reproducibility of the crosslinking process is low. In order to increase cohesion and water resistivity, Hercules Polycup resins were evaluated. These are water-soluble, polyamide epichlorohydrine (EPC)-type materials effective as crosslinking agents for carboxylated acrylic PSAs or CSBR latexes, carboxymethyl cellulose (CMC), PVA, etc. It was proposed to use 0.5% solids of the latex-based crosslinker in order to increase the modulus. A hexamethylol melamine resin, commercially available from American Cyanamid Co. under the trade designation Cymel 303, together with 0.5 mole p-toluene sulfonic acid catalyst also may be used. The fluid heat-reactive resin Cymel 301 was proposed as a tackifier for neoprene latex, yielding very high bond strength upon the application of heat. Polyvinyl alcohol (PVA)-stabilized dispersions may be crosslinked with Quilon (Hercules).

Improvement of Emulsion Properties

Hot-melt PSAs and solvent-based systems are stable systems with unlimited shelf life. Dispersed systems and aqueous or solvent-based dispersions may coagulate as a result of shear forces, temperature, or chemical interactions. Therefore stabilizing agents (mostly surface-active agents) are added in order to improve their mechanical and/or shelf stability. These additives ensure the dispersion of the PSAs before and during coating, its coatability in the optimum case having only a minor effect on the end-use properties of the products. As dispersion additives to improve the stability of PSA dispersions surface-active agents (emulsifiers) or protective colloids may be added. The polymerization is carried out with emulsifiers, but poststabilizing surface active agents can be added as well. Sedimentation depends on the particle diameter, the density difference between solvent and solute, and the electrical potential of the particles.

Stabilizing agents. Stabilizing agents are emulsifiers (surface-active agents) used during the polymerization or thereafter in order to ensure shelf stability under static and dynamical conditions (storage, formulation, coating). Generally, stabilizing agents are recipe specific and it is not the task of formulators to select them and to specify their level. Carboxylated latexes have very good inherent colloidal stability and need no additional stabilizers. Many emulsifiers are commercially available for emulsion polymerization and may be classified as anionic, cationic, or nonionic. Generally, anionic emulsifiers are preferred, but nonionic emulsifiers also are efficient [120].

The common anionic surfactants are alkyl or alkyl ether sulfates such as sodium lauryl sulfate, sodium lauryl ether sulfate, alkyl oxyl sulfonates (e.g., sodium dodecyl benzene sulfonates) and alkyl sulfosuccinates (e.g., sodium hexyl sulfosuccinates) (2–3%). Surface-active agents, like surfactants, are used in dispersions for stabilizing and for wettability. A detailed discussion of their effect on the properties of the coated adhesive will be presented in Section 3.1 of Chapter 10.

Protective colloids. Protective colloids are necessary for dispersions with particle sizes larger than 1 μm (because of the large particle surface) [178]. The use of protective colloids allows better mechanical, electrolytic, and storage stability, as well as neutral pH, higher particle size, narrower molecular weight distribution, and higher surface tension [179]. Using a polymeric colloid which is a low molecular weight alkali-soluble polymer, surfactant-free adhesives may be prepared [91]. On the other hand, a high viscosity, a lower tack, and higher water sensitivity are characteristics of protective colloid-stabilized latexes. In general, only a few PSAs dispersions, mostly based on EVAc or neoprene, contain PVA as stabilizer.

Wetting agents. Inherent problems common to water-based systems include substrate wetting, foaming, foam and levelling deficiencies, as well as a possible reduction in adhesive properties. The first requirement for the bonding process is that the adhesive must establish intimate contact with the adherent; it must be capable of wetting it [180]. For virtually all uses of PSAs this is the case; the surface energy of the adhesive is much lower than that of the adherent, which is a condition for good wetting. The wetting process is far from being the instantaneous phenomenon one would like to believe it to be. Using radioactive tracers, Toyama, Ito and Nakatsuka [181] showed that the adhesive was present in discrete parts. The quantity percentage depends on the dwell time of the adhesive.

In order to improve the coating properties (i.e., wetout), surfactants are added to the formulation. Surfactants are defined as substances which significantly reduce the surface tension of liquids at very low concentrations. The degree to which these molecules accomplish this depends on the balance of their hydrophilic and hydrophobic regions. When properly balanced, surfactants will concentrate at the liquid/air interface, causing a compressive force to act on the surface [182]. For a given solid, liquid, or fluid medium, a unique value for the contact angle of a triplet line would be expected, but in practice a given system often displays a range of contact angles between lower levels (receding angle) and an upper limit (advancing angle); this difference is often referred to as hysteresis. While the static reduction of the surface tension can give the formulator a quick idea of the efficiency of the surfactant, many industrial applications never reach equilibrium. Therefore it is important in processes where surfaces are generated at a rapid rate, that the surfactants migrate rapidly to the interface so as to prevent film retraction and other surface defects; hence the dynamic surface tension is a very important parameter. Good dynamic surface tension means that the surfactant migrates rapidly to the interface so as to prevent film retraction. The formulator has to employ the wetting agent(s) leading to good wetting on the most difficult surfaces at the lowest viscosity level and necessary defoamer level, and at the lowest wetting agent concentration. Surface tension reduction and dynamic surface tension characterize the activity of the wetting agent, leaving only a limited range of available wetting agents (independent from the economic aspects). On the other hand, water sensitivity and migration, and the latter's influence on the peel and removability, limit the level of the wetting agent to be used. In order to select the surfactants, their chemical nature, their surface tension reducing effect, and their influence on the coated adhesive layer should be determined. The choice of the surfactants is not so simple because the required surface tension depends on the characteristics of the web to be coated (face stock or release liner) as well as the coating parameters (coating device). Aqueous dispersions usually have a built-in level of surfactants that are required for dispersion stability during and after the polymerization or for handling purposes.

These dispersion inherent surfactants may be polymerization related or postadditives. Postadditives may be micromolecular (surfactants) or macromolecular (protective colloids). Their level depends on the end-use of the PSAs [183]. Quite different surfactants are used for primary dispersions (elastomer dispersions: acrylic PSAs, CSBR, EVAc) or secondary ones (resins or other tackifiers, antioxidants, etc.). On the other hand, the different chemical base of the elastomer dispersions requires different emulsifiers. Generally, surfactants (not protective colloids) are used for PSAs (as compared to viscosity, water sensitivity), but poly(vinyl alcohol)-stabilized carboxylated neoprene latexes make an excellent base for PSAs [10]. The choice of emulsifiers used for secondary dispersions depends on the nature of the additive, but its molecular weight is important also. During the emulsification of polybutenes, surfactants with different hydrophilic-lipophilic balances (HLB) are used as a function of the molecular weight of polybutenes; the optimum HLB for polybutenes increases with increasing molecular weight. Regarding the choice of surfactants for primary dispersions, it depends on the nature (original surfactant level and surface tension level) of the elastomer dispersion and of its formulation (tackified or not).

The surface tension of most neoprene latexes and their compounds is relatively low (usually around 30–40 dyne/cm as compared to water at 72 dyne/cm); for example, Neoprene Latex 102 has a surface tension of 58 dyne/cm at 20°C [179]. A polyvinyl acetate latex has a surface tension of 38

Table 8.26 Surface Tension Values of Water-based PSAs

Batch number	Surface tension value (dyn/cm)
01	42.4
02	41.9
03	41.6
04	40.0
05	40.8
06	40.2
07	40.8
08	43.2
09	39.2
10	44.0
11	40.8
12	39.0

Water-based PSA: Acronal V 205

dyne/cm, whereas styrene butadiene latexes possess a surface tension of 45–60 dyne/cm [184]. Dispersions of the same composition may have quite different surface tensions (Table 8.26); moreover the same dilution of a dispersion may lead to very different surface tensions (Figure 8.16).

The use of wetting agents has been reviewed [22]. The choice of the wetting agent depends on the efficiency and the influence on the adhesive end-use performance (i.e., the activity to improve wetting and coatability). Their influence on the adhesive performance characteristics concerns the change of the adhesive properties, the optical properties, chemical resistance, and aging of the coating. Their versatility to be used (ease of compounding, influence on the viscosity, drying speed) has to be considered. For example, some sulfosuccinates have only a limited solubility in cold water; in order to improve their solubility alcohol should be added. Some sulfosuccinates are sensitive to pH changes and thicken with the increase of the pH. Sulfosuccinates are sometimes hygroscopic and therefore cause bleeding and migration; they can reduce the drying speed, or cause adhesive residues upon removal of the label from a substrate (i.e., they are not recommended for removable PSAs). The most common wetting agents are sulfosuccinate derivatives [8, 185]; dioctylsulfosuccinates display better properties than other esters (Table 8.27).

The selection criteria for wetting agents include the ability to provide good coverage on difficult-to-wet surfaces, a reduced foam level, no adverse effect

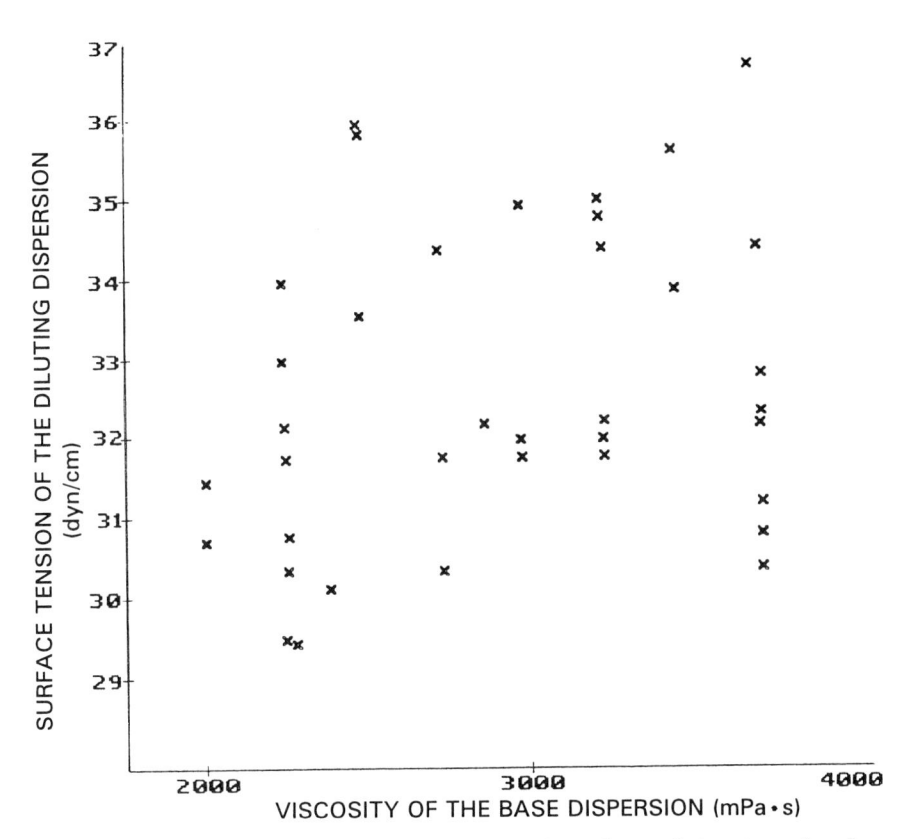

Figure 8.16 Dilution of water-based PSA. Interdependence of viscosity and surface tension.

Table 8.27 Common Sulphosuccinate Derivatives Used as Wetting Agents for Water-Based PSAs

	Code	Supplier	Solids content (%)	Chemical basis
1	Disponil SUS-65	Henkel	40	Natrium, fatty alcohol, ethylene oxide, sulfosuccinate
2	Triton GR-5	Rohm	60	Natrium, dioctyl sulfosuccinate
3	Rewopol SBFA 50	Rewo	30	Di-natrium, fatty alcohol, polyglycolether sulfosuccinate
4	Fenopon	GAF	—	—
5	Lankropol GR-2-3	Münzing	—	—
6	Aerosol A 102	Cyanamid	30	Di-natrium, fatty alcohol, sulfosuccinate
7	Aerosol OT	Cyanamid	—	Natrium, dioctylsulfosuccinate
8	Aerosol 18	Cyanamid	35	Di-natrium, N-octadecyl sulfosuccinamate
9	Rewopol SB-MB-80	Rewo	80	Di-isohexylsulfosuccinate
10	Rewopol SBFA 30	Rewo	40	Di-natrium, fatty alcohol, polyglycolether sulfosuccinate
11	Lumithen IRA	BASF	—	Natrium, dioctyl sulfosuccinate
12	Aerosol MA	Cyanamid	—	Natrium, dihexyl sulfosuccinate

on adhesive properties, and a reduced water sensitivity. The best ability to provide good coverage on difficult-to-wet surfaces are displayed by sulfosuccinates. Fluorinated surfactants are superior (Table 8.28) but not economical [186]. In some cases better results are obtained with acetylenic diol surfactants alone or in mixtures [182].

Surfactants. The presence of surfactants can lead to foaming. The evaluation of the foam level is discussed in Chapter 10 (Test Methods). Very critical remains the lack of adverse effect on the adhesive or end-use properties. In order to limit or avoid this, it is necessary to determine the influence (mechanisms) of surfactants on the PSA layer.

Surfactants exert a chemical and a physical influence on PSAs. Surfactants interact mainly with the bulk polymer, with components used as reactants for the bulk polymer or with other dispersion additives. Similar interactions are chemically possible between the surfactants and the other components of the laminate (face stock or liner). Physical processes occuring with the aid of the

Table 8.28 Wetout Improvement, Using Acetylenic Diol Surfactants in a Blend with Ethoxylates and Sulfosuccinates

	Surface active agent characteristics				
	Formulated surface active agent				
Code Supplier	Genapol X80 Hoechst	Surfynol 61 Air Products	Lumithen IRA BASF	Global surface active agent level (wt%)	Wetout index (subjective evaluation)
	50	50	—	2	4
	50	50	—	4	4–5
	33	66	—	3	4
	66	33	—	3	3
	25	75	—	4	2
	—	50	50	2	2
	—	66	33	1.5	2
	—	80	20	1.5	2

A water-based acrylic PSA was coated on SL release. The wetout quality index is defined as follows: 1, excellent wetout; 2, excellent wetout, stationary for 5 sec; 3, wetout not stable, liquid film shrinks after 3 sec; 4, wetout not stable, film shrinks after 3 sec, pinholes and craters; 5, no wetout, film shrinks instantaneously, craters.

surfactants may influence the properties of the PSA laminate. A surfactant in an adhesive tends to migrate to the surface, where it often reduces the bond strength [8,91,187]. Surfactants also may react with chemicals added in the formulation in order to interact with the reactive groups of the bulk polymer; an interference with certain crosslinking agents used as modifiers also may occur [8].

The amount of nonadhesive additives in water-based dispersions influences the adhesive properties of the PSAs; thus the effect of surfactants should be taken into account. The adhesive performance is influenced to a larger degree than expected, by the surfactants and other additives in the tackifier dispersion [57]. Midgley [13] found that surfactants reduce the tack, but with the exception of a slight change in rolling ball tack, there is a little difference between the same water- or solvent-based adhesive. Emulsion polymers contain surface-active agents which cause a decrease in tack and adhesion [110]. For CSBR different, optimum tack values are achieved, when one reference latex is used with different emulsification systems [66]. Low levels of potassium rosin acid soap may even improve the tack and peel adhesion of CSBR latexes. Similar effects are obtained by using ethoxylated tetramethyl-decyne-diol (Surfynol 465), and the nonionic soap, octyl-phenoxy-polyethoxy-ethanol (Igepal CA 630, GAF) [133,188]. Lauroyl sulphates migrate rapidly to the surface and reduce tack [133, 188]. Ethoxylated octylphenol affects (reduces) the peel after aging [189]. Water-based PSAs contain surfactants which complicate the formulating process because of the interference with certain crosslinking agents [8]. Furthermore a surfactant in an adhesive tends to migrate to the surface where it often reduces bond strength. The presence of the surfactant reduces the wet bond strength of the laminate [91]. Furthermore, surfactants reduce the peel of PSAs [190]. For special applications an emulsifier is added into the adhesive formulation in order to avoid that the adhesive at low temperatures acts like a rubber (i.e., in this case the emulsifier is acting like a plasticizer). The influence of the surfactant on the peel also has been summarized by Makati [191].

Interaction of Surfactants with Other Layers in the Laminate. In the particular case where the adhesive is to be used to form self-adhesive laminates and where at least one of the surfaces is a printed surface, the presence of any residual surfactant can lead to discoloration and bleeding of the ink [14]. This is a recognized problem in applications such as overlaminated books or printed labels when the purpose of the film is to preserve the integrity of the printed surface.

Generally, the hardness of a clear film obtained by evaporation of a latex is largely dependent on the polymer composition and the temperature. The surfactants and colloids have the same effect, with regard to the humidity.

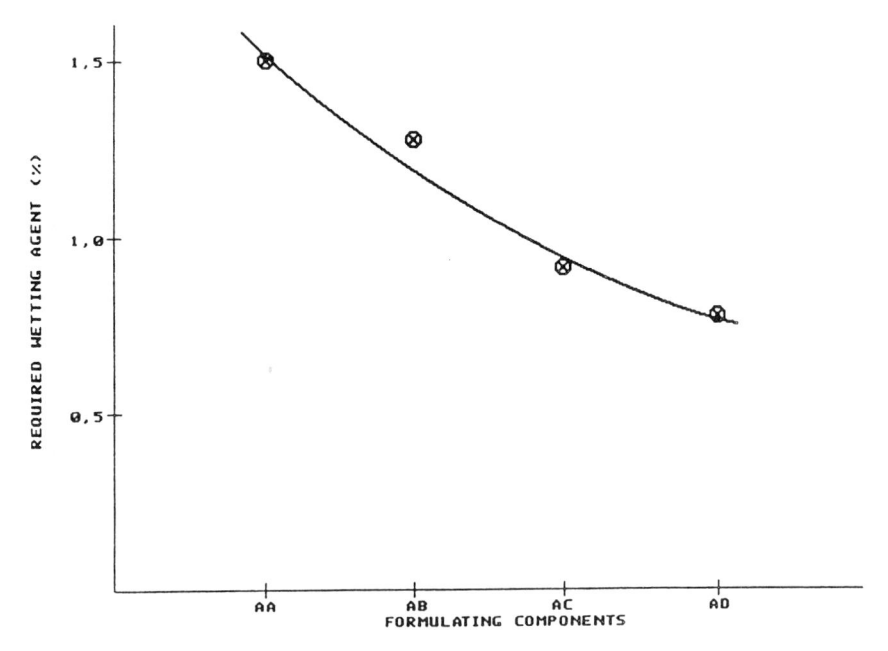

Figure 8.17 Dependence of the required surface active agent level on the nature of a PSA dispersion. A, B, C, and D are different water-based acrylic PSAs.

The presence of the surfactant reduces the water resistance of the laminate [91,110]. Water absorbtion in acrylic latex films is influenced by the storage conditions/time or thermal coalescence, which reduce water absorption [192]. Addition of emulsifier for latex stabilization increases the starting rate of water dissolution in acrylic films; an increased amount of surfactant reduces the drying rate under given conditions [66]. Ethoxylated nonylphenol derivatives bring a high viscosity increase, but poor cold water solubility [83]. Anionics are strongly hydrophylic and are particularly undesirable in wet environments [189]. The water resistance of the surfactants depends on their chemical composition. The resistance to rewetting is good for fatty acid derivatives, poor for anionic, and fair for nonionic materials.

Generally, polymers have different surface tensions. Normally ready-to-use systems are blends of several batches. The same amount of wetting agent is not sufficient in order to ensure the same surface tension for PSAs from different sources (i.e., the required surfactant level depends on the base polymer) (Figure 8.17).

Generally, the nature and the level of the surfactant used depend on the base polymer of the PSAs. For ethylene vinyl acetate dioctyl maleinate copolymers the amount of the emulsifying agent varies by a factor 1 to 10, preferably about 2–8% of the amount of monomers used [94]. Water-based acrylic PSAs possess a lower surface tension than common EVAc or CSBR dispersions. Current surface tension values are 34–40 dyn/cm while EVAc dispersions show more than 45 dyn/cm; rubber latexes have a surface tension of 45–60 dyn/cm [193]. Water-based acrylic PSAs generate more coagulum (grit) than CSBR dispersions and exhibit a better mechanical stability than CSBR.

Generally, PSA properties improve as the anionic surfactant concentration increases. All major properties can be altered by changes in surfactant nature and level. The hydrophilic-hydrophobic balance of a surfactant is another of the physical parameters available to enable an easier choice of the emulsifiers. Relating the water-soluble and water-insoluble part of the surfactant is a useful guide to the solubility of the product itself and to the solubilizing ability which it possesses.

Surface-active agents were used in synthetic latexes in amounts of 1–6% [194]. Natural latexes contain 1% surface-active agents. Some natural latexes include acids as an emulsifying agent [104]. In practice 0.75–1.5% (wet/wet) sulphosuccinates and 1 to 3% ethoxylates are used; the surfactant level also depends on the pH. The level of the water sensitive wetting agents may be lowered using the special fluor-based emulsifiers or the volatile ones (Surfynol S 61), but actually they remain too expensive. Table 8.29 illustrates the influence of the concentration of fluorinated surfactants on the surface tension.

The viscosity of many dispersions depends on the pH. As wetting out is a function of the viscosity, the level of the wetting agent necessary in order to achieve a good wetout depends on the pH (Figure 8.18). As seen in Figure

Table 8.29 The Influence of Fluorinated Surfactants on the Surface Tension of a PSA Formulation

Fluorinated surface active agent	Concentration (%)	Surface tension (dyn/cm)
Zonyl, FSK	0	35.7
	0.2	36.8
	0.5	30.7
Zonyl PT 448	0	40.8
	0.2	41.7
	0.5	42.8

Arylic PSA: Acronal V 205.

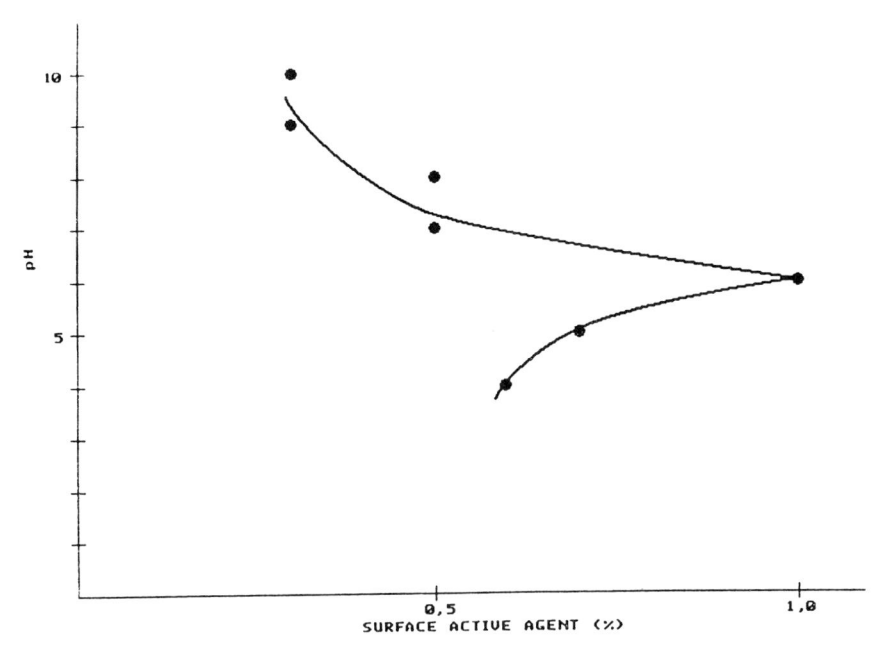

Figure 8.18 Dependence of the surface-active agent level required for good wetout on the pH of water-based acrylic PSAs.

8.18, the amount of the sulfosuccinate active agent used for wetout depends on the pH; a minimum level is required in the high acid domain.

Adjustment of Viscosity. The wetout is a function of viscosity [195], surface tension of the water-based adhesive, and foaming. The viscosity of the formulation depends on its basic components. Elastomeric dispersions exhibit a different viscosity, depending on their chemical composition (the presence of water-soluble monomers), dispersion characteristics (which depends on the surfactants used during their synthesis) and solids content. Generally, water-based acrylic PSAs possess a higher solids content and a higher solids content/viscosity ratio than common CSBR/EVAc dispersions as well as a better dilution response (Table 8.30). This is very important when the viscosity has to be lowered for coating (machinability) purposes. In some cases, the running viscosity has to be increased again (before or after dilution); there are certain special applications where reducing the viscosity appears required.

With regard to the influence of the viscosity and the surface tension on the wetting behavior, Hansmann [196] stated that the decrease of the viscosity

Table 8.30 Solids Content and Viscosity of Water-Based Acrylic PSAs Versus EVAc and CSBR PSAs

Water-Based PSA				Emulsion characteristics			
Brand name	Code	Supplier	Base	Method	Viscosity (mPa·s)	Shear rate (s^{-1})	Total solids (%)
Dilexo	N600	Condea	Acrylic	DIN 53019, 20°C	400–800	57	55
Dilexo	AK610	Condea	SAcrylic	DIN 53019, 20°C	800–1500	57	55
Latex	3703	Polysar	CSBR	Brookfield LVF, S2, 30 rpm	250	—	50
Latex	3958	Polysar	CSBR	Brookfield LVF, S2, 30 rpm	500	—	53
Mowilith	130	Hoechst	EVAc	Brookfield RVT S5, 20 rpm	5000–11000	—	50
Mowilith	131	Hoechst	EVAc	Brookfield RVT S5, 20 rpm	5000–13000	—	50
Mowilith	132	Hoechst	EVAc	Brookfield RVT S5, 20 rpm	4000–1000	—	60
Acronal	4D	BASF	Acrylic	DIN 53019, 23°C	15–40	250	50
Acronal	80 D	BASF	Acrylic	DIN 53019, 23°C	80–150	250	50
Acronal	V 205	BASF	Acrylic	DIN 53019, 23°C	800–1600	250	69
Nacor	314	National	Acrylic	Brookfield	> 200	—	52
Nacor	525	National	Acrylic	Brookfield	> 200	—	53

influences the wetting more than the surface tension. The rheological shortcomings of latex PSAs with conventional surfactants arise at least in part from the need to incorporate thickeners, such as hydroxyalkyl cellulose or polyacrylic acid. Without thickening the conventional low viscosity latex will not completely wet a silicone-coated release liner. This is a necessity for transfer coating which is preferred for heat sensitive face stocks [91]. Addition of thickeners (alginates) may avoid stripes [97]. Migration through face stock paper may be avoided using thickeners as well [198]. To improve surface coverage the viscosity can be increased with the addition of cellulosic or polyvinyl alcohol (PVA) thickeners. However, problems with machinability, air entrainment, and moisture sensitivity may arise. Thixotropy can be introduced by adding different components like silica, ricinus oil, titanium, or zirconium chelates. [75]. As thickeners methylcellulose, PVA, starch, polyacrylate salts, alginates, dextrine, and ethoxylcellulose also can be used [199, 200]. Copolymers of methylvinyl ether and maleic anhydride are used as thickeners for natural rubber and synthetic latexes [201].

According to their activity as thickeners, these agents may be classified as pH regulators, surfactants, protective colloids/polyelectrolytes, water-soluble polymers, fillers, crosslinking agents, or solvents. It is evident that most PSA emulsions containing polar groups incorporated in the polymer backbone should modify their viscosity through a pH change. For this purpose, alkaline additives (mostly fugitive ones) are used. In some cases discoloration (by ammonia or alkaline hydroxides) was observed and organic amines are preferred. The change of pH is required for some polyelectrolyte thickeners (polyacrylic acids) (i.e., these act mostly in the alkaline domain). Aqueous solutions of carboxylated polymers have low viscosities. The polymer is believed to be tightly coiled and only slightly ionized; as the pH is raised, more carboxyl groups become ionized, the polymer chain is further uncoiled, and, as a result, the viscosity increases [86]. Changes in the flow curves of acrylic polymer dispersions incorporating acrylic acid during a treatment with ammonium hydroxyde were described with the Cross theory of aggregation of dispersed particles [202]. If the surface tension value of the material to be coated is too high and the formula is not thickened properly, defects in the film will appear. Cratering, pinholing, and crawling are all evidence of poor substrate wetting, of a too high surface tension value and/or of an improperly thickened system. Viscosity adjusters are agents used to lower or increase the viscosity of the water-based systems.

High speed rotogravure machines are running with 100–500 mPa·s dilute dispersions. On the other hand, good wetting requires a higher viscosity or/and wetting agents. Some wetting agents increase the viscosity also, thereby thickening the dispersion. The high running speeds assume higher solids contents associated in several cases with an inadequate rheology of the semidry coated

Table 8.31 Common Thickeners Used for WB PSA Formulations

Code	Supplier	Basis	Usual concentration (wt%)	Comments, ref.
Borchigel L75	Borchers	Polyurethane	< 1.5	—
Viscoatex 46	Coatex	Acrylic	< 2	—
Coatex BR 100	Coatex	Polyurethane	< 1.5	—
Cyanamid A-370	Cyanamid	Modified polyacrylamide	< 2	FDA approval
Kelzan M	Lancro	Polysaccharide, natural sugar derivative	0.2–0.5	Tested for acrylics with good results
Viscalex V430	Allied Colloids	Acrylic	0.1–0.5	Current industrial use
Polysar Latex 6100	Polysar	Acrylic	< 2	Proposed for CSBR [203]
Rohagit SD15	Rohm & Haas	Acrylic	< 1.5	Current industrial use
Latekoll D	BASF	Acrylic	< 1.5	Proposed for neoprene latex [204]
Jaguar CHMP	—	Natural polysaccharide	< 1	
Kelset	Langer	Natrium calcium alginate	< 2	
Collacral VL	BASF	Acrylic	< 2	Recommended for film coatings

Manufacture of Pressure-Sensitive Adhesives

layer, resulting in an unsmooth coating. In this case dilution of the base dispersion and rethickening it is necessary. Summarizing, two kinds of viscosity adjustments are needed: thickening and lowering of the viscosity; pH regulating agents are not really thickeners, but they are added more in order to increase the dispersion stability or to change certain adhesive properties. Table 8.31 lists some common thickeners used for water-based PSAs.

Soft hydrophobic and hydrophylic comonomers (butyl acrylate/methacrylic acid, BuAc/MAA) produce thickening latices (their viscosity increases more than 1000% in the alkaline region) which makes the penetration of the alkali (natrium ions) in the polymer particle possible, and thus change the polymer volume and structure. The influence of the pH regulating agent was discussed in Section 2.2 of Chapter 2.

Some surfactants also act as thickeners. This behavior is very pronounced for certain sulfosuccinates and may differ for the "same" compound from different suppliers. Care should be taken whether surfactants should be used as thickeners. PSA dispersions possess a certain surfactant level as a result of the polymerization process; some surfactant amount can come from the face stock material. All these may interfere with the adhesive and end-use properties (see Section 1.1 of Chapter 6). The hydrophobic segment of a soap can migrate to the relatively nonpolar rubber adhesive and alter release, tack and readhesion properties. Contamination can come not only from the migration of certain species in the release coating, but also from species in the face stock such as plasticizers from fibers or the soaps in paper saturants. Calendered PVC (emulsion PVC, EPVC) made from 0.1 μm diameter particles, has a thin layer of emulsifier on its surface; this layer behaves like an adhesive [205].

Protective polyacid-based colloids are preferred, together with polyamides, because they are less sensitive to microbial attack and cause less decrease of the tack than natural derivatives.

Polymer thickeners for emulsions are copolymers of (meth)acrylic acid (25–45%) and alkyl (meth)acrylates (25–65%), esters, ethylene glycol derivatives (1–40%), polyethylene unsaturated monomers (0–11%), and hydrophyllic ethylene monomers < 10% [206]. Water-based polymers (e.g., polyvinyl ethers) improve the adhesive characteristics also, but display water and aging sensitivity. Polyvinylpyrolidone in PSAs imparts a high initial tack, strength, and hardness; water-remoistenable PSAs cast on paper reduce the usual tendency to curl [207]. N-vinylpyrolidone-ethylene methacrylate (EMA) copolymers may be used as thickeners also [208]. Fillers generally decrease the adhesive properties and have to be used in concentrations below 5%. Cross-linking agents (e.g., metallic salts) interact with the polar monomers from the bulk polymer or from the protective colloid. They improve the water resistance, but decrease the adhesive properties. Cohesion and heat resistance can be increased further through crosslinking. The divalent metal ion/acid functionality approach used with acrylate PSAs also applies to new systems such as ethylene/vinyl acetate/maleinate copolymers. Solvents may be added to water-based adhesives and can improve their anchorage 7–30% [200]. Depending on their solubility in water, solvents increase the viscosity of water-based PSAs. The choice of an adequate thickener depends mainly on the raw dispersion and the end-use requirements. The thickening power on polymer latices is influenced by the type of emulsifier, particle size (i.e., surface area), and presence of mineral fillers. High molecular weight water-soluble polyelectrolytes are readily absorbed by polymers or mineral particles. The degree of absorbtion depends on the type of latex, emulsifiers, thickeners, or mineral filler employed. For a poorly stabilized latex, the absorbtion is the greatest and the thickening power the most pronounced [209].

In some dispersions the viscosity may be too high, even though no conventional thickener is present. The easiest way to reduce the viscosity is to dilute the emulsion with water, if a reduction in solids content can be tolerated. If a high solids content must be maintained, the addition of alkaline

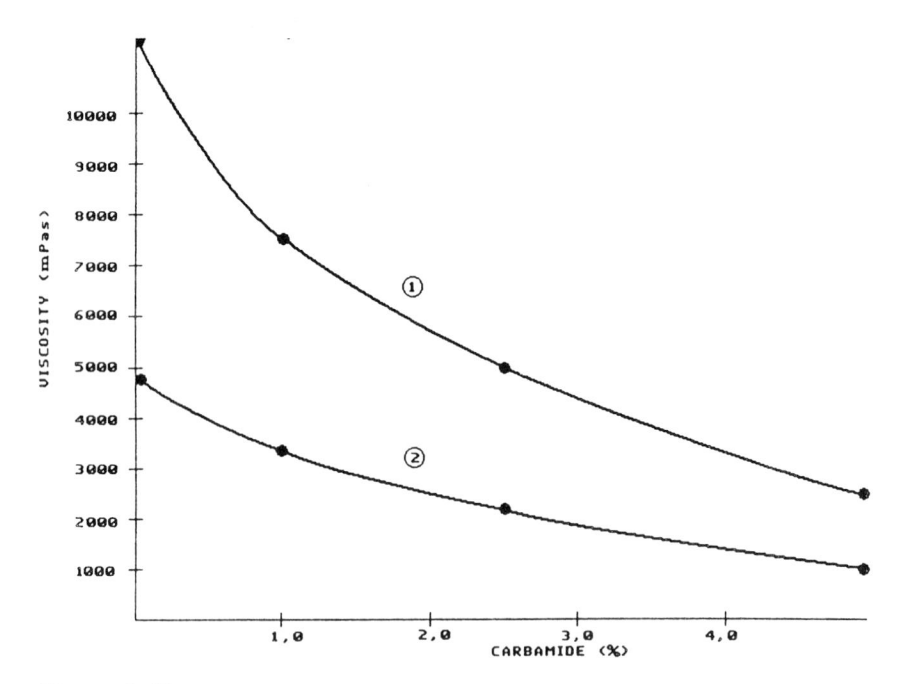

Figure 8.19 Viscosity decrease through formulating. The use of carbamide as viscosity reducing agent. The Brookfield viscosity of the formulation after: 1) a storage time of 2 days; 2) instantaneously.

Table 8.32 Agents for Lowering the Viscosity of PSAs

	Water-based PSA		Viscosity		
Viscosity reducing agent	Code	Supplier	Before	After	Refs.
Dicyandiamide	1360	Hoechst	5000	2400	[212]
Carbamide	V 205	BASF	5000	3600	—
Polyvinyl pyrrolidone	Natural latex	—	—	—	—

materials (e.g., sodium or potassium hydroxide, sodium polyphosphate, ethylene diamine, diethylamine, etc.) may reduce the viscosity. Urea (up to 6% wet weight) may be used to decrease the viscosity of natural latexes [210]. The viscosity may be decreased by special additives like polyvinylpyrolidone [211].

Table 8.32 lists some viscosity reducing agents; Figure 8.19 illustrates the influence of carbamide addition on the viscosity of a common water-based PSA.

Coatability and Foaming

By virtue of lowering the surface tension, any surfactant reduces the energy needed to create foam; however, what really dictates the foam volume achieved is the stability of the foam. Surfactants form a dense, ordered, viscous film on the surface of aqueous solutions in which bubbles of air can be encapsulated. The mechanism apparently involves the hydrophylic polyethylene oxide chains forming closely packed hydrated layers that are crosslinked by interstitial water molecules. In contrast, defoaming copolymer surfactants form low viscosity, mobile, open surface films that form areas of weakness in the surface of a bubble, eventually leading to bubble rupture [213]. There is no generally valid theory concerning the mechanism of defoamers [214].

Generally, if more surfactants are used there is a competitive adsorbtion of the nonionic and anionic surfactants on the latex particles. The addition of a surfactant may cause the desorbtion of the other one. In practice, this can cause problems such as foaming [215]. Foaming also depends on the surface quality; on porous surfaces more foam is left [214]. In stirred systems foam builds up if the strain τ is greater than the capillary pressure p [216]. The strain depends on the particle diameter ϕ according to the following equation:

$$\tau = \rho \cdot 1.9 \cdot (\varepsilon \cdot \phi)^{2/3} \tag{8.2}$$

where ρ denotes the density of the particles. The particle diameter is a function of the surface tension and therefore foaming is a function of the surface tension.

Foaming on the coating device depends on the depth of the immersed coating cylinder. Deeply immersed coating cylinders cannot carry much air into the liquid, therefore low foaming results. The rotation direction of the coating cylinder influences foaming also. Reverse cylinder sense (against the web direction) is suggested [217]. Temperature (related to the choice of the surface-active agent) also may influence foaming [189]. Typical low foaming nonionic surfactants foam less at temperatures above their cloud points because they become insoluble.

Defoamers are often used because the polymers are present as emulsions (or latexes) in water and the hydrophobic nature of the polymers implies that substantial amounts of surfactants are necessary in order to maintain the stability of the dispersion. Conventional defoamers (e.g., mineral oils and waxes) or silicones can be typically used in an amount of about 0.25 wt% of the wet coating mix. Defoamers are added into the dispersion in order to reduce foam formation. Their use is well known from the synthesis of water-based adhesives, where defoamers are necessary to aid free monomer reduction when heavier reflux develops. Different kinds of defoamers (e.g., silicones, fatty acid derivatives, or alcohols) in an amount of about 0.15% are suggested [217]. Defoamers absorb into the polymer, especially at elevated temperatures, and thus have a short life and require frequent replenishment. When foaming during the handling or coating is encountered, regular additions of defoamer should be made. Many additives were found to minimize or eliminate the formation of foam, and a large number of proprietary defoaming agents are on the market. None is completely satisfactory for all compounds; conditions of use are such that only by trial and error one can identify the most efficient agent or combination of agents for a given compound and process. With any antifoaming agent the minimum amount required should be used since such agents tend to cause localized coagulation, poor wetting, and/or "fish eyes" in the film. Fish eye formation can sometimes be prevented by adding a small amount of a chelating agent.

Besides the many antifoam agents, there are other readily available materials which function as defoamers. Higher alcohols (such as 2–ethylhexanol), water-soluble oils, and tributylphosphates also have been used for this purpose [218]. Equipment and devices used for testing foam formation are discussed in Chapter 10; as defoamers, silicones, fatty acid derivatives and caprylalcohol are used. A blend of tributylphosphate/pine oil (3/1) also may be used as defoamer [199]. There are many antifoam agents and they are usually classified as silicones and nonsilicones [185]. Silicone fluids and emulsions are common defoaming agents. The silicone fluids most commonly used in the printing ink industry are based on polydimethyl siloxane where the chain length may vary from 1 to 1000. Antifoam agents must be used with care, since the resulting adhesive performance characteristics are concentration dependent. An excess can cause pinholing and poor roll-to-roll ink transfer. Silicone derivatives are used in a concentration of 0.05–0.2% [199]. Defoamers based on mineral oil may contain emulsifiers and other defoaming agents like fatty acid esters. It is often preferable to use a nonsilicone type to avoid film scission. Polyethylene/polypropylene glycol ether derivatives of amines and alcohols were suggested. Propylene glycol, diethylene glycol, and carbamide also have been used [217].

Defoamers for labels should not influence the printing properties of the face material. They must also meet FDA and BGA requirements. Foaming is a problem mostly encountered with medium level viscosity formulations. Solvents (used as defoamer) may swell the dispersion particles in the coating, giving a more continuous dry film. Hoechst stresses the role of solvents to improve water resistivity, especially with regard to water whitening. Table 8.33 lists some commonly used defoamers.

Adjustment of pH

Generally, water-based dispersions are stable above a pH of 2; acrylic dispersions are usually delivered with a pH of 5–6, vinyl acetate copolymers at a pH of 4–5, and natural rubbers and neoprene dispersions in the alkaline pH range. Dispersions of polymers bearing carboxylic groups (CSBR or rosin acids) are acid; their neutralization may change their stability, adhesive and end-use properties (e.g., tack, shear, water sensitivity, aging resistance). In order to avoid an ionic shock, formulation components have to display the same pH range (i.e., they need neutralization agents). Nonvolatile neutralization agents may change the water absorption/desorbtion of the polymer particles (i.e., drying properties) or may interact with other components of the laminate; they also may change the viscosity of the dispersion or destroy ester-based surfactants (e.g., sulfosuccinates).

The modulus, peel adhesion, and water resistance of the adhesive depend on the nature of the neutralization agent used [219]. For neoprene latices diethanol-amine (DEA) is recommended [220]. Polychloroprene latices are supplied with a pH of about 7. It is advisable to adjust the pH upwards to obtain a long storage stability and to ensure good cohesive strength and film aging. The adjustment of the pH in order to modify the balance between adhesive tack and cohesive strength is just one of several formulation variables the adhesive manufacturer has available with the carboxylated chloroprene latexes [219]. For CSBR it is desirable to adjust the pH upwards in order to obtain a long storage stability and good film aging [179]. In some cases the use of thickeners makes the change of pH necessary.

Table 8.33 Common Defoaming Agents

Name	Code	Supplier	Base	Comments
Polymekon	1488	Goldschmidt	Silicone derivative	—
—	EOPK-53	Goldschmidt	Silicone derivative	Not BGA-FDA approved.
NOPCO	8034	Diamond Shamrock	—	One of the most recommended defoamers.
BYK	040	BYK	—	—
Bevaloid	581 B	Erbslöh	—	—
Dehydran	G8	Henkel	—	—
Drewplus	T-4211	Drew Ameroid	—	One of the most common defoamers used for PSA. FDA approved.
Lanco-Foamex	VP-345	Langer	—	—
Lumithene	ES	BASF	—	—
Surfynol	104	Air Products	Acetylenic alcohol	Wetting agent
Polymekon	81	Goldschmidt	Silicone derivative	—

Solvents

Solvents are added to adhesives in order to improve the anchorage on a substrate [221] or to plasticize them. The drying of latex adhesives is particularly susceptible to environmental conditions (temperature, humidity). What is not so obvious is that these also affect the surface of the dry film. Therefore coalescence aids can be added to counteract this problem. Generally, up to 20% solvents may be added. The acceleration of the bonding process (green strength) that occurs with polyvinyl acetate adhesives cannot be observed for PSAs [54]. In the case of water-based adhesives, the addition of coalescing solvents, plasticizers, etc., depends to a large extent on their water solubility. Additives which are water soluble should be added at 20°C, preferably diluted with an equal quantity of water. If added at too high a temperature or concentration, they can cause destabilization of the latex by removal of water. If the additive is insoluble in water, it can be added while hot. The higher the addition temperature, the faster the absorbtion of the solvent into the polymer particles and the less the effect on the particle size.

Addition of toluene to an aqueous PSA produces a swelling of polymer particles; therefore thickening occurs and no other thickeners are needed [105]. Solvent addition can lead to a better adhesion/anchorage of the adhesive to special surfaces. The addition of 10–20% solvent to latex under adequate stirring is possible [222]. Like pure solvents, rubber or adhesive solutions (20%) may be added as well [223] in order to get higher solids content. The reverse procedure—the addition of rubber latexes in adhesive solutions—has been practiced also [223]. Solvents may improve the coalescence and water resistance of the coated adhesive. Generally, care should be taken with the use of water-soluble solvents.

Special Additives

Special additives are necessary in order to impart some specific end-use properties to the dispersion. The most important of these products are the waterproofing agents and the wet-adhesion agents. Special labels for wine bottles (mostly champagne) need prolonged water stability (in cold water). Normal water-based formulations do not meet this requirement. Because of the tension (strain) induced in the web, hot-melt or solvent-based PSAs do not completely meet these requirements and there is a need to improve the water resistance of the water-based adhesive. On the other hand, labeling of wine-cellar stored, recycled, and freshly washed bottles, and deep-freeze labeling require wet adhesion (on water condensation or a water layer).

Special adhesives with a noninflammable character need special flameproof additives. Generally, these are the same as used for solvent-based PSAs, but their formulation is more difficult because of the low viscosity of the aqueous PSAs (e.g., Genomoll P, Stibium oxide, Al silicate).

The role of solvents as special solubilizing agents for the wetting agent should be noted. As an example, diluted Acronal V 205 does not wetout solvent-based silicone release liners at a concentration of 1.2% (wet/wet) Lumithen IRA; 5% ethyl alcohol improves the wetout.

Solubilizing Agents

Solubilizing agents ensure solubility of the water-based adhesive layer after storage at room temperature or elevated temperatures, and in cold or hot water. These are special additives used for wash-off labels, used mainly for recycled bottles. Theoretically, solubilizing agents are micromolecular (wetting agents) or macromolecular (protective colloids). From the class of protective colloids used, there are only a few that provide good water solubility without a loss of the tack. The most important ones are polyvinylethers (Lutonal M, Gantrez NM), polyvinylpyrolidone copolymers (Antarez) and polyacrylic acid copolymers. Their use always depends on the balance water solubility/tack. All these products are sensitive to bleeding, thus care should be taken when selecting these products (solubility/bleeding); for example, PVP-K-15 gives migration (formulated in Vinamul 4512) while PVP-K-90 does not cause migration.

Polyvinyl ethers may be used as aqueous solutions or dispersions. They were originally employed as tackifiers, especially for film coatings on soft PVC because of their good plasticizer resistance. They are water soluble (Lutonal M) and thus they impart good water solubility to the formula. Unfortunately, they lead to strike-through (bleedthrough) and have a low aging resistance.

Fungicides

Fungicides/bacteriocides can be added to the emulsion to prevent biodeterioration. Antimicrobical agents like p-hydroxyheptanphenol or p-hydroxycyclohexylphenone may be used [224].

Compounding of Water-Based PSAs

The compounding of water-based adhesives should be carried out in a tank at low stirrer speeds to minimize foam formation. The solids content of both latexes and resin emulsions should be closely monitored to prevent deviations from the intended polymer/resin ratio and to ensure consistency of adhesive properties. Generally, the dispersion additives (surfactants, defoamers) are fed in at the end of the formulation process. Because of the absorbtion phenomena of defoamers, they should be added stepwise. In some cases a separate preparation of the surfactant solution, antioxidant solution (dispersion), or filler slurry is needed. The main suppliers of elastomer/tackifier dispersions are providing technical information concerning the storage/use of their products [225,226]. The main characteristics of the equipment used are given below.

Tanks. Either vertical or horizontal designs are acceptable. Filament wound, glass fiber-reinforced polyester tanks are recommended for their relatively low cost, easy installation, and excellent resistance to chemical attack. Other construction materials also are acceptable. Stainless steel tanks are more expensive than other types but have proven durability and inertness to polymer emulsions and are easily cleaned. Aluminum tanks may be used for some emulsions, but are darkened by others. Some acrylic emulsions (mostly tackified ones) must be agitated continuously to ensure homogeneity. A mixer entering on the side (300–500 rpm) or a vertical shaft agitator (with a peripheral velocity below 200 ft/min) may be used. The space above the emulsion must be humidified to prevent the formation of skin over the surface of the liquid as well as on the walls of the tank. In some cases steam is injected for 3–5 min every 8 hr [227].

Pumps. Low shear pumps should be used. Diaphragm pumps exert the lowest shearing force. They are particularly suited to transfer highly loaded formulated emulsions containing resins or other additives. Pumps should always be operated under conditions of flooded suction. Generally, minimum agitation, turbulence, or pulsation is recommended.

Pipelines and filters. Pipelines of steel or rigid PVC may be used. All pipes should be easily accessible for cleaning. Outside lines should be insulated and heated. When polymer emulsions are withdrawn from storage tanks, skin and foreign matter must be removed before they pass through pumps, meters, and other equipment. A basket strainer having about 1 ft^2 of filtering area provides adequate capacity for this task. Removable inserts are convenient for cleaning purposes. Stainless steel screens (14–20 mesh) suffice for coarse straining. For still finer filtration, cartridge filters with disposable inserts are employed. Generally, filters with filter bags (like nylon) should be used. Working pressure for 200 mesh filters is about 20–30 psi (to start).

2.10 Technological Considerations

Formulating solvent-based or water-based PSAs needs no special technological equipment. Formulating water-based dispersions should be carried out taking into account the shear sensitivity of the dispersed system. When formulating hot-melt PSAs one should consider the temperature and aging sensitivity of these materials.

There are different technical possibilities to compound the tackifier with the raw PSA materials. The tackifying technology depends on the tackifier, PSA, and adhesive end-use properties of the laminate. Hot-melt PSAs require solid and/or liquid tackifiers, as do solvent-based PSAs; water-based PSAs are usually formulated in combination with water-based tackifiers. Each of the current tackifying technologies show some advantages and disadvantages.

Tackifying with Water-Based Tackifiers

From a technical point of view this remains the most simple technology. Demixing problems during feed-in can be avoided. No special knowledge for the converter is needed. The coating properties (wetout, mechanical stability) are the best; no flammable agents are used. Current dispersion coating machines may be used without special precautions.

On the other hand, aqueous resin-tackifier dispersions show some weaknesses also. These can affect the properties of the final coating in areas such as a lower water resistivity of the coating, a lower anchorage of the adhesive, a higher penetration or bleeding of the adhesive, lower shear and poorer die-cutting. The water-based resin dispersions are secondary dispersions made with surface-active agents (2–14%). This additional amount of surface-active agent in the final coating increases its solubility and hygroscopy. Concerning the influence of the emulsion character of the tackifier on the adhesive properties, there are divergent data in the literature. It is assumed that the adhesive properties of the emulsion-tackified PSAs are inferior to those of solvent-based resin-tackified PSAs. This hypothesis is illustrated by the limited amount of tackified water-based PSAs for film coating. In this case the loading of the tackifier remains limited (< 10%). Therefore solvent-based tackifiers are used more often. On the other hand, it was shown that with the exception of a slight change in rolling ball tack, there is little difference between the same adhesives coated with water and solvent carriers.

Tackifying with Solvent-Based Tackifiers

Solvent-based tackification is an old technology enjoying some renaissance because of the superior water resistivity and (perhaps) tack of the solvent-based resin-tackified coatings. The anchorage of the directly-coated solvent-based resin tackified coatings also appears superior. Unfortunately there are some technological disadvantages associated with solvent-based resin tackifying. First, the manufacture of the concentrated (70–80%) resin solutions causes difficulties for a converter without chemical knowledge. Furthermore the formulation (feed-in) technology of the resin solution in the aqueous dispersion requires special knowledge and equipment. Because of the low level of surface-active agents the mechanical stability of the solvent-based resin-tackified dispersion is low; the wetout of the dispersion is inferior and generally does not allow transfer coating on solventless silicones. Theoretically, no special converting machines (e.g., explosion proof) are required for solvent-based resin-tackified formulations; practically, some precautions are needed.

Except for the water resistivity, no improvement in the peel/tack/shear balance for solvent-based resin tackified water-based acrylic PSAs is observed.

Tackifying with Liquid Resins

Tackifying with liquid, unformulated resins is the most simple method to feed-in the tackifier (no pH or viscosity regulation necessary). Unfortunately the amount of the liquid tackifier which can be added, remains limited because of the strike-through of the tackifier. Loading more than 10% liquid tackifier causes migration or bleeding because of the pronounced decrease of the cohesion. Different liquid resins were tested (Resamin, Cymel, Polyethylenimin, Hercolyn, etc.); the best results were obtained with a level of 5–10% hydrocarbon-based resins, mainly with regard to the tack.

Tackifying with Molten Resins

There are only a few aqueous dispersions with sufficient heat stability (at the boiling temperature); some copolymers of chlorine-butadiene (e.g., DuPont CR-115) meet these requirements. Thus it is possible to add the solid high softening point tackifier as a (molten) liquid into these dispersions; these formulations display a higher shear strength. Unfortunately these special polymers are very expensive. On the other hand, special equipment is required for correctly dispersing the molten resin. Experience shows that more than 30–40% loading (by weight) of Neoprene CR-115 (Dow) is required for current acrylic PSAs in order to raise the cohesive strength.

Special Features

The age of the formulated (tackified) water-based acrylic PSAs influences the adhesion/cohesion balance. Polyethylene peel increases as a function of dwell time (Table 8.34). A dwell time of several hours (optimally 24 hr) is suggested.

2.11 Comparison Between Solvent-Based, Water-Based, and Hot-Melt PSAs

In Chapter 6 the properties and performance characteristics of rubber-resin-based and acrylic-based PSAs were compared. Next, a comparative study of the PSAs with a different physical state (i.e., solvent-based, water-based, and hot-melt PSAs) evaluates the formulating features of different raw materials, taking into account their adhesive and other properties.

Solvent-Based Formulations

Solvent-based acrylic PSAs have the following strengths [228]: a high water resistance, an improved clarity for polyester overlays, and an extremely high

Table 8.34 Peel Improvement After Longer Storage Time of the Formulated Adhesive

Adhesive characteristics		Storage time		
		1 hr	1 day	1 week
Coating weight (g/m^2)		24	28	25
Peel (N/25 mm)	Glass	25 PT	26 PT	20 PT
	PE	18	25 PT	20 PT
Shear (min)		45	46	31

Formulation: Acronal V 205/Snowtack CF 52; Face stock: paper; room-temperature shear measured on stainless steel.

temperature resistance. The drawbacks of solvent-based acrylic PSAs include the following:

- lack of compliance with environmental regulations;
- need for incineration or solvent recovery with accompanying capital expenditure;
- fire hazards/static electricity;
- higher insurance rates;
- future availability of solvents.

Solvent-based adhesives generally have been the preferred choice of converters because of their ease of application and their desirable balance of PSA performance properties. However, there remain significant problems associated with these solvent systems. These include high costs, environmental and toxicity concerns, poor shelf life, and very real fire hazards because of the solvents. Both hot-melt PSAs and water-based PSAs were developed to overcome these concerns [229].

Of the two alternatives, hot-melt PSAs (particularly rubber-based) have generally been plagued by relatively poor oxidation stability, marginal PSA performance (especially at elevated temperatures), and high initial application replacement costs. In contrast, emulsions are water-based systems which present no major environmental or toxicity concerns; they also pose no fire hazard associated with the volatiles present. Emulsion systems possess a further advantage in their ability to be coated on coating equipment previously used for solvent-based products. Because of the fire and explosion hazards, there is a need for the use of special ventilating and explosion-proof equipment when coating solvent-based PSAs. In the large scale use of solvent-based adhesives the removal of organic solvent during drying requires special solvent recovery equipment to avoid problems of air pollution. Solvent emission restrictions, increasing energy costs, the potential decreased availability of energy, rising solvent prices, and more stringend health and safety regulations make the availability of safer, less energy intensive alternatives urgent [46].

Solvent-based adhesives generally display good adhesion to a broader range of film substrates than water-based coatings [230]. Solvent-based rubber-resin PSAs have the following benefits: as fully commercial products they possess agressive adhesion, good cohesion, and water resistance. The limitations include poor aging, a low plasticizer resistance, low temperature adhesion and solvent emission. Solvent-based rubber PSAs coat easily and fast at lower coating weights. Due to the usually low solid content, high coating weights require a reduction of the coating speed. The viscosity of solvent systems cannot easily be changed which implies that the coating head needs to be adjusted to the actual viscosity. Explosion hazards remain real and a solvent recovery installation is required. Wetting is not a problem, fish eyes seldom occur, and the resulting coating is smooth. Die-cutting and slitting may be critical since the adhesive is often somewhat elastic, soft, and exhibits stringing.

Solvent-based acrylic PSAs wet the web easily, do not form fish eyes, and lead to smooth, transparent coatings. They allow fast machine speeds, but at higher coat weights (as in tapes) the speed should be reduced in order to prevent explosion; explosion-proof equipment and solvent recovery are required. The finished products are relatively hard, so slitting or die-cutting normally present no problems. Solvent acrylic PSAs have excellent aging characteristics and resistance to elevated temperatures and plasticizers; they also offer the best balance of adhesion and cohesion, as well as an excellent water resistance. Acrylic PSAs are harder than rubbers; this can be seen in the less agressive tack and slower build-up of peel strength. Lower adhesion to nonpolar polyolefins is caused by the polar chemistry of acrylic PSAs. Solutions of acrylic-based PSAs have the following advantages:

- no degradation in air, ozone, sunlight, heat;
- elevated temperature strength;
- low temperature flexibility retention and bonding;
- resistance to oil, fuel and chemicals;
- resistance to plasticizer migration (from PVC);
- good adhesion to metals, plastics, and other substrates.

The weaknesses include rather low quick stick values and softening by oxygenated solvents (glycols, alcohols).

In general, solvent coatings are free of migrating species of hydrophilic components which can cause humidity induced variations in release properties. Solvent coatings also dry faster because of the low boiling point or the

Table 8.35 Comparison of the Properties of Water-Based, Solvent-Based, and Hot-Melt PSAs

	PSA label	Adhesive technology		
Application field	Properties	Water-based	Solvent-based	Hot-melt
Permanent	Label converting	E	G	E
	Adhesion	E	G	E
	Application temperature range	E	E	G
	Formulation flexibility	G	E	F
Removable	Label converting	G	F	F
	Adhesion	G	G	F
	Application temperature range	G	E	F
	Formulation flexibility	F	E	F
Cold temperature	Label converting	E	F	F
	Adhesion	G	F	F
	Application temperature range	G	G	F
	Formulation flexibility	E	G	F
Clear film	Label converting	G	E	P
	Adhesion	G	E	F
	Application temperature range	G	G	F
	Formulation flexibility	F	G	F

E = Excellent; F = Fair; G = Good.

low heat of evaporation of the solvent used. Solvent rubber-based adhesives may cause problems when printing and die-cutting of labels due to legging [76]. Normally, solvent coatings contain 40% or less solids because of viscosity constraints. Equipment clean-up requires the use of solvents. Contrary to conventional wisdom, the drying costs are greater for solvent coatings. While the heat of evaporation of hydrocarbon solvents is much less than that of water, a greater amount of heated air must be passed through the oven in order to operate below the lower explosion limit.

Table 8.35 summarizes comparatively the most important properties of solvent-based, water-based, and hot-melt PSAs [120]. In Table 8.36 the advantages and disadvantages of water-based and solvent-based adhesives systems are listed. The comparison of the technical and economical characteristics of PSAs labels as a function of the physical state of the adhesive is summarized in Table 8.37 [231].

Table 8.36 General Comparison of Solvent Acrylic and Emulsion Acrylic PSA

	Acrylic PSA	
Properties	Solvent-based	Emulsion
PSA performance	Good to excellent	Good to excellent
Aging properties	Excellent	Excellent
UV stability	Excellent	Excellent
Thermal stability	Excellent	Excellent
Oxidation stability	Excellent	Excellent
Bleeding/migration	Good	Fair
Edge ooze	Good	Fair
Solvent resistance	Good	Good
Water resistance	Good	Fair to poor
Crosslinkability	Excellent	Fair
Clarity	Excellent	Good
Coating properties		
Foaming	Not applicable	High
Wetout	Excellent	Poor

Table 8.37 Technical and Economical Performance of PSA Labels as a Function of the Physical State of the Adhesive

		Physical state of the PSA		
Properties		Solution	Emulsion	Hot melt
Adhesive properties	Tack	Excellent	Good	Excellent
	Peel	Excellent	Excellent	Good
	Shear	Excellent	Good	Poor
End-use properties	Aging	Excellent	Good to fair	Poor
	Versatility	Excellent	Good	Poor
	Bleeding	Good	Fair	Fair to poor
Converting properties	Die cutting	Good	Fair	Poor
	Printing	Excellent	Fair	Good
Cost		Very high	Moderate	Low

Water-Based PSAs

The replacement of solvent-based adhesives with aqueous latexes is a continuing trend in the industry, for reasons of air quality, safety, and economics. However, the transition has not been simple and is by no means complete. Aqueous adhesives have found limited acceptance, being regarded as generally inferior to organic solution adhesives either in performance or coating rheology or both [91].

The deficiencies of common aqueous PSAs may be summarized as follows [228]: the inferior balance of tack, peel, and shear properties, the low water and humidity resistance, the bad plasticizer migration resistance, the limited heat resistance, and concerns about machinability and/or coatability. The poor coating rheology typical of conventional latex adhesives is manifested in several forms: first, there is a tendency in roll application to form ridges parallel to the coating direction. Ridges adversely effect adhesive performance by reducing the effective area of contact with the substrate, but more substantially, destroy the optical appearance (transparency), particularly when a clear film is used. Other disadvantages include the shear rate and water sensitivity. High relative humidities may cause liquid components to migrate through label papers, so that staining appears, a common characteristic of several water-soluble adhesives [138].

Special emulsion polymerization techniques allow properties which cannot be acquired through other systems (e.g., solvent-based or hot-melt). The core/shell microstructure of the polymer particles (core/shell structured particle blends) produces a latex which shows evidence of the presence of a copolymer in its films, although the particles do not contain any copolymer in the emulsified state [232].

As far as adhesive properties of the water-based PSAs are concerned, they do not offer any advantages versus solvent-based PSAs. The typical room temperature shear values from the solvent-borne adhesives are significantly higher than the typical values for water-borne adhesives [233]. The shear of solvent-based CSBR PSAs is a little higher than that of water-based ones [234]; theoretically, water-based systems can yield superior shear performance related to solvent-based systems [3].

The emulsion polymerization does not impose viscosity/molecular weight restrictions, because the polymers are formed as discrete particles suspended in the medium. The high molecular weight achieved during emulsion polymerization provides suitable performance characteristics for a wide variety of PSA applications without necessitating further modification [138]. An inherent advantage of the emulsion polymerization method is that crosslinking can be built into the polymer at the time of polymerization. A number of difunctional crosslinking monomers may be used such as 1,6–hexanediol diacrylate, eth-

Manufacture of Pressure-Sensitive Adhesives 441

ylene glycol diacrylate, butanediol dimethacrylate, N-methylol acrylamide, and triethylene glycol dimethacrylate. In practice, the performance of water-based PSAs is reduced by the composite structure.

Performance disadvantages of water-based acrylic PSAs technology include the need to improve the machine coatability, the lower adhesion to plastic substrates, the lack of clarity of the adhesive film, and the relatively low water resistance. The advantages of water-based coatings are evident; lower dry weight costs, easy clean-up, and environmental acceptance. Furthermore, they cause no pollution, require reduced energy for drying, cost less than solvent types, and hence they displace some rubber-based polymers. Water-based systems may be dried a high speeds when the web allows high temperatures (e.g., release paper); explosive hazards do not exist. However, wetting of the web is often a tricky problem; much work is carried out in the area of wetting agents to improve the wetout and minimize effects on performance. Most adhesives are fairly hard and cause no problems during slitting or die-cutting, unless they are highly tackified. An advantage of the water-based adhesives is their high solids content. Rubber solutions generally possess a minimum of 15–20% solids, water-based latexes have a solids content of at least 35%, while solution polymers generally have a narrower MWD than emulsion polymers.

The difficulties involved in the transition to water-based coating include [235]: wetting, coalescence, and technological problems, as well as concerns about the properties of the coating. The surface tension of water is 72 dyne/cm, while a typical solvent has a surface tension of 25–30 dyne/cm. With many webs, having a surface energy of 30–40 dyne/cm wetting becomes problematic. Film forming from aqueous media requires coalescence. A switch from solvent-based products to water-borne products often results in longer ovens and the corrosion of piping, pumps, reservoirs, etc. Furthermore, the dried films have an inherent water sensitivity; also, the high surface tension of water results in higher power consumption, an increased machine wear, and waste streams which cannot be incinerated or dumped into a sewage system.

Therefore the following general requirements towards aqueous dispersions [236] can be stated:

- Dispersing high molecular weight polymers can achieve a higher solids content than solution adhesives, thus maximizing the cohesive strength.
- Dispersions should have the viscosity and general flow properties necessary to allow efficient film coating with minimal striations on conventional coating equipment.
- There should be adequate wetting and no crosslinking of the film between the coating head and the drying tunnel.

- There should be low viscosity and maximum flow coupled with the highest possible solids content for optimum drying efficiency.
- High mechanical stability of the adhesive dispersion is an absolute necessity for efficient and trouble-free machine coating.

The concerns include the tendency to foam and the presence of migrating components. Care must be taken in the selection of defoamers which can exert a negative effect on adhesive and/or release properties.

Hot-Melt PSAs

Hot-melt PSAs offer all (or most) advantages of solution PSAs (i.e., no vapor hazard nor fire hazard); they require a low processing energy and no special storage conditions, and allow unlimited coating possibilities (speed, thickness). However, they also possess some weaknesses, such as the need for specially designed equipment, an elevated application temperature with higher processing costs, and process sensitivity; they remain inferior in heat-sensitive coating applications and display the lowest heat resistance.

REFERENCES

1. E. Djagarowa, W. Rainow and W. L. Dimitrow, *Plaste u. Kaut.*, (2) 100 (1970); in *Adhäsion*, (12) 363 (1970).
2. *Coating*, (6) 175 (1972).
3. J. Kendall, F. Foley and S. G. Chu, *Adhesives Age*, (9) 26 (1986).
4. A. Zawilinski, *Adhesives Age*, (9) 29 (1984).
5. *Coating*, (7) 184 (1984).
6. R. Hinterwaldner, *Adhäsion*, (3) 14 (1985).
7. U. Fiederling, *apr*, (16) 463 (1986).
8. J. P. Keally and R. E. Zenk (Minnesota Mining and Manuf. Co., USA); Canadian Patent, 1224.678/19.07.82 (USP. 399350).
9. J. A. Schlademan, *Coating*, (1) 12 (1986).
10. J. C. Fitch and A. M. Snow, *Adhesives Age*, (10) 24 (1977).
11. *Die Herstellung von Haftklebstoffen*, T1.2.2; 15d, BASF, Ludwigshafen, Nov. 1979.
12. D. Satas, *Handbook of Pressure Sensitive Technology*, Van Nostrand-Rheinhold Co., New York, 1982.
13. A. Midgley, *Adhesives Age*, (9) 17 (1986).
14. EP 0251672.
15. C. W. Koch, *J. Polymer Sci.*, (C3) 139 (1963).
16. P. J. Counsell and R.S. Whitehouse, in *Development in Adhesives* (W.C. Wake, Ed.), Vol. 1, Ed. Applied Science Publishers, London 1977, p. 99.
17. A. Bull, *Shell Bulletin*, TB, RBX/73/8/6.
18. R. E. Downey, *Adhesives Age*, (3) 35 (1974).
19. *Coating*, (1) 16 (1984).

20. A. W. Bamborough, 16th Munich Adhesive and Finishing Seminar, 1991, p. 96.
21. *Coating*, (1) 13 (1984).
22. *Coating*, (18) 240 (1972).
23. G. Ruckel, *Coating*, (1) 18 (1984).
24. P. A. Mancinelli, *New Development in Acrylic HMPSA Technology*, p. 165.
25. R. Mudge, Ethylene-Vinylacetate based, waterbased PSA, in *TECH 12, Advances in Pressure Sensitive Tape Technology, Technical Seminar Proceedings*, Itasca, IL, May 1989.
26. E. G. Ewing and J. C. Erickson, *Tappi J.*, (16) 158 (1988).
27. *Adhäsion*, (4) 24 (1985).
28. *Adhäsion*, (9) 352 (1966).
29. J. Class and S. G. Chu, *Org. Coat. Appl. Polymer Sci., Proceed*, **48**:126 (1989).
30. R. Sattelmeyer, *Adhäsion*, (10) 278 (1976).
31. C. A. Dahlquist, in *Handbook of Pressure Sensitive Technology* (D. Satas, Ed.), Van Nostrand Rheinhold Co., New York, 1982, p. 82.
32. *Coating*, (7) 186 (1984).
33. J. Villa, *Adhäsion*, (10) 284 (1977).
34. L. Heymans, European Tape and Label Conference, Exxon, Brussel, April 1989, p. 113.
35. G. W. Drechsler, *Coating*, (3) 98 (1987).
36. F. H. Wetzel, *Rubber Age*, (82) 291 (1957).
37. C. W. Koch and A. N. Abbott, *Rubber Age*, (82) 471 (1957).
38. H. Hultsch, *Farbe u. Lack*, (77) 11 (1971).
39. E. De Walt, *Adhesives Age*, (3) 38 (1970).
40. *Adhesives Age*, (12) 35 (1987).
41. G. W. Drechsler, *Coating*, (2) 52 (1987).
42. *Adhäsion*, (5) 162 (1971).
43. R. Houwink and G. Salomon, in *Adhesion and Adhesives*, Vol. 2, Elsevier Publ. Co., Amsterdam, 1967. Chapter 17.
44. M. Sheriff, R. W. Knibbs and P. G. Langley, *J. Appl. Polymer Sci.*, **17**:3423 (1973).
45. G. Kraus and K. W. Rollman, *J. Appl. Polymer Sci.*, (23) 3311 (1977).
46. T. G. Wood, *Adhesives Age*, (7) 19 (1987).
47. *Coating*, (7) 184 (1984).
48. S. W. Medina and F. W. Distefano, *Adhesives Age*, (2) 18 (1989).
49. EP 0244997
50. P. Green, *Labels and Labeling*, (11/12) 38 (1985).
51. British Petrol, Hyvis, Technical bulletin (1985).
52. P. Dunckley, *Adhäsion*, (11) 19 (1989).
53. *Adhäsion*, (12) 390 (1969).
54. *Adhäsion*, (12) 430 (1972).
55. T. G. Wood, *Adhesives Age*, (7) 19 (1987).
56. M. F. Tse and K. O. McElrath, European Tape and Label Conference, Exxon, Brussel, April 1989, p. 91.
57. *Adhesives Age*, (11) 40 (1988).

58. *apr*, (16) 462 (1986).
59. T. Yoshida, European Tape and Label Conference, Exxon, Brussel, April 1989, p. 69.
60. O. Cada and P. Peremsky, *Adhäsion*, (5) 19 (1986).
61. R. Schuman and B. Josephs (Dennison Manuf. Co., USA), PCT/US86/02304/ 25.08.86.
62. I. J. Davis (National Starch, Chem. Co., USA), U.S. Pat., 4.728.572/01.03.88.
63. J. M. Avenco, *Resin Review*, (3) (1972).
64. J. Lin, W. Wen and B. Sun, *Adhesives Age*, (12) 22 (1985).
65. A. L. Bull, *Shell Elastomers, Thermoplastic Rubbers*, Technical Manual, TR 8.12, p. 13.
66. R. W. Rance, E. Lazarus and F. Wilkes, *Labels and Labeling*, (2) 16 (1988).
67. J. F. Kwiatek, *Adhesives Age*, (11) 28 (1988).
68. R. Jordan, *Coating*, (10) 278 (1982).
69. R. Dörpelkus, *apr*, (16) 456 (1986).
70. W. Drüschke, *Adhesion and Tack of PSA*, AFERA Meeting, Edinburgh, October 1986.
71. W. Blume, *Adhäsion*, (7/8) 20 (1983).
72. L. C. Broggs, *Adhesives Age*, (5) 19 (1983).
73. M. E. Ahner, H. L. Evans, S. G. Hentges and M. R. Gerstenberger, *Coating*, (10) 330 (1986).
74. *Coating*, (10) 304 (1969).
75. S. Pila, *Defazet*, (2) 54 (1974).
76. H. Müller, J. Türk and W. Druschke BASF, Ludwigshafen, EP 0.118.726/ 11.02.83.
77. *Movilith DM 56*, Technical Data Sheet, Hoechst.
78. *Adhesives Age*, (10) 36 (1977).
79. *Coating*, (1) 14 (1987).
80. *Coating*, (2) 46 (1970).
81. *Movilith DM 104*, Kunstharze Hoechst, Hoe, KGM, 3070d.
82. B. Henzel, *Plast-Europe*, (3) 86 (1982).
83. *Coating*, (3) 79 (1974).
84. Firestone, *Technical Service Report*, 6110, August 1986.
85. L. Krutzel, *Adhesives Age*, (9) 21 (1987).
86. *Converting and Packaging*, (3) 24 (1986).
87. C. Massa, *Coating*, (7) 239 (1989).
88. R. Wilken, *apr*, (37/38) 1042 (1988).
89. C. Zang, U.S. Pat., 3.532.652.
90. H. Salmen and M. Htun, *apr*, (35) 28 (1986).
91. C. P. Iovine, S. J. Jer and F. Paulp (National Starch Chem. Co.), EP 0.212.358/ 04.03.87, p. 3.
92. C. T. Albright, D. S. Culp, (Avery Internat. Co., USA) EP 0099087B1/ 23.12.87.
93. G. R. Hamed and C. H. Hsieh, *J. Polymer Physics*, **21**:1415 (1983).
94. R. P. Mudge (National Starch Chem. Co., USA), EP 0.225.541/11.12.85.

95. J. Lechat, *The pressure sensitive labelstock market in Western Europe*, Finat World Congress, Monaco, 1989.
96. M. Mazurek (Minnesota Mining and Manuf. Co., USA), EP 4.693.935/15.09.87.
97. J. Lin, W. Wen and B. Sun, *Adhäsion*, (12) 21 (1985).
98. L. Jacob, 8th Munich Adhesive and Finishing Seminar, 1983, p. 42.
99. *Coating*, (10) 366 (1988).
100. *Vinnapas, Eigenschaften und Anwendung*, 7.1. Teil, Anwendung, Wacker, München, 1976.
101. K. Goller, *Adhäsion*, (4) 101 (1974).
102. T. J. Bonk and S. J. Thomas (Minnesota Mining Manuf. Co., USA), EP 120.708.31/03.10.84.
103. *Adhesives Age*, (5) 54 (1987).
104. DE-AS 2407484/EP 0118726.
105. *Die Herstellung von Haftklebstoffen*, Tl.-2, 2-14d, Dec. 1979, BASF, Ludwigshafen.
106. J. L. Wacker and P. B. Foreman, *CAS, Colloids*, **1**:5 (1988).
107. Silikone, *Sitren*, 532276, Technical Information, Th. Goldschmidt A.G., VK-Silikone.
108. *Adhäsion*, (7) 242 (1972).
109. *Coating*, (9) 247 (1985).
110. D. Satas, *Adhesives Age*, (8) 28 (1988).
111. J. R. Creasney, D. B. Russel and M. P. Wagner, *Am. Chem. Soc., Rep.*, 35, 1968.
112. J. Hansmann, *Adhäsion*, (10) 360 (1970).
113. B. Mayer, *Coating*, (4) 111 (1969).
114. R. Hinterwaldner, *Adhäsion*, (7) 244 (1972).
115. F. Hartmann, *Coating*, (1) 7 (1969).
116. *Coating*, (11) 338 (1973).
117. J. Young, *Tappi J.*, (5) 78 (1988).
118. G. Grove, *Adhesives Age*, (6) 39 (1971).
119. S. Tadashi, M. Noboyuki, I. Yoshimide and T. Toyokichi, *CAS Crosslinking Agents*, **14**:6 (1988), 108.222762n.
120. T. H. Haddock (Johnson & Johnson, USA), EP 0.130.080 B1/02.01.85.
121. P. J. Moles, *Polym. Paint. Colour J.*, **178**:4209 (1988).
122. M. Hiroyasu, T. Kitazaki, Y. Truma, M. Tetsuaki and K. Yunichi, DE 354486861/18.12.85.
123. US Pat., 3.691.140/18.09.1972.
124. Eastman Kodak, US Pat., 359.943; in *Adhäsion*, (5) 328 (1967).
125. M. Hasegawa, US Pat., 4.460.634/17.07.84; in *Adhesives Age*, (3) 22 (87).
126. P. Tkaczuk, *Adhesives Age*, (8) 19 (1988).
127. EP 213860.
128. F. W. Brown, W. St. Paul and L. E. Winslow, US Pat., 4.629.663; in *Adhesives Age*, (5) 15 (1987).
129. Kimberley Clark Co., US Pat., 799.429 in *Coating*, (1) 9 (1970).
130. *Adhäsion*, (9) 349 (1965).

131. M. Ulrich, US Pat., 24.906, in D. Satas, *Handbook of Pressure Sensitive Technology*, Van Nostrand-Rheinhold Co., New York, 1982, p. 203.
132. *Adhesives Age*, (10) 16 (1977).
133. EP 010046
134. J. Harrison, J. F. Johnson and W. Rossyates, *Polym. Eng. and Sci.*, **(14)** 865 (1982).
135. S. Mitton and C. Mak, *Adhesives Age*, (1) 15 (1983).
136. *Offset-Technik*, (8) 11 (1986).
137. *Coating*, (8) 240 (1972).
138. *Adhesives Age*, (11) 28 (1983).
139. W. Wittke, *Coating*, (9) 334 (1987).
140. W. Wittke, *Coating*, (7) 249 (1987).
141. W. Wittke, *Coating*, (8) 206 (1987).
142. US Pat., 3351.572.
143. J. Paris, *Papier u. Kunststoffverarb.*, (9) 53 (1988).
144. R. M. Friedrich, European Tape and Label Conference, Exxon, Brussel, April 1989, p. 181.
145. H. J. Fricke, *Allgem. Papierrundsch.*, (24/25) 720 (1983).
146. F. King, *Coating*, (2) 59 (1970).
147. R. Jordan, *Coating*, (2) 37 (1986).
148. D. Krüger, *Kaut., Gummi, Kunstst.*, **(6)** 549 (1988).
149. Shell Elastomers, Technical handbuch, TR 5.1.1. (G) p. 6.
150. Shell Elastomers, Cariflex TR, Merkblatt TR 5.2.1. (G) p. 2.
151. E. Diani, A. Riva, A. Iacono and E. Agostinis, 16th Munich Adhesive and Finishing Seminar, 1994, p. 77.
152. R. A. Fox, *Adhesives Age*, (10) 35 (1977).
153. A. Patel and R. Thomas, *Tappi J.*, (6) 166 (1988).
154. H. Hadert, *Coating*, (7) 203 (1970).
155. D. De Jaeger, 11th Munich Adhesive and Finishing Seminar, 1986, p. 87.
156. *Coating*, (12) 344 (1984).
157. *Adhesives Age*, (8) 35 (1983).
158. Celanese Resins Systems, Technical Information, LHM, 991.
159. *Coating*, (3) 68 (1985).
160. C. N. Clubs and B. W. Foster, *Adhesives Age*, (11) 18 (1988).
161. *European Adhesives and Sealants*, (9) 3 (1987).
162. F. C. Jagisch, Recent Developments in Styrene Block Copolymers for Tape and Label PSAs, European Tape and Label Conference, Exxon, Brussel, 1993.
163. F. Jagisch and L. Jacob, *Advances in Styrene Block Copolymer Technology*, PSTC Technical Seminar, Tech. XIV, May 2, 1991.
164. D. De Jager and J. B. Bortwyck, *Thermoplastischer Kautschuk*, Technisches Handbuch, TR 8.11 (G), Shell Res. B.V.
165. *Papier u. Kunststoffverarb.*, (9) 48 (1990).
166. R. Hinterwaldner, *Coating*, (7) 176 (1980).
167. H. Röltgen, *Coating*, (11) 400 (1986).
168. R. Hinterwaldner, *Coating*, (7) 250 (1991).

169. G. Auchter, J. Barwich, G. Remmer and H. Jäger, *Adhäsion*, (1/2) 14 (1993).
170. S. J. Piaseczinsky, *Adhesives & Sealants Council*, 4/89, Sonderdruck.
171. R. Jordan, *Coating*, (2) 27 (1984).
172. F. V. Distefano, S. W. Medina and B. R. Visayendran, *Sampe J.*, (7/8) 27 (1988).
173. H. Yang, L. Jacob and L. Heymans, 15th Munich Adhesive and Finishing Seminar, 1990, p. 92.
174. R. Hinterwaldner, *Coating*, (5) 198 (1991).
175. *Technische Informationsblätter, Kautschuk Latices*, Nr 11/2, Aug. 1977, Bayer, Leverkusen.
176. M. Takuhiko, *Setchaku*, (9) 389 (1987); in *CAS* **7**:5 (1988).
177. J. A. Fries, *New Developments in PSA*, p. 27.
178. *Coating*, (4) 11 (1974).
179. Celanese Chemical Company, *Acrylate Esters*, 4M/5-71, Technical Bulletin.
180. J. Johnston, *Adhesives Age*, (12) 24 (1983).
181. H. Toyama, T. Ito and H. Nakatsuma, *J. Appl. Polymer Sci.*, (17) 3495 (1973).
182. W. R. Dougherty, 15th Munich Adhesive and Finishing Seminar, 1990, p. 70.
183. H. Saxen, *Etiketten*, (4) 9 (1994).
184. *Coating*, (5) 121 (1971).
185. *Converting and Packaging*, (3) 26 (1986).
186. *Adhäsion*, (11) 481 (1967).
187. C. L. Zao, Y. Holl, T. Pith and M. Cambia, *Colloid Polym. Sci.*, (9) 823 (1987).
188. *Product and Properties Index*, Polysar, Arnhem, 02/1985.
189. *Surfynol Technical Bulletin*, Air Products and Chemicals Inc. (1991).
190. M. H. Tanashi, K. Yasuaki, M. Tetsuaki and K. Yunichi (Nichiban Co, Japan), OS/DE 3.544.868 A1/15.05.85.
191. A. C. Makati, *Tappi J.*, (6) 147 (1988).
192. J. R. Snuparek, A. Bidman, J. Hanus and B. Hajkova, *J. Appl. Polymer Sci.*, **28**:1421 (1983).
193. A. Zosel, *Colloid Polymer Sci.*, **263**:541 (1985).
194. *Adhesives Age*, (9) 15 (1976).
195. L. A. Rutter, *Adhäsion*, (6) 180 (1971).
196. J. Hausmann, *Adhäsion*, (4) 21 (1985).
197. E. Böhmer, *Norsk Skogindustrie*, (8) 258 (1968).
198. *Adhesives Age*, (12) 36 (1986).
199. *Coating*, (4) 118 (1969).
200. R. Pfister, *Coating*, (6) 171 (1969).
201. *Coating*, (6) 177 (1969).
202. O. Quadrat, *J. Appl. Polymer Sci.*, **35**:1 (1988).
203. *Coating*, (7) 120 (1983).
204. *Coating*, (7) 20 (1984).
205. G. Meinel, *Papier u. Kunststoffverarb.*, (10) 26 (1985).
206. *Coating*, (11) 4 (1988).
207. L. Becher, D. Lorenz, H. L. Lowd, A. S. Wood and N. D. Wyman, in *Handbook of Water Soluble Gums and Resins* (R. L. Davidson, Ed.), McGraw-Hill, New York, 1980, Chapter 21.

208. *Coating*, (5) 122 (1974).
209. *Viscalex, Polyelectrolyte Thickening Agents*, TPD/60/10, Allied Colloids.
210. I. B. Portnaja, N. O. Kazacinskaja, O. A. Stenina and R. M. Panic, *Kautschuk Rezina*, (6) 25 (1986).
211. *Adhäsion*, (9) 266 (1965).
212. *Coating*, (2) 120 (1970).
213. P. H. Gamlen and R. M. Lane (ICI), *Research and Technology*, p. 9.
214. W. Heilen, H. F. Fink, O. Klocker and G. Koener, *Coating*, (9) 338 (1987).
215. P. Stenius, J. Kuortti and B. Kronberg, *Tappi J.*, (5) 56 (1984).
216. R. Hölinger, *Chemie-Anlagen & Verfahren*, (11) 19 (1983).
217. *Coating*, (2) 39 (1970).
218. *Mowilith*, Farbwerke Hoechst A.G., Frankfurt, 5 Aufl., 1970, p. 131.
219. *Adhäsion*, (10) 307 (1968).
220. *Adhesives Age*, (10) 24 (1977).
221. *Coating*, (6) 188 (1969).
222. H. Hadery, *Coating*, (2) 17 (1970).
223. *Coating*, (12) 390 (1969).
224. *Coating*, (5) 122 (1974).
225. *Kunstharze Hoechst, Verpackung u. Lagerung von Kunststoff Dispersionen und Kunstharzen*, Hoechst, Aug. 1982, GSA 0010.
226. *Kunstharze Hoechst, Verpackung u. Lagerung von Kunststoff Dispersionen und Kunstharzen*, TR 1, Hoechst, Aug. 1982, GSA 008.
227. *Emulsion Polymerization of Acrylic Monomers*, Rohm and Haas, CM-104A/cf.
228. R. Lombardi, *Paper, Film and Foil Conv.*, (3) 74 (1988).
229. C. L. Mao and W. Medina, presented at the PSA Workshop Univ. of Southern California, LA, February 1986.
230. R. D. Bafford and G. R. Faircloth, *Adhesives Age*, (12) 24 (1987).
231. R. Hinterwaldner, *Coating*, (5) 172 (1987).
232. K. Pathamanthan, J. J. Cavaille and G. A. Yohari, *Polymer*, **29**: 311 (1988).
233. D. G. Pierson and J. J. Wilczynski, *Adhesives Age*, (8) 52 (1980).
234. *Adhesives Age*, (9) 22 (1986).
235. D. J. Zimmer and J. Smurphy, *Tappi J.*, (12) 123 (1989).
236. *Adhesives Age*, (10) 26 (1977).

9
Manufacture of Pressure-Sensitive Labels

The pressure-sensitive label is almost always manufactured and used as a two component laminate (i.e., pressure-sensitive adhesives are coated onto face stock materials, the adhesive-coated face stock material will be covered with a release liner); therefore manufacturing pressure-sensitive labels includes a coating and laminating operation. In some special cases (electron beam or UV polymerization) PSAs are actually manufactured during the coating (i.e., the polymerization of the special raw materials, monomers or oligomers, is carried out after or during the coating).

A laminate is a composite structure that is obtained by assembling two or more materials with the help of a bonding agent. Different lamination processes include wet lamination or dry lamination. Lamination with PSAs is a dry (adhesive) lamination process where the adhesive-coated web is dried before meeting the other laminate component (i.e., the release liner). Labeling also may be considered as laminating. In this case, "cold" PSA lamination, as its name implies, involves the use of PSA-coated paper (or film) with a treated removable paper or film liner. The system includes a laminator which unwinds the film, presses it into the substrate and rewinds the liner. In order to obtain a stable laminate there is a need for a bonding agent (i.e., to laminate it is necessary to coat the web with adhesive). In the present context, laminating can be taken to mean the production of a layered material from two or more ready made webs, using a bonding medium. Laminating is always the second step of the manufacturing process. In the first stage the bonding agents of the laminate (i.e., the adhesive) are coated on one of the laminate components (i.e., on the face stock or on the release liner).

1 COATING TECHNOLOGY

Pressure-sensitive adhesives are coated either directly or by transfer. When direct coating, PSAs are coated directly on the final web (face stock) material. When transfer coating, PSAs are deposited on a web, dried and/or crosslinked, and finally transferred onto the face stock material. This method is necessary in order to avoid destruction of or damage to the face stock material during drying or to avoid the formation of air bubbles. The choice between direct or transfer coating influences the penetration and anchorage of the adhesive as well as the release force [1]. Figure 9.1 displays a PSA-coating machine which is able to coat either directly or by the transfer method (i.e., via a presiliconized paper web) [2].

In the case of transfer coating it is very important to adjust the adhesive characteristics with respect to the silicone so that a perfect wetout will be obtained. The use of a precoated (siliconized) web in transfer coating of PSAs led to the development of tandem systems where in-line coating with PSAs is carried out. Figure 9.2 represents a classical tandem machine for the production of label stock.

The siliconized substrate is produced in the first part; subsequently, the adhesive coating takes place. In the second laminating section, the adhesive layer is finally transferred onto the paper or film carrier material. In some cases, especially when removable adhesive-coated substrates are produced, first the application of a primer onto the face stock will be necessary (see Chapters 2 and 6); Figure 9.3 shows such a combination.

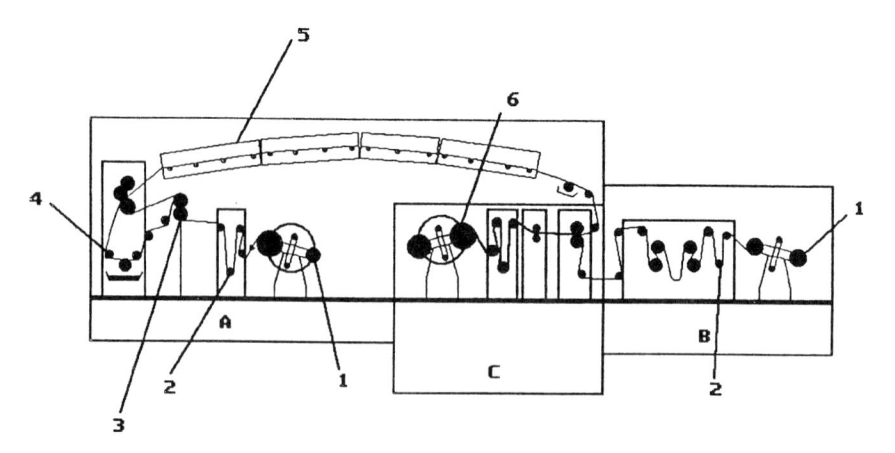

Figure 9.1 Coating machine for direct and transfer coating. A) Coating station; B) Secondary web; C) Laminating 1) unwinder; 2) automatic tension control of the web; 3) press roll; 4) coating device; 5) drying tunnel; 6) rewinder.

Manufacture of Pressure-Sensitive Labels 451

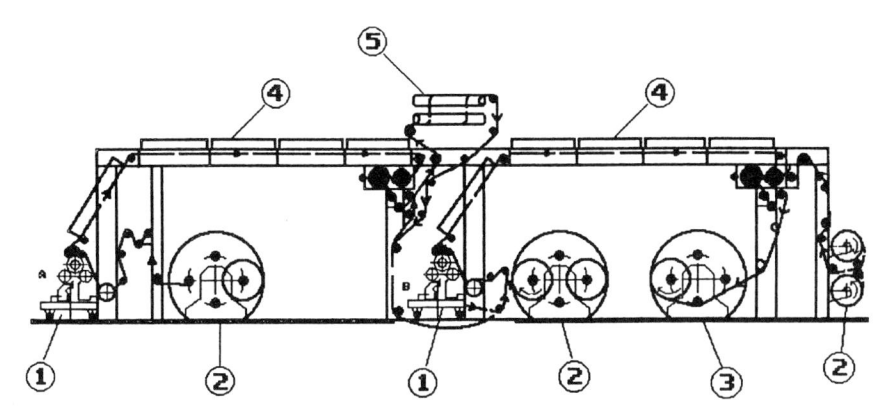

Figure 9.2 Classical tandem machine for the production of label material. Siliconizing and PSA coating. 1) coating device; 2) unwind; 3) rewind; 4) drying channel; 5) turnbars.

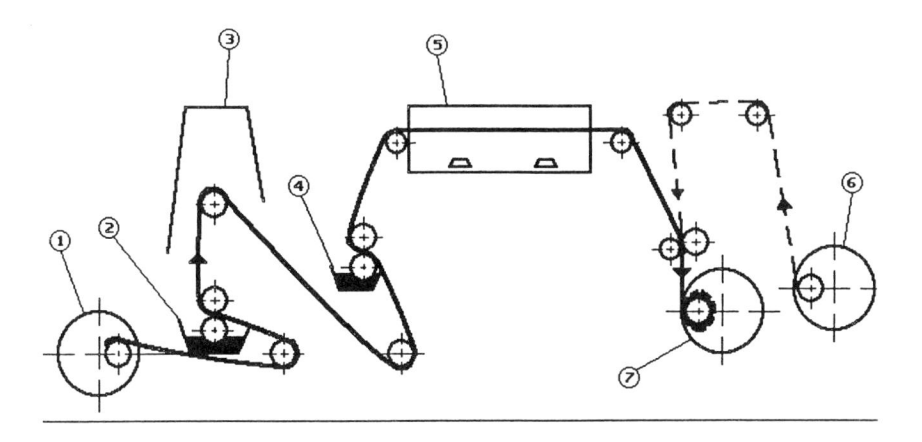

Figure 9.3 Coating machine for primer and adhesive coating. 1) automatic unwinder; 2) coating device for primer; 3) drying tunnel; 4) coating device for PSA; 5) drying tunnel; 6) unwinder; 7) rewinder.

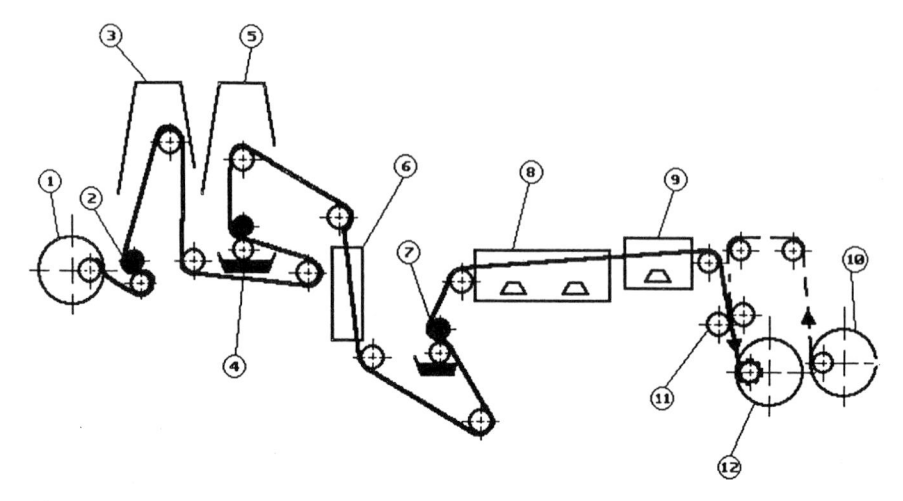

Figure 9.4 In-line coating/laminate plant. 1) automatic unwinder; 2) printing unit; 3) drying tunnel; 4) primer coating; 5) drying tunnel; 6) cooling and humidifying; 7) PSA coating device; 8) drying tunnel; 9) cooling, humidifying; 10) unwinder; 11) laminating unit; 12) rewinder.

In some cases the silicone-coating machine has to be equipped with an additional flexoprinter for the printing of a logo on the backside of the web [3]. Thus the coating/laminating machine is (or may be) a complex laminating/printing device carrying out consecutive coating processes involving an adhesive, an adhesive primer, and printing ink (Figure 9.4). In an ideal case, this process occurs on an in-line machine. This machine integrates all necessary converting processes to produce label stock, as for example:

- backside printing;
- silicone coating with curing;
- selfadhesive coating with drying tunnel;
- rehumidification of the siliconized and adhesive-coated label material;
- primer coating with dryers;
- laminating;
- cutting system.

A high performance coating and laminating machine is controlled by a process control system [3]. The individual process parameters include drying temperatures, air velocities in the dryers, chilling roll temperatures, web tensions, roll pressures for press rolls, applicator roll speeds, and steam feed of jetsteamers. They are controlled by independently working analog regulators which also can be adjusted manually [3,4].

According to this scheme the initial coating step is followed by the laminating step. The coating technology may be an adhesive or nonadhesive one. The adhesive formulations may be applied to the liner by any one of a variety of conventional coating techniques (e.g., roll coating, spray coating, curtain coating, etc.) by employing suitable conventional coating devices designed for such coating methods.

Next, an overview of different adhesive technologies for PSAs (i.e., a comparison of performance, coating properties, and applications) is presented. Today a wide range of adhesive technologies is used in the production of PSAs. The first PSAs were based on rubber-resin solutions, but newer developments have gained a significant part of the market and have paved the way to new applications and end-uses, like the development of acrylic PSAs in solution and later in water. In addition acrylic PSAs provide excellent aging and weathering resistance; they are used as a pure polymer (mostly in solvent-based acrylic PSAs) or with addition of tackifiers (many water-based acrylic PSAs). Environmental problems, and energy considerations are the main reasons for the worldwide boom of the use of water-based systems for very different applications. Both adhesive systems (solvent-based and water-based) require the evaporation of the carrier liquid (solvent or water) in a drying oven. Hot-melt PSAs also have been developed for pressure-sensitive uses. They are based on different thermoplastic polymers with addition of tackifiers and other ingredients, and provide ease of coating because of the lack of a carrier liquid (i.e., no additional drying is needed). On the other hand, all these adhesive systems depend on the adhesive state, require different coating viscosities and show a dependence of the viscosity on temperature and shear (i.e., they require different coating geometries/machines). The "TA Luft" regulation in Germany and similar legislation in other countries forces producers to review emission levels of solvents and to take appropriate measures to limit emission levels. Therefore, the drying systems for solvent-based or water-based systems differ; on the other hand, similarities exist between coating technologies for adhesive and nonadhesive systems.

2 COATING MACHINES

A short review of the coating machinery is presented next. Technical developments in this area continue at a rapid pace. As early as 1971, 2.40 m wide coating machines were being built [5]; in 1988, the running speed of the new machines was already the double of the older machines. On the other hand, as a function of the nature of the adhesive medium and of the desired coating weight (which has to be reduced) quite different coating systems (devices) have to be used. For example, a primer with a coating weight of 0.5 g/m^2 may be deposited via gravure printing [7,41]. Dispersions with 50% solids, may be coated by the use of reverse gravure coating, down to 4–6 g/m^2. Migration may produce coating defects; they may be avoided using high coating speeds and thickeners [7,42].

Different coating systems were designed for different kinds of PSAs. For solvent-based PSAs reverse roll coating, for water-based adhesives an air knife system was proposed [9]. Tape coating requires a one- or two-station machine, whereas label stock coating lines may need up to four coating stations. After coating the material is wound into jumbo rolls for subsequent finishing on an off-line slitter-rewinder. Label stock is manufactured for further converting and/or printing operations. A release paper protects the adhesive and separates the layers or sheets. All label stock and double-sided pressure-sensitive tapes use a release sheet (liner). Label stock manufactured on an one-station coater might require two or three passes to complete the manufacturing process. The first pass applies the primer, if required; in the second pass the adhesive is applied with subsequent lamination to the release liner. A separate pass is needed to apply the silicone coating on the release paper. A two- or three-station line could perform all of these steps in a single operation.

The number of the components of a coating machine may be reduced and, depending on the nature of the liquid laminate component to be coated, the differences arise from the application of an adhesive, release, or primer coating.

2.1 Adhesive Coating Machines

The main component for the three different technologies are described next. A solvent-based PSAs coating machine includes the following main parts:

- automatic unwinder;
- automatic web guide;
- automatic dual infeed tension nip for prestretch and stabilization of the web;
- primer coating unit;
- release coating unit;
- automatic tension nip;
- self-adhesive coating unit;
- drying tunnel;
- chill rolls;
- rewinder.

A hot-melt PSA machine has the following sections:

- unwinder;
- primer unit;
- tension nip;

- slot die or other coating head geometry;
- chill roll;
- rewinder.

A coating machine for water-based adhesives has the following parts [10]:

- unwinding cylinders;
- coating device (smoothing bar);
- drying oven (IR heaters);
- winding cylinders;
- humidifier;
- laminating station;
- unwinding cylinders (secondary unwinder);
- rewinding.

A common web width is 1500 mm; the coating weight can vary between 10 and 250 g/m^2.

There are common parts used for coating/laminating machines for different kinds of PSAs. Moreover, some parts like winders [11–15] for the laminating station are well known from the general practice of wet or dry adhesive lamination. On the other hand, remoisturization of the paper web also is necessary in paper laminating or printing [16]. Web control techniques for PSA coating are well known from the practice of laminating or extrusion [17–22]. The most important part of the coating machine remains the coating head. Its choice is influenced by the rheology of the adhesive (and vice versa), so the formulation must be compatible with the coating device.

2.2 Coating Devices/Coating Systems

From a theoretical point of view, a coating device system is based on a knife (or blade), a rotating cylinder, or a slot die. Possible geometries of the coating head are shown in Figure 9.5. The following coating systems are possible [23]:

- direct transfer—metering by knife-over-roll;
- direct transfer—metering by slow, rotating roll-knife (roll blade coater);
- kiss-coating—premetering by knife or reverse roll;
- multiroll coating (in the direct sense)—premetering in direct or reverse sense;
- reverse roll coating—premetering by knife-over-roll or reverse roll.

A slot-die coating for dispersions also remains possible. This is used in the case of air sensitive adhesives, or when strip coating (ungummed areas) is required [10]. Reverse roll coating allows varying coating weights (i.e., an easy adjustment of the coating weight). Unfortunately, there is a lower limit of the coating weight for this procedure [24]. An air knife also can be used;

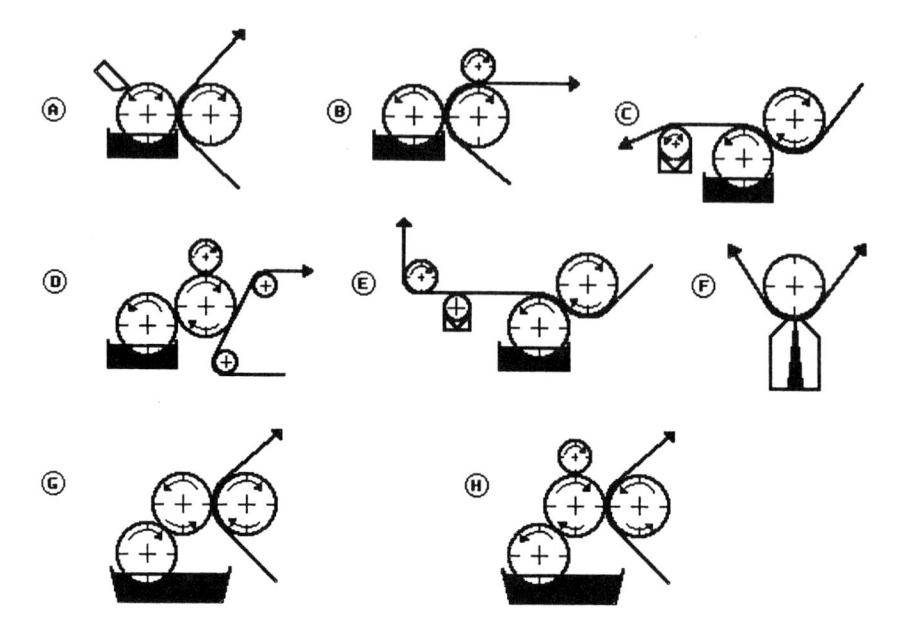

Figure 9.5 Coating device. Possible constructions A) direct transfer, knife over roll; B) direct transfer, knife over web on the roll; C) direct transfer, knife over the web; D) kiss coating; E) direct transfer knife on the web and air knife; F) direct transfer with slot die; G) multiroll coating in direct sense; H) reverse roll coating.

this older system has lost its importance over the last several years. In this case a one-cylinder coating device picks up the adhesive and coats it on the web; the premetering occurs via another cylinder, but the final metering occurs via the air knife [25]. There are many papers describing coating devices [11, 26–30]. Actually, the most common ones are cartridge style coaters [29]. The most important coating systems are described next.

Meyer Bar

The main components include a pan, a chrome-plated cylinder with a two-direction drive, and a Meyer bar that is adjustable both vertically and horizontally and is driven by a stepless speed charge gear motor. The amount of coated adhesive is controlled by a wire wound rod.

Direct Gravure

The main components include a pan, an engraved roll, a doctor blade, and an impression roll. The engraved roll is driven by a direct current motor. The coating weight is determined by the depth of the engraving.

Direct Rotogravure/Fountainless

The coating is applied with an integral doctor blade and applicator system. This system is especially designed for higher speed coating processes.

Offset Gravure

The main components comprise a pan, an offset roller, a back-up roll, and a doctor blade. The engraved and offset rolls are equipped with a direct current drive. The coating weight is controlled by the depth of the engraving and by the roll speed.

Multiroll Coating Heads

The advantage of multiroll coating heads is that there is no wear out of blades or engraved cylinders; all rolls are driven. The coating weight is controlled by the nip pressure, rubber hardness, and roll speed.

The properties of the PSAs to be coated are function of the coater; parameters such as running speed, viscosity, antifoam requirements, shear stress, reflux requirements, wetout difficulty, and the general application influence the usability of a coating device for a certain adhesive [31] (Figure 9.6). The self-adhesive coating unit differs as a function of the adhesive technology (solvent-based, water-based, hot-melt).

Knife and Blade Coaters

Various types of doctor blade systems and roller systems for coating onto flexible substrates such as textiles, synthetic leather, etc., and their systems and technological uses were described by Patermann [32]. The schematic construction of a metering device with knife-over-roll (static or rotating) has been reviewed [33].

Knife and blade coaters obtain their nomenclature from their application systems. These machines are employed to deposit thick coating layers with at least a 50% solids content. A scraper is used to perform the final metering, coating, and smoothing operations, since both methods are a form of the excess coating system. Knife coaters employ a rigid blade that has sharp or rounded edges, while the blade coater utilizes a thin springlike doctor blade. Both machines are similar but they have different coating reservoirs, tension, and drying and wiping systems. They also are called applicator rods, Meyer bars, equilizer bars, or coating rods [34]. The web passes over the metering rod, which may be stationary or which may be rotating slowly; the direction of the rotation may be the same as the web or opposite. In contrast to the scraper (blade, knife) systems, roll application (rotating cylinder) or slot-die coating devices do not use an excess of adhesive. In the case of rotating cylinders the fine (gravure) surface structure of the roll regulates the adhesive coating weight. In the case of a slot die the opening of the slot die regulates

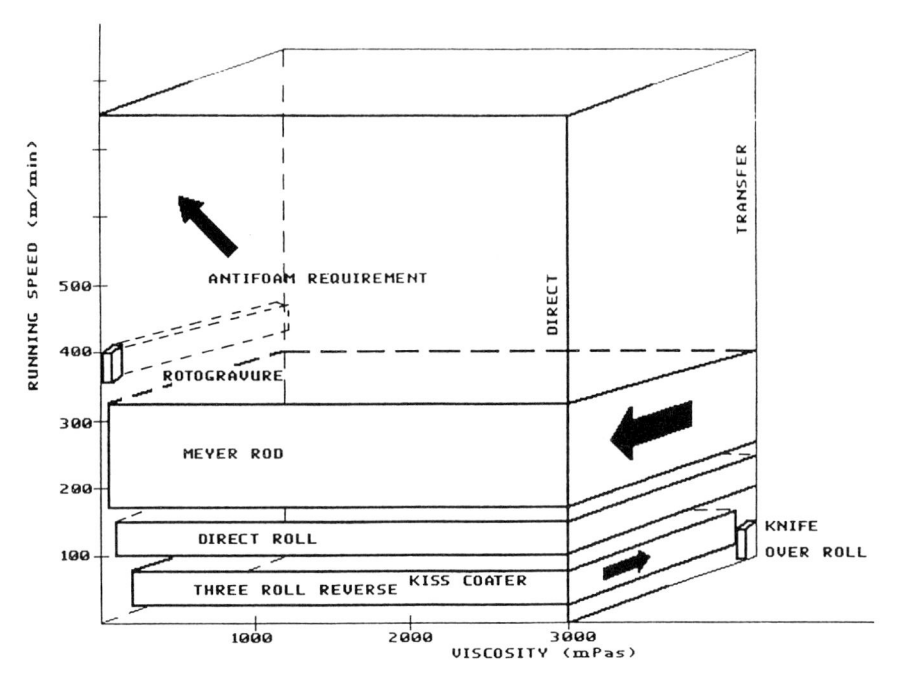

Figure 9.6 Parameters influencing the usability of the coating device.

the coating weight. Subsequent development of both systems resulted in more complex metering devices. The use of many frictional cylinders or a knife on the cylinder (or round knife, roll) allows a fine adjustment of the coating weight. In a similar manner, the combination of a slot die with a pump (gear-in-the-die) assures a finer coating weight regulation. The coating of PSAs onto a web may be carried out with different coating techniques for different adhesives. For solvent-based systems a Meyer bar or reverse-roll system is proposed; water-based systems tend to use a Meyer bar, reverse-roll coater, roll bar, or air brush. Hot-melt PSAs are coated via roll systems or slot-die systems, eventually with the aid of an extruder.

Generally, one should differentiate the knife/blade coaters according to the location of the knife. Theoretically, it is possible to wipe off the excess of the adhesive with a knife that is in contact with the web driving cylinder (Figure 9.7) or with the web (Figure 9.8).

Knife-Over-Roll. This is a coating device with a knife on the cylinder (doctor blade); this method allows the use of adhesives with viscosities up to 40,000 mPa·s; commonly used viscosities average 5000–15,000 mPa·s [35].

Manufacture of Pressure-Sensitive Labels

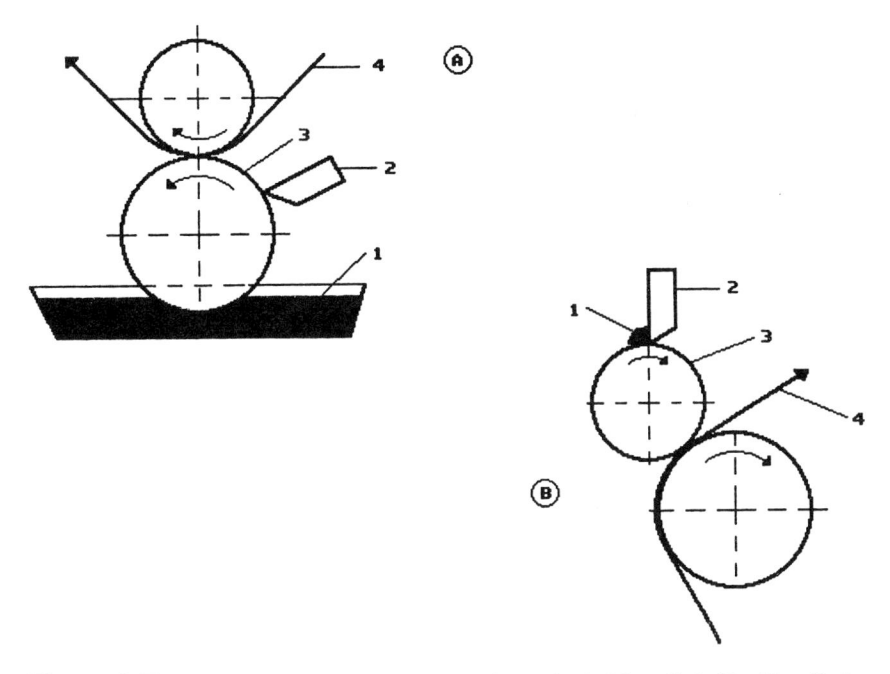

Figure 9.7 Coating device with knife on the roll. 1) PSA; 2) knife; 3) cylinder; 4) web.

In this case the web runs with the same speed as the driving cylinder and the stationary knife wipes down the excess of PSAs from the top of the web. The quality of the knife (shape, material, etc.) determines the coating weight and appearance (Figure 9.9). Other requirements for good quality include the absence of blade damage and dried adhesive residues, a high viscosity (lower limit above 2000 mPa·s) and a high lower limit of the coating weight. The method is sensitive towards web thickness and cylinder tolerances. Generally, speeds of 220 m/min are achieved [36,37]. The following types of knife are possible:

- floating knife;
- roller knife;
- rubber blanketed knife.

The thickness of the coated film is generally higher than the opening of the blade, and depends on the pressure before and after the blade [38]. In production the web travels through a wetting station, then to the metering rod where a metered thickness of the coating is allowed to pass between the wires, and

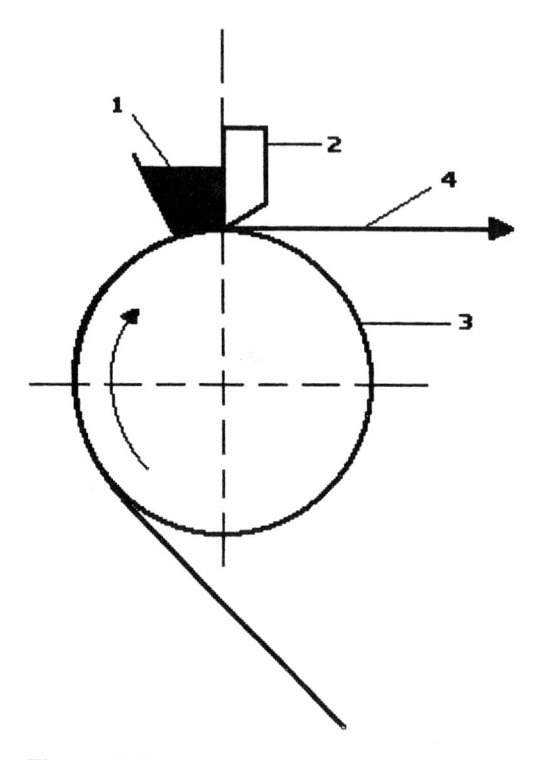

Figure 9.8 Coating device with knife on the web. 1) PSA; 2) knife; 3) cylinder; 4) web.

the excess of liquid retrieved into the holding tank. The coated liquid may be applied at the wetting station by several different methods. The web can be immersed directly into a tank or an applicator can be rotated in the reservoir to transfer the liquid to the web at the top of its rotation. It is important to apply an excess of coating liquid at this station in order for the metering rod to do its job. The types of knifes (blades) known [39] are either smooth, wire wound, or machined. There are different constructions for holding the knife, namely:

- clip type holder;
- magnetic holder;
- champion type;
- Mylar insert holder;
- heated rod holder;
- water lubricated rod holder.

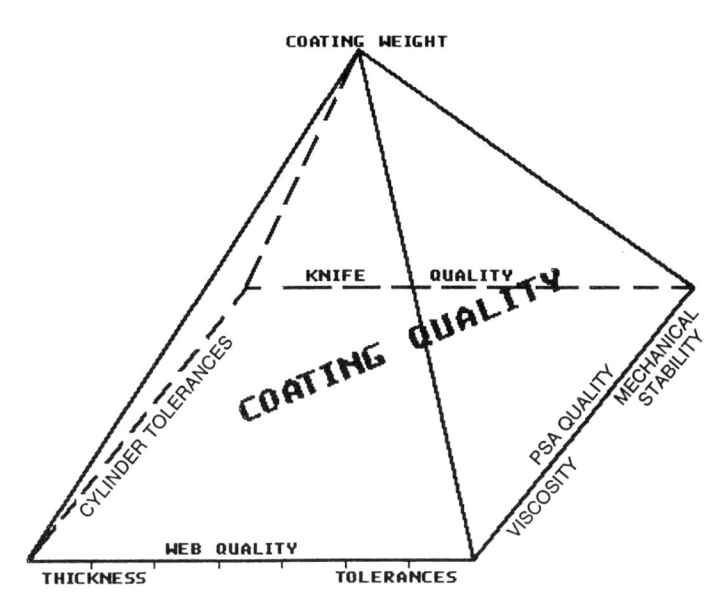

Figure 9.9 Coating device with knife over roll. Parameters influencing the coating weight and coating quality.

The blade or knife may be used as a polishing bar, inserted knife, knife-over-roll, metering rod, etc. The web passes over the metering rod, which may be stationary, or may be rotated slowly. A smoothing rod is a finely polished metering rod without wire and is used in several different ways. It can be installed in the web path after coating with a gravure cylinder to eliminate the etched pattern and enhance the coated surface. It also can be used after a wire-wound metering rod in those cases where the viscosity of the coating liquid prevents it from flattening out because of low surface tension [40]. In applications where a large amount of coating liquid must be metered off the web, two wire-wound rods are used in tandem spaced together or apart. The first rod has a larger size wire for doctoring off most of the excess liquid, and the second, smaller wire rod, controls the finished coating thickness. A typical coating station may use one or two metering rods, and also have a position for a smoothing rod.

Rod Coating (Meyer Bar), Wire Wound Rod Coating

Rod coating or Meyer rod coating has been in existence since the early 1900s [40]. These machines used equalizer bars or doctor rods; the rods are made of steel wound with different sizes of wire. In the last several years, rods with threads instead of being wound with wire also have been developed. The coater applies an excess of coating to the web with an applicator roll. The applicator roll usually is supplied with edge wipers, which wipe the coating off the roll at the edges. The applicator roll is usually driven by a variable speed drive which follows line speed at an adjustable ratio. The web passes over the applicator roll where it picks up an excess of 3–10 times the desired final coating weight; it then passes over a wire wound rod whose wire size determines the final coating weight. The coating weight is determined by the diameter of the wire and by the speed difference between the web and the rod. Metering rods may be smooth, wire wound, or machined [41]. The new ISO box-rod coaters have an extremely wide coating weight range, ease of operation, and precise coating weight regulation. For classical wire wound rods the maximum viscosity is 400 mPa·s (6-mm rod diameter, 0.075–0.75 mm wire) [41].

The thickness of the coating is governed by cross-sectional areas of the gravures between the wire coils of the rod. The geometry of this system creates a wet film thickness which is directly proportional to the diameter of the wire used [34]. Wet film coating thicknesses accurate to 2–3 μm can be achieved easily using metering rods. Mathematically, the average thickness of the area between the wires (i.e., the coating thickness) is 0.1073 times the wire diameter. In production coating there are other physical factors influencing the coating weight. The most important of these is the phenomenon of shearing the liquid. In fact, not all of the coated material passes through the gaps; some liquid adheres to the surface of the wire. The impact of this is usually small, but can be significant when small wires are used or when viscosities are high. The web speed, the web tension, and adhesive penetration in the face material will also affect the coating thickness. The sum of all these variations seldom implies a 20% difference from the theoretical coating weight.

The coating weight depends on the diameter of the wire as follows; for a wire diameter of 0.3, 0.5, 0.7, and 0.9 mm the coating weights are 10, 14, 20, and 30 g/m^2, respectively [42]. The maximum coating weight which can be achieved with this system amounts to 50 g/m^2. In order to obtain a constant coating weight, this system needs a lower dispersion viscosity, a low foaming level, a speed as high as 150 m/min, and a regulated web tension [42]. With a Meyer bar, the wire wound bar regulates the adhesive layer thickness on the web. With wire diameters of ranging from 0.1 to 0.9 mm, coating weights of 15–100 g/m^2 are possible. Adhesives with viscosities of 2000–25,000 mPa·s may be coated with a ±1 g/m^2 tolerance. In order to determine the choice of an adequate metering rod, tables or diagrams describing the dependence of the coating weight on the rod size are used (Figure 9.10).

The knife-over-roll coater may generate mechanical stress, and foam; however, a price is paid not only in line speed but also in the difficulty to achieve a good reflux [44]. When an applicator roller is used in a rod coating system,

Manufacture of Pressure-Sensitive Labels

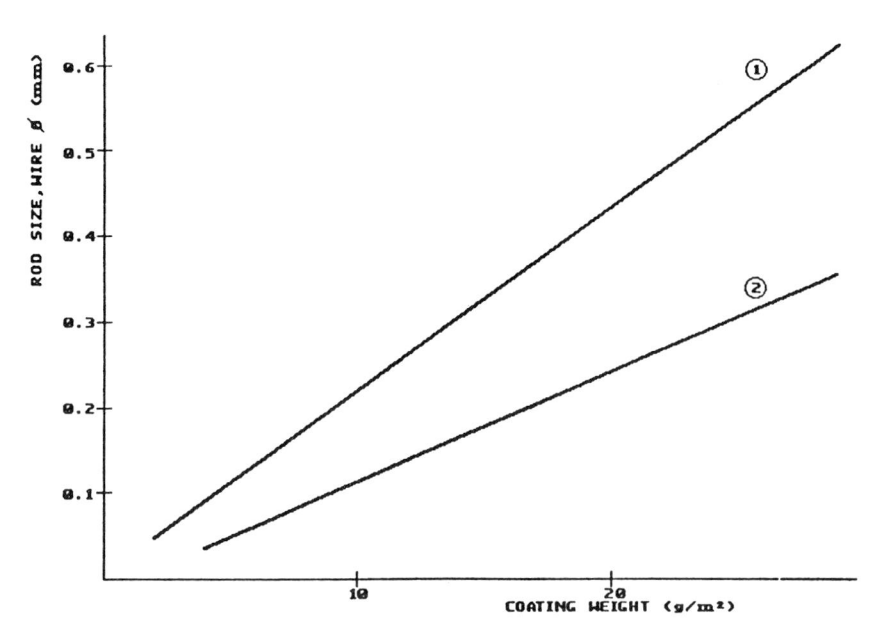

Figure 9.10 Rod coating. Dependence of the coating weight on the rod size. 1) Enol, 2) one different PSA.

the speed of the applicator is not a critical factor. Other systems using gravure or knife coating methods require precise roller speed adjustments. In a rod coating system the machine operator can adjust the applicator roller speed within a wider range and can do it while the machine is running. In rod coating the actual thickness of the coating can be affected by web speed, viscosity, and other factors. A gravure cylinder, although expensive and limited in its range, provides an extremely accurate coating (appearance, weight) almost independently from the web speed. Depending on the application, rod coating speeds are usually limited to 300 m/min, although some coaters claim web speeds up to 600 m/min [34]. The critical factor in the web speed of a rod coater is the time for the striations formed by the rod to flatten out and become smooth; the web speed must be controlled to allow time for flattening before the web is dried. The design of a rod coating station should ensure that the web makes intimate contact with the wires of the metering rod. "The wrap angle," the angle of the web direction as it approaches the rod and its direction as it leaves the rod, should be 15° for a heavy web tension, or up to 25° for a light web tension. Web tension is a critical factor in the design of a rod coating station. With a wrap angle of 15–25°, the web must be tight enough to ensure intimate contact with the metering rod, yet not so tight that the web is deformed by the wires.

Adhesives can solidify between the wire windings of the rod whenever the coater is stopped. Many coating machines have a "throw-off" feature, a mechanical means of separating the web from the rod automatically whenever the machine is turned off. When a coating liquid is allowed to extend to the edges of the web, it can spill over to the dry side and contaminate the idle roller. There are several methods used to maintain a dry edge, according to the coating process used. In gravure coating, the most common method is to undercoat the backing roll at the edges; disadvantages include the need for a wide assortment of back-up rollers. In roll coating the spillover is usually controlled by specially constructed dams which are placed in direct contact with the coating roll. In a rod coating system the dry edge is controlled by wipers or deckle straps on each edge of the applicator roll [44]. Because the wipers are easily moved their position can be adjusted while the coater is operating, with no downtime at all. The next step in the development of knife/blade scrapers is the use of "round blades" or stationary rolls (Figure 9.11).

In this case the blade has a "break-down" edge; a direct mass transfer of the PSAs on the web occurs, as well as metering by the aid of a stationary round blade. In a subsequent development the stationary blade is replaced with a rotating cylinder; in this case one would have a metering roll (roller coater).

The direction of rotation may be the same or the opposite to that of the web,. The most common procedure is to rotate the rod slowly in the opposite direction to the movement of the web. The rotation works twofold: it flushes the coating material between the wires, keeping the wire surfaces wet, and it prevents setting up and hardening of some liquids. The rotation will also distribute any abrasive wear evenly over the wires [34].

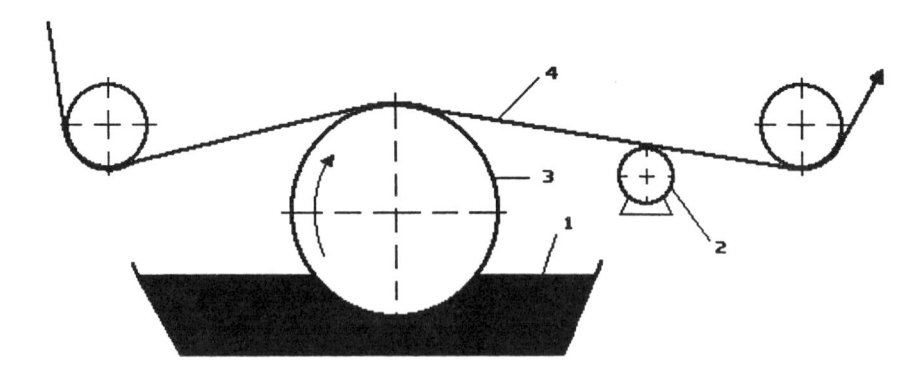

Figure 9.11 Coating device with stationary roll. 1) PSA; 2) stationary roll; 3) cylinder; 4) web.

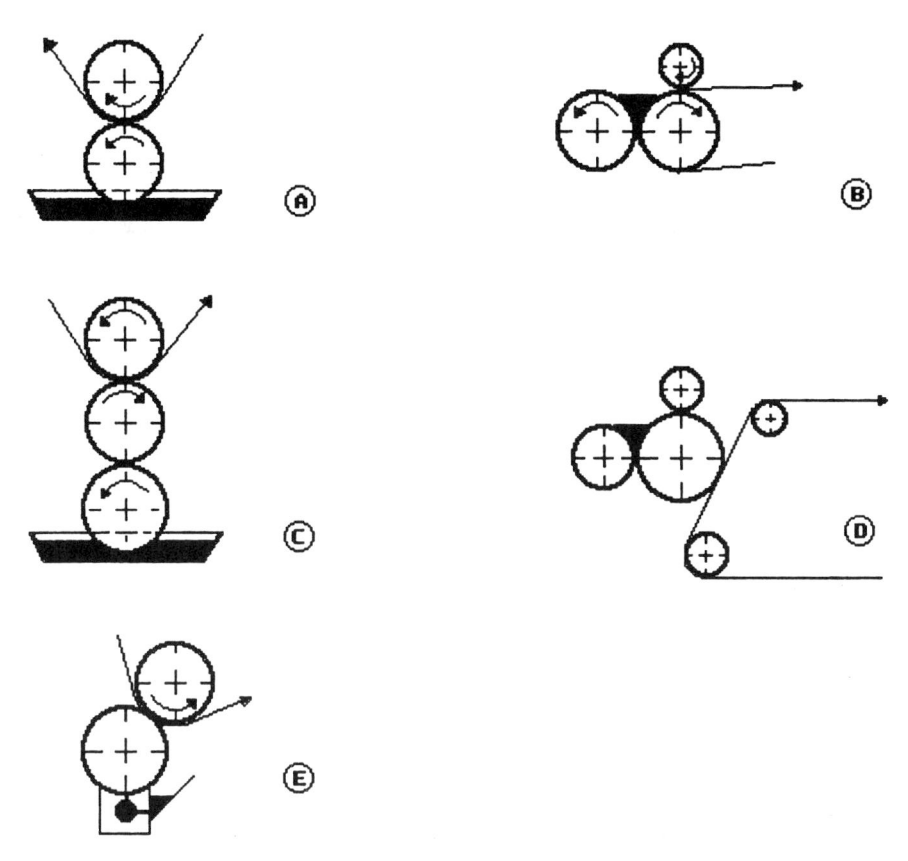

Figure 9.12 Roll coating systems with direct/indirect mass transfer on the web. A) direct gravure; B) direct gravure with reverse roll; C) offset gravure; D) kiss coater; E) slot die, duplex coater.

There are different roll coating systems for solvent-based or water-based PSAs, with low or high viscosities; they differ according to the direct or indirect character of the adhesive mass transfer onto the web (Figure 9.12).

Roll Coating Devices/Rotogravure

In a manner different from the scraper (blade, knife) systems, roll application (rotating cylinder) or slot-die coating devices do not use an excess of adhesive. In the case of rotating cylinders the fine (gravure) surface structure of the roll regulates the adhesive coating weight. In the case of a slot die the opening of the slot die determines the coating weight. Subsequent developments of both systems led to more complex metering devices. The use of several

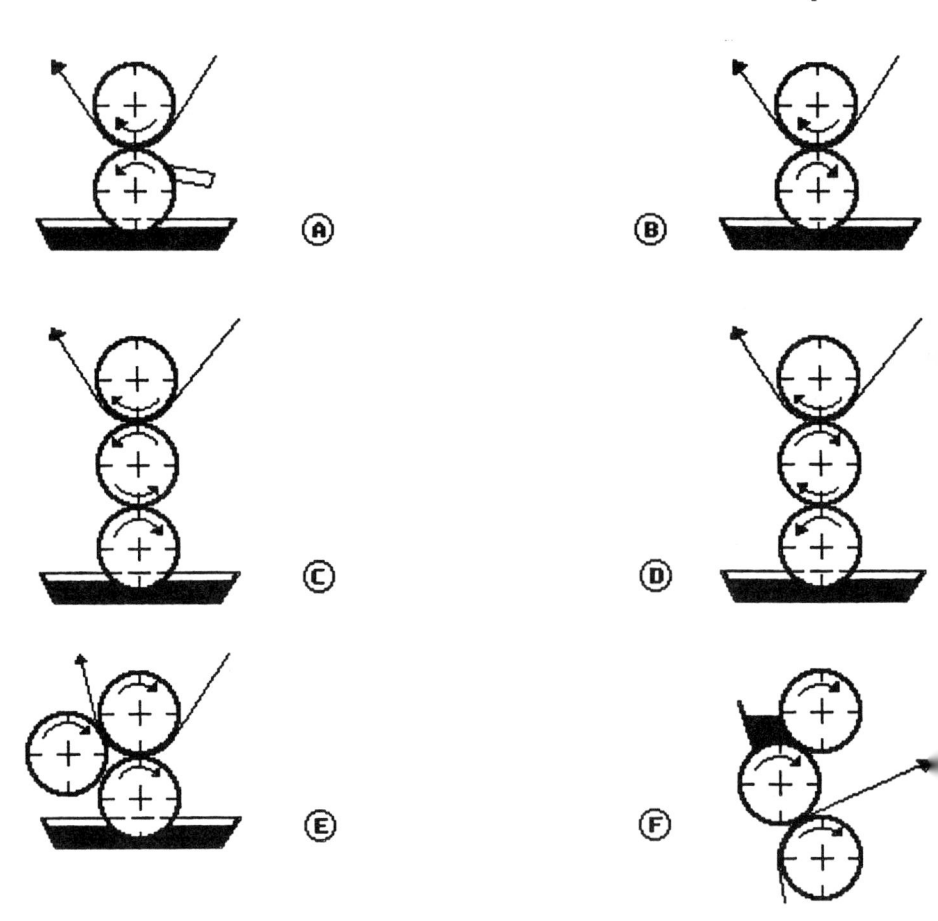

Figure 9.13 The main coating methods used for roll coating. A) direct gravure; B) reverse gravure; C) direct offset gravure; D) reverse offset gravure; E) three-roll reverse pan fed gravure; F) three-roll reverse nip fed gravure.

frictional cylinders, or a knife on the cylinder (or round knife on a roll), on the cylinders allows a fine adjustment of the coating weight. In a similar manner the combination of the slot die with a pump (gear with die) ensures a finer regulation of coating weight. In order to define the choice of the adequate coating device, classification charts were developed [45]. Figure 9.13 summarizes the most commonly used coating geometries.

Theoretically, it is possible to use a direct gravure when direct contact between the engraved cylinder and the web occurs, and indirect gravure when the adhesive is transferred from the engraved cylinder to the web via intermediate

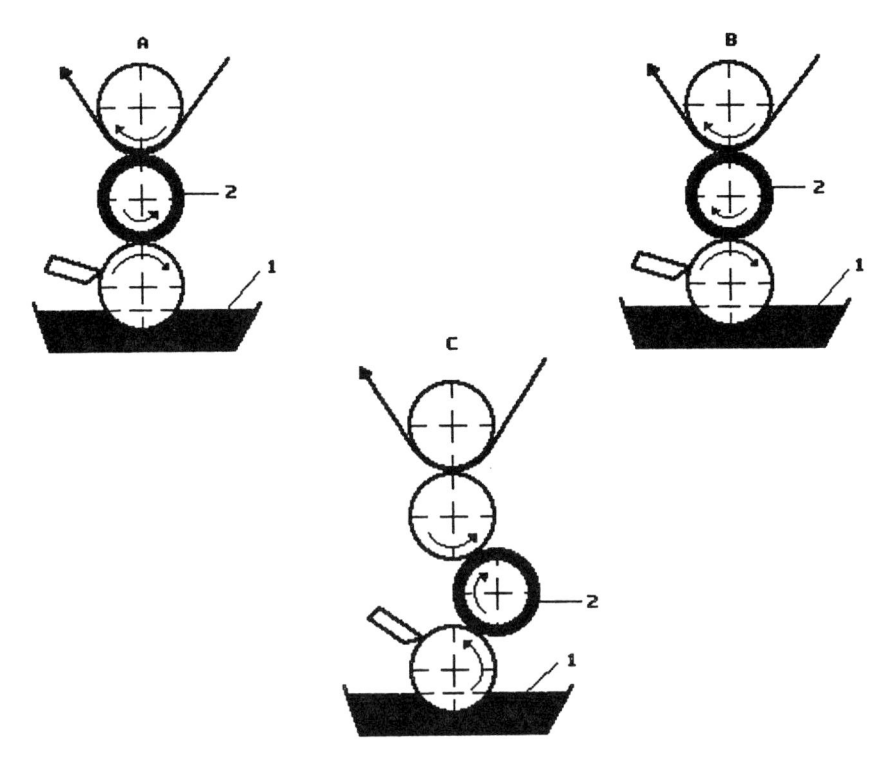

Figure 9.14 Roll coating. The basis offset gravure configurations. A) vertical, two-roll offset gravure, direct; B) vertical, two-roll offset gravure, reverse; C) angular three-roll offset gravure.

cylinders. In principle, the engraved cylinders depth (gravure) may be zero (i.e., a smooth cylinder also can be used). Moreover, according to the rotation of the engraved cylinder one can distinguish direct or reverse roll coaters. In many cases reverse gravure is combined with reverse roll coating. A reverse rotation of a Meyer bar also is used in order to regulate the coating weight.

Indirect gravure or offset gravure coating employs one or more transfer rollers between the gravure cylinder and the substrate. Indirect coating is offset from the gravure cylinder to the transfer roller and than applied to the web. A coating is offset when the adhesive is transferred to a dry surface and then applied to another dry surface. The intermediate rollers improve the wet coating characteristics, reduce or remove the adhesive pattern, and enable the unit to apply a uniform and thinner coating film. There are two basic offset gravure configurations (i.e., with vertical and angular orientations) (Figure 9.14).

An offset gravure coater is an extension of the direct gravure system, where the rubber pressure roll becomes a driven offset on the transfer roll, and a third roll (steel) is added [34]. In operation the coating is first transferred to the intermediate offset roll before being applied to the web. The advantages of offset gravure include the following:

- The process is suitable for running with rough web surfaces.
- The engraving pattern of coatings with poor flow is eliminated.
- A more complete removal of the coating from the cells is possible (suction effect of rubber).

A virtually self-cleaning system can be obtained; it is recommended for 100% solids. More shearing of the coating is obtained and pattern coating is possible. A variation of this method can be used for two-sided coating. The essential distinction between differential offset gravure and normal offset gravure is that coating weight variations may be obtained by independently altering the speeds of the compression roll as well as the offset roll in their relation to the speed of the coating roll or web itself.

The vertical offset gravure configuration is best suited for liquids with slow to medium evaporation rates. The transferred wet coating remains on the rubber-covered offset roller through an angle of rotation of 180° before it is deposited. The angle of rotation through which the coating remains on the wetted cylinder before being transferred or deposited is normally referred to as the wet angle. The angular offset gravure configuration (Figure 9.14) reduces the wet angle by almost one half. The angular design can be used with faster evaporating diluents and when more than two transfers are required. The coating is applied to the rubber-covered transfer roller by the gravure process, after which the coating pattern has a tendency to flow together, since the rubber covering repels the liquid. The coating then is transferred by the film splitting process to the chromed steel application roller, which in turn deposits the coating onto the web.

Two-roll gravure coaters may have a reverse direction of the engraved roll. In reverse gravure the engraved cylinder rotates in the opposite direction as the web being coated. In this case the wiping action of the opposite rotation of the etched cylinder tends to obscure the cell pattern that sometimes remains on the surface of the applied coating. This is advantageous for coatings with poor flow-out properties. Because the web is not nipped against the etched cylinder, as in direct gravure, the speed of the etched cylinder can be varied; this offers the ability to vary the coating weight. The doctor blade can be relocated to the opposite side of the coater, which is not easy in direct gravure coaters. An alternate procedure would be to keep the blade mechanism in its original location and adapt a flooded blade technique [34]. In the reverse gravure it is not possible for the etched cylinder and the impression cylinder to be pressed tightly together to form a tension contact nip. Here, the tensile strength and elasticity of the web are important factors for predicting the need for a driven backup roll. A negative factor on reverse gravure is the increase in the rate of gravure cylinder wear. Another design of reverse gravure is able to vary the speed of the etched cylinder. The relation between the speed of the etched roll and the speed of the web running in the opposite direction has a decisive influence on the overall quality of the coating. The etched roll speed must be at least as high as that of the web or even higher. The indirect reverse gravure may be used for webs having quite different thicknesses [46]. In gravure coating the most common method is to undercut the backing roll at the edges. The disadvantage is the need of an assortment of backup rollers, one for each width of web the user handles. Changing the backing roller for each width also requires more time.

Coating Weight

Generally, when coating with cylinders, the adhesive is taken from an adhesive pan (kiss coating). The coating weight depends on the pressure of the web on the cylinder, the viscosity of the fluid, web speed, and cylinder speed [43]; coating weights of 2–50 g/m^2 are possible. Viscosities of 20 DIN sec to 12,000 mPa·s are used, with coating weight tolerances of ±1 g/m^2. Generally, the web width is limited to 1600 mm; the system is open and tends to dry. Continuous pumping (recirculation) of the material content is necessary. Generally, for gravure cylinders, the coating weight depends on the geometry of the gravure. The factors influencing the ink hold out for printing also influence the coating weight of the adhesive. These factors depend on the engraved cylinder rolls, on the machine parameters, and on the adhesive. Figure 9.15 summarizes the parameters influencing the coating weight (e.g., line screen number, angle, depth, velocity, viscosity).

One of the most important factors influencing the hold out is the line screen number (Figure 9.16). The adhesive holdout decreases with the line number. It also depends on the depth and angle of the engraving (Figure 9.17). The adhesive holdout also depends on the running speed (Figure 9.18) and increases with the viscosity (Figure 9.19).

There is a (theoretical) choice between gravure depth and angle (Figure 9.20). The relative importance of the different parameters influencing the adhesive holdout may be combined in a formula:

$$H = f\left[\left(\frac{9C \cdot 1.3V}{1.1S}\right) \cdot (e \cdot N)\right] \tag{9.1}$$

where H is the holdout, C is the coating cylinder, V is the viscosity, S is the speed, N is the nature of the adhesive, and e is a coefficient that is dependent on N; the measured coefficients (9, 1.3, 1.1, or e) may differ according to

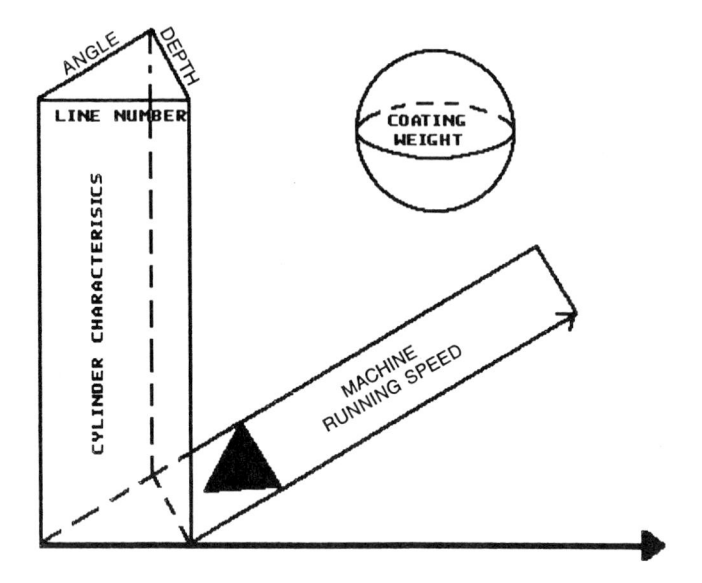

Figure 9.15 Machine and running parameters influencing the coating weight. Gravure coating.

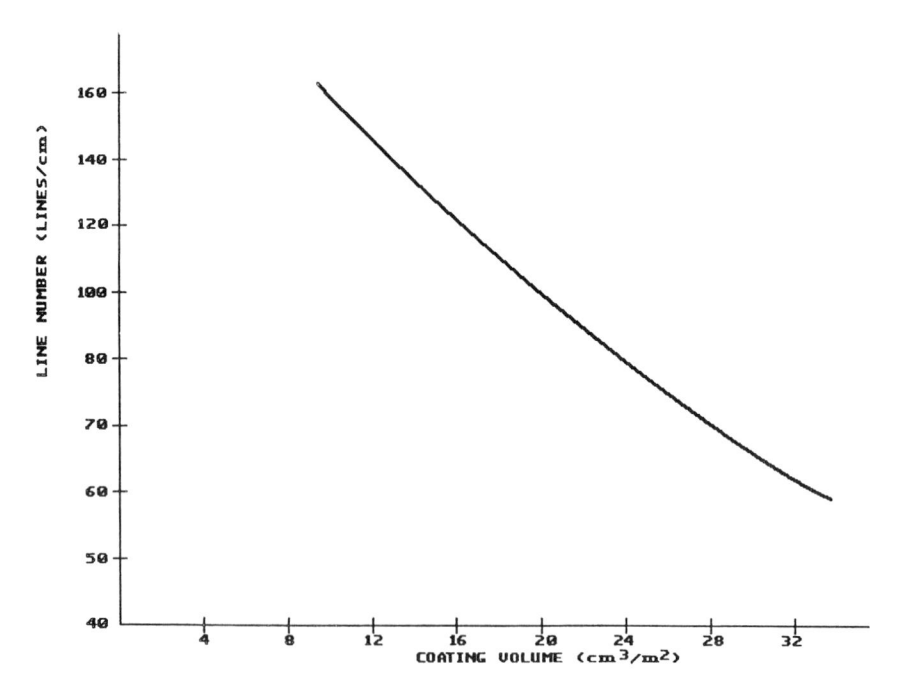

Figure 9.16 Gravure coating. The influence of the line number on the adhesive hold out. Angle 80°.

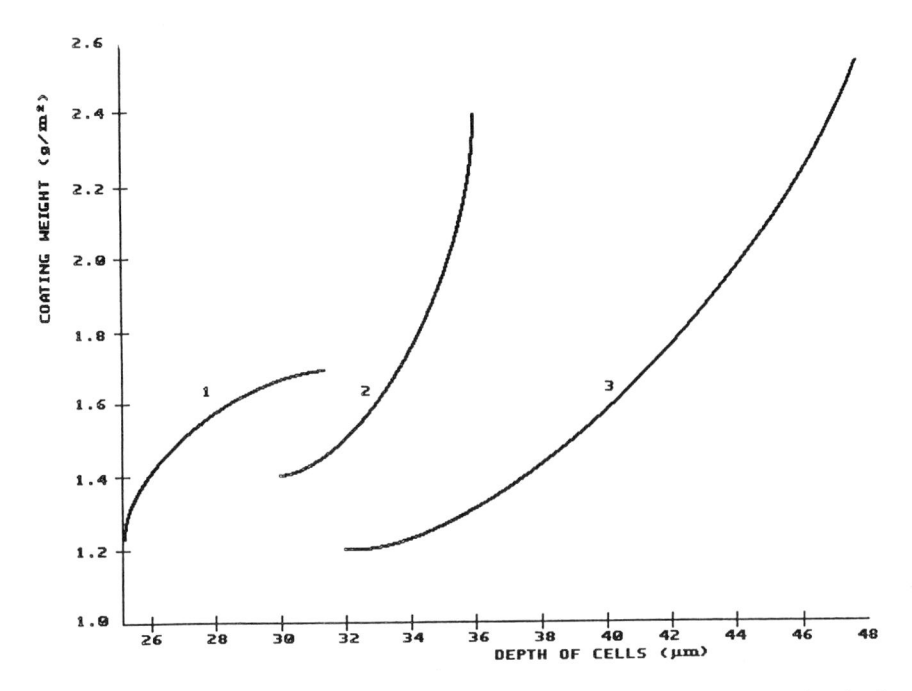

Figure 9.17 Gravure coating. Dependence of the adhesive hold out on the depth and angle of the engraving. Line number 70; 1) 140° angle; 2) 130° angle; 3) 120° angle.

the experimental conditions. As can be seen from the above equation, the most important parameter influencing the coating weight is the cylinder. The influence of the viscosity also remains important; the influence of the speed is relatively small:

$$C \gg V > S \tag{9.2}$$

The nature of the adhesive may cause changes of 200% in the holdout (coating weight). Concerning the cylinder characteristics, the most important parameter remains the depth of the gravure (up to 500% change). The line screen has a decisive influence too (about 50%). The holdout also increases with the engraving angle (Figure 9.17). The coating weight for gravure cylinders depends on the gravure number, form, depth, and cavity ratio [47–49]. Gravure cylinders are used for coating adhesives with viscosities of 100–3000 mPa·s, and an ensuing coating weight of 10–50 g/m² [48]. Different line screen numbers are recorded for different coating weights; the gravure characteristics (quadra gravure, free-flow, etc.) depend on the adhesive character-

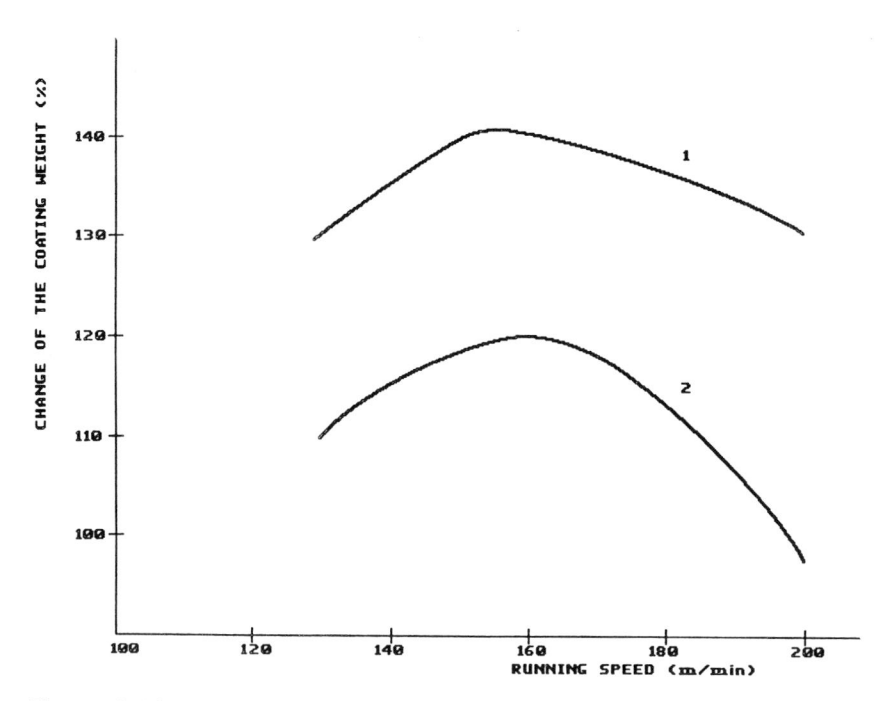

Figure 9.18 Gravure coating. Dependence of the adhesive hold out on the running speed. 1) line number 70, cell depth 48 μm; 2) line number 80, cell depth 37 μm.

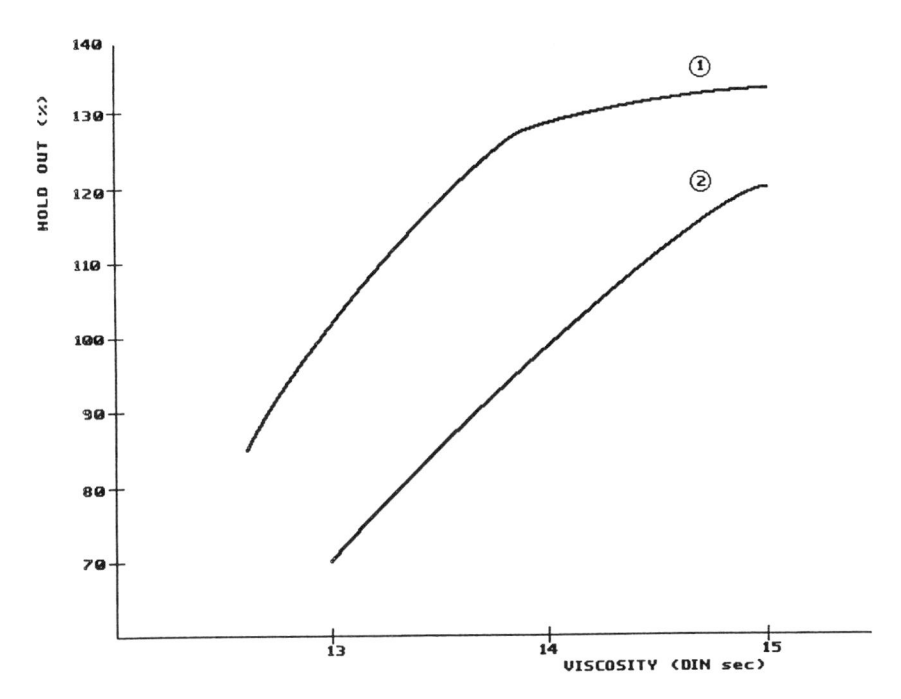

Figure 9.19 Gravure coating. Dependence of the adhesive hold out on the viscosity. 1) cell depth 27 μm; 2) cell depth 30 μm.

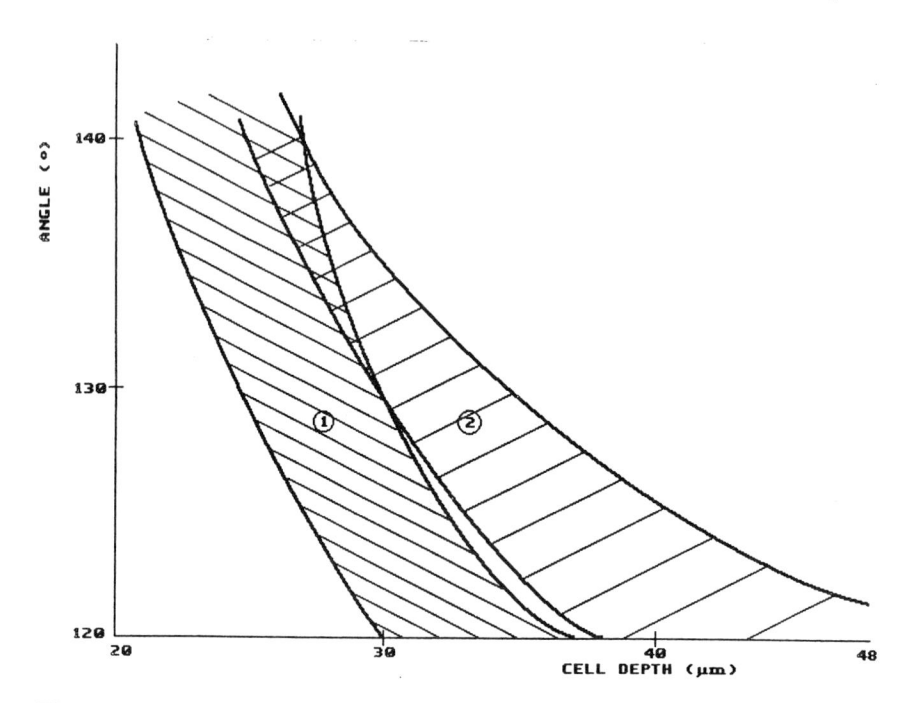

Figure 9.20 Gravure coating. The influence of the gravure cell depth/angle on the coating weight. The choice of these parameters for the same coating weight; 1) line number 80; 2) line number 70.

istics [51,52,60]. For example, a 40-line screen is recommended with a 70-μm depth for a coating weight of 5–7 g/m^2 [50]. The choice of a cylinder is made easier using tables or diagrams concerning the coating weight dependence on the cylinder geometry/line number (Figure 9.21).

Roll Coaters with Doctor Blade/Mixed Systems

Reverse gravure with a closed chamber and a doctor blade [53] in the coating station also is used. A coating weight of 0.2–30 g/m^2 is possible; viscosities from a low number of DIN seconds up to 5000 mPa·s can be used. Changes in coating weight are obtained by changing the cylinder. The blade and the cylinder cause shear of the adhesive [43]. Changes in web width are made via changes in web covering or by changing the engraved cylinder.

Conventional or reverse angle blades can be installed [34]. In the reverse angle doctor blade arrangement the excess coating is sheared from the surface of the engraved roll rather than being squeezed from it. The system eliminates the need to increase blade pressure for higher viscosities and reduces the

Manufacture of Pressure-Sensitive Labels

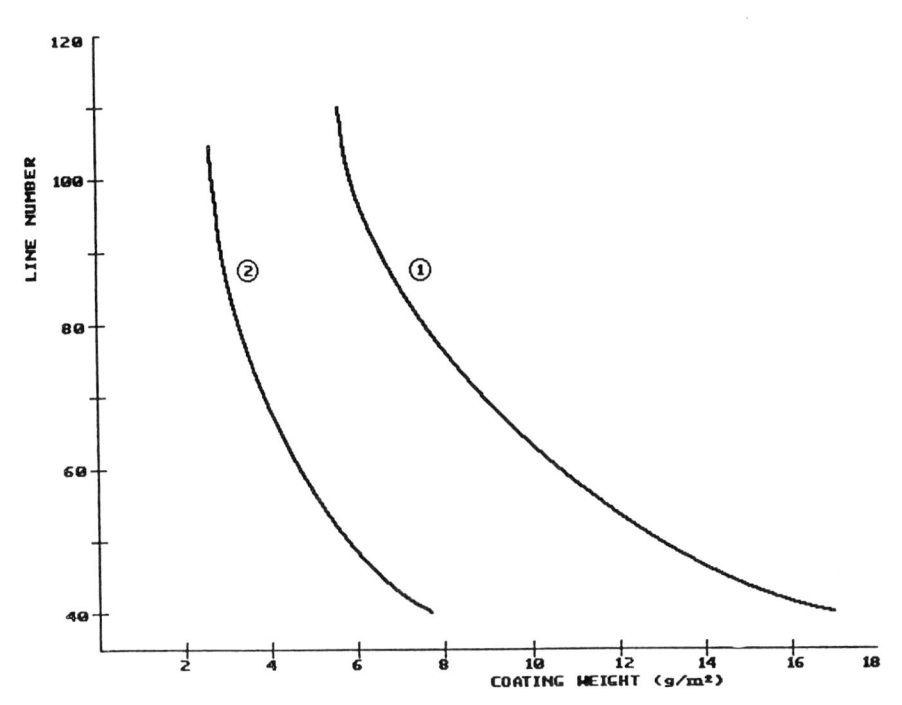

Figure 9.21 Roll coating, gravure coating, coating weight. Dependence on the cylinder geometry. 1) solid content 45%; 2) solid content 75%.

turbulence in the coating fountain. A fountainless applicator with a coating chamber also may be used. Knurled gravure rolls in combination with a doctor blade may be used to coat primers with a low viscosity at 25°C with a thickness of less than 1 μm [54]. Among the possible coating systems (e.g., MEC roll, reverse roll, fixed doctor blade system, etc.), the most used one appears to be the MEC roll [23]. This system is not only a simple configuration, but it also is easy to set up in order to obtain a uniform coating with the desired coating weight. The system is designed to overcome a series of problems, for example:

- correct ratio between coating cylinder diameter and doctor (metering) roll diameter;
- angular position of the doctor roll;
- mechanical tolerances of the order of a few microns concerning excentricity, surface uniformity;
- the doctor roll speed has to be adjustable, and to run dependent on the coating cylinder speed by an electronic master/slave system;

- adjustable compensation for the doctor roll flexure;
- fine adjustment of the gap between coating cylinder and doctor roll, independently adjustable parallelism on operator and drive side.

There are six common designs used for gravure coating systems [55]. Gravure coaters are normally constructed in two-roll or three-roll configurations. Two-roll constructions may be direct, reverse, and differential reverse. Three roll-machines may be offset, reverse offset, and differential reverse offset. For direct gravure in a two-roll design, the roll is partially submerged in a coating pan and as it turns it brings the coating up. The excess coating is wiped from the surface with a doctor blade leaving an amount predetermined by the total carrying capacity of the etched or knurled cells. Therefore, for a selected cylinder the only way to change the coating weight is to change the solids content in the coating itself. The doctor blade is usually of the oscillating type and the upper web back-up roll is made essentially of an elastomer-covered sleeve. The web runs between it and the engraved roll. The advantages of direct gravure include [55]:

- widely used and excepted;
- provides a consistent and uniform coating;
- low to medium equipment cost;
- independent from operator skills.

The disadvantages of the direct gravure include:

- the pattern of the engraved roll may be evident, if the coating does not have adequate flow;
- the doctoring of viscous coatings at high speeds is difficult;
- high foam generation at higher speeds.

There are several ways to reduce the disadvantages of direct gravure. They include heating of the coating or using a polishing bar. The latter can be variable speed driven, in either direction and cooled or heated.

2.3 Choice of Coating Geometry

Generally, the choice of a coating device depends on the following parameters [43]:

- desired coating weight range;
- viscosity;
- changes in coating weight;
- uniformity of the coating weight;
- abrasion of the metering devices;
- changes in coating (web) width;
- changes in coating medium.

There are general requirements for coating devices, such as environmental and energy considerations, short operating time, and ease of operation [6]. New cartridge-style coaters and quick changeover rubber sleeve rolls allow a full change of the gravure roll and supply system in 5 min or less [56].

The coating of PSAs on a web may be carried out using different coating techniques. For solvent-based systems Meyer-bar or reverse-roll-coat systems are proposed; water-based systems use Meyer-bar, reverse-roll-coater, roll-bar, or air-brush systems [2]. Hot-melt PSAs may be coated via roll systems or slot-die systems, and eventually through extrusion.

Roll Coating of Solvent-Based PSAs

The cylinders used for PSA coating generally have a 180–400 mm diameter, a regulated drive, and are explosion proof [57]. Coating machines for solvent-based PSAs vary as a function of the viscosity of the adhesive. For low viscosity systems, a driving cylinder brings the PSAs from the pan and coats it on the web (Figure 9.22). A reverse roll regulates the coating weight (metering roll). For highly viscous systems, two metering rolls bring the adhesive on the web (reverse rolls) (Figure 9.23) and each is continuously regulated.

Common roller coaters with accugravure operate with solvent-based PSAs having a solids content of 38–40%, at speeds of 300–400 m/min. Slot-die coating devices may be used for solvent-based PSAs also [57]; viscosities up to 1–2 million mPa·s are possible and coating speeds of 150 m/min were achieved. Solvent recovery systems remain an absolute necessity.

Coating Machines for Water-Based Systems

Whereas for the solvent-based and hot-melt PSAs coating the operating principles, the coating systems, etc., are widely known, the same is not generally valid about water-based systems. Generally, reverse gravure and kiss coating were used since the 1970s [59]. Film forming of water-based adhesives was described [60]; a wide range of requirements must be fulfilled when using a PSA emulsion coater.

To keep foaming at the lowest possible level the coater must have a closed supply system (i.e., overflow should be avoided, which precludes an open pan feed and roll gap feed system). Since the emulsion-based PSAs were introduced for label products the coat weight range varied between 20 and 24 g/m^2 (dry). To keep the coat weight tolerance within maximum ±5%, the viscosity of the emulsion must be kept at a constant level (i.e., the temperature must remain under control). To keep the intervals between cleaning as long as possible (coagulation of an emulsion cannot be avoided) the coating head must be easily accessible for quick cleaning to get rid of streaks; good access is necessary even during operation.

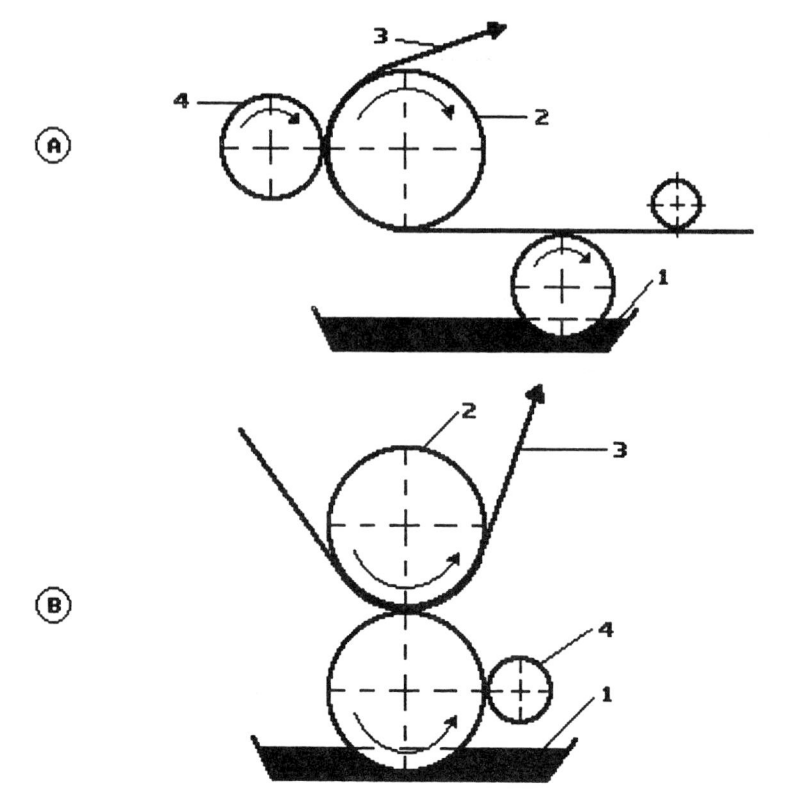

Figure 9.22 Roll coating for a solvent-based PSA. Coating systems for low viscosity solvent-based PSA. 1) PSA; 2) driving cylinder; 3) web; 4) rotary doctor bar.

The following coating devices are used for water-based PSAs [42]:

- reverse roll;
- gravure with direct adhesive transfer;
- knife-over-roll;
- reverse gravure.

These methods use a premetering system with the aid of a multicylinder system. On the other hand, there are some methods using an excess of adhesive, namely roll blade, air brush, and knife-over-web. The systems with knife-over-roll or knife-over-web may be used for high coating weights, and medium to high viscosities. Manufacturing speeds higher than 100 m/min are possible; a low foam level is ensured. The shape of the knife does not have any influence when solvent-based adhesives are used, but for dispersion-based

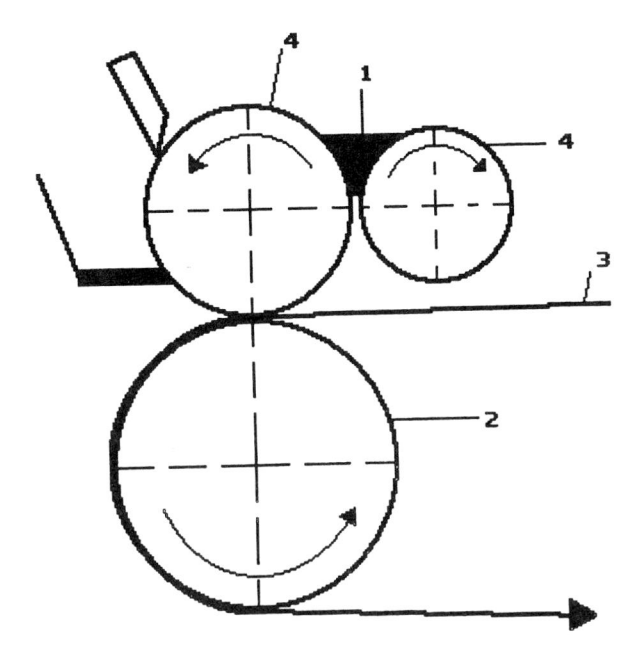

Figure 9.23 Roll coating for a solvent-based PSA. Coating system for high viscosity solvent-based PSAs. 1) PSA; 2) driving cylinder; 3) web; 4) metering roll.

ones there is a need for a "break down edge", as a stationary metering roll is not possible. In Europe today, two coating methods for emulsion PSAs actually dominate their fields (i.e., the metering bar systems for tapes and the reverse gravure system for labels). The coating machines for PSAs tapes using water-based PSAs have been described [53].

Figure 9.24 shows the main components of a reverse gravure system. A closer look at the reverse gravure system reveals why it also is considered more and more for coating tapes. The most outstanding advantages of this coating method are its ease of operation and cleaning, as well as the very smooth surface of the adhesive. The method is not sensitive towards circular roller tolerances [4]. Unfortunately, changing the coating weight requires the change of the coating cylinder since it is not possible to cover a complete coating weight range with a single gravure size. Other disadvantages include the difficulty to do pattern coating and to control the coating weight at low speeds (below 50 m/min).

Different kinds of reverse roll coatings [35] include: the nip-fed system, the pan-fed system, and the Dahlgren system. The nip-fed and pan-fed systems use a metal cylinder; the Dahlgren system requires a special hydrophylic

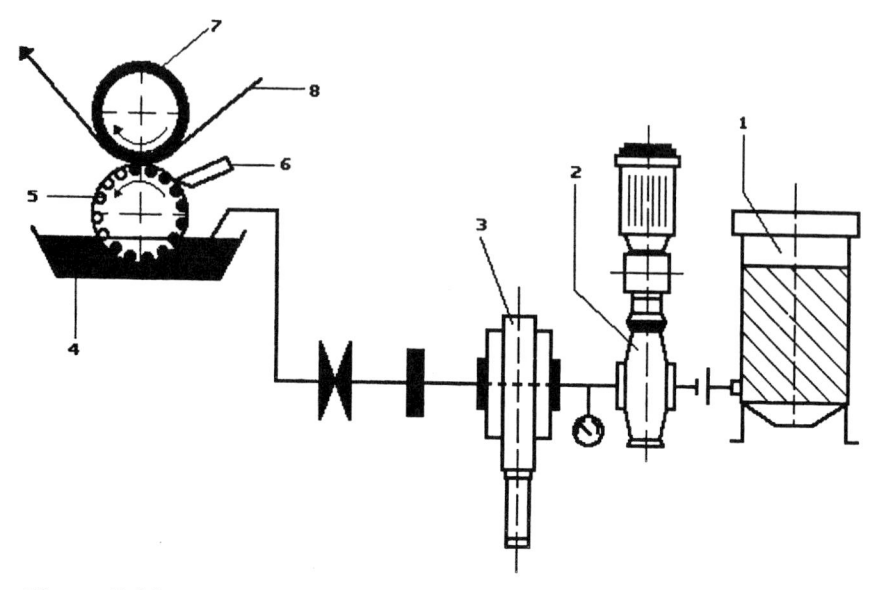

Figure 9.24 The main components of the reverse gravure system. 1) supply tank; 2) pump; 3) filter; 4) feed box (pan); 5) gravure roll; 6) knife; 7) backing roll; 8) web.

cylinder (rubber). Reverse roll coaters were developed for adhesives with a high viscosity [49]. Generally, the coating device is built up from the following parts: coating cylinder, metering cylinder, transfer cylinder, and a falling blade. The transfer cylinder runs at 100% of the maximal speed of the machine. The coating cylinder runs in a reverse sense to the transfer cylinder, at a speed of –15 to +200% of the coating cylinder [49]. Metering cylinders run in a reverse sense to the coating cylinder, with a rotation speed difference of 10% with respect to the coating cylinder. In order to achieve a smooth film, the cylinder speed should be 1.3–1.5 higher than the velocity of the web (i.e., for a production speed of 200 m/min, the cylinder should run at 300 m/min) in order to avoid adhesive defects [42].

Conventional roll coaters designed for solvent-based adhesives can be used successfully including techniques such as gravure, gravure offset, reverse kiss, direct roll kiss, air knife, etc. Because of the basic differences between solution adhesives and polymer dispersions there are considerable differences in rheology and it is necessary to take extra care when coating dispersions in order to optimize coating efficiency. For example, a greater clarity and transparency of the final film structure and smoothness of adhesive coating are ensured if a smoothing roll is installed just after the coating head. Suggested

Table 9.1 Performance of a Reverse Gravure Coater for Water-Based PSAs

Coater characteristics		Coating characteristics, coating weight (g/m^2)
Line number	Speed (m/min)	
10	50	36
20	200	22
40	50	15
50	50	8
70	50	6
80	50	4

Gravure angle: 45°; viscosity range: 800–1500 mPa.s

specifications of such a smoothing roll include a 1.5–2.0-in. diameter, a polished or plated surface, a reverse direction drive, a speed of rotation which is 4 times the linear speed, and at a distance from the coating head of about 20 mm; the track angle is determined through trials [61].

With a given gravure size it is possible to adjust the coat weight within a range of approximately 10%. This can be achieved by changing the pump pressure, the speed of the gravure roll, and the pressure of the doctor blade, and by the appropriate choice of the type of doctor blade (thickness of the polyester blade and the shape of the blade tip). High speed (short contact time between web and gravure) implies a low coating weight, whereas a low dispersion viscosity means high coating weight. For water-based PSAs on a reverse gravure coater, it is ideal to have a high solids content with a low viscosity. The range of the solids or the viscosity is dictated by the surface tension of the dispersion; production speeds of up to 300 m/min with a coating weight tolerance of ±3% are possible. The capabilities of a reverse gravure coater for emulsion PSAs are summarized in Table 9.1.

Reverse roll coaters coat PSAs with a viscosity of 12,000–50,000 mPa·s [49]; the coating weight ranges are 0.5–800 g/m^2. The reverse roll coater was developed for viscous media; the shear forces reduce the viscosity during the coating. The coating weight is determined by the distance between coating and metering cylinder, and by the relative speed of the two cylinders respective to the web. Generally, there is a speed difference between the coating and metering cylinder of 10%. In order to avoid a discontinuous adhesive coating and to obtain a fine, smooth adhesive surface, there are some special requirements the PSAs need to fulfill [35]:

- The metering roll has to be fixed or to run so rapidly that no drying of the adhesive on the roll can occur.
- The diameter of the metering roll has to be small in comparison to the application roll.
- A low viscosity is needed.
- An adhesive without thixotropy is needed (i.e., the emulsion PSAs must be designed for the reverse gravure method).

Given these characteristics, production speeds of up to 300 m/min and coating weight tolerances of ±3% may be achieved [62]. For reverse gravure a speed of 200–300 m/min with a coating weight of 18–25 g/m^2 is possible [42,76]. Generally, 65–160 lines/cm, an angle of 40–90°, and a cell depth of 22–85 μm are used for engraved cylinders with a diagonal screen and the calotte shape of cells is 45°; for water-based dispersions high line numbers and less gravure depth are required.

Having designed water-based acrylic adhesives with a desirable rheology, wetout behavior, low foaming, and good drying characteristics, adhesive suppliers find that almost every converter operates a different coater [63]. This means further formulation refinements to tailor make water-based PSAs for each type of coater configuration (Table 9.2). The higher speed rotogravure coaters require emulsions with higher mechanical stability [63]. They require low viscosity dispersions when wetout is difficult; therefore more surfactants are added in the formulation, which in turn leads to higher foaming. The reverse roll coating is suggested for direct and transfer coating (Table 9.3) as the choice of a coater depends on the basic formulation.

Table 9.2 Dependence of the Formulation of Water-Based PSAs on the Coater Configuration

Viscosity range		
Value (mPa·s)	Method	Coater type
100–1000	Brookfield, 20 rpm	Wire wound rod (Meyer bar)
1400–3400	Brookfield, 12 rpm, spindle 3	Gravure
100–1500	Brookfield, 20 rpm	Dahlgren
3000–10000	Brookfield, 20 rpm	Knife over roll
50–500	Brookfield, 20 rpm	Reverse gravure
1000–5000	Brookfield, 20 rpm	Nip fed
200–3000	Brookfield, 20 rpm	Pan fed
1000–10000	Brookfield, 20 rpm	Slot die

Table 9.3 Dependence of the Choice of the Coater on the Basic Formulation

Coater type	Formulation characteristics			Running characteristics					End-use characteristics				Refs.
	Physical state	Viscosity (mPa.s)	Anti-foam requirements	Speed (m/min)	Shear stress	Reflow requirement	Wet-out difficulty	General use	Coating weight (g/m^2)	Face stock paper	Film	Others	
Knife Meyer rod	SB, WB	100–1000 L	M	H	M	L	H	D/T	0.5–50	x	x	—	
Chambered Doctor blade	HM, WB	M	L	M	M	L	M	D/T	5–50	x	x	—	
Knife over roll	SB, WB	> 3000	M L	L	M	H	L	T	0.5–35	x	x	—	
Direct roll	SB, WB	M	M	M	M H	M	M	D/T	20–200	x	x	x	[31, 203–206]
Reverse roll	HM, SB, WB	1000–5000 300–500	M L	H	M L	M H	M L	D/T	2–50	x	x	x	
Rotogravure	HM, SB, WB	50–500 VL	H	H	M H	L	H	D	1–40	x	x	x	
Slot die	HM, WB	1000–10000	L	M	M L	M	H	D/T	1–200	x	x	x	

SB, solvent-based; WB, water-based; HM, hot-melt
D, direct coating; T, transfer coating;
VL, very low; L, low; M, moderate; H, high

2.4 Other Coating Devices

Air Knife

It appears easier to coat an adhesive when the web is flooded (air knife) with PSAs before doctoring occurs, rather than receiving coverage via a theoretically precise layer transfer from a metering roll. The air knife often allows the passage of a particle of solid matter which could cause continuous lines when caught under the blade. Spray from the continual blast of air onto the wet coating can be a nuisance and foaming will only be made worse, but the introduction of predoctoring, either by roller or wire-wound rod, can reduce air knife pressures and volumes, and hence eliminate potential problems.

In the air brush method the coating weight is regulated via air flow. Coating weights of 7–11 g/m^2 are possible at a viscosity of about 200 mPa·s, with a tolerance of 1 g/m^2 [43]. Air knife systems operate with a 1500-mm water pressure. In the case of kiss-roll coating with air brush, the latex is coated using a smooth cylinder, and the excess of the latex is blown away by an air brush; a thin air stream with a pressure of 6.9×10^{-3} N/mm^2 should be used.

Slot-Die Coaters

Simple slot dies may have coating weight tolerances of ± 1.5 g/m^2 [65]. The slot-die coater and its use for dispersions have been discussed in a detailed manner [66]. Bolton-Emerson developed a gear-in-die slot orifice die. This has a liquid pumping section built into the die; the pump is designed to cover the complete die length. Advantages of this new die coating method include higher solids, better accuracy, no foaming, and low shear [67]. The original and patented tube extrusion die for full and strip coatings was invented by George Park; over the years it was substantially further developed [27]. The die evolution went from the original simple tube die to the coathanger configuration, from there to the gear-in-die (GID), and lately to today's Duplex coater. The Duplex system lends itself not only to aqueous, but also to two-component crosslinking adhesives. The shear rate in the slot die is lower than in roll coaters, thus dispersions which need a high shear rate to be coated (for film forming) are not adequate for slot-die coating [26].

A coating speed of 250 m/min is achieved for hot-melt PSA coatings [6]. Film coatings may show ribbing patterns that are generated by a poor adhesive system. Ribbing seen on the applicator roll will result in ribs in the film. Foaming can be caused by the recirculation of the adhesive emulsion from the pick-up roll into the pan. The main coating defects and the methods to avoid them are listed in Table 9.4.

Table 9.4 Coating Defects and Methods to Avoid Them

Coating defect	Cause/Solution
1. Foaming	*Cause*: Pumping, stirring speed too high; viscosity too low; wetting agent level too high, inadequate.
	Solution: Reduce speed; increase the viscosity; add defoamer.
2. Wetting problems	*Cause*: Viscosity too low; surface active agent inadequate, surface active agent level too low; release liner inadequate; treatment level (films) too low.
	Solution: Increase viscosity; add adequate surface active agent, increase the coating speed and shear; increase the drying speed, change the release liner; treat the film again.
3. Bubbles in the adhesive layer	*Cause*: Air speed too high; drying speed too high.
	Solution: Reduce air blow level, reduce drying temperature in the first drying step.
4. Migration of the adhesive in the face stock material	*Cause*: Viscosity too low; drying speed too low; paper porosity too high; low molecular water soluble components in the formulation; low molecular tackifier/plasticizer in the formulation; surfactant level too high.
	Solution: Increase viscosity; decrease surfactant level; increase the molecular weight of the low molecular weight components in the formulation; use primer coating, make transfer coating.
5. Dry adhesive layer is too soft, legging	*Cause*: Adhesive is not dried enough; surfactant level too high.
	Solution: Improve drying; use higher temperature and air volume (WB).
6. Adhesive layer is not smooth enough, shows lines	*Cause*: Viscosity too low/high; grit; coating device is not clean, mechanical stability of the PSA is not sufficient; too high shear during transport.
	Solution: Change the rheology of the product; filter the PSA, change the shearing conditions, change the coating device.
7. Release too high/low	*Cause*: Drying is not adequate; release liner is not adequate; release liner is too fresh/old.
	Solution: Improve drying conditions; change release liner quality and age.
8. Coating weight too high/low	*Cause*: Viscosity is too low/high; solid content is too high/low; coating speed is too high/low.
	Solution: Change the rheology of the formulation; change the solid content of the formulation; change the coating speed, coating geometry.

3 COATING OF HOT-MELT PSAs

The composite laminate is typically manufactured by coating the hot-melt PSA in a molten state at a temperature above about 130°C, from a slot die or roll coater onto a release liner. The coated release liner then is laminated to the face stock with a nip roll using pressure between a rubber and a steel roll. The coating machines and devices used for hot-melt PSAa have been described [68–70]. Coating devices used for hot-melt PSAs are similar to coating devices used for solventless silicones. Evidently, other coating systems like slot die roll and reverse roll coating geometries were proposed [71].

3.1 Roll Coaters for Hot-Melt PSAs

Coating devices using engraved cylinders move the adhesive from the nip with the aid of a cylinder. The adhesive layer is smoothened by a knife (pressure and angle regulated). The molten adhesive is transferred directly to the web or indirectly through the use of an elastical cylinder. The coating weight depends on the line numbers and geometry of the cylinder, and the viscosity, adhesion, and cohesion of the adhesive; the viscosity is limited to 5000 mPa·s. A reverse roll system permits the use of 150–20,000 mPa·s. In this case, a metallic cylinder removes the adhesive from the nip. The excess adhesive is controlled by a reverse roll metering cylinder (cleaned by a knife) [68].

The advantages of roll coating systems for hot-melt PSAs [71] are a low sensitivity towards dirt, low tolerances in coating weight (1 g/m^2 at 60–350 m/min), rapid changes of hot-melt PSA formulations, and a rapid switch on/off and stop mode is possible. The disadvantages of the roll coating systems for hot-melt PSAs can be summarized as follows [71]: a direct contact between cylinders should be avoided, a coating weight range of only 18–100 g/m^2 is possible, a fine ridging on the adhesive layer may be observed, and the system is only half closed and thus it is not protected from oxidation.

It may be concluded that depending on the desired coating weight, either the slot-die system (10–16 g/m^2) or the roll coating (16–100 g/m^2) should be selected [71]. A roller coating system with four cylinders is able to operate up to 5000 mPa·s, and coats up to 250 g/m^2 [72]. Thus the main disadvantage or limiting factor in the use of roll coaters for the hot-melt PSAs remains the viscosity [72]. In the kiss coat procedure the web is going over the application cylinder (one or two cylinders) and is coated in normal or reverse roll sense. The coating weight is strongly influenced by the contact web/cylinder, web tension, and rotating speed of the cylinder. No viscosity above 920 mPa·s can be successfully coated. Continuous metering of the adhesive is possible using a smooth cylinder coating device, where the web travels between two or more cylinders. The speed of the web, the viscosity, temperature, clearance knife, web angle, cylinder synchron, etc., influence the coating weight. This system

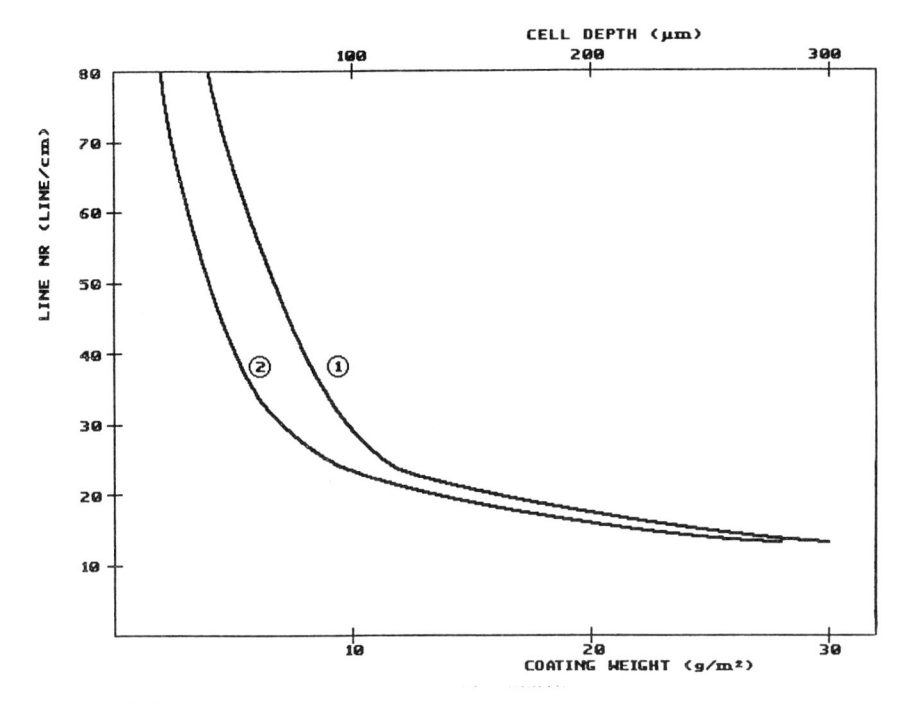

Figure 9.25 Gravure coating for hot-melt PSAs. The choice of a hot-melt coating cylinder as a function of the coating weight on the line number. Working temperature 120°C. 1) and 2) are two different cell depths.

may be used for hot-melt viscosities of 300–500 mPa·s [68]. In a manner similar to solvent-based or water-based adhesives, diagrams or tables are used in order to select the adequate engraved cylinder (Figure 9.25).

For hot-melt coatings a running speed of 250 m/min can be achieved [73].

3.2 Slot-Die Coating for Hot-Melt PSAs

The justification for purchasing hot-melt slot-die coaters is based on economic considerations. The use of highly concentrated coating media and the necessity to avoid solvents has created the demand for coating heads which can handle highly viscous materials. The roller coating methods have steered new development into different directions. Through the combination of reverse-roll-coating with the knife-over-roll coating, a new coating method was born which allows the handling of a wide range of viscosities and which is applicable to existing equipment and machinery. Above a viscosity of 30,000 mPa·s it is not possible to use a metering roll; in this case a slot die is required. Higher viscosities are used for special products like electron beam-curing of hot-melt PSAs. In this case viscosities up to 1,000,000 mPa·s are possible, but only in the final stage [73]. Coating systems for hot-melt PSAs with slot die (1300–1500 mm) are able to coat 18–40 g/m^2 at a manufacturing speed of 15–270 m/min [74]. The gear-in-die system used for solvent-based and hot-melt PSAs may operate in a temperature range of room temperature to 250°C, with viscosities of 200–1,500,000 mPa·s.

The slot-die system for hot-melt PSAs was tested in 1964 by George Park (Park Coater) [76]. The advantages of the die system may be summarized as follows:

- high viscosities are allowed;
- it is a simple coating device;
- coating weight adjustment is easy;
- there is a low oxidation level (closed system) [76];
- coating weights as low as 10 g/m^2 are possible;
- a wide web possible [77].

The disadvantages of the system are the following:

- no changes in the recipe are possible (without cleaning);
- frequent cleaning of the die is necessary [76];
- there is a sensitivity towards changes in the thickness of the web [77];
- there is a sensitivity towards fine adjustments of the die;
- there may be lines in the adhesive layer.

For hot-melt PSA coatings where usually coating weight and viscosity are rather high, the system is entirely heated by diathermic oil and consists of a slot die and rewinding bar, an interconnection hose and hot-melt PSA supply gear pump that are electronically adjusted to line speed in order to apply the set and targeted coating weight at all line speeds. This system possesses good reliability, coating thickness and uniformity, and adjustment and repeatability of the production conditions. If the coating device is fed by a continuous extruder, the residence time of the molten PSA is not more than 2–6 min (200–800 kg/hr). Viscosities of 120 Pa·s at 175°C are possible and the energy input of the system is 7–12 kWhr/kg [78].

Metering devices for GID and Duplex systems, used mainly for hot-melt and high solid solvent-based coatings (mainly for tapes), show the following advantages [43]:

- lower costs for drying and solvent recovery;
- higher productivity (30%) at a maximum of 420 m/min;
- lower installation costs;

- lower pollution;
- simpler cleaning;
- less space required;
- higher accuracy;
- higher solids;
- lower shear;
- no foaming.

The GID device possesses an interspace between the die and the coating roll; the adhesive is fed into this interspace with or without the pump (Duplex/GID). Advantages include the absence of turbulence, foam, and residues on the die. The system was originally developed to be used for all coating media (i.e., emulsion, hot-melt, and solvent-based PSAs) [67]. It has an entirely new slot orifice die, with a liquid pumping section built into the die. The pump is designed to cover the complete die length. The coating die is hinged for quick opening and easy cleaning; adhesive transfer to the coating slot is given via an hydrodynamically acting rotor over the entire width, with rotor speed adjustable according to media viscosity and required coating weight, and no pressure builds up in the coating die (thus there is no overflow and build-up of adhesive on the edges).

4 DRYING OF THE COATING

After coating solvent-based or water-based adhesives the liquid adhesive carrier has to be removed (i.e., a drying channel or tunnel is required). Recent developments for drying tunnels have been discussed [80–88]. The main component of a dryer is the adhesive drying tunnel operating in the most cases with heated air. The web travels through the tunnel supported by a conveyor belt, idlers, or air (contact free).

4.1 Adhesive Drying Tunnel

The design of a drying tunnel and its parameters depend on the nature of the adhesive to be coated and dried. The choice of the conveyor belt and length of the drying tunnel is of fundamental importance and is selected depending on whether the application uses solvent-based or water-based adhesives, the desired dry coat weight, the solids content, the operating speed, the type of web, etc. The various sections with independent air flow rate, temperature, and exhaust/recycle features can be selected and/or combined; among them:

- standard rollers;
- rollers with adjustable air over/underneath the web;
- air flotation;

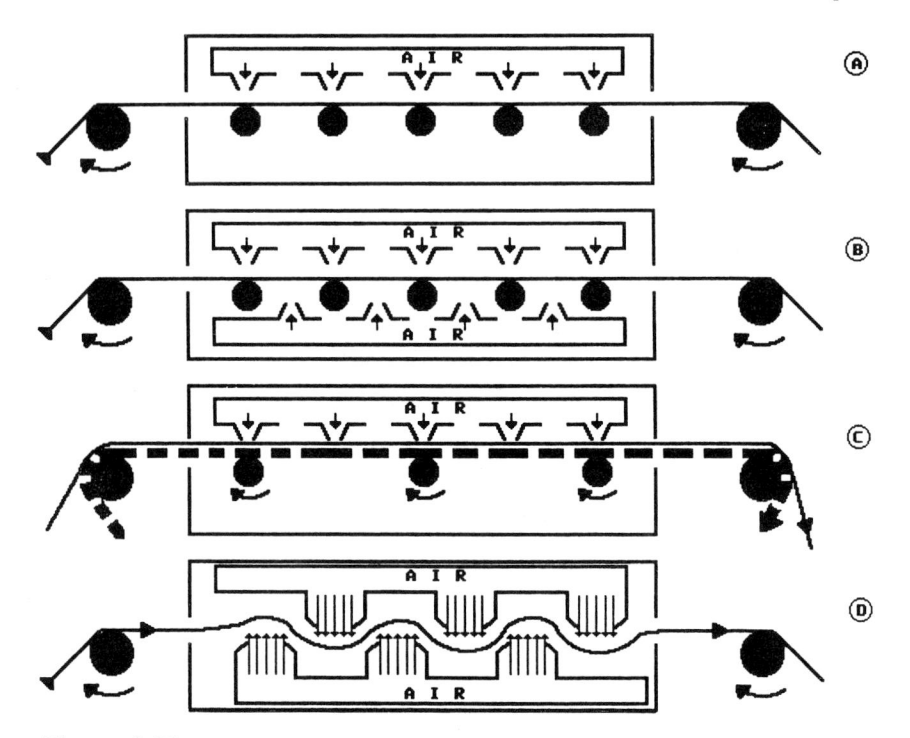

Figure 9.26 The main constructive parts of the drying tunnel. A) drying tunnel with rollers, with air jets over the web; B) drying tunnel with rollers, with air jets over and under the web; C) drying tunnel with rollers and conveyor belt; D) drying tunnel with air flotation.

- conveyor belt;
- festoons.

Figure 9.26 shows the main types of a drying tunnel.

The choice of drying system for solvent-based or water-based adhesives (or both) must take into account the relative evaporation curves of the solvent or water through the various sections. In the case of solvents an increased amount of solvent should be evaporated at the beginning of the tunnel. For water-based coating the water evaporated is maximum in the middle of the path. The drying conditions for water-based PSAs differ from those of solvent-based ones. Water-based systems are more efficiently dried by air volumes rather than air temperature. The temperature gradient throughout the drying oven should also be different [89] (see also Figure 7.2). Energy requirements for solvent-based coatings are 170–200 cal/kg and for water-based ones 400–

500 cal/kg [90]. On the other hand, safety requirements require more air volume than necessary from a thermal point of view. For example, drying a water-based dispersion requires 2.2×10^5 kcal/hr compared to 7.61 kcal/hr for a solvent-based one. A common mistake is the application of too much heat [31]. This leads to the "flash off" of the upper water layer and skinning occurs. For aqueous PSAs drying is generally more efficient with multizone ovens programmed for gradually increasing temperatures. Typically, a first zone temperature of 160°F may be utilized with subsequent gradients up to 200°F. Such designs have eliminated the skinning and coating defects [44]. If an air-flotation oven is used, the distance between the coating surface and the nozzle is quite critical. Adjustments of as little as 12 mm can create a totally new drying environment as this gap defines the actual drying temperature [31]. Drying channels have slot dies for the air (8–10 cm between the dies), situated 2–4 cm from the web; the air flow rate amounts to about 60–80 m/sec [91]. At the entrance of the channel, the temperature should be high enough in order to evaporate large water quantities. After a decrease of the temperature in the middle, the temperature should increase again at the end in order to eliminate the remainder of the humidity from the adhesive [91].

Different drying methods are actually used (hot air, IR, radio frequency, UV, electron beam) [92,93]. Generally, the drying time depends on the adhesive and on the drying conditions. The drying parameters depend on the coating weight, temperature, volume, and speed of the air. The air flow rate depends like other aerodynamical parameters on the machine construction. For aqueous dispersions drying also depends on the pH.

There is a lot of literature data concerning the drying of adhesives in general and PSAs in particular [90,93,94]; solvent-based and water-based adhesives require quite different drying conditions. Concerning the speed of the air, the blow off (blow away) of the coating should be avoided through a limited air speed (i.e., 25–45 m/sec) [93,95].

Infrared Drying

Infrared heaters generally are used in plastics processing and coating [96,97]. Wavelengths are selected in function of the product to be dried [10]. Infrared heaters used as preheating displace 10–30% of the humidity from the coating. Infrared drying also may be used for paper coating [98,99]. Infrared drying may avoid adhesive penetration [100]. Short wavelength devices have a 50–300 kW/m^2 output, 80% IR heat output, and fast on/off switching. The wavelength of IR dryers for water-based adhesives should be in the domain where maximum IR absorption by water occurs [91]. Infrared drying and equipment has been described [10]; heating element parameters, air knives, exhaust and cooling, and processing considerations are reviewed and an example of an IR dryer calculation is given. Infrared radiation with a wavelength of 0.76 µm to 1 mm should be used; the highest energy transfer is given at a wavelength shorter than < 2 μm [102]. The advantages and disadvantages of IR/UV drying have been discussed [103]. Radiofrequency drying may be used also [104]. Broad-band infrared radiant driers also may be used. The main benefit claimed is that they greatly speed up drying, cutting the residence time by at least half. Another drier, the air knife (originally developed for use with water-based coatings), blows thin curtains of very hot air across freshly coated web at speeds up to 1400 m/min. Another system utilizes convection to achieve high heat transfer rates without the web being lifted from the nozzle or disturbing the coated surface. New developments offer cassette-based drying systems to provide the best drier for each type of web with rapid changeover.

Flying Dryer (Flotation Dryers, Contact-Free Dryers)

The construction of a flying dryer has been described [39]. The distance between web and dies is about 3 mm [105–107]. The air speed amounts to about 50 m/sec, the temperature 250°C, and the dryers length 3–30 m. The primary advantage of using a flotation drying system is the elimination of damage to the web or coating because there are no idle rollers. The flotation system allows full web coating coverage and the web must be run with much lower tensions. Cleaning downtime also is reduced [14]. Flotation driers for paper were discussed by Drechsler [39].

Radiofrequency Drying

Radiofrequency (RF) drying is used mostly for PSA-coated paper material [83,84]. It is a characteristic known as the "dielectric loss factor" which determines how readily a given material will respond to RF energy. Typically for tap water the loss factor is 100 at an RF value of 10 MHz (i.e., at a frequency of 10 million oscillations per second); for paper it is less than 1. Thus, wet areas of a product are selectively heated and dried while dry areas are unaffected [108].

Ultraviolet and Electron-Beam Driers

The use of UV predominates when drying inks, coatings, and adhesives in printing and converting operations [109]. Heat-sensitive substrates may run at maximum speeds without damage or change of moisture content of the web.

UV and electron-beam driers are much more energy efficient than thermal driers, as they do not depend on heat. The high capital cost of electron-beam driers realistically makes them only suitable for use as a final cure system. In flexo printing using an electron beam-curable ink on a PSA hot-melt laminate, the electron beam radiation provides at the same time cure of the print and after cure of the hot-melt PSAs. The advantages of high speed electron-radiation curing include the absence of solvents or monomers in the converted product, the improved web profile achieved by "cold" running, short web paths, economical operation, easy cleaning, quick changeover, and the absence of the explosion risk [109].

During drying the liquid carrier (solvent or water) is evaporated; recycling them is very important. The recycling of solvents has been discussed [110–118]. Today, solvent-based adhesives are facing a continuing attack of environmental regulations. For many years there were very satisfactory means of recovering solvents and preventing emissions into the air; various alternatives are available:

- *Absorbtion by active carbon and dissolution by steam.* This the most common method; this system is suitable for hydrocarbon solvents, which covers most rubbers and some acrylic PSAs.
- *Recovery by condensation.* This system is more expensive, but suitable for all solvents and yields the solvent or solvent blend without contamination [118].
- *Afterburning.* As selling recovered solvents becomes more difficult and complicated solvent blends become more prevalent in the manufacture of high performance acrylic adhesives, many manufacturers are changing over to afterburners on the drying line. When the solvent vapors are concentrated in nitrogen before burning, the afterburning can yield an energy output that can be used to heat the drying air [120].

Biological cleaning of the air is used also [121]. Solvents that are easily recoverable are used routinely; products are also polymerized in toluene and hexane. Generally, the following solvents are used: toluene, hexane, ethyl acetate, and isopropyl alcohol.

The two most prominent processes in the recovery of solvents from the process air are absorbtion and condensation. Absorbtion is accomplished on a bed of activated carbon, where the solvent is trapped by the carbon and is recovered by steam stripping. In condensation the solvent is recovered by lowering the temperature of the process air to a point necessary to condense the solvent and lower the saturation concentration of the process air. The presence of reactive materials (e.g., isocyanates, phenol, free monomers, and ketones such as methyl ethyl ketone) can be troublesome. Ketones, which have a higher heat of absorbtion than many other solvents, are of particular concern and generally their presence is forbidden in solvent-recoverable adhesives. The water insolubility factor plays a most important role in the recovery of the solvent from the carbon bed. If esters of chlorinated solvents are present, corrosion problems may occur.

5 SIMULTANEOUS MANUFACTURE OF PSAs AND PSA LAMINATES

5.1 Radiation Curing of PSAs

Chapter 5 (Chemical Composition) discussed some special cases where the monomers to be used for manufacturing the adhesive are not polymerized, or polymerized only partially and coated as such; in this case, the PSA is manufactured during the coating. These are special materials with the ability to polymerize through radiation (electron beam or UV). As an intermediate step, there are special "classical" PSA formulations which can be dried or cured using radiation as well. Although an intensive effort was carried out in this field, electron beam- or UV-cured PSAs or PSA-related materials are still considered exotic.

Pressure is growing on convertors using solvent-based materials to look for methods which are environmentally acceptable [122]. Here the field of chemistry can offer new groups of products, namely systems with a low solvent and a high solids content (i.e., water-based systems, coatings with solid resin in powder form, and 100% prepolymerized solid systems). The following feasible solutions as alternatives for solvent-containing formulations are known for PSAs: formulations on an aqueous basis, formulations on a hot-melt basis and radiation-curable and crosslinkable formulations. Numerous papers were published in the last several years in this field, some of which could serve as a basis for industrial applications. Techniques which do not include the disadvantages of thermal drying are the various methods of radiation drying, in particular those including ionizing radiation such as ultraviolet- and electron beam-curing. In this case the energy is transferred without contact. Energy transfer can take place in a vacuum or through a vacuum section. Electron beams belong to the directly ionizing beams at energy levels above 3 eV. Some of the advantages of electron beam processing for PSAs include higher productivity, compliance with regulatory restrictions, enhanced profitability, and increased product development capacity [124].

S-I-S block copolymers can be crosslinked to become PSAs by means of reactive monomers under radiation [125–127]. Electron radiation crosslinking of PSAs has enjoyed industrial applications for several years now making PSA tapes. Furthermore Huber [129,130] was able to present new radiation-curable copolyesters. These are saturated copolyesters containing reactive acrylate end groups [128]. Nitzl and Förster [123,126,131] studied hot-melt PSA formulas based on block copolymers; the base polymer used was Kraton D 1320X, a branched styrene butadiene copolymer with a functionalized end group in combination with hydrocarbon resins. The various formulas were crosslinked by electron beam exposure at doses of 2 and 4 Mrad. The available results show that there is no clear dependence of the adhesive properties on the formulation [123].

As postulated by Seng [132], the most important application fields of radiation curing is the curing of silicones and PSAs. Release-coating curing is done mainly for labels, while curing of the adhesive layer mostly for labels and tapes; a postcuring via electron beam is made for tapes also [128]. In all these domains, both techniques present advantages and disadvantages [132]. Electron beam-curing is faster, up to 100–500 m/min versus 100–200 m/min for UV, but needs a nitrogen atmosphere and may damage paper. UV displays the disadvantages of rest monomers, a limited thickness of the layer, and too low penetration of the radiation.

The state of development of radiation curing (electron beam and UV) has been discussed [133,135]. The first application of electron beam-curing for PSA tapes was described in 1954 [136]. Silicone acrylic PSAs were used by Hurst in 1983 as release liner [137]. An in-line UV-curing hot-melt PSA/silione release manufacturing line was proposed in 1982 [137]. Other papers discuss the problems of radiation curing for lacquers, printing inks and adhesives [138–142]. The coating and printing of radiation curing systems has been discussed in a detailed manner [143–147]. The technology of UV curing for varnishes and pigments in the printing industry ("Face stock printing for labels") has been discussed [148–152]. Radiation curing possibilities for plastics and paper coating were covered by Röltgen and others [153–158]. Guidelines for the formulation of radiation-cured adhesives were reviewed by Hinterwaldner [159–160]. The electron beam-curing and drying of varnishes and coating was described by Strohner and Fuhr [161,162].

5.2 Siliconizing Through Radiation

An emerging coating application which lends itself to UV/electron beam-cure technology involves silicone release coatings. There are several reasons for the interest in these new techniques. Because UV and especially electron beam-curing occurs at very low temperatures, these kinds of silicone release coatings can be applied and used on heat-sensitive liners (e.g., PET, polypropylene, or polyethylene films) and polyolefin-coated papers without distorting such liners. For the same reason the application of radiation-curable release coatings to Kraft paper liners eliminates the need for remoisturization of the coated paper. The complete, immediate cure of radiation-curable release coatings eliminates "blocking" of rolls of finished product and provides greater assurance of uniform release characteristics from the outside of a roll to its core [163]. Electron beam-curing enables simultaneous curing of both sides of two-sided coated release liners. UV/electron beam technology is a considerably more efficient means of curing than thermal drying; the energy cost is less than that of conventional (i.e., thermally dried) release coatings and the need for large ovens disappears. The practice of in-line mixing of release coatings with a relatively short pot life could be eliminated with these types of coatings. These fully formulated products typically have shelf lifes of 6 months to 1 yr. Theoretically, the "hot" UV system, the "cold" UV system, and electron beam-curing can be used [164]. In some cases solvent-based silicones were postcured with electron beam [164]. The coating weight for electron beam-cured silicones amounts to 0.7–1.5 g/m^2 with a curing time of 0.3 sec. In some cases the release properties of electron beam-cured silicone papers improve with storage [165]. The performance characteristics of electron beam-cured silicones and the advantages/disadvantages of the procedure were summarized by Brus [167].

Radiation-cured silicones have to be coated as a continuous layer, at a coating speed of 300 m/min, a coating weight of less than 1 g/m^2, and with a viscosity below 1500 mPa·s. They contain at least 90% dimethylsiloxane. Radiation-cured silicones are based on prepolymers; radiation-cured PSAs also contain reactive monomers [155]. The most common radiation-cured silicone systems are based on acrylic (silicone derivatives) which are crosslinkable via electron beam or UV radiation. Another system is based on mercapto-modified silicones; these are not air sensitive and may be used for UV curing. The initial tests were carried out with UV-cured silicones, mostly in combination with hot-melt PSAs. In comparison with electron beam-cured PSAs, electron beam-cured silicone liners need no more than a 0.5 Mrad dose. The highest efficiency is achieved coupling hot-melt PSA coatings with in-line siliconizing. In this case about 30% of the energy costs may be saved [155].

Electron Beam Curing

Electron beam-cured silicone release papers exhibit different release forces depending on the paper thickness [168]. A 67-g/m^2 paper displays a higher release peel force than a 90-g/m^2 paper (84–98 mN/cm) [168]. Electron beam-curing may damage the paper and increase its stiffness; PVC may undergo yellowing. Furthermore, electron beam-cured silicone release liners have limited storage resistance. In this application the electron beam-curable silicone release layer is coated via rotogravure (1 g/m^2) [169]. For release liners manufactured via electron beam-curing a coating weight of 0.8–1.2 g/m^2 and 0.4 g/m^2 is suggested for paper and for films, respectively, at a radiation dose of 1.5–2.0 Mrad [169]. An inert atmosphere is required for this application; the oxygen concentration should be lower than 50 ppm [170].

The advantages of radiation curing include the following [135]:

- The paper is not dried.
- Thermal sensitive films can be used.
- No drying oven is necessary.
- High coating weights are possible.

- High running speeds are possible (up to 900 m/min) [140].
- Two-sided in-line coating is possible.

On the other hand, some technological disadvantages need to be mentioned. The prepolymer/diluting agent blend is highly viscous, so smootheners are necessary [158]. Another problem is the toxicity of the used monomers [139].

The average energy consumption for solvent-based or water-based siliconizing is about 0.14 kWh/m^2; a UV-cured release liner needs 0.0143 kWh/m^2 [154].

In general, electron beam technology provides manufacturers of PSAs laminates with the following benefits [171]:

- New products can be developed by electron beam-curing/crosslinking using a variety of substrates, coatings, and adhesives, with different property combinations.
- A higher productivity is provided because modern electron beam units operate at higher speeds and process wider webs with minimum waste.
- Electron beam users are in compliance with environmental regulations because the systems use solventless adhesives, inks, and coatings.
- Profitability is enhanced because electron beam technology can generate new product and market opportunities while improving the performance of existing products.
- Reproducible product uniformity is achieved because electron beam units ensure a uniform cure.

Electron beam-curing/crosslinking can enhance the shear resistance, the high temperature resistance, and oil or solvent resistance of thermoplastic rubber hot-melt PSAs. Integrating the electron beam processor in line with the hot-melt coating equipment enables single operation processing prior to slitting.

"Dual cure" is a method where curing is initiated via radiation and completed through a thermal method [172]. There are special S-I-S block copolymers suitable for electron beam-curing [173,174].

There are two main types of electron beam generators: partial beam and scanner [172]. The electrocurtain process was described by Lauppi [175,176].

UV Curing

Limited space requirements and a low thermal loading of the face stock material are the advantages of UV-cured silicones [177]. The coating weight for UV crosslinked silicones may be adjusted within a tolerance of ±5%. UV-curing siliconizing machines run at a speed of 400 m/min [177]. UV-cured PSAs possess the following advantages [178]: rapid curing, low energy requirements, limited machine preparation costs, absence of solvents, and a long shelf life. They may be coated using screen printing or roll coating technology.

6 MANUFACTURE OF THE RELEASE LINER

6.1 Nature of the Release Liners

Generally, the PSA label is manufactured as a laminate composed of a face material laminated to a release liner with a PSA. In many cases the adhesive is first coated on the release liner and not on the face stock. There are two methods of depositing the PSAs onto the face sheet: direct and transfer coating. The former deposits the adhesive onto the face material. In the transfer method the adhesive is deposited on the release paper and is subsequently transferred to the face stock during the laminating process. It is evident that especially in transfer coating, the performance of the release liner will also influence the performance of the final PSA label. On the other hand, the properties of the release liner depend on its chemical composition and manufacturing conditions. Many converters manufacture PSAs and release coatings and there are some common features for both technologies. Although it is not the aim of this work to discuss the chemistry and manufacturing of release liners its influence on the label (and label technology) are reviewed next.

Other materials than silicone may be used as a release liner. The release surface will normally be provided by a layer of silicone resin or a material providing similar release properties coated on the liner. Silicone-free vinyl acetate-based emulsion polymers were suggested as release coating [179]. Derivatives of polyvinyl carbamate also are used for protective films and tapes, but the release effect of silicones remains superior. Where a coating of silicone polymer is used, the coating weight will typically not exceed 2 g/m^2 and a liner of 60–100 g/m^2 should be used [180].

Paper and nonpaper materials are used as liners. As printing, converting, and label application speeds are increasing, web breaks and V-tears in the liner cause unacceptable scrap loss and downtimes. Hence plastic films are now competing with paper-based liners. Today solvent-based, solventless, and emulsion (water-based) silicone coatings are routinely coated.

The liner is the packaging for the adhesive, the carrier material for die-cutting, and sometimes the transparent support for the label [140]. Generally, (as shown earlier in Chapter 7) carrier materials for PSAs are flexible webs, supplied as rolls which are coated, dried, and rewound [3]; silicone derivatives are used as coating materials. Their function is to separate the adhesive from the release paper, to offer an easy but controlled and reproducible separation of the face paper from its adhesive layer (or vice versa). A very low coating weight of silicone is deposited and a very smooth defect-free surface should be obtained.

The weight of the release liner depends upon the method of label converting, the type of die cutting and tools, and the mechanical strength of the sheet. The release coating is usually a silicone formulation and is designed to provide a variable release level, again depending on the method of converting and whether the adhesive is cooled over the whole web width or is pattern or zone coated. For years the only release liners used in pressure-sensitive label applications were calendered kraft papers in the 40–50 lb weight range, with one side silicone coated. As processing speeds increased, web breaks and V-tears in paper liners became a growing concern. The use of 150-g siliconized PET as a release base resolved the problems associated with high speed dispensing applications [181]. Paper and polyester are just two of the commercially available release liner systems. Oriented polypropylene and polyolefinic blends also are used for high speed application systems. Another technology employs a liner structure of 30-lb kraft paper extrusion coated with 14 lb of polypropylene; these materials offer a greater margin of security during die cutting [182]. Polyethylene has been used as a release liner since 1970 [183].

Face stock materials and release liner materials were reviewed in Chapter 7. Their nature is important for quality control purposes because of their influence on the release (peel) force. Papers used as release liner must display superior mechanical properties, little thickness variation, and end-use tailored properties [6]. The release force for pigmented papers depends on the nature of the clay as well as the level and nature of the resin [184]. The curing method also influences the choice of the release paper [6]. The main factors influencing the release force include the silicone formulation control release agent (CRA) level, the coating defects present, the mechanical destruction of the silicone layer, and the possible incompatibility between release and adhesive formulation. Mechanical defects can be avoided using softer rubber-coated cylinders (60 shore hardness, SH) and the offset process. The adhesive and release properties may be influenced by the interaction of the adhesive and silicone. This problem may be avoided by using another silicone catalyst, but the activation temperature may increase. High peel forces are necessary to separate water-based acrylic PSA labels from Sn-catalyzed silicones. Only addition polymerization with a Pt catalyst allows lower release forces [185].

Generally, the most difficult problem is the control of the silicone coating weight. The silicone coating should not show any pores and display a good anchorage to the liner [6]. Appropriate amounts of solvent-based silicones are $0.5–0.8$ g/m^2, and for solventless silicones, $0.7–1.5$ g/m^2 [184]. The reactive groups in the silicone or PSAs, their mobility, and the mechanical influences determine the release properties [186].

Usually the release liner displays the minimum resistance to debonding. However, this low release force is inadequate in certain applications as greater resistance may be required. In these cases special silicone resins (control release agents) are used. These agents must be compatible with the base silicone resins and be capable of being blended with them in any ratio; various base resins are used as control release agents:

- control release resins for polycondensation in solvent systems;
- control release emulsions for polycondensation systems;
- control release resins for polyaddition systems with and without solvent.

Polycondensation siliconizing is not used in Europe [6]. In 1988 in Europe, 50% of the silicones were solvent-based, 39% solventless, and the rest water-based. Since the 1980s, solventless silicones are being used at an increasing pace [186].

6.2 Coating Machines for Silicones

The same coating machines as the ones used for PSA coating are suitable for siliconizing. Coating machines for PSAs have been extensively discussed in the literature and were described in a detailed manner in Section 1.2 [37,187]. Silicone coating machines are similar to the adhesive coating machines. Different coating devices should be selected for different applications [37]:

- in line manufacturing: release-Corona-primer-PSA, Corona-primer-PSA;
- manufacture of very thin layers, like solventless silicones;
- manufacture of self-adhesive medical products;
- manufacture of self-adhesive special papers (crêpe);
- manufacture of self-adhesive woven and nonwoven materials;
- manufacture of self-adhesive tamper-proof and protective films;
- manufacture of pattern coatings.

The rheology of PSAs differs only slightly from that of the silicones (coating technology), hence similar coating systems may be used for both systems. For solventless silicones a two–cylinder coating device was used at 120 m/min running speed [188]. A four–cylinder geometry deposits 0.5–2 g solventless silicone [189]. A five–cylinder nip fed coating device deposits 0.3–1.5 g/m^2 [37]. The machine speed for siliconizing ranges from 50–600 m/min [190]. Siliconizing at 400 m/min ensuring a ±5% weight tolerance appears possible [6,191].

6.3 Technology for Solvent-Based Systems

Different coating geometries were designed for different silicone/release types [192]. Machine widths of 80–250 cm and speeds of 50–200 m/min are indicated. The most important coating devices include gravure roll, roll blade systems, and kiss coating with knife-over-roll. Generally, silicone coating machines possess a coating, drying, and rehumidification station [193].

6.4 Technology for Solventless Siliconizing

A machine for solventless silicone coating is equipped with fully automatic unwinders and rewinders, as well as with a rehumidification unit [194]. Contrary to solvent-based or aqueous systems, nearly ten times or less liquid mass must be applied on the substrate (i.e., a quantity of 0.7 g/m^2 of paper normally suffices, generally less than 1 g/m^2) [188]. Solvent-free silicone preleased coatings vary from 1.1 to 1.4 g/m^2 [194]. Today coating quality tolerances are guaranteed within ±5% in longitudinal and cross directions. The silicone is applied with a five-roll system running at different speeds and provided with distinct surface structures [194]. Today speeds of up to 400 m/min with a working width of 2.3 m are reached and speeds of 500 m/min are conceivable. Usually, thermal systems are used for silicone curing, with single- or double-sided airfoil dryers, depending on the type of the substrate.

A combination of roll and blade coaters was suggested for 100% solventless silicone coating systems [95]. Totally solventless systems are cured with platinum or rhodium catalysts as addition reactions of Si-H groups and vinyl groups occur.

Siloxane microemulsions also may be used as release agent [196]. A release agent with excellent mechanical stability contains an organopolysiloxane microemulsion obtained by emulsion polymerization of an organopolysiloxane. Water-based silicones have been known since 1986 [197]. In 1988 20% of the silicone release liner production was based on water-based technology [198]. The advantages/disadvantages of the solvent-based/aqueous release coating are summarized in Table 9.5.

Table 9.5 Advantages/Disadvantages of Solvent- and Water-Based Release Coatings

	Advantages	Disadvantages
Solvent-based	Anchorage to variety of liners	Cost
	Absence of migrating species	Flammability
	Less sensitive to humidity variations	Environmental regulations
		Low solids content
	Faster drying	Clean-up
		Energy costs
Water-based	Cost	Foaming
	Nonflammable	Poor anchorage to certain liners
	Cleanup	Migrating species
	Nonpolluting	Humidity sensitivity

7 REHUMIDIFICATION/CONDITIONING

Due to the high thermal web strain caused by the generally high crosslinking temperatures of the silicone, humidity is extracted from the web. Therefore, this lost humidity must be added to the oven-dried web after the coating or curing process in order to eliminate the tendency for curling or for flatness correction purposes. For this purpose the paper is led through one or more jetsteamers where a highly saturated steam/air mixture is generated. Several papers review rehumidification [193,199–202]. In some cases, in order to increase the hygroscopy of the web, additives like carbamide, glycerine, or calciumnitrate are used [91].

REFERENCES

1. A. Dobman and J. Planje, *Papier u. Kunststoffverarb.*, (1) 37 (1986).
2. *Die Herstellung von Haftklebstoffen*, T1.-2, 2-14d, Dec. 1979, BASF, Ludwigshafen.
3. *Adhäsion*, (1/2) 27 (1987).
4. *Coating*, (4) 123 (1988).
5. *Paper, Film and Foil Conv.*, (10) 39 (1971).
6. J. Paris, *Papier u. Kunststoffverarb.*, (9) 53 (1988).
7. *Coating*, (3) 56 (1984).
8. E. Böhmer, *Norsk Skogindustrie*, (8) 258 (1968).
9. *Coating*, (6) 156 (1974).
10. W. Schaezle, *Coating*, (9) 314 (1987).
11. H. Hadert, *Coating*, (6) 175 (1969).
12. H. Klein, *Adhäsion*, (1) 15 (1976).
13. H. Klein, *Coating*, (11) 406 (1987).
14. J. R. Martin, *Paper, Film and Foil Converter*, (9) 81 (1989).
15. *Papier u. Kunststoffverarb.*, (11) 25 (1985).
16. *Coating*, (4) 23 (1969).
17. H. Klein, *Coating*, (10) 378 (1988).
18. H. Klein, *Coating*, (1) 18 (1986).
19. H. Klein, *Coating*, (12) 336 (1983).
20. H. Klein, *Coating*, (3) 60 (1988).
21. H. Klein, *Coating*, (4) 122 (1987).
22. H. Klein, *Coating*, (2) 49 (1987).
23. C. Massa, *Coating*, (7) 239 (1989).
24. G. Renz, *apr*, (24-25) 960 (1986).
25. *Coating*, (7) 240 (1987).
26. G. W. Drechsler, *Coating*, (7) 251 (1987).
27. *The Duplex Coater*, Bematec, 1992, Lausanne
28. H. Hardegger, *Coating*, (6) 194 (1992).
29. H. Klein, *Coating*, (2) 42 (1988).

30. D. Percivalle, European Tape and Label Conference, Exxon, Brussel, April 1993, p. 133.
31. *Adhesives Age*, (11) 28 (1983).
32. H. D. Patermann, *Kunststoffberater*, (1/2) 39 (1988).
33. *Coating*, (2) 49 (1988).
34. D. M. McLeod and P. M. Fahrendorf, Tappi, Polymer Laminating Conference Proceedings, September 1988.
35. National Starch Chem. Co., *Durotak*, Pressure Sensitive Adhesives, Technical Bulletin, 4/86.
36. H. Klein, *Coating*, (2) 42 (1988).
37. H. Klein, *Coating*, (10) 372 (1988).
38. H. Baumeister, *Coating*, (10) 8 (1982).
39. W. Drechsler, *Coating*, (7) 253 (1987).
40. R. B. Kohler, P. H. Lathrop and D. M. McLeod, Tappi, Polymers, Laminating and Coating Conference Proceedings, September 1988, p. 205.
41. *Coating*, (3) 87 (1969).
42. H. Türk, *Papier u. Kunststoffverarb.*, (10) 22 (1985).
43. P. Herzog, *Coating*, (3) 73 (1988).
44. F. T. Sanderson, *Adhesives Age*, (12) 19 (1983).
45. *Adhesives Age*, (8) 24 (1988).
46. *Coating*, (2) 48 (1988).
47. F. Renk, *Coating*, (4) 146 (1988).
48. H. D. Patermann, *Coating*, (3) 96 (1988).
49. H. D. Patermann, *Coating*, (6) 224 (1988).
50. F. Weyres, *Coating*, (4) 110 (1985).
51. T.G.I., Twentse Graveerindustrie, Glanerbrug, *Rostermalzen*, Technical bulletin.
52. Technical bulletin, Pamarco Inc., Roselle, NJ.
53. Mario Benzi, *Coating*, (4) 120 (1991).
54. T. J. Bonk and S. J. Thomas (Minnesota Mining Manuf. Co., USA), EP 120.708.81/03.10.84.
55. P. M. Fahrendorf, *Paper, Film and Foil Conv.*, (9) 106 (1989).
56. J. R. Martine, *Paper, Film and Foil Conv.*, (9) 81 (1989).
57. H. Klein, *Coating*, (12) 142 (1984); (10) 372 (1988).
58. *Coating*, (12) 365 (1984).
59. *Coating*, (2) 50 (1978).
60. H. Klein, *Coating*, (7) 225 (1981).
61. Croda Adhesives, *Water Based Laminating Adhesives*, Newark (U.K.).
62. Backofen & Meier AG, Bülach, *Reverse Gravure Coating of Emulsion PSA*.
63. *Adhesives Age*, (3) 87 (1969).
64. *Coating*, (3) 87 (1969).
65. J. Paris, *Papier u. Kunststoffverarb.*, (9) 44 (1990).
66. H. Hardegger, *Coating*, (5) 153 (1992).
67. *Adhesives Age*, (7) 36 (1986).
68. H. Bohlman, *Papier u. Kunststoffverarb.*, (3) 12 (1979).
69. *Adhäsion*, (1/2) 23 (1993).

70. *apr*, (1) 18 (1978).
71. G. Kupfer, 11th Munich Adhesive and Finishing Seminar, 1986, p. 105.
72. H. M. Maschke, *Coating*, (6) 186 (1970).
73. *Coating*, (12) 344 (1984).
74. H. Klein, *Coating*, (11) 390 (1986).
75. *Coating*, (11) 393 (1986).
76. *Coating*, (8) 399 (1987).
77. G. Kupfer, *apr*, (16) 428 (1986).
78. *Coating*, (7) 238 (1989).
79. *Coating*, (7) 222 (1989).
80. R. Wimberger, *Coating*, (7) 258 (1991).
81. *Coating*, (7) 258 (1991).
82. R. Dehnen, *Der Siebdruck*, **32**:36 (1986).
83. A. Tawn, *Farbe u. Lack*, (5) 416 (1977).
84. J. J. Koch, *Der Siebdruck*, (3) 32 (1987).
85. H. Dickerhoff, *Coating*, (6) 187 (1993).
86. H. Klein, *Coating*, (7) 255 (1991).
87. E. Klas, *Coating*, (7) 254 (1991).
88. H. Klein, *Coating*, (3) 24 (1993).
89. *Adhesives Age*, (3) 41 (1985).
90. G. Kupfer, *Coating*, (1) 6 (1984).
91. Solvay & Cie, *Beschichtung von Kunststoffolien mit IXAN, WA*, Prospect Br. 1002d-B-0,3-0979.
92. *Coating*, (8) 311 (1982).
93. H. Hadert, *Coating*, (3) 54 (1973).
94. H. Wittke, *Coating*, (9) 290 (1986).
95. F. Krizek, *Coating*, (9) 316 (1987).
96. G. Menges, E. H. Stroh and D. Weinand, *Papier u. Kunststoffverab.*, (9) 68 (1986).
97. *Papier u. Kunststoffverarb.*, (9) 38 (1986).
98. R. A. Grainwille, *Paper*, (12) 763 (1973).
99. *Der Siebdruck*, (3) 240 (1987).
100. W. Barnscheidt, *Wochenbl. f. Papierf.*, (2) 63 (1974).
101. *Coating*, (12) 334 (1985).
102. *Kaut., Gummi, Kunstst.*, (10) 899 (1983).
103. J. Lawton, *Paper*, (12) 769 (1973).
104. T. Frecska, *Screenprinting*, (6) 68 (1987).
105. *Coating*, (6) 208 (1988).
106. *Coating*, (6) 211 (1989).
107. *Coating*, (7) 250 (1988).
108. *Coating*, (6) 156 (1974).
109. *Converting Today*, (8) 13 (1991).
110. W. Wittke, *Coating*, (7) 249 (1987).
111. W. Wittke, *Coating*, (3) 85 (1987).
112. W. Wittke, *Coating*, (6) 2100 (1987).

113. W. Wittke, *Coating*, (5) 168 (1987).
114. *Coating*, (12) 345 (19874).
115. *Coating*, (7) 180 (1981).
116. *Druck-Print*, (2) 60 (1989).
117. W. Wittke, *Coating*, (9) 334 (1987).
118. H. Lubberich, *Papier u. Kunststoffverarb.*, (5) 114 (1986).
119. *Coating*, (4) 120 (1992).
120. *Druckwelt*, (12) 44 (1987).
121. D. Eitner, *Chemie-Umwelt-Technik*, (4) 50 (1992).
122. P. Holl and E. Föll, European Tape and Label Conference, Exxon, Brussel, April 1989, p. 151.
123. K. Nitzl, G. Peter, T. Horna and W. Hasselbeck, Radiation curable PSA, 15th Munich Adhesive and Finishing Seminar, 1990.
124. *Adhesives Age*, (7) 38 (1986).
125. D. J. Clair, *Adhesives Age*, **23**:30 (1980).
126. K. Nitzl, T. Horna and U. Hoffmann, 16th Munich Adhesive and Finishing Seminar, 1991, p. 100.
127. M. Dupont and M. De Keyzer, 19th Munich Adhesive and Finishign Seminar, 1994, p. 120.
128. *Adhäsion*, (4) 4 (1985).
129. H. Huber and H. Müller, *European Adhesives and Sealants*, (3) 120 (1987).
130. H. Huber and H. Müller, 13th Munich Adhesive and Finishing Seminar, 1988.
131. A. Förster, *Thesis*, FH München, FB 05/1988.
132. H. P. Seng, *Coating*, (7) 245 (1992).
133. *Coating*, (2) 50 (1986).
134. M. Schmalz and W. Neumann, European Tape and Label Conference, Brussels, 1993, Exxon, p. 143.
135. R. Seidel, *apr*, (16) 432 (1986).
136. S. O. Hendrichs (Minnesota Mining and Manuf. Co., USA), US Pat., 2.956.904 cited in ref. 135.
137. *Coating*, (6) 154 (1983).
138. A. R. Hurst, *Production of EBC Silicone Release Paper and Film*, Radcure 82, Lausanne.
139. *Der Polygraph*, (5) 366 (1986).
140. J. Paris, *Papier u. Kunststoffverarb.*, (6) 12 (1986).
141. H. Röltgen, *Coating*, (11) 399 (1986).
142. H. G. Müller, *Kunststoff. J.*, (9) 8 (1986).
143. *Coating*, (11) 292 (1982).
144. *Coating*, (12) 332 (1982).
145. F. T. Birk, *Coating*, (9) 238 (1985).
146. R. Hinterwaldner, *Coating*, (3) 73 (1985).
147. *Coating*, (6) 154 (1983).
148. K. F. Roesh, *Coating*, (12) 353 (1983).
149. K. D. Schröter, *Coating*, (11) 331 (1984).
150. R. P. Cawthorne, *Papier u. Kunststoffverarb.*, (10) 56 (1985).

151. *Coating*, (2) 45 (1986).
152. P. P. Zylka, *Der Siebdruck*, (3) 60 (1986).
153. H. Röltgen, *Coating*, (12) 331 (1985).
154. H. Röltgen, *Coating*, (4) 102 (1985).
155. H. Röltgen, *Coating*, (5) 124 (1985).
156. T. Birk, *Coating*, (12) 278 (1985).
157. R. Hinterwaldner, *Adhäsion*, (3) 14 (1989).
158. G. A. Kupfer, *Coating*, (12) 258 (1985).
159. R. Hinterwaldner, *Adhäsion*, (10) 24 (1985).
160. R. Hinterwaldner, *Adhäsion*, (12) 13 (1985).
161. G. Strohner, *Adhäsion*, (1/2) 24 (1985).
162. K. Fuhr, *Defazet*, (6/7) 257 (1977).
163. *Tappi J.*, (9) 27 (1987).
164. *Coating*, (12) 344 (1984).
165. W. Graebe, *Papier u. Kunststoffverarb.*, (2) 49 (1985).
166. *Coating*, (11) 396 (1987).
167. H. Brus, C. Weitemeyer and J. Jachmann, *Finat News*, (3) 27 (1987).
168. P. Holl, *Verpackungsrundsch.*, (3) 220 (1986).
169. K. Nitzl, *Adhäsion*, (6) 23 (1987).
170. D. A. Meskan, *Coating*, (7) 238 (1992).
171. R. Kardaghian, *Adhesives Age*, (4) 24 (1987).
172. *Adhäsion*, (12) 8 (1988).
173. *Coating*, (5) 177 (1988).
174. *Coating*, (4) 121 (1987).
175. V. Lauppi, *Coating*, (7) 188 (1984).
176. V. Lauppi, *Coating*, (1) 8 (1984).
177. *Chemische Industrie*, (8) 64 (1989).
178. J. Lin, W. Wen and B. Sun, *Adhesives Age*, (12) 22 (1985).
179. R. A. Bafford and G. E. Faircloth, *Adhesives Age*, (12) 234 (1987).
180. EP 0251672
181. A. W. Norman, *Adhesives Age*, (4) 35 (1974).
182. J. M. Casey, *Tappi J.*, (6) 151 (1988).
183. L. C. Fehrmann, *Adhesives Age*, (6) 48 (1970).
184. *Coating*, (11) 396 (1987).
185. R. Hinterwaldner, *Coating*, (4) 87 (1984).
186. R. Hinterwaldner, *Coating*, (5) 172 (1987).
187. G. Kupfer, *Coating*, (3) 25 (1986).
188. H. Klein, *Adhäsion*, (7) 185 (1976).
189. H. Zander, *Adhäsion*, (11) 37 (1988).
190. R. Thomas, *Allg. Papierrundschau*, (16) 437 (1986).
191. *Adhäsion*, (6) 6 (1987).
192. R. Thomas, 4th PTS, *Adhesive Seminar*, Munich, 1984, p. 33.
193. R. Pagendarm, *Coating*, (5) 182 (1988).
194. H. U. Hermann, 2nd International Cham Tenero Meeting of RSM Industry, Locarno, March 29th, 1990, p. 21.

195. *Adhesives Age*, (7) 36 (1986).
196. *CAS, Emulsion Polymerization*, (10) 3 (1988), 108.152233f.
197. *Coating*, (1) 3 (1986).
198. A. Soldat, A. Fall and G. Lechtenböhmer, 18th Munich Adhesive and Finishing Seminar, 1993, p. 145.
199. *Coating*, (4) 242 (1988).
200. H. Hadert, *Coating*, (6) 175 (1969).
201. *Coating*, (19) 277 (1982).
202. E. Mandershausen, *Coating*, (19) 271 (1974).
203. Air Products, *Airflex*, Polymer Chemicals Technical Report, p. 5.
204. *Papier u. Kunststoffverarb.*, (2) 16 (1979).
205. ARC Europe, Dynamic Flow Chamber, Chambered Doctor Blade System, Technische Information, *Vergleich verschiedener Kammerrakelsysteme*, June 1994.
206. B. Voges, *Klebrohstoffe, Beschichtungstechnikum*, BASF, Ludwigshafen 1993).

10
Test Methods

As in the case of many other finished products a wide range of test methods were developed for PSA-based labels, tapes, and coatings. A number of organizations such as FINAT, AFERA, PSTC, and TLMI have established (standard) test methods which are widely used in the industry, although there remain significant differences between the various methods used [1]. These methods provide a good basis for the evaluation of adhesives, but some modifications or additional tests are required when testing materials for specific applications. Therefore, the principal PSAs manufacturers and convertors have developed their own methodology. It is not the aim of this book to discuss the standardized PSA testing methods in detail. Here it appears of more interest to describe the special methods worked out for special end-uses or for specific PSA applications. It is evident that for a label manufacturer the properties of the liquid adhesive, its coating behavior, and the PSA label performance characteristics are very important. From the point of view of methodology, these areas differ quite a bit.

1 EVALUATION OF THE LIQUID ADHESIVE

Pressure-sensitive adhesives are coated in the liquid state (i.e., as a dispersion, as a solution, or as a molten adhesive). Therefore, the evaluation of the adhesive starts with the examination of the characteristics of the liquid PSAs, generally by testing the flow properties. Evidently, the flow properties of the solvent-based PSAs (stable systems), water-based PSAs (shear-sensitive systems), and hot-melt PSAs (highly viscous, temperature-sensitive systems) differ. The properties of the liquid adhesive to be tested include the solids content, the viscosity, the flow properties, and the free monomer content [2]. Later, the following properties of the dried PSA coating should be tested, namely the 90° and 180° peel adhesion, the cohesion (shear), the tack, the Williams plasticity, the shrinkage (plasticizer influence of the PVC) or dimensional stability behavior, the storage properties, the temperature resistance, the peel from the release liner (release force), as well as the aging on the substrate.

1.1 Hot-Melt PSAs

The most important processing parameter of hot-melt PSAs is their viscosity and thus this parameter should be tested according to common methods. Not only the viscosity but also its time/temperature dependence (aging) appears very important. Melt viscosities are often determined in the temperature interval range of 140–200°C using a Brookfield viscometer (spindle LVF4) [3]. Recent developments in the measuring techniques/devices, using cone and plate geometries, allow a more rapid determination of the viscosity curves at different shear rates and temperatures requiring a minimal amount of adhesive.

1.2 Solvent-Based PSAs

Generally, the processability of solvent-based PSAs depends on their viscosity and solids content. Coating devices are designed for a given viscosity; whereas the coating weight and the drying speed depend on the solids content. In general, the solids content and viscosity should be measured for formulated compositions. Because of the nonideal flow behavior of the adhesive solutions (dispersions), viscosity measurements at different flow rates should be made. However, the most common industrial method is the cup-flow method where

Table 10.1 Test Methods for Viscosity Measurement of Water-Based PSAs

PSA supplier	Apparatus	Type	Spindle	RPM	Temperature (°C)
			Test conditions		
AKZO	Brookfield	RVT	3	50	20
Polysar	Brookfield	RVT	2	30	—
Rhône-Poulenc	Brookfield	RVT	—	50	23
BASF	Contraves	STV C111	—	—	23
Hoechst	Brookfield	RVT	5	20	23
	Contraves	MS-C11	—	—	23
Rohm & Haas	Brookfield	LVT	3	30	—

Test Methods

a one point viscosity measurement (flow time versus cup orifice) is carried out; Ford cup or DIN cup 3 or 4 are common [4]. Table 10.1 summarizes the most commonly used test methods for viscosity measurements.

1.3 Water-Based PSAs

Generally, the properties of aqueous dispersions to be tested [5] include the viscosity and its stability in time, the thinning (diluting) characteristics, the solids content, and the temperature stability. With regard to the processing properties of the dispersion, there are several requirements: the absence of foaming, no drying on the rolls, the wetout, the mechanical stability when being pumped, and the lack of odor. The liquid flow properties of water-based PSAs are crucial for their processability. Aqueous dispersions are shear-sensitive systems, where mechanical influences during storage, handling, and coating operations may cause the formation of grit or foam; therefore the examination of the mechanical stability remains very important. On the other hand, water-based adhesives are difficult to coat because of their low viscosity and high surface tension, and because of the low surface tension of the release liner or face stock web. The density and particle size of the dispersion may influence its stability, the coating properties, and drying ability as well. Foam properties determine the processing speed and coating quality/appearance.

The particle size and particle size distribution influence the viscosity, surface tension, and mechanical stability of the dispersion. Mechanical stability influences the viscosity; viscosity and surface tension determine the coatability. For the PSA conversion, coatability remains the main property of the liquid adhesive because the wettability also leads to the choice of the adequate coating geometry, of the face stock and release liner.

Coating Properties

Wettability, the desired coating weight under required conditions (running speed, temperature), and optical appearance are factors of the coatability behavior. The viscosity and surface tension influence the wettability: they should be tested a priori as should the viscosity (thinning, diluting response) and mechanical stability.

The methods for measuring the viscosity were discussed earlier (see Section 1.1). For water-based PSAs the viscosity (Brookfield RVT, spindle 4, 20 rpm) at room temperature should be tested immediately after compounding [6]; in some cases the thixotropy index also is measured. The thixotropy index is the ratio of the Brookfield viscosity at 0.5 rpm to the Brookfield viscosity at 50 rpm. This index shows an increase of the low speed viscosity of 5% for conventional polyacrylic PSAs used as thickener, as compared to 33% for some special products [6].

Wetting Characteristics

The wetting characteristics of a dispersion are examined either directly or indirectly. The direct evaluation of the wetting characteristics implies the subjective examination of the wetout of the original or slightly diluted dispersion on a release liner or on a face stock material. The indirect characterization of the wetting properties refers to the measurement of the surface tension of the dispersion or of the wetting angle.

Direct Examination of the Wetting Out. The evaluation of the wetout should be carried out on the surface of the carrier material which is to be coated. Depending on the nature of the coating (direct/transfer), either the face stock or the release liner should be used as test surface. The test of the wetout on a face stock material is mostly used for film carriers.

The liquid typical PSA first is applied onto a release liner before transfer application onto the face stock; both solvent-based and solventless silicone liners should be included in the evaluation. After drying a comparison of the coatings should be made, based on the visual inspection of the presence (or absence) of defects (cratering, pinholes, fish eyes, etc.) [7].

Vinyl Wetting Test. A film of the experimental latex is cast on the PVC sheet using the draw-down bar. The film coating is inspected for fish eyes and cratering. The fewer fish eyes and craters the film contains the better the vinyl wetting of the latex.

Indirect Examination of the Wetout. The evaluation of the wetting out and the existing norms have been described [8,9]. The evaluation of the surface tension according to Union Carbide or ASTM yields no real information about the adhesion [10]. This test is made more in order to characterize the coating performance of an adhesive or of a surface to be coated. The coating performance and wetting out depend on the surface tension, but the surface tension measurement alone is not sufficient to characterize the actual behavior of an adhesive on a coating machine. In the coating of a latex at high speeds the dynamic aspect of wetting also appears critical. The measurement of the contact angle of a latex on a face stock material or the surface tension of the latex with different surfactant systems determines the choice of the surfactant [11]. The wettability of adhesives also may be examined using a wettability balance [12]. Independent from the analysis of the statical wetout, dynamic tests under shear also are necessary. Surface tension tests were standardized according to the FINAT methods. Similar test methods were published by DIN and ASTM. Surface tension measurements and (wet) film-weighing machine measurements allow the evaluation of adequate surfactants, their diffusion velocity, the reduction of the surface tension, and evaluation of the surfactant concentration required [13].

The evaluation of the surface tension, with the aid of a wettability/wetting angle measuring device has been described [14]. A simple method to test the dynamic surface tension is the use of a stalagmometer (capillary tube). This method—the pendant drop method [15]—involves the measurement of the surface tension of the aqueous dispersion (as compared to water) using a stalagmometer to determine the number of liquid drops. The dynamic surface tension is given by the ratio of the number of water droplets (multiplied by the surface tension of water) to the number of the emulsion droplets.

Wetting Angle. The method of the wetting angle may be used to test the face stock, the release liner, or the PSAs. The same method applies in order to determine the nature (solvent-based, water-based, or hot-melt) of an unknown PSAs (or label). In this case it is assumed that the water-based adhesive always leaves traces of surface agents on the release liner. Therefore, a delaminated release liner which had been originally coated with a water-based adhesive allows a better wetout for pure water. In order to determine the water-based or solvent-based nature of an unknown PSA sample, the wetting angle of water on the delaminated release liner should be measured (deionized water should be used). For example, typical wetting angle data for a solvent-based rubber-resin and water-based tackified acrylic PSA coated release liner are 101–103° and 91–93°, respectively.

Wettability of a Release Liner. In the evaluation of the wettability of an adhesive on a release liner care should be taken because of the change of the wettability of the siliconized paper with time (Table 10.2). Following a small increase after a few days, it decreases with storage time (i.e., the wettability increases with increasing age of the siliconized paper).

Wetting Angle for Adhesive Extract. Extracting with water the water-soluble surface-active agents from a PSA-coated laminate (adhesive nature unknown) and comparing the wetting angle of the extract with those of a known hot-melt or solvent-based PSA, it is possible to identify the nature of the adhesive (Table 10.3).

Foam Content

The most important tests used for the characterization of the foaming ability of a dispersion consist of the following:

1. Stirring and measuring the foam density.
2. High speed stirring (42,000 rpm) and measuring the foam height.
3. Foam generation by air bubbling.
4. Foam dynamics as a function of time.

Table 10.2 Wettability of Siliconized Paper as a Function of Age

Characteristics of silicone coating (wetting angle, °)

Age of the coating (days)	Nature of the silicone coating	
	Solventless	Solvent-based
0	103	109
	103	107
	100	113
	96	109
1	108	87
	107	89
	103	106
	103	105
2	108	107
	106	108
	110	110
	107	111
3	106	111
	104	108
	105	111
	101	110
7	106	104
	108	105
	102	105
	103	106
30	98	106
	98	106
	100	107
	98	106

Table 10.3 Test of the Water-Based Nature of PSAs. Wetting Angle for Aqueous Laminate Extracts

Adhesive nature	Surface active agent level in the recipe (%)	Wetting angle (°)
Tackified water-based acrylic	1.5	62.0
Solvent-based rubber/resin	0	94.9
Solvent-based rubber/resin	0	84.8
Tackified Water-based CSBR acrylic	0.8	77
Water-based acrylic	5	47

Generally, tackified dispersions display a higher foam content than untackified ones. The evaluation of the foaming behavior is carried out using different stirring devices and measuring the volume, height, or weight of the foam column [16]. The foam number after 5 min is used for foam characterization [17]. Byk [18] showed that the best results are obtained stirring the diluted dispersion 1 min with a high speed mixer. A volume of 50 ml of the stirred, foamed, aqueous dispersion is weighed. The weight of the liquid (dispersion/air mixture) is an index for the foaming characteristics. Surfactant solutions (or water-based formulations) should be subjected to high speed agitation at 25–60°C for 3 min in a blender, and the foam volume should be measured [19].

Procedure for Testing the Foam Weight [20]

- The emulsions are diluted to a solids concentration of 7%; 100 g of the diluted latex is poured into a blender.
- The content is stirred for 30 sec at 8000 rpm and subsequently for 150 sec at 15,000 rpm.
- The generated foam and remaining liquid are poured into a 1000-ml graduated beaker. The volume of foam and liquid is reported as foam height (ml).

Another method is:

- First, a well-defined sample volume is poured into a 1000-ml graduated cylinder.
- The dispersion then is stirred at 4500 rpm for 30 sec. The volume is recorded and a 5-min time period is started.
- Volumes should be recorded at the 15 sec mark and then each minute thereafter.

- With the aid of the weight/volume correlation density values can be calculated. With this method the sample contains both entrained air and froth foam, and the lowest density value corresponds to the case of the highest amount of foam. Errors related to the transfer of the foam into another vessel are eliminated.

Mechanical Stability (Sieve Residue)

The term mechanical stability refers to the ability of the emulsion polymer to resist coagulation under the influence of shearing stresses as encountered during agitation or pumping. Shear under machine conditions depends on the running speed and viscosity, and the shear characteristics of the coating device. In the most frequently employed test the polymer is stirred at high speed in a blender for a specific time (typically 10 min). The amount of coagulum (instability) resulting from this treatment is determined by filtering the latex through a 80-mesh screen [21] whereas a 100-mesh screen also is sometimes used [22].

A rapid measure of mechanical stability is obtained by filtering a sample through a 100-mesh screen, then stirring it at high speed in a blender for 5–10 min. The emulsion is again passed through the screen and washed through with water until only solid, if any, remains on the wire. If there is no solid residue the emulsion is considered to possess very satisfactory mechanical stability. If so desired the amount of solid can be weighed after drying the screen in a circulating draft of vacuum oven at 80–100°C.

The Hamilton Beach blender test is another way of testing the mechanical stability. The percent residue created by subjecting the emulsion to high speed agitation for a set period of time is a measure of the stability of the emulsion (or the lack thereof).

Sieve Residue (Tackifying Resin). This is a measure of how much deposit the user can expect to find in the filters. Normally, the specification allows a maximum of 0.05%; a more realistic figure, however, is a maximum of 0.015%.

Grit Content. The grit represents the coagulum which did not pass through a 200-mesh screen [23]. It can be tested visually, either by coating a thin (2–5 g/m^2) adhesive layer on a transparent face stock material or by using a screen (gravimetrically).

Drying Ability and Conditions

While screening different water-based PSAs with similar adhesive but different processing conditions it appears very important to estimate their drying ability. The drying ability is the drying speed (loss of water with time) under real conditions. Unfortunately, there is no standard method nor apparatus for

Table 10.4 Drying Conditions

	Coating characteristics		Drying conditions		
Face stock	Coating weight (g/m^2)	Nature	Temperature (°C)	Time (min)	Ref.
PET	—	Solvent-based	90	45	[24]
—	—	—	100	5–10	[25]
—	—	Water-based	120	10	[23]
PET	—	Water-based	130	3	—
PP	25	Water-based	220	0.5	[22]
—	—	—	93/204	—	[26]
Paper	—	—	70	5	[27]
PET	—	Solvent-based	RT/90	15/5	[28]

this test. Preliminary screening tests may be carried out by the weight loss determination of the coating in time (see Figure 6.5).

Tests with PSA laminates require a dry adhesive coating. Different companies use quite different drying conditions as can be seen from Table 10.4. ASTM has developed methods of determining "drying, curing, or film forming of an organic coating at room temperatures" (TE5–1740–65T).

There are some properties which allow a screening of water-based PSAs from the point of view of the coating properties. Parameters such as solids content, particle size, and density yield indirect information about the machine properties of the adhesive.

Particle Size

Different methods used for the determination of the particle size and distribution are known; either simple or more sophisticated pieces of equipment are used. Particle size is related to mechanical stability, water resistance, viscosity, and surface tension. The measurement of the particle size can be carried out with a stalagmometer, by determining the number of liquid drops for the pure dispersion and the modified one, upon adding an increased surfactant level to the dispersion. The method is:

- The dispersion to be tested should be diluted (1:1.5) with water.
- Increasing quantities of surfactant (in fact a 10% wet weight surfactant solution) in portions of 1–2 cc should be added and the surface tension should be measured. The surface tension/concentration of surfactant plot

for the unknown dispersion is to be compared with the plot of a known dispersion (using the same surfactant). The slope of the surface tension versus surfactant concentration plot is equal to the moles of soap absorbed per gram of polymer. If the same surfactant is used, as in the synthesis of the emulsion, and the original amount of soap per gram of polymer solids is known, the surface area per gram of polymer can be calculated, and hence the average surface radius of the polymer particles.

For emulsions, the particle size is a measure of the quality of the dispersion. Particles with a diameter of 0.25–0.5 μm can be observed with a microscope. Other methods for measuring particle size and particle size distribution exist (i.e., laser light scattering).

Solids Content

When measuring the solids content of resin emulsions it is very important that the sample remains in the oven for only 40 min, as the base used for emulsification is volatile. Karl Fischer water content measurements tend to indicate a solids content of about 1% higher than solids determined in an oven.

Density

This is a measure of how much air is contained in the emulsion. A low density (i.e., high air entrainment) causes problems through dehydratation of the emulsion at the surface (especially if the temperature exceeds 30°C).

2 EVALUATION OF THE SOLID ADHESIVE

Many different test methods were established in order to characterize the adhesive, converting, and special properties of the solid (i.e., dried) PSAs. As these depend on the coating weight, the measurement of this parameter is discussed next. The adhesive properties of the solid PSAs are discussed in Chapter 6.

2.1 Test of Coating Weight

A simple gravimetric method can help to keep coating weight specifications within ±0.5 g/m^2 resulting in considerable savings. The coating weight is tested by dissolving the adhesive layer in a solvent and determining the weight difference between coated and uncoated face material.

Apparatus:

- Balance with a precision of 0.001 g.
- Die-cutting device for 100-cm^2 samples.
- IR lamp (300 W).

Method:

- Prepare a 100-cm^2 sample by die-cutting the laminate and weigh it with a 0.001-g precision.
- Wash-off the adhesive layer with a common solvent for the adhesive (e.g., toluene).
- Dry the cleaned face material under the IR lamp for 20 sec at 20 cm distance from the lamp.
- Check the weight of the cleaned, dried face material again.
- The weight difference indicates the coating weight.

Generally, for paper/PET coatings, this method yields more reliable results than comparing the weight of the coated and uncoated face material.

In-line gravimetric measurements of the coating weight can be carried out by inserting siliconized paper or aluminum foils in the laminating station. Methods based on beam penetration and absorption (IR, β-rays,. etc.) also are often used in-line.

2.2 Other Properties

Other properties of solid (dry) PSAs or raw materials to be examined consist of the softening point of the resin, the Williams plasticity/modulus, the nature of the adhesive, the FDA/BGA compliance, the tensile strength, and the cold flow.

Softening Point of the Resin

Tackified dispersions differ through their softening (melting) point. The tack/shear balance of an adhesive is related to the softening point of the tackifier resin (see Chapter 6). Softening points as measured by the Mettler device are 5–10°C higher than ring-and-ball softening points. Other, subjective methods for the examination of the softening behavior of a coating also have been developed; an example is given next.

The softening point of the adhesive coating is determined subjectively (adhesive transfer, cohesive break) by peeling off (90°) the laminate affixed on a heated substrate (test-bank) in different temperature ranges:

Apparatus: Koffler heating stand

Method:

- A 1 × 30-cm adhesive strip should be affixed to the heated stand.
- The strip should be peeled off instantaneously at a 90° angle.
- The temperature of the heated zone giving adhesive transfer should be noted.

The softening point provides information about the nature of the adhesive (hot-melt/solvent-based, competitive samples) and the formulation (tackifier amount and nature).

Williams Plasticity

The Williams plasticity number indicates the deformability of the adhesive mass under a static load. Williams plasticity has recently regained favor for PSA testing purposes. It was used extensively in the 1950s for following molecular weight changes in reactions during polymer production, but was replaced by the Mooney viscosity which is normally faster and more convenient to measure [29].

The Williams plasticity is used for screening hot-melt PSAs [30]; it is run at 100°F for 15 min [31]. The sample coated on a release liner is dried for 24 hr, weighed (to 2.0 g), conditioned 15 min at test temperature, the load is applied, and the plasticity measured 10 min after the application of the load [32]. In some cases the Williams plasticity is evaluated at different temperatures (38–200°F). Some screening values for the Williams plasticity number are summarized in Table 10.5.

In this test method the thickness of the sample (in mm) is measured while subject to a constant load (5 kg) at 100°F (38–37.8°C) [35].

Table 10.5 Screening Values for Williams Plasticity Number and Interdependence of the Williams Plasticity/Modulus.

Williams plasticity number ($\times 10^{-2}$)	Rheological characteristics			Adhesive nature	Ref.
	Storage modulus ($\times 10^{-5}$)	Loss modulus ($\times 10^{-5}$)	tan δ		
1.9	1.4	0.7	−0.51	—	[33]
5.1	—	—	—	Solvent-based acrylic	[31]
3.2	—	—	—	Solvent-based acrylic	[31]
5.3	—	—	—	Solvent-based acrylic	[31]
2.2	1.6	0.82	0.46	Water-based acrylic	[33]
2.2	1.7	0.82	0.48	Water-based acrylic	[33]
1.9	1.4	0.7	0.50	Water-based acrylic	[33]
1.9	1.4	0.6	0.43	Water-based acrylic	[33]
2.4	2.1	0.9	0.44	Water-based acrylic	[33]
1.6	1.7	1.0	0.59	Water-based acrylic	[33]
1.0	—	—		Hot-melt acrylic	[34]

Test Methods

Apparatus:

- Plastometer with gauge (Williams Plastometer, Scott Testers Inc.) in accordance with ASTM-D-926.
- An oven.
- Release paper.

Method:

- A wet film of the adhesive is coated on siliconized release paper, so as to produce a dry film. It is dried at room temperature and further dried 5 min at 135°C in a circulating air oven. The adhesive is removed from the silicone paper, and a pellet, (about 2 g) is formed in the shape of a ball.
- Condition the ball-shaped sample (placed between two silicone papers) for 20 min at 38°C.
- With the plastometer and sample at 38°C, place sample (between two pieces of release paper) into plastometer, and record the thickness of sample after 14 min.
- The Williams plasticity number (PN) is the thickness of the pellet in mm after 14 min compression at 37.8°C in the plastometer under a 5-kg load.

This method yields valuable information about dried (coated) materials. Some companies use this method; others consider it as inadequate for dispersions.

Cold Flow

In this test the dimensional change of an adhesive sample after storage under load during a long time (1 week) is measured [36].

Apparatus:

- Glass slides
- A 1000-g weight.

Method:

- Prepare a 0.3 g spheric sample of the adhesive.
- Place the sample to be tested between two glass slides with a 1000-g weight on top of the glass sandwich.
- After 1 week at room temperature the diameter of the squeezed adhesive disc is measured (cm).
- Calculate the ratio of the diameters of the experimental and standard adhesive. The cold flow resistance is equal to the diameter of the squeezed disc of the experimental adhesive divided by the diameter of the squeezed disc of the standard adhesive. Cold flow also may be evaluated using a load of 3 psi at a temperature of 120°F [37].

Tensile Strength

The method (ASTM D-638) is used for the evaluation of adhesive films. For this purpose, 3.2 ± 0.4-mm thick and 19 ± 0.5-mm wide dumbbell-shaped specimens are prepared and strained in a tensile tester until break occurs. The tensile force at break is registered and divided by the original cross-section area of the narrow section of the specimen.

Apparatus: Tensile tester

Method:

- A 500-μm adhesive layer is coated on a siliconized release liner.
- After drying (at 70°C for 30 min) the adhesive layer is converted into a test specimen. The adhesive sample should be conditioned 24 hr at 21°C and 55% relative humidity (RH). The sample should be strained at a rate of 300 mm/min in a tensile tester until break occurs.

The test of the tensile strength at break yields unrealistic values (too high) for the cohesion; the tensile strength at 300% elongation provides data in good agreement with measured shear values.

FDA and BGA Compliance

Generally, PSAs should comply with the following FDA and BGA regulations:

FDA 21 CFR[*]	175 105 (adhesives)
	176 170 (aqueous food)
	176 180 (dry food)
	178 3400 (emulsifier)
BGA	Proposal XIV for polymer dispersions, 01.08.85 170.
	Mitteilung Bundesgesundheitsblatt 28, 305 (1985).
	Prüfung auf Physiologische Unbedenklichkeit.

* Code of Federal Regulations

3 LAMINATE PROPERTIES

Most properties of PSAs may be examined when coated (i.e., in the PSA laminate). All PSA samples should be examined for their main characteristics; those manufactured for special purposes should also be examined for their special application related properties.

3.1 General Laminate Properties

Independent from their end-use, all PSA laminates possess an inherent degree of tack, peel, and shear (i.e., an adhesion/cohesion balance as well as some converting properties). The coating weight influences all these properties (see Chapter 6). On the other hand, the adhesive and converting properties depend on the aging response of the adhesive.

Adhesive Properties

The first step in the evaluation process of the adhesive (or other) properties involves the preparation of the specimen. Generally, the PSA label specimen to be tested is manufactured in the laboratory when tests concern the PSA, the face stock material, or the release liner (i.e., when screening the laminate components).

Quality control tests are carried out on specimens taken from the industrial production. Generally, laboratory samples are prepared via laboratory (direct/ indirect) coating, drying, and conditioning of the coated sample. However, the laminate components, the equipment, and the drying and conditioning parameters used by different PSA suppliers or converters can vary quite a bit.

Test methods have been established in the United States (ASTM) or in Germany (DIN). European Norms (EN) also have been defined. Several adhesive manufacturing associations have developed test procedures that are internationally accepted and are being used in the trade as reference methods. The test procedures for PSAs generally examine three key aspects: the tack, the peel adhesion at 90° and 180°, and the shear resistance (cohesion). Procedures for measuring adhesion of self-adhesive materials to a test surface normally rely on a peel test [38]. Such test procedures were published by a number of organizations such as FINAT, AFERA, and PSTC. They also have been incorporated in buying specifications of the German military procurement office (BWB) [39].

Standard test methods for adhesives were described by MacDonald [40] and Symietz [41]. Test methods for PSAs are derived from those used originally for gummed papers [42]. Shear, tensile, and peeling tests were conducted on gummed papers in order to obtain the performance characteristics with respect to the three classical stress modes. The basic types of stress to which a joint may be subjected fall into three categories: tensile, shear, and peel [43]. For the evaluation of the adhesive properties, reference materials like tapes were used in some cases [44]. Thus as a reference ordinary office-grade Scotch tape was selected; it has a rolling ball tack of about 180 mm whereas rolling ball tack data of masking tape typically is 75–130 mm. On the other hand, it is difficult to compare the shear strength of an ordinary adhesive tape because the backing of this commercial product will not support the test weight [44]. However, as a reference value the masking tape shear resistance is 1–17 hr under PSTC-7 conditions.

Preparation of the Specimen. The preparation of the specimen includes the coating of the adhesive on a solid state material, the drying of the coated material, and in some cases laminating it on the release liner/face stock material. The preparation of the specimen also includes in most cases the conditioning of the samples.

A layer of liquid adhesive (or molten in the case of hot-melt PSAs) thick enough so as to produce 20 g/m^2 of dry adhesive film is spread on a Mylar film by means of an adjustable adhesive film applicator and subsequently dried. A sheet of silicone paper is applied on the dry (cold) adhesive film. Test specimens are cut before removing the silicone paper. In some cases the indirect or transfer coating method is used. For hot-melt PSAs coatings are sometimes made from solutions. Generally, different kinds of laboratory equipment are used in order to coat the adhesive. Hand operated devices like the Bird, Erichsen, Drage, Biddle applicators, etc., may be used [16, 45].

Face Stock and Substrate Used. A series of performance tests can be carried out on experimental adhesives using essentially the end-use substrates. For reasons of comparison most tests are made on standard plates. The largest single factor in the measurement of bond strength is the test plate. However, normalized test procedures differ in their definition of the surfaces required. FINAT uses float process plate glass; AFERA suggests brushed steel, while PSTC and BWB propose polished steel. FINAT specifies glass as the most suitable material; the surface smoothness is well defined and the chemical nature of the material is stable and constant [38]. Other parameters also influence the results [38] (e.g., drying of the test plate after cleaning). Different procedures and surface agents will leave different deposits on the test plate, which may influence adhesion of the self-adhesive material to the plate.

Drying Conditions. Drying conditions influence the content of liquid components in the adhesive layer, the composite structure (coalescence), and the roughness of the adhesive layer. They can also affect the mechanical properties of the solid state laminate components. In most cases different coating weights, different test surfaces, and different drying conditions are used.

Performance data are determined from laboratory prepared test laminates composed of 1 mil dry adhesive coated on 2-mil polyester protected with siliconized release paper [31,34]. Different kind of adhesives (thermoplastic, thermosetting, self-crosslinking, and emulsion PSAs) are dried at different temperatures and oven times. Hence, an exposure to ambient air for 15 min and 2 min oven time at 200–250°C are suggested for solvent-based adhesives; a 1 min ambient air exposure followed by 3 min oven (212°F) exposure is suggested for emulsion polymers. A tackified CSBR coated on paper is tested by coating it on release paper, dried 5 min at 70°C, then 2 min at 100°C, and then transfer-coated into paper (1 ±0.1 mil dry). Samples on PET are prepared by directly coating adhesives on chemically treated films and drying for 5 min at 158°F and 2 min at 212°F. The adhesive then is laminated onto a

Table 10.6 Coating-Drying Conditions of Laboratory Samples of Water-Based Coatings

	Drying steps				
	1st Step		2nd Step		
	Temperature (°C)	Time (min)	Temperature (°C)	Time (min)	Ref.
1	70	5.0	100	2.0	[27]
2	115	3.0	23	1440.0	—
3	RT	30.0	105	5–10.0	—
4	220	0.5	—	—	[22]
5	105	2.0	—	—	—
6	70	10.0	—	—	—
7	RT	15.0	90	5	[28]

release liner and conditioned a minimum of 2 hr at standard temperature and humidity before testing [32]. Drying at 90°C during 3 min followed by a conditioning at 23 ±2°C and 50% RH for 16 hr also is used in practice. The adhesive is coated on the face stock material with a Meyer bar so as to yield a 20–25 g/m^2 dry adhesive after 2 min in a ventilated oven at 105°C [47]. Then a protective liner is placed on the adhesive film and conditioned at 20°C and 50% RH before testing. There are quite different recommendations concerning the coating/drying conditions of laboratory samples (Table 10.6).

Sample preparation is carried out as follows [21]: a wire rod or wire-wound Meyer rod is used to apply 25 g/m^2 (dry weight) of the adhesive on a 25-μm polyester film. The coating then is dried for 3 min at 115°C and conditioned for 24 hr at 23°C and 65% RH before being tested. Coatings on polyethylene should be dried at 60°C on soft PVC at 45–50°C.

Test Conditions/Conditioning. A specification for materials testing always has to start with a definition of environmental conditions. Methods will normally specify 23°C at 50% RH, but there are considerable differences in accepted tolerances (Table 10.7) [38,49,50]. If the laminate to be tested contains paper, broad tolerances in the testing environment will raise considerable problems. One should therefore aim at achieving environmental conditions according to DIN ISO 137 (1982) [51].

Some test equipment allows the measurement not only at a constant peel rate but also with a constant force. From the point of view of application

Table 10.7 Test conditions

	Air temperature (°C)	Relative humidity (%)	Standard
1	23±1	50±3	DIN/ISO
2	23±2	50±6	BWB-TL
3	23±2	50±5	FINAT
4	23±1	50±5	AFERA
5	23±2	50±2	PSTC
6	23±1	50±2	TLMI

technology, this mode of measurement is often of greater interest. The PSA-coated films should be conditioned under different conditions [52]. Fresh films should be stored 24 hr at 23°C and 50% RH with no contact with the release paper. Aged films should be conditioned at 40°C for 14 days and then 24 hr at 23°C. For water-based PSA formulations the coalescence (film forming) occurs slowly as compared to hot-melt PSAs and solvent-based PSAs (interpenetration of resin and dispersion).

Measuring the Tack

Generally, the test procedure to measure tack consists of two steps: bond formation and bond separation. Some of the test methods (quick stick or loop tack and Polyken tack) simulate the real bond formation conditions; others, like rolling ball tack, measure the tack under quite different experimental conditions. Table 10.8 summarizes the different test methods for the characterization of the tack as suggested by suppliers of PSAs or raw materials for PSAs.

The first and least sophisticated test for the tack is to put the thumb on the adhesive that was coated onto a sheet of paper. The next level of complexity involves sticking a loop of adhesive-coated paper to different substrates. Each time the loop is pulled from the surface to which it sticks, the tester forms an opinion about how sticky the adhesive is [53]. The normalized test methods are used to compare different adhesives [54]. The data are not absolute values, but may be used to compare the various products.

Quick Stick. Quick stick (PSTC 5) measures the instantaneous adhesion of a loop of adhesive-coated material using no external pressure to secure contact. Because of the special loop form of the specimen quick stick also is called loop tack. According to another definition the quick stick tack value is the force required to separate at a specific rate a loop of material which was brought into contact with a standard surface (FINAT test method 9 or

Table 10.8 Test Methods for the Tack Measurement

	Principle of the method				
	Stress nature	Procedure	Name/variants	Norm	Unit
1.	High speed, repeated dynamical bonding/debonding	Slowdown of a rolling item	Rolling ball Rolling cylinder	PSTC 6	in, cm in, cm
2.	Low speed, dynamical bonding/debonding	Debonding: via peel, tensile forces	Loop tack Quick stick	PSTC 5 TLMI L-IB1 TLMI L-IB2 FTM 9	g/in^2 lb/in; oz/in^2 oz/in; $N/25$ mm N/in; $N/25$ mm
		Debonding: via tensile forces	Polyken tack Probe tack		gm; N, gm/cm^2

FINAT FTM 9). The FINAT FTM 9 quick stick method differs from AFERA (4015) and the PSTC 5 quick stick method as the latter measures peel at 90° without making a loop. The quick stick method is relatively simple and may be carried out using common tensile strength test machines [55]. Unfortunately, the contact time is long and depends on the area, the contact area differs, and when using the FINAT method the peeling angle is not constant.

Quick stick has the advantage of allowing tack to be measured on a wide range of substrates such as stainless steel, glass, polyethylene, and paper. It can, however, result in high tack readings on smooth substrates such as glass for adhesives with a low finger tack on rough surfaces [1]. The EPSMA method for double-sided tapes varies in that it specifies that a loop of PET film is brought into contact with the tape on a panel; this procedure discounts the effect of the material rigidity.

Loop Tack. Loop tack is a measure of the force required to remove a standard adhesive-coated film loop from a standard stainless steel plate after a short contact of the test strip with the steel plate in the absence of pressure [35,36]. A 0.5 × 4-in. strip of one mil Mylar film coated with the sample adhesive is formed into a loop with the adhesive on the outside; the loop is applied to the test plate until the PSA loop contacts 0.5 in.^{2} of the surface area on the plate. The loop is then removed from the plate at a rate of 12 in./min and the loop tack is the maximum force observed. According to

Table 10.9 Loop Tack Test Methods

Face material	Substrate	Test conditions Contact surface (cm^2)	Speed (mm/min)	Ref.
PET (50 µ)	Stainless steel	2.54 × 1.27	—	[35]
PET	Stainless steel	—	200	[47]
PET	Stainless steel	1.27 × 1.27	300	[56]
Paper	Stainless steel	—	300	—
Paper	Stainless steel	2.54 × 2.54	200	[56]

another method, a loop formed from a 2.54-cm wide strip of coated paper is lowered by a tensile testing machine onto a glass plate. The force required to remove the tape measured at a rate of 300 mm/min at 20°C is the measured loop tack. Tack (measured as loop tack) depends on the substrate; the following values were obtained for different substrates (in N/in) [17]:

Stainless steel	2.8
Glass	3.5
Polyethylene	2.7
PVC	2.5

In a modified PSTC 5 loop tack (quick stick) test polished stainless steel is used as substrate [47]. When a contact surface of about 1 in.^{2} is achieved, the loop is separated from the panel at an 8 in./min separation speed. Quick stick or loop tack is the average of five test values. In a further modification (modified PSTC 5) a 200 mm/min rate may be applied. In some cases a 2.54 × 12.7-cm adhesive-coated strip also is used [56]. Tack should be tested in N/20 mm on chromed steel plate [35]. Quick stick to stainless steel and to kraft paper also are measured [37]; peel build-up and restick also are tested. Loop tack also can be tested through the test method ATM 136–84 [17]. Table 10.9 summarizes the test methods concerning loop tack.

A test plate of 40 × 135 mm should be used for loop tack [46]. The nature of the joint failure should be characterized as follows:

A = adhesive break (no adhesive on the plate)
B = cohesive break (adhesive traces on the plate and carrier)
A_o = adhesive transfer (adhesive on the plate)

The intermediate steps are characterized in percent.

Loop tack is probably the most common test method [1]. The Polyken probe method is the most well known among the probe tack tests [37].

Polyken Tack. Polyken tack is a measure of the tackiness of PSAs. The Polyken probe tack testing machine provides a means of bringing the tip of a flat probe into contact with PSA materials at controlled rates, contact pressures, dwell times, and subsequently measuring the force required to break the adhesive bond in g/cm^2. A Polyken Probe Tack Tester (Kendall) according to ASTM has been described [58]. Here a 120-g force and a 0.5-cm sample diameter are used. Polyken tack is tested with a $100\text{-}g/cm^2$ contact pressure, 1-sec dwell time, and a 1-cm/sec test rate [59]. Since this method implies a very small contact area, small imperfections in the adhesive film can lead to disparate values.

The use of a probe to test tack is an extension of thumb tack testing, a probe test method having been used in the 1940s [60]. Kamagata [61] tried to control conventional thumb tack testing by using a small balance [26]. Another earlier method was the Ball Tack Test using a steel ball as contact surface [61]. In the 1950s Wetzel [62] proposed a probe testing device that could be fitted in a standard extensiometer. Later this method was redesigned into the Polyken Probe Tack Tester. The probe is attached to a force gauge; the sample to be tested is attached to another weight to control the applied pressure which then is lowered onto the probe and pulled off. The equipment applies a $100\text{-}g/cm^2$ pressure with a 5-mm diameter stainless steel probe, a contact time of 1 sec, and a rate of removal of 1 cm/sec. There also is the Kendall probe tack tester for research investigation purposes. The Polyken Probe Tack Tester and the test method also have been suggested for the evaluation of re-adhering adhesives [58,61,63].

A more precise version of the probe tack method is the one developed by Zosel [55] enabling the use of contact times as short as 0.01 second. The probe tack method proposed by Wetzel [62] and developed by Hammond [58] has the disadvantage that a special apparatus is necessary and the contact area is too low (≤ 0.2 cm^2) [55].

Other Tack Test Methods. Another method proposed by Bull [64] (in a manner different from the rolling ball method where the adhesive surface is static, and the adhesiveless contact surface rotates) has a polished rotating adhesiveless cylinder in contact with a rotating adhesive-coated cylinder. This method measures the force (resistance) necessary to move the cylinders. Unfortunately, a special apparatus is needed. Like in the rolling ball tests, debonding is influenced by the compression of the adhesive by the cylinder.

Tack Measured with the Toothed Wheel Method. Druschke [55] summarized the different factors influencing the adhesion (and the test method); they are;

- the surface tension of the adhesive, face stock, and adherent;
- the mechanical (viscoelastic) properties of the adhesive, face stock, and adherent;
- the surface structure of the adhesive, release liner, and adherent;
- the thickness of the adhesive, face stock, release liner, and adherent;
- the test conditions.

Taking these parameters into acount, Druschke improved the rotating drum method. A further development of the rotating drum method is the pitch-wheel method [65]. In this method the polished wheel used in the rotating drum method is replaced by a pitched wheel, and the contact force is measured using a common tensile strength test machine. Druschke [55] proposed the toothed wheel method which permits very short contact times, as does the rotating drum principle. In the new method, a toothed wheel rotating freely around its axis is fixed in the upper clamp of a tensile tester with a liner. A rotating drum is pushed against the wheel until an almost pressureless contact is achieved. The tensile tester measures the force acting against the rotating drum with the adhesive layer.

Rolling Ball Test. The rolling ball tack (RBT) test measures tack as a function of the distance travelled by a steel ball on an adhesive-coated substrate. The rolling ball tack, according to PSTC Method 6 is measured as follows: a stainless steel ball is allowed to run down a slope from a point down the inclined plane onto the adhesive tape; the distance travelled along the sample is measured in centimeters. In this test an 11-mm diameter steel ball rolls down a plane having a length of 18 cm and inclined at an angle of 21°30′, the adhesive thickness being at least 25 µm. In a modified rolling ball test, according to PSTC-6, an inclined angle of 21°30′, a necessary length of 12.5 cm, and a ball weighing 7.59 g with a diameter of 1.229 cm (0.484 in.) is used. In a different manner from other tack testing methods, the rolling ball method does not measure the debonding force [30]. The apparatus used is relatively simple, but unfortunately the test is influenced by adhesive residues on the ball or by the viscosity and thickness of the adhesive layer. Table 10.10 summarizes the characteristics of the tack test methods by rolling ball.

Rolling Cylinder Tack. The rolling cylinder tack (RCT) method is a modified version of RBT, known as "Douglas tack test." Here one uses a stainless steel cylinder with a diameter of 24.5 mm, a length of 19.05 mm, weighing 75 g, and a travel path 203 mm long with an angle of 5° [3].

The reproducibility of the RCT values is better than the reproducibility of the RBT values; the running path is shorter and less material is necessary (as

Table 10.10 Rolling Ball Tack; Characteristics of the Test Method

Method	Travel path		Rolling ball characteristics		Ref.
	Length (mm)	Angle (°)	Diameter (mm)	Weight (g)	
—	173	25	14	4	[67]
PSTC 6	180	21°30'	11.1	—	[68]
PSTC 6	125	21°30'	12.29	7.59	[47]
—	180	21°30'	—	—	[47]

compared to competitive samples). For paper laminates values of 1.5–2.5 are very good and values of 7.5–8.0 fair. No more than 15 cm can be accepted. For film laminates values of 2.5–3.5 are very good, values of 4.5–5 are fair, and values above 7–8 cm are not acceptable. Rolling ball tack is more sensitive towards the coating weight than rolling cylinder tack (Table 10.11).

Evaluation of the Peel Adhesion

A combined method (an assembly of different test methods) was proposed for the evaluation of different adhesive characteristics of hot-melt PSAs [69].

Table 10.11 Dependence of the Tack on the Coating Weight; Rolling Ball Versus Rolling Cylinder

Coating weight (g/m^2)	Tack values, cm	
	Rolling ball tack	Rolling cylinder tack
16	10.4	4.5
	10.7	4.8
	10.0	4.2
19	4.5	4.5
	5.0	5.0
	5.0	4.0
27	3.4	3.0
	3.1	2.9
	3.1	3.0

Peel adhesion should be tested according to PSTC 1, shear according to PSTC 7, and tack according to PSTC 6 (rolling ball), Polyken, or quick stick tests. Melt viscosity is measured with Brookfield RVT spindle 7, 20 rpm. The softening point should be determined with the ring-and-ball method (ASTM E 2867) at 180°C [69]. A computerized system for measuring the peel adhesion interfaces the slip peel tester to a personal computer. This system measures peak, valley, and average peel forces in various geometries [70]. These examples illustrate that although there are standard methods for measuring the peel, really quite different "home made" methods are used to measure different adhesive performance characteristics. Kaelble [30] showed that peel depends on the modulus and thickness of the adhesive, the solid state component of the label, and on the peeling angle. The best known test of the adhesion is the method used for laquers according to DIN 53151 and ISO 2409. Adhesive strength (peel) also may be determined according to ASTM D 1000. Here the peel or tensile strength measurement should be used for the evaluation of the bond strength. This was a question in the first years of the development of PSAs [46]. For a long time tensile strength measurements and peel measurements were carried out in parallel [42]. Kemmenater [71] proposed a tension measuring device for the peel. FIPAGO accepted a pendulum device [72, 73]; in this test, the adhesive strength was tested via peel work in mm/kp [73]. Today the most common method for testing the adhesive strength of PSAs is the peel test. Different versions of this method using different peel angles, peel rates, and substrates are used. The history of peel testers was described by Liese [73]. FIPAGO recommended the adhesive strength tester of PKL used for gummed papers. The Werle Tack tester measures adhesive strength in a similar manner based on the peel adhesion [72]. A systematic study of the test methods for wet adhesive tapes was reviewed during the FIPAGO meeting in 1964.

For different end-uses, labels for different applications, varying peel test angles are used. For a flexible and a rigid adherent a 180° peel or 90° peel, for peel between two flexible adherents a 180° T peel is carried out (Figure 10.1). Common angles are 90° and 180°; for label stock the 180° geometry according to FINAT FTM 1 has found wider application than the 90° geometry [38].

As a result of the viscoelastic properties of PSAs and the mechanical properties of the face material, the adhesive bond strength depends on the peel rate (see Chapter 2). In most test equipment single point measurements at 5 mm/sec are specified. Test results show that the suitability of a range of products may vary with the peel rate [38]. This should be considered when practical application conditions differ strongly from test conditions. A similar statement holds true for release forces from siliconized release paper. Dwell times (the contact time between adhesive and substrate) also influence the

Test Methods 533

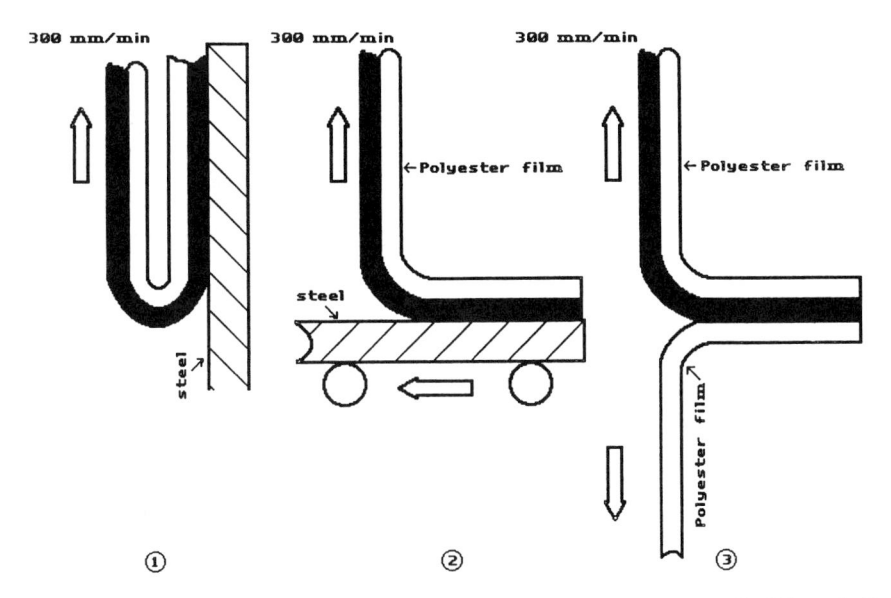

Figure 10.1 Test methods used for peel evaluation. 1) 180° peel; 2) 90° peel; 3) T-peel.

peel value. This is very important especially for peel tests on difficult surfaces. In some cases (for single-coated tapes) the 180° angle PSTC 1 method is modified in order to allow 20 min or 24 hr contact of the adhesive with the test panel [35].

Peel adhesion is determined in accordance with ASTM D-3330–78, PSTC-1, and FINAT FTM 1 and 2, and is a measure of the force required to remove a coated flexible sheet material from a test panel at a specified angle and rate of removal. According to American norms, a 1-in. wide coated sheet is applied to a horizontal surface of the clean, stainless steel test plate with at least 5 in. of the coated sheet material in firm contact with the steel plate. A rubber roller is used to firmly apply the strip and remove all discontinuous and entrapped air. The peel adhesion of the sample then is measured in a tensile tester [56].

According to the 180° peel adhesion AFERA 3001 method, one 10-mm wide specimen of PSA film (prepared at least 24 hr before running the test) is applied on a stainless steel plate. Ten minutes later the specimen is peeled off by means of a tensile tester at a speed of 304 mm/min. The measurement is carried out at 23 ± 2°C. In fact, tests according to PSTC 1 and AFERA 4001 PB are carried out with a 25-mm wide sample, peeled off at 300 mm/min speed, and the peel force is given in g/25 mm [74]. FIPAGO uses the measurement of the peeling work for the evaluation of the peel adhesion. A modified PSTC 1 method [75] prescribes a polished stainless steel panel as a substrate. After 1 min dwell time the tape is peeled away from the panel at an angle of 180° and at an 8 in./min separation speed. The 180° peel adhesion is the average of five test values and is expressed in g/in. width of tape. Another modified PSTC 1 method on a 2.5 × 25 × 125-mm glass plate uses a tape applicator. The peeling force is registered at 12 mm intervals and given in N/25 mm. Alternatively, a strip of 2.54-cm wide coated paper is bonded to a glass plate by applying a constant pressure [57]. A tensile testing machine then is used to measure the force required to peel the paper strip at an angle of 180° at a rate of 300 mm/min at 20°C. The testing speed for film laminates should be 300 mm/min and 75 mm/min for paper [46]. Peel adhesion should be tested according to FINAT FTM 1, but at a peeling rate of 200 mm/min; 180° peel adhesion should be checked.

Apparatus:

- Tensile testing machine according to DIN 51220, Class I, and DIN 51221 (tensile strength).
- Glass slides (2.5 × 45 × 125 mm).
- Tape applicator D-4271 (Sandes Place Research Institute).
- Cleaning solution

Method [66,76]:

- Clean the glass slides 24 hr in cleaning solution; flush with deionized water and then dry the glass slides.
- A strip of coated material (25 × 250 mm) is bonded to the glass plate by applying a constant pressure (by tape applicator or roll).
- The noncontacted portion of the sample then is doubled back (nearly touching itself) so that the angle of removal of the strip from the plate will be 180°. The free end of the test strip is attached to the adhesion tester. The test plate then is clamped in the jaws of the tensile testing machine.
- The peel force should be rated at each 12 mm interval.

90° Peel Adhesion. The 90° peel adhesion is measured as the force required to remove an adhesive-coated material from a test plate at a specific speed and at an angle of 90° [74]. A sample of PSA-coated material 1-in. (2.54 mm) wide and about 4-in. (about 10 cm) long is bonded to a target substrate surface (glass unless otherwise stated) [35]. One end of the adhesive-coated sheet is peeled off the target substrate and held in a plane perpendicular to the target substrate which is restrained in its initial plane. The coated sheet is pulled away at a constant linear rate of 300 mm/min and an angle of 90°. The peel adhesion at 90° is twice the value of 180° peel [21]. For 180° peel measurements the results depend on the face stock material [78]. For thicker and stiffer materials a higher peel value will be obtained than at 90° [77]. Especially for soft PVC, 90° peel gives more exact values [78]. A special case of 90° peel adhesion test is the butt tensile test. This test (90° adhesive test) restricts the evaluation to a single narrow area, in contrast to a standard peel test on which a continually new adhesive part is being examined [79]. Special peel adhesion measurements at different angles and temperatures (between 45° and 90°, at 50°C and 110°C), as well as hot shear gradient (SAFT) measurements were proposed.

T-Peel Adhesion. The T-peel adhesion test evaluates the "unwind" characteristics of the wound tape [22]. The tape is bonded to the other piece of the same film. The ends of the two films are inserted into the opposing jaws of a tensile machine which then pulls the two films apart at an 180° angle. T-peel adhesion may be measured according to ASTM D-1876 or DIN 53281. This method is intended primarily for determining the relative peel resistance of adhesive bonds between flexible adherents or when standard peel values are too low. The adherents shall have such dimensions and physical properties to permit bending them through any angle up to 90° without breaking and/or cracking.

Influence of the Peeling Rate. Peel and tack increase with increasing test speed, but only up to a limit [80]. At very high measuring speeds the tack increases again. An increase of the temperature gives increased molecular mobility resulting in increased tack and reduction in shear resistance. A high stress rate is equal to a drop in temperature. Indeed, universal testing of a roll of tape occurs at 12 in./min, while modern slitting and unwinding machines, or tape application processes, may run at several hundred feet per minute. At harder unwind, adhesive pick-off or transfer may occur [79].

High Speed Peel Tester. This equipment allows a 10–300 m/min testing speed for release force measurements [81]. With this method the separation force of the PSA-coated face stock from the release liner is evaluated at speeds similar to those typically used to convert and dispense the laminated material. It therefore provides a better characterization of the material than FINAT FTM 3. Very low peel values (from the release liner) may create label-fly during conversion or application; high values may produce web break when skeleton stripping or dispensing failure during automatic application occurs. The test is carried out like a 180° peel test at jaw separation speeds between 10 and 300 m/min. In order to ensure good contact between the release paper and the adhesive, the sample is placed between two flat metal or glass plates and kept for 20 hr at 23°C ±2°C under pressure. There is an influence of the way of stripping on the high speed release test results (i.e., face from liner or liner from face). Generally, higher results are obtained when removing the face material from the release liner.

The peel force measurement is usually made at a single pull rate and ambient temperature [82]. Generally, the PSA is polymer-based; consequently, it behaves like a viscoelastic medium. The bond strength is not constant for varying strain rates and temperatures. The rate of separation and temperature are related to the adhesive strength in a complicated manner, and these factors can influence the nature of the failure mode (adhesive or cohesive). This can lead to a situation in which an adhesive is selected on the basis of its performance at one set of strain rate/temperature conditions, but ultimately fails because in use it experiences a different set of loading conditions. Rather than judging adhesive strength at a single separation rate and temperature, surface contour diagrams should be used to illustrate the adhesion over a wide range of conditions [82]. This technique allows a comparison of different PSAs while examining the maxima and minima of their rate/temperature graphs as well as how a PSA laminate meets or exceeds a given peel value. A temperature range of –20 to 158°F may be selected, a range of separation rates of 0.50–50 in./min was suggested.

The time/temperature influence on the peel test was discussed in Chapter 6. Druschke [55] summarized the different dwell times used by AFERA (10 min), PSTC (1 min), and FINAT (20 min and 24 hr) for peel testing purposes.

Influence of the Substrate. The influence of both components of a joint on the debonding is well known (see Chapter 2). Peel adhesion should also be examined as a function of the substrate (adherent). For example, peel was tested on stainless steel and HDPE [32]. Samples of 20 × 280 mm on a stainless steel plate with a bond length of 100 mm were evaluated. For peel from HDPE, 0.25-cm thick, high density polyethylene panels were used.

The most important factor in measuring the bond strength is the test plate [38] which can be float process plate glass (FINAT), brushed steel (AFERA), or polished steel (PSTC and BWB). Glass remains the most suitable material [38]. The FINAT Test Method 1 proposes the use of float process plate glass as substrate or test material. All peel test methods except FINAT suggest stainless steel plates as the adherent [55]. Standard tests of tack and peel are made on steel (PSTC-5 and 1), but measurements on polyethylene, polypropylene, cellulose, PET, PA, and glass are often suggested. Adhesion on different substrates (aluminum glass, nylon, acrylic sheet, polycarbonate, rigid polyethylene, rigid polypropylene, polystyrene, stainless steel) was tested [83]. Generally, peel adhesion should be evaluated with and without the use of a primer [84]. The surface energy of the adherent, the compatibility of the surface with the adhesive, and its pH also influence the adhesion [79].

Peel on Cardboard. It is essential that PSAs are evaluated against the final intended surface, namely the National Bureau of Standards reference material 1810 [79]. Because cardboard has a grain that is dependent on the direction of the manufacturing process, this should be determined prior to any testing so that the same direction is used in any measurement. An adhesive tape can cause 100% delamination of the cardboard when stripped in one direction, but may only cause 5–10% fiber tear when stripped in the perpendicular direction. In the cardboard fiber tear test, a strip of tape is applied to the cardboard substrate, affixed with a 2-lb roller and removed immediately by a pulling force extended on a tape at a 90° angle to the substrate surface. The adhesion properties of a product vary directly with the properties of its surface covered by fiber [22].

Peel Adhesion of Plasticized PVC Film. A PVC film/1 mil PSA layer/ siliconized paper laminate is proposed for the evaluation of the adhesion of plasticized PVC film [59]. After specified aging times, the siliconized paper is removed and the adhesive-coated PVC film then is adhered to a stainless steel panel. After a 24 hr dwell time the 180° peel strength is determined according to PSTC test method 1.

Polyethylene Peel on Polyethylene Plate. This method tests the 180° peel adhesion on polyethylene using a rigid polyethylene surface (plate). The test method is the same used for polyethylene foil but with HDPE plates instead; the cleaning of the plates should be carried out with n-hexane. Adhesion to HDPE is lower than to LDPE [85]. There are different bond failure modes:

1. Face failure: a 100% transfer of adhesive to the substrate.
2. Clean release failure: 100% adhesive release from the substrate.
3. Cohesive failure: the adhesive splits between the face and the substrate.

The nature of the failure is very important for removable PSAs. Papers with a different surface treatment yield different peel values for the same adhesive. Permanent adhesives should cause destruction of the paper (paper tear); partial destruction is noted as paper strain [88] which generally occurs at a force of 1500 g/in.

Measuring the Shear

As discussed in Chapter 6, the cohesion of an adhesive may be estimated as its resistance to shear forces. Similar to tack and peel measurements, there are standard and special methods for the evaluation of the shear properties. There are a lot of different methods providing information about the cohesive strength. These include shear or holding power, shear at a 20° angle, heat distortion temperature, heat resistance, 90° peel, shear resistance versus humidity, etc. [89]. Different methods for measuring shear based on the examination of shear, peel, and tensile strength were described [96]. A tensile strength measurement also may be used for the evaluation of the cohesion [78]. On the other hand, shear measurements may characterize other adhesive properties. Shear tests are used for the characterization of the mechanical properties of adhesives [91]. The temperature resistance via 45° shear at 70°C has been measured [78].

There are a lot of different methods based on quite different test parameters for shear testing purposes. The most important method (FINAT FTM 8 and PSTC 7) measure holding power or shear adhesion, and record the time to failure at room temperature. Other methods are carried out at 65°C (Shell), 50°C (Ashland), or 70°C (Rohm). Some tests involve temperature measurement (shear adhesion failure temperature, heat distortion temperatures). Here the temperature is gradually increased and the temperature at bond failure measured; for the shear adhesion failure temperature (SAFT) a 40°F/hr gradient is generally used. The slip distance is measured by Wacker Chemie GmbH (70°C, 5 min). Others measure the temperature at which a change from adhesive failure to cohesion failure occurs.

The test conditions differ also. Shear should be tested on 5 cm^2 samples, at room temperature or 50°C with a 1-kg load, and conditioned for 12 hr [46]. Holding power is tested statically according to different norms like ɾAFERA 4012, TL7510–011, or U.S. Federal Test 20554 (PPT-T-60-D) [92]. These examples illustrate that there are many different shear measurement methods although standardized ones were defined.

Standard Tests for Shear Strength. Shear is determined in accordance with ASTM D 3654–78, PSTC 7, and FINAT Test Method 8, and is a measure of the cohesiveness (internal strength) of an adhesive. It is based on the time required for a statically loaded PSA sample to separate from a standard flat surface in a direction essentially parallel to the surface to which it was affixed with a standard pressure.

Statical Shear—Room Temperature Shear. Each test is conducted on an adhesive-coated strip applied to a standard stainless steel panel in a manner such that a 0.5 × 0.5-in. portion of the strip is in fixed contact with the panel with one end of the strip being free [56]. The steel panel with the coated strip attached is held in a rack such that the panel forms an angle of 178–180° with the free end tape extended; the latter then is loaded with a force of 500 g applied as a hanging weight from the free end of the test strip. The elapsed time required for each test strip to separate from the test panel is recorded as shear strength. According to FINAT FTM 8 (a 178° shear adhesion test) a 25-mm wide specimen of PSA film (prepared at least 24 hr before running the test) is applied to one end of a stainless steel plate in order to obtain a contact area of 25 × 25 mm. Five to ten minutes after applying the specimen on the plate, the assembly is suspended from a plate holder, which maintains the plate at an angle of 2° from the vertical. A 1-kg weight is immediately attached to the clamp. The time recorder is automatically switched off by the weight as it falls. According to AFERA 4012–P1 (modified) a 25 × 12.5-mm sample is used in combination with a 1000-g weight [86]. Besides the standard, normalized shear test methods there are several modified versions. In another test method, a strip of coated paper is fixed to the edge of a glass plate so that an area of 6.45 cm^2 (1 in.^{2}) is in contact with the glass. The force required to remove the strip at an angle of 2° is measured at a separation of 1.0 mm/min and 20°C [57].

As in the case of peel measurements the influence of the contact time (dwell time) on the shear test results should be considered. Shear adhesion testing according to PSTC 7 (1000 g/in.^{2} load) implies a dwell time of 24 hr. Shear should be measured after a 20 min dwell time, using a 25 × 25 mm or 25 × 12.5 mm area, with a load of 5 or 10 Newton [2]. As test substrates, stainless steel or glass should be used. Tests should be carried out at 25°C and 50% RH. Failure time or slip distance should be defined as shear resistance. The influence of the sample dimensions should also be taken into account. Room temperature shear should be measured using an increased contact surface in order to avoid errors by coating defects. For a modified version of FINAT FTM 8, a load of 2 kg and a contact area of 12.5 cm^2 on aluminum test plates are used. A strip of PSA-coated material (20 × 150 mm) should be affixed with a light pressure on an aluminum test plate (previously cleaned with toluene) so that the contact length is 62.5 mm. At least three measurements should be performed.

The basic problem with the standard lap shear test is that the average measured shear strength is not a material property that characterizes the adhesive uniquely. Instead, it is a rather vague quantity that also is strongly dependent on the geometry of the joint being tested [93]. For instance, if the overlap were doubled and all other variables left unaltered, the strength of joint would not be doubled. Similarly, if the joint geometry and adhesive are kept constant and the adherent material changed, the apparent strength of the adhesive will change dramatically [93]. From the point of view of the measurement efficiency, shear tests are more difficult and less efficient than tack or peel measurements. Static, normal temperature shear possess low reproducibility and requires long measurement times. In order to get precise results in a short time, hot shear measurements were introduced (i.e., shear tests are carried out at elevated temperatures). Another possibility to accelerate shear tests is the use of the dynamical shear test. For adhesive characterization purposes not only the time, but the nature of the failure also is very important.

The mode of bond failure is described by shear tests with different codes such as P, C, PS, and SPS, where P means panel failure, C means cohesive failure, PS failure with panel staining, SPS means slight panel staining [94]. Modes of shear failure include:

1. Face stock failure: 100% transfer of the adhesive to the substrate (adherent).
2. Clear release failure: 100% adhesive release from the substrate (adherent).
3. Cohesive failure: (zip effect) adhesive splits between the face stock and the substrate. This results as a direct consequence of an imbalance between cohesive and adhesive properties and is an adhesive weakness.

Room temperature shear tests (0.5 × 0.5 cm, 1000 g) and elevated temperature shear tests (with a 4.4 psi load) were used [32]. To force the cohesive failure rather than substrate or panel failure, the polymer was tested on aluminum foil as substrate, after a 72-hr dwell time and a 20 psi load. Values of 1500 min to more than 3000 min were recorded.

Hot Shear (Holding Power). This test is a modification of PSTC 7. A 1.27 × 2.54-cm specimen of the adhesive is mounted on a 7.5 × 20-cm stainless steel panel; an aluminum foil is added as reinforcement for the face material. The panel then is positioned with its longitudinal dimension so that the back of the panel forms an angle of 178° with the extended piece of tape, the 2.54-cm dimension of the adhesive extending in the vertical direction [66]. The assembly then is placed in a 70°C oven and a 1-kg weight is attached to the free end of the tape. The time required for the adhesive to fail cohesively is reported in minutes. There are many different variations of hot shear methods. Shear resistance at elevated temperatures is measured at 200 and 250°F (2.2 psi) [95]. High temperature shear at 400°F, based on a 5 kg/in^2 bond, 0-min dwell time was evaluated [66,96]. Shear at 70°C is reported in [66] and 50°C creep is determined [97]. The method is similar to PSTC 7; only the weight is reduced from 500 g to 250 g and the temperature is increased from 25°C to 50°C.

Shear Adhesion Failure Temperature: Dead Load Hot Strength Test. The shear adhesion failure temperature (SAFT) determines the temperature at which a pressure-sensitive specimen delaminates under a static load in a shear mode [98]; it is a method of determining the resistance to shear of a tape under constant load under a rising temperature. A variation of the hot shear test method can be achieved by mounting the laminate construction in an oven and suspending a weight from the sample in such a way that a vertical shear stress is applied to the bond. The oven temperature may be raised gradually to define the temperature at which the bond fails [99]. When measuring the SAFT the sample is subjected to a 30 min dwell time at room temperature. The load is attached, the sample conditioned 20 min at 100°F; then the temperature is increased 1°F/min until shear failure occurs up to a maximum of 350°F [99]. The 178°C (modified PSTC 7) test is a measure of the ability of a PSA tape to withstand an elevated temperature rising at 40°F/hr under a constant force [99]. For a rubber-resin-based adhesive tackified with Wingtack a temperature range to failure of 78–87°C was found [100].

Hot-melt PSAs based on Kraton D display SAFT values (on Mylar) of 60–90°C [101]. The test assembly is supported by one end in a vertical position, with a weight attached to the other end and heated in an oven for periods of 15 min; the temperature starts at 30°C and is raised progressively by 5°C increments after each 15 min period.

The SAFT value is $(T - 5)$°C, being T the temperature at which the bond failure occurs. The adhesive tape is applied to a glass plate (25 × 25-mm overlap) and a 2-kg roller passed over the tape, once in each direction [3]. The glass plate is placed in a temperature-programmed oven and clamped at an angle of 2° to the vertical. A load of 500 g is hung from the tape and the oven temperature increased by 4°C/min. The temperature at which the tape drops off the glass plate is recorded.

The validity of the time/temperature superposition principle is demonstrated by the parallel use of dynamic shear tests (controlled forces) and SAFT (controlled temperature). Practically, it is easier and faster to use dynamic shear [57].

Dynamic Shear. This method tests the shear properties of the adhesive in a tensile tester under an increasing load (force). Current static-shear test methods use a constant load at longer test times; they show (related to the nature of the test) poor reproducibility and need very long test times. Following lamination and appropriate conditioning, the samples are placed in a typical tensile strength testing machine, and shear adhesion can be defined by applying shear stress at a given rate (dynamical shear).

Apparatus:

- Tensile tester
- Glass plates (see peel test)

Method:

- A strip of coated material is fixed to the edge of the glass plate so that an area of 6.45 cm^2 (1 in.^{2}) is in contact with the glass.
- The force required to remove the strip at an angle of 2° is measured at a rate of 1 in./min and 20°C.

Other variants of the method exist [57].

Dynamic Lap Shear. Here a 3 × 1-in. aluminum/aluminum assembly is used with an adhesive coating weight of 3 mils. The aluminum plates are pulled apart in a 180° configuration at a speed of 1 in./min [102]. The standard shear method is essentially a static test and therefore no upper value is defined. A dynamic test method is preferred, using a machine that has the capability of running extremely slowly (e.g., 0.02 in./min) [79]. A new dynamic method at a testing speed of 2.5–100 mm/min (as a preliminary test) and 50 mm/min as common testing speed, a laminating pressure of 154 kpa, and temperatures of 23, 40, and 70°C (allowing a rapid assessment of holding power) was developed [92]. A strip of the PSA-coated material is adhered by its adhesive layer to a primed polyethylene substrate with an adhesive contact bond area of 1 × 0.5 in. When testing is carried out at an elevated temperature, this substrate is first reinforced with aluminum foil to impart rigidity.

Twenty Degree Hold. The 20° hold test is similar to a standard shear test except the test plate is inclined 20° from the vertical direction [31]. This test measures a combined peel and shear strength of the adhesive mounted on a 1 mil Mylar film when applied under standard forces to a corrugated cardboard substrate [56]. Shear from cardboard according to PSTC-7 is tested with a 200-lb bursting strength cardboard. The tape is applied in the corrugation direction [22]. With regard to carton sealing properties, cardboard shear, cardboard adhesion (% fiber tear), and high humidity aging (80° RH/125°F, 3 days) are evaluated [37].

Apparatus:

- Shear tester block
- 500-g weight

Method:

- Samples of adhesive-coated PET are applied on a cardboard substrate such that a 1.5-in. edge of the sample is aligned parallel to the corrugated flutes (ridges) of the substrate.
- After application, the sample is rolled with a 4.5-lb rubber roller one time parallel to the 1.5-in. edge of the sample at a roller speed of 12 in./min.
- The sample is mounted in a shear tester at an angle of 20° to the vertical.
- A 500 g weight then is affixed and a timer is started.
- Hold values are reported in minutes.

The shear test shows the following disadvantages: the information obtained remains limited, the reproducibility is low, and the time required too long. More reliable results are given by creep tests (measuring a creep time slope in order to determine the viscous and the elastic components) or by measuring the creep (cold flow) directly.

Test Methods

Table 10.12 Experimental Conditions Used for Shear Measurements

	Sample dimensions (mm × mm)	Substrate	Weight (g)	Temperature (°C)	Method	Ref.
1	25×12.5	Glass	1000	RT	AFERA 4012–P1 modified	[86]
2	25.4×12.7	Stainless steel	1000	70	PSTC 1 modified	[86]
3	25.0×25.0	Stainless steel	1000	40	FTM 8	—
4	12.5×12.5	Stainless steel	500	—	ASTM D-3654–78	[57]
5	12.5×12.5	Stainless steel	500 1000	RT	PSTC 7	—
6	12.5×12.5	—	1000	RT	—	[29]
7	24.5×24.5	—	500 1000	20 65	PSTC modified	[47]
8	12.5×12.5	—	500	RT	PSTC 7 modified 24 hr dwell time	[87]
9	10×50	—	1000	5	—	—
10	645	Glass	—	RT	Dynamic 1 mm/min	[57]

Automotive PSAs Shear. This test measures the slip after a given time, using a 500 g/in.^{2} test sample surface at 158°F, according to the Fisher Body Procedure TM-45–134 [28]. Table 10.12 summarizes the experimental conditions used in shear measurements; Table 10.13 lists the various test methods and cohesive strength values. A complete evaluation of the adhesive properties is possible after measuring all these performance characteristics (i.e., tack, peel, and shear). There are quite different graphical representations allowing the comparison of the above properties as a function of the formulation and in relation to one another. Figures 10.2, 10.3, and 10.4 present the main types of diagrams used for the evaluation of the adhesive properties.

Figure 10.2 shows that the adhesive properties can be evaluated separately (one property) for different formulations (A) or different properties for a given formulation (B). Figure 10.3 presents an overview of the performance characteristics for different chemical compositions (C) or for a given formulation (D). Figure 10.4 shows that the properties can be plotted separately as a function of the chemical composition and a base characteristic of the main components, in a three-dimensional (E) or two-dimensional graph (F).

Table 10.13 Test Methods and Values for Shear

	Test conditions								
	Temperature (°C)	Angle (°)	Dimensions of the sample (mm)	Weight (g)	Face stock material	Value	Units	Norm/method	Ref.
1	RT	178	6.2×6.2	500	PET	120	min	PSTC-7	[103]
2	RT	180	12.5×12.5	500	PET	>100	hr	FTM-8	[104]
3	RT	180	25×25	10 N	PET	0.7–168	hr	FTM-8	[2]
4	RT	180	—	—	PET	250	min	FTM-7	[21]
5	RT	180	—	—	Paper	36–162	min	FTM-7	[105]
6	20	180	25×25	1000	PET	—	—	—	[47]
7	65	180	25×25	500	PET	—	—	—	[47]
8	70	180	25×25	1000	PET	>30	hr	PSTC-7	[106]
9	RT	180	25×25	1000	PET	>85	hr	—	[2]
10	RT	180	25×25	1000	PET	5–10	hr	—	[89]
11	70	180	—	—	PET	0.5–2	hr	—	[107]
12	RT	180	25×25	1000	PET	>200	hr	—	[108]
13	70	180	25×25	1000	PET	>100	hr	—	[109]
14	RT	180	25×25	1000	PET	>200	hr	—	[110]
15	RT	180	25×25	1000	PET	>5000	min	—	[109]
16	70–90	180	25×12.5	—	—	—	—	—	[86]
17	RT	180	6.45 cm2	—	—	80–150	N	Dynamic	[57]
18	—	180	25×25	1000	PET	75–92	°C	SAFT	[109]

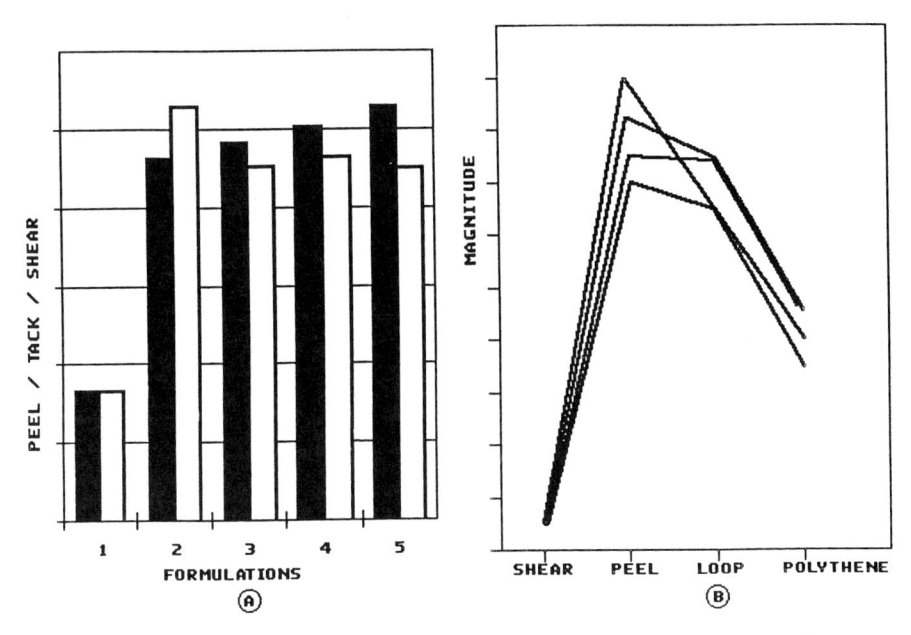

Figure 10.2 The main types of diagrams used for the evaluation of the adhesive properties. A) Adhesive characteristics for different formulations (peel, tack, and shear). B) Adhesive characteristics (peel, tack, and shear) for a given formulation.

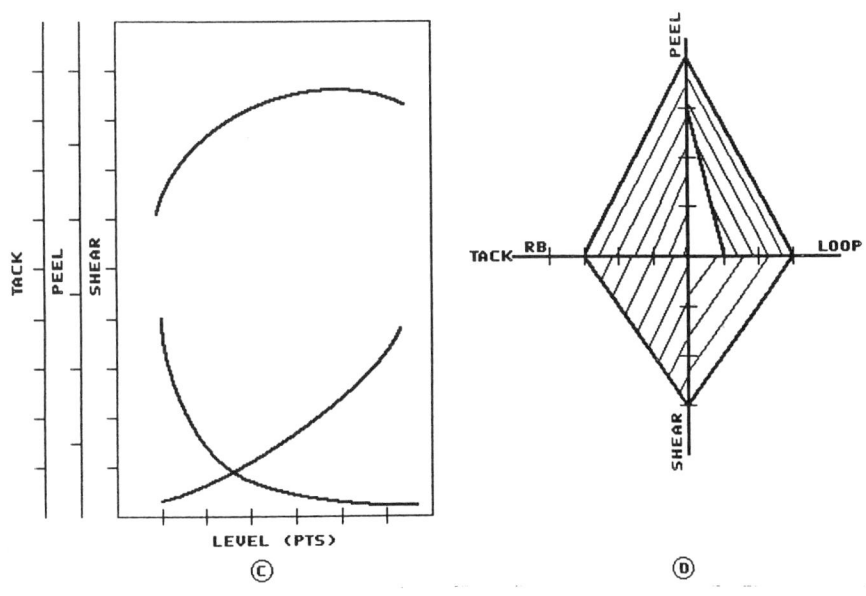

Figure 10.3 The main types of diagrams used for the evaluation of the adhesive properties. C) All performances as a function of the chemical composition. D) All performances for a given chemical composition.

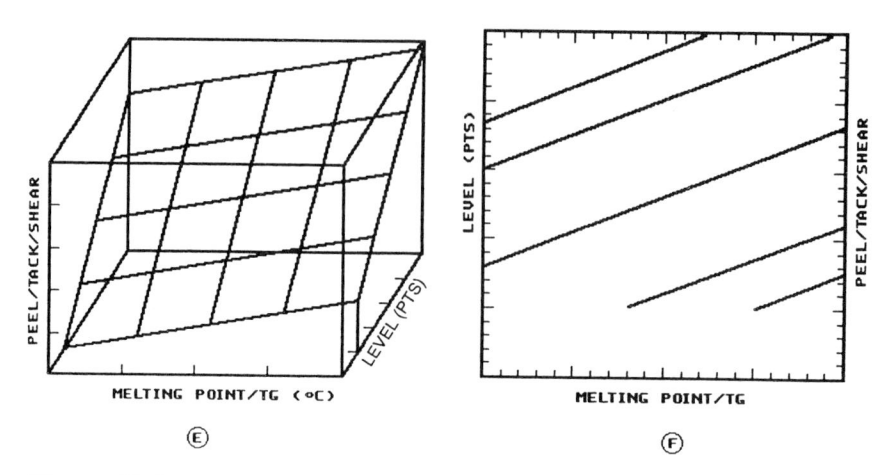

Figure 10.4 The main types of diagrams used for the evaluation of the adhesive properties. E) Space-diagram of the adhesive characteristics (peel or tack or shear) as a function of the chemical composition and component characteristics; F) Plane diagram of the adhesive characteristics.

Other Properties

There are some properties of the coated adhesive which influence the adhesive performance; they include the aging properties of the adhesive and the coating weight. On the other hand, the adhesive and converting properties together determine the application field of the label; therefore the converting properties should also be tested.

Coating Weight. The coating weight influences the adhesive, converting, and end-use properties. The dependence of the adhesive properties on the coating weight was extensively discussed in Chapter 6. Industrial in-line measurement and control of the coating weight, as well as spot checks of the coating weight are carried out. The coating weight may be checked using β rays or IR radiation [100]. For laboratory purposes the gravimetric method is generally used. In order to avoid the discrepancies due to the sensitivity of the face stock material towards the liquid adhesive, aluminum foil may be used. The coating weight is measured by dissolving the adhesive layer and determining the weight difference between the coated and uncoated face material. The method was described in Section 2.1.

Aging Properties. Aging tests are carried out on liquid and dried PSAs. For liquid PSAs these are mostly limited to testing the viscosity of molten hot-melt PSAs. For the dried PSAs or laminates aging tests are carried out in order to check the solid components of the laminate or the PSAs (shrinkage, migration). These aging tests use one of the standard methods for the characterization of the adhesive properties (or their combination) after the storage of the adhesive (laminate) under well-defined conditions (temperature, light, humidity) for a defined time. The most important adhesive characteristics evaluated after aging are the peel adhesion and the tack; generally, peel increases with time and tack decreases. A more complex test concerns the aging of removable PSAs, where "clean" removability should be examined also. In some cases aging tests are carried out in order to make a choice of the adequate adhesive [111]. Aging tests for hot-melt PSAs are carried out (melt viscosity stability) at 177°C for 4 days [111].

The test of the aging characteristics of plastics have been summarized [111–113]. In a similar manner adhesive-coated laminates are exposed to weathering tests and their adhesive properties are tested subjectively or with a mechanical device. Generally, the weaknesses of special classes of adhesives are well known; hence aging test methods are developed to examine these shortcomings after storage. Consequently, the conditions proposed for accelerated aging (storage) can vary quite a bit. The following environments were proposed for aging tests of PSA laminates [111]: –65°F and 50% RH, +73°F and 50% RH, +150°F and 50% RH, and +73°F and 95% RH. Temperature and humidity cycling (MIL STD 331) and diurnal cycling from –65 to +165°F at 95% RH are carried out. Test specimens are examined after 0, 8, 16, 27, and 40 weeks exposure. Aging at 70°C for 1 day corresponds to 0.5 yr of natural aging [106]. Aging tests are carried out at room temperature, 54°C, or 71°C for 1 to 4 weeks [114].

For aging tapes, a storage at 80% humidity at 66°C for 6 days and at 50% humidity at 29°C for 4 hr according to PSTC-9 was proposed [115]. Exposed film tack may be evaluated by leaving tapes in a laboratory environment and periodically checking the finger tack. Aging for 1 day or 1 week at room temperature, 8 days at 88°C and 8 days under a sunlamp is done for chloroprene latices [116]. The aging of S-I-S-based hot-melt PSAs was tested after storage at 150°C for 24, 48, 72, and 100 hr [113]. Tests were carried out at 70°C after 1, 2, 3, and 4 weeks [113]. Pure S-I-S films were aged at 95°C for 150 min [116]. Adhesive build-up is tested for hot-melt PSAs [117]. No significant increase in 180° peel adhesion to stainless steel was noted from initial testing through 24 hr. With hot-melt PSAs, peel decreases after aging. A hot-melt PSA with a 180° peel adhesion of 8.50 lbs/in. (control) displays only 3.70 lbs/in. after 7 days at 120°F, and 4.70 lbs/in. after 7 days at 150°F [37]. Aging of S-I-S-based hot-melt PSAs is performed at 95°C, on polyester, and in a dark environment and dry air [113].

Table 10.14 Test Conditions for Aging

Temperature (°C)	Time	Ref.
70	7 days	[100]
67	7 days	[37]
50	3 days	[27]
50	30 days	[78]
70	1, 2, 3, 4 weeks	[119]
54, 71	1, 2, 3, 4 weeks	[114]
40	2 weeks	[120]
22, 67	1, 3 weeks	[120]

The aging behavior may be estimated by measuring peel or tack changes. Peel strength should be tested initially and after 14 days at 40°C, as well as shear strength and rolling ball tack. Exposed film tack after days at 70°C may be measured also. The UV stability of hot-melt PSAs is tested after 0, 1, 3, and 5 hr via 180° peel, shear, tack, color, and density measurements [118]. Table 10.14 summarizes the test conditions for aging.

The aging test is carried out by storing a laminate sample, 100 × 100 mm, in an oven, between two wooden plates, under a load of 1 kg/100 cm^2, and at 70°C for 4 days. The degree of the adhesive degradation during aging, and the migration (bleeding) are examined.

- Evaluation of the adhesive degradation (finger tack test). The degree of degradation is rated as follows:

    D_0 = OK, no changes in adhesive properties;
    D_1 = Slight legging, may be used however;
    D_2 = About 1 mm legs, adhesion lowered;
    D_3 = Fiber yield, adhesive transfer on the fingers;
    D_4 = Smearing, fluidity of the adhesion, loss of cohesion;
    D_5 = Dry, nontacky adhesive.

- Evaluation of the migration (bleedthrough): after peeling off the liner, coated and uncoated strips of the face material should be compared. The quality is rated as follows [27]:

    M_0 = No difference, very good;
    M_1 = Coated surface generally lighter, fair;
    M_2 = Migration, pointlike sites, poor;

M_3 = Migration, numerous pointlike penetration sites, pinhole-like or more;
M_4 = More than 50% of the surface covered by bleeding.

Aging Test (not FTM). This method tests subjectively the adhesion of the coated PSAs layer after aging under light and temperature exposure. For current, industrial control tests, a "30 min aging time" should be used.

Apparatus:

- Sun tester
- Oven, without air blow cooling (weatherometer), substrate temperature 60–70°C.
- Glass plates (slides) 2.5 mm thin, 25 mm wide.

Method:

- A strip of PSA-coated material (25 × 100 mm or longer) is applied to the glass slide, with the coated adhesive side against the light, with a 20-mm portion at the end of the slide left exposed, the rest being covered with release paper/aluminum foil.
- After 20 min aging of the exposed part, the next 20 mm portion should be exposed and so on.
- After aging the adhesive quality (finger tack) of the aged material should be subjectively evaluated.

D_0 = OK, no changes;
D_1 = No major changes, good-fair, slight legging;
D_2 = 1 mm legs, the adhesive is softer
D_3 = Legging, adhesive transfer;
D_4 = Smearing, loss of the cohesion;
D_5 = Loss of the tack.

Accelerated Weathering Testing. One mil films of adhesive coated on one mil PET film are exposed directly (adhesive side) to the light source in a UV device. A straight UV cycle is imposed (no moisture) [118].

Aging Test of the Release Force. Accelerated aging before carrying out a high speed release test can be carried out by placing a set of self-adhesive strips between two flat metal or glass plates and keeping them for 20 hr in an air circulating oven at 70°C ±5°C. The strips then should be removed and conditioned for at least 4 hr. The environment's effect on the release characteristics of the release paper is measured.

Aged siliconized paper is applied to freshly prepared adhesives and the force required to remove the paper measured in the 180° mode [111]. Adhesive film aging may be tested at 150°F and 15% RH at 4 and 12 days intervals.

Table 10.15 Shrinkage Values for Different Face Stock Materials

Face stock characteristics		Test conditions		Shrinkage (%)		
Nature	Thickness (μm)	Temperature (°C)	Time (min)	MD	TD	Ref.
PET	30–350	190	5	3.0	2.0	[122]
PET	—	—	—	3.3	3.5	—
LDPE	30–200	70	60	—	1–3.0	—
HDPE	—	80	15	1.0	2.0	—
CPP	100	90	5	1.0	—	[123]
BOPP	50–75	130	5	< 8	< 4	[122]
BOPP	—	135	7	5.0	5.0	[124]
PVC	40–200	70	60	—	0–2.0	—
Paper	—	—	—	0.8–1	—	[125]

Exposure under these conditions for four days according to ASTM D-3611–77 represents two years of natural aging.

Aging of PVC Face Stock. Aging tests on PVC (coated with PSAs) may be carried out after 1 week at 158°F [121]. This test appears equivalent or more severe than 7.5 days exposure at room temperature [59].

Dimensional Stability. Dimensional stability testing was standardized according to FINAT FTM 14. Shrinkage values for different face stock materials are summarized in Table 10.15 (see also Table 6.42).

Plasticizer Resistance. Different practical tests are described in the literature that deal with the plasticizer resistance of common adhesives [126]. Most of them use a PVC film (soft PVC with a well-defined plasticizer content) and store the PVC between the adhesive surfaces to be tested. The plasticizer migration is estimated as a weight difference [127,128]. Most plastic films are compounded materials, and contain additives. There are possible interactions between micromolecular components of the film (e.g., PVC) and the adhesive. Plasticizers and emulsifiers from the film can migrate into the adhesive. On the other hand, monomers, oligomers, and surface agents from the adhesive can migrate, causing stiffening of the face stock, or its shrinkage, and the loss of the adhesive properties. The insufficient aging resistance of the rubber-based adhesives excludes their use on plastic films. On the other hand, several EVAc-based adhesives show high shrinkage. Thus, the most suitable adhesive materials for film coating are the acrylic PSAs.

The shrinkage of a plasticized vinyl film with a 1.0 mil transfer coated adhesive film, conditioned 24 hr at standard humidity and temperature, is measured as the change in length and width; then it is heat aged on the liner, reconditioned, and measured again. Shrinkage on vinyl is tested after 7 days at 158°F. The plasticizer migration resistance of the polymer is acceptable for vinyl applications if the shrinkage, when heat aged on a release liner, is in the range of 0.3–0.9% [32].

In the mounted shrinkage test a PSA-coated PVC face stock is bonded to a test surface, often stainless steel, aluminum and/or release liner. The laminate then is exposed to elevated temperature accelerated aging, typically 70°C for one week, and the percentage shrink back of the PVC film from its original dimensions is measured. Both machine direction (MD) (subject to greatest stress-induced elongation) and cross-machine direction (CMD) measurements are carried out [37]. In another test, a 4 × 4-in. coated sample is adhered to a formica panel and aged for 7 days at 110°F [32]. Evaluation criteria include shrinkage in the machine direction less than 0.5%, and shrinkage in the cross direction also below 0.5%.

Plasticizer migration is tested by 180° peel after 24 hr dwell time, 24 hr aged peel, and 1 week aging on liner at 158°F. The peel retention after 7 days at room temperature and 158°F, the shrinkage on the liner (6–8%), and mounted shrinkage (11–17%) were measured (24 hr at 158°F) [95].

Migration. When pressure-sensitive adhesives were coated onto porous substrates such as paper, an assessment of the tendency of the adhesive to migrate or bleed through the paper should be made (see also Chapter 7). The migration of the adhesive can discolor the face surface and reduce the effective adhesive film thickness, thus changing various adhesive properties. On the other hand, the migration may dilute the adhesive and decrease the adhesive properties. Migration is tested by storing the label samples at increased temperatures and then examining it against a black cardboard [129]. Suggested storage temperatures are 60°C and 71°C for hot-melt and solvent-based PSAs, respectively.

There are different methods to test the bleeding through of an adhesive. The test can be run by placing samples of coated label stock, based on paper of a certain quality (e.g., litho paper), into an oven kept at elevated temperatures (minimum of 70°C). The test samples are inspected for discoloration at weekly intervals [130].

Face stock penetration is tested using a 60-lb white paper. The adhesive is coated on siliconized paper and laminated with the Kromecoat litho paper [59]. The adhesive laminate is stored in a 158°F oven during specified times (5 days). Bleedthrough should be tested under a load of 1 lb/in.^{2}. No migration at 140°F on standard label stock should occur [37].

Edge Ooze. This property refers to the flow characteristics of the coated adhesive.

Method:

- A sample (a stack of sheets, having a mean height of 3 cm) should be stored at room temperature for 24 hr.
- After storage the sample should be guillotined through the middle.
- The appearance of the cut surface should be evaluated subjectively.
- The evaluation of the cuttability (edge ooze) should be repeated after 9 succesive cutting operations of the sheet stack.

Rating:

1 = very good, no blocking, sheets can be moved (from the sheet stack);
2 = slight bonding on the cut surface, sheets can be moved only after separating the cut pieces;
3 = pronounced adhesive residue on the cut surfaces, slight gumming of the cut and cutting surface, sheets cannot be moved (from the sheet stack);
4 = very pronounced flow-out of the adhesive, very pronounced gumming on the cut and cutting surface.

3.2 Special Laminate Properties

In this chapter the test methods designed to evaluate the special features of PSA labels used in quite different application fields (e.g., removable film, deep freeze) are covered.

Removable PSA Labels

The most important special features of removable PSA labels are the removability, the migration/bleeding behavior, and the edge lifting properties.

Removability. In the automotive industry special removable tapes have to display very high removability. Thus, selected tapes were removed cleanly from enameled lacquered or melamine test surfaces after 1 hr at 150°C or 30 min at 121°C [131]. For common PSA labels the experimental conditions are not so severe, but generally instantaneous and removability after aging (room temperature and high temperature aging) are carried out. Removable adhesives can be formulated so as not to display adhesion build-up (i.e., the adhesion to the substrate does not increase to the point where the label cannot be removed cleanly even after exposure to heat) (see also Chapter 6). The most important tests covering removability are the following:

- removability from different surfaces, after aging at elevated temperatures over a period of time, evaluated subjectively (see Table 6.18);
- removability from glass, after aging at elevated temperatures, evaluated as peel from glass;
- mirror test, residue free removability from mirror glass, evaluated subjectively.

Removability of Hot-Melt PSAs. The removability is tested by aging the stainless steel panels with 1 × 6-in. strips of label stock at room temperature and elevated temperature (48°C), cooling 1 hr at room temperature, followed by peel adhesion measurements of the label stock [132]. Other test cycles include 24 hr at room temperature, 24 hr at 48°C, 1 week at room temperature and 1 week at 48°C. For hot-melt PSAs the initial peel from stainless steel should be generally below 1.0 lb or 16 oz/in. The resulting peel (after build-up) should be below 2.5 lbs or 40 oz [132]. For removable hot-melt PSAs the initial peel, 24 hr peel at room temperature, peel adhesion at 48°C, 1 week peel at room temperature, and 1 week peel after aging at 48°C are measured [47]. Tack values for removable hot-melt PSAs (90° quick stick) are 0.5–1.3 lb/in., rolling ball tack values are 1–3 in, and peel values are 0.9–2.6 lb/in; the SAFT amounts to 128–156°F [132].

Not only the peel value but also the failure mode is important in the removability test of PSAs. Therefore, the amount of adhesive transfer for peelable tapes or labels is evaluated after storage at 40°C for 7 days [17].

Removability of Water-Based PSAs. The evaluation of removable water-based acrylic PSAs is carried out after 1 week storage at 70°C on PET and HDPE [2].

Adhesive Residue. When the removability test (described earlier) is performed, the surface (substrate) underneath the label sample is usually inspected to determine the amount of adhesive residue left on the surface of the adherent. There are different, subjective rating scales. In some cases each sample is assigned a numerical rating from 0 to 5 based on an arbitrary scale. In other cases adhesive transfer is estimated with A, L, and SL ratings, where A means adhesive transfer, L corresponds to legging, and SL denotes slight legging or stringing. High temperature resistance may be tested as the percentage adhesive transfer after 30 min at 250°F [37]. Removability should be evaluated on glossy stain-resistant acrylic enamel after aging at elevated temperatures and peeling at an angle of 45°.

Apparatus:

- Test panel covered with glossy stain-resistant acrylic enamel paint (automotive paint).

- Samples of labels (tapes) 1.27 × 10.16 cm adhered at room temperature to the test surface.
- Ventilated oven.

Method:

- Samples are fixed to the panel.
- A 2-kg rubber-coated metal roller is run twice over the samples.
- Samples are left 1 hr at 128°C or 150°C in a ventilated oven.
- Samples are peeled back at an angle of 45° at an approximate rate of 1.9 m/min.
- The panel is removed from the oven and examined. A nontacky deposit is reported as residue.

Repositionability. With repositionable adhesives the adhesive-coated paper may be readily lifted and removed from the contact surface and reapplied at least 8 additional times to paper surfaces without reducing the adhesive properties [133]. Measurements done with a 7-g/m^2 adhesive coat weight, on a 70-g/m^2 paper as face stock material, yield the following peel adhesion values on newspaper (test conditions: 180° peel, 300 mm/min): 110 g initially, 85 g after 50 repeated peel tests, and 75 g after 100 repeated peel tests [63].

Lifting (Mandrel Hold). Different lifting tests are carried out in order to check if the peel/shear resistance of the adhesive balances the lifting forces of the face stock (deformation of the face stock material on a curved surface). In a test called "Curved Panel Lifting Test at 150°C," an aluminum panel with a radius of curvature of 23 cm and a length of 35.5 cm in the curved direction is used [131]. The tapes to be tested are applied to the aluminum panel in its curved direction. The assembly bearing panel then is put into an air-circulating oven at 150°C for 10 min, allowed to cool, and examined for failures. A rating of "pass" means no lifting has occured; any lifting at either end of the strip is noted as "end failure" and the total length of tape which has lifted is rated.

This method should be used as a practical test for the peel of removable PSAs. The peel is subjectively evaluated on the basis of lifting or flagging of the sample affixed on round, cylindrical items.

Apparatus:

- H-PVC and polyethylene cylinders (21 mm);
- Glass cylinders (16 mm).

Method:

- Two samples (labels 15 × 50 mm for PVC, and 15 × 40 mm on glass) are applied in such a manner that the machine direction of the face is either parallel or perpendicular to the axis of the cylinder.

- The samples are stored at room temperature.
- The lifting (delaminated) length of the samples is measured at regular intervals (i.e., 1 day, 1 week).

Lifting (winging) also is evaluated after 5 hr on vertical surfaces [63].

Filmic PSA Labels

In film-coating of PSAs, a range of special requirements must be fulfilled. The first screening of the adhesive (and laminate) covers a lot of special tests carried out in the laboratory or on a pilot coater (Figure 10.5). Laboratory tests carried out with water-based PSAs estimate the coatability and the adhesive properties of PSAs. An additional aging test completes the laboratory screening. In the coatability test the wetout is checked for minimum viscosity and minimum wetting agent level. In the preliminary test of the adhesive properties, the peel, tack, and shear balance are estimated. The preliminary evaluation of the adhesive properties includes the estimation of the shrinkage, color, and storage behavior. In the pilot phase of the test, the coatability, convertability, adhesive, and aging properties of the laminate and its processability are examined. The coatability should be tested by wetout, through the rheology, and optical appearance. The convertability of the dispersion should be checked as a function of the speed, foam formation, and behavior on the metering roll. The discrepancies between laboratory-measured and pilot-manufactured label material should be evaluated with regard to the adhesive and aging properties. The processability of the laminate should be examined as far as cuttability, release, and storage are concerned. Figure 10.6 summarizes the steps of the preliminary screening on a laboratory and pilot scale for PSAs for film coating, including: coagulum, transparency wetout, shrinkage, tack/polyethylene peel, water whitening, wet anchorage, and wet adhesion.

For water-based PSAs for transfer coating onto soft PVC, the following properties are to be tested for their emulsion and adhesive properties, as well as for their water whitening and water resistance.

Emulsion Properties. The relevant emulsion properties are:

- Viscosity: the viscosity should be measured with a DIN 53211 Ford Cup 4 mm, at a temperature of 20°C or with a Brookfield viscosimeter, LVT, Spindle 3, 12 rpm. Target values are 17.5 sec ±1 sec or 150–200 mPa·s.
- Solids content: the solids content should be measured according to DIN at 140°C, over a period of 20 min, through weight difference measurements. The target value should exceed 50% ±1%.
- Wetout: the aqueous PSAs should be coated with an Erichsen knife coater (70 μm) on solventless silicone release paper at a coating weight of 20

TEST RANGE

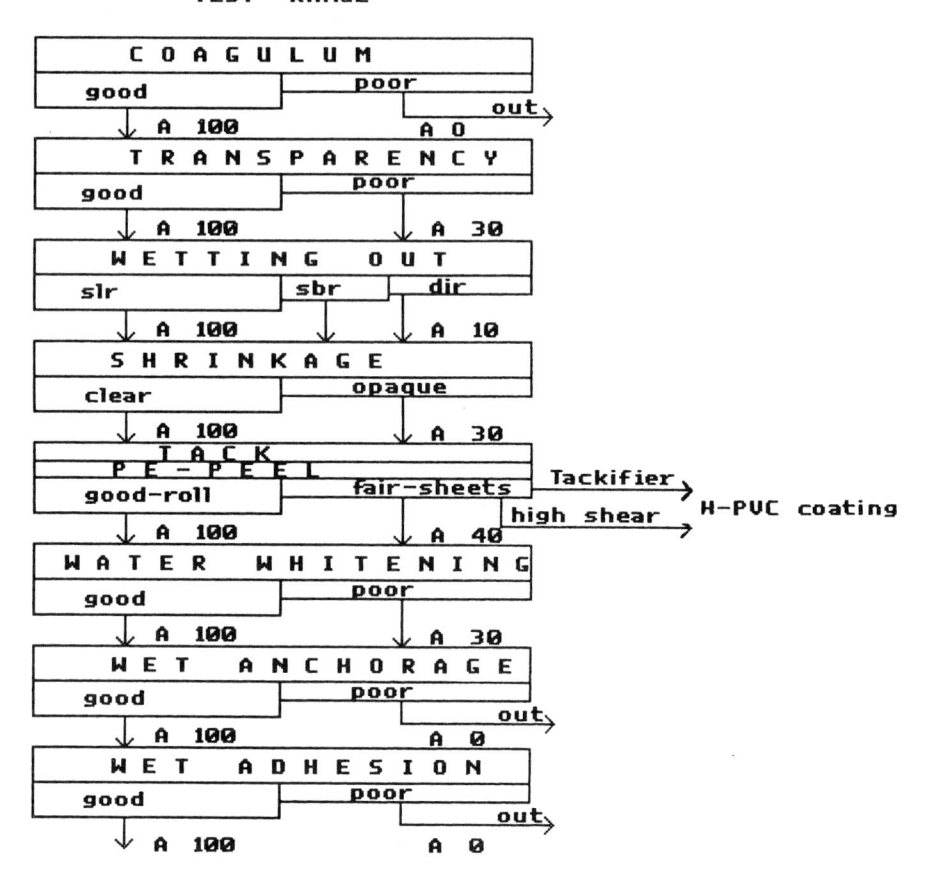

```
┌─────────────────────────────────────────┐
│           C O A G U L U M               │
├──────────────────┬──────────────────────┤
│     good         │        poor          │
└──────────────────┴──────────────────────┘ out →
        ↓ A 100              A 0
┌─────────────────────────────────────────┐
│        T R A N S P A R E N C Y          │
├──────────────────┬──────────────────────┤
│     good         │        poor          │
└──────────────────┴──────────────────────┘
        ↓ A 100           ↓ A 30
┌─────────────────────────────────────────┐
│          W E T T I N G   O U T          │
├──────────┬──────────┬───────────────────┤
│   slr    │   sbr    │       dir         │
└──────────┴──────────┴───────────────────┘
        ↓ A 100      ↓         ↓ A 10
┌─────────────────────────────────────────┐
│          S H R I N K A G E              │
├──────────────────┬──────────────────────┤
│     clear        │       opaque         │
└──────────────────┴──────────────────────┘
        ↓ A 100           ↓ A 30
                    T A C K
┌─────────────────────────────────────────┐
│          P E - P E E L                  │
├──────────────────┬──────────────────────┤
│   good-roll      │    fair-sheets       │  Tackifier →
│                  │                      │  high shear → H-PVC coating
└──────────────────┴──────────────────────┘
        ↓ A 100           ↓ A 40
┌─────────────────────────────────────────┐
│      W A T E R   W H I T E N I N G      │
├──────────────────┬──────────────────────┤
│     good         │        poor          │
└──────────────────┴──────────────────────┘
        ↓ A 100           ↓ A 30
┌─────────────────────────────────────────┐
│      W E T   A N C H O R A G E          │
├──────────────────┬──────────────────────┤
│     good         │        poor          │
└──────────────────┴──────────────────────┘ out →
        ↓ A 100              A 0
┌─────────────────────────────────────────┐
│       W E T   A D H E S I O N           │
├──────────────────┬──────────────────────┤
│     good         │        poor          │
└──────────────────┴──────────────────────┘ out →
        ↓ A 100              A 0
```

Figure 10.5 Screening test for PSA for film coating. Evaluation of the suitability (in %); A, adequacy; Slr, solventless release; Sbr, solvent-based release.

Test Methods

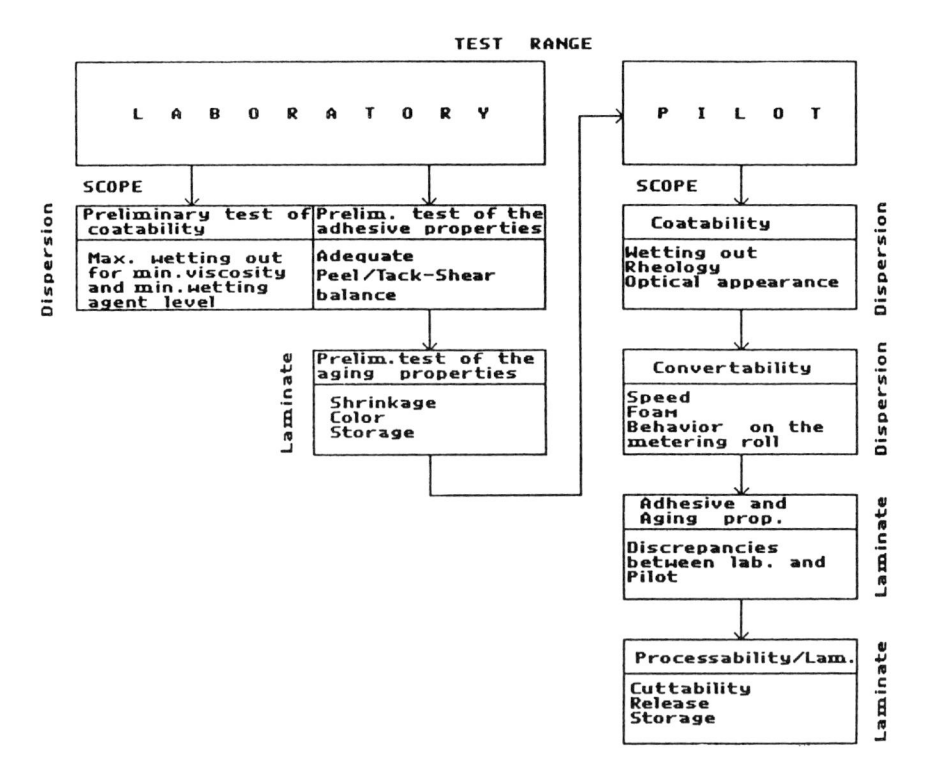

Figure 10.6 Screening test used on laboratory and pilot scale for the evaluation of the general characteristics of a PSA film laminate.

g/m^2 or less dry. Wetting out is adequate if the wet PSA layer displays no transversal shrinkage within 10–15 sec.

Adhesive Properties. The most important adhesive properties include:

- Peel from glass: the PSAs layer is coated using an Erichsen 70-μm knife coater on a 100-μm clear PVC film and dried 10 min at 70°C. The target value is 10 N/25 mm.
- Peel from polyethylene: the strip should be affixed on a polyethylene film laminated onto a test panel; the target value is 6 N/25 mm.
- Rolling cylinder tack: as discussed in Chapter 6, the rolling cylinder tack test measures tack as a function of the distance traveled by a steel cylinder on an adhesive-coated substrate after it has rolled down an inclined plane. The target value averages 2.5 cm ±1 cm, for a coating weight of 20 g/m^2 (dry).

- Hot shear: the hot shear should be measured according to a modified FINAT FTM 8. Face stock materials include PVC, PE, PP, or PET. For PVC materials the film should be reinforced with paper in order to avoid elongation phenomena. As equipment a special heated panel should be built, tilted 2° from the vertical to which the steel plate with the sample on it can be attached. When the temperature reaches 50°C, the free end of the suspended sample should be loaded with a 2-kg weight. The longer the sample holds, the better the hot shear resistance. The target value is a minimum of 20 min.
- Water whitening:
  - The PSA is coated on a 100-μm film by the aid of an Erichsen knife coater (70 μm). After 30 min room temperature drying supplemental drying should be carried out at 70°C during 10 min. The sample is then covered with release paper. For comparison purposes the face stock should possess ungummed areas also.
  - A coated (partially ungummed) strip of the film is immersed in distilled water at room temperature. The moment color changes appear (whitening of the adhesive layer) should be noted. The longer the transparency persists, the better the water whitening resistance. Target values are rated as follows: 10 sec, fair; 15 sec, good; and 17 sec, very good.
- Loss of transparency: PSA-coated film can suffer water whitening upon immersion in water. Water whitening is characterized by the speed of the loss of transparency (the water-whitening test just described) and by the absolute value of the loss of transparency, as compared to the transparency of a dry coating.

  The film laminate should be covered with a transparent (clear) PVC film and a bubble-free film/adhesive/film laminate should be prepared. Another part of the same film coating (clear film + adhesive) should be immersed in distilled water at room temperature for 7 min. After 7 min the immersed sample should be laminated with the same transparent (clear) film as used for the dry laminate. Transmittance (transparency) of both samples (film-dry adhesive-film and film-wet adhesive-film) should be measured comparatively with a colorimeter. The differences in transparency between dry and wet laminate) are measured; target values are $< 8\%$ = fair and $< 3\%$ = good.
- Wet anchorage: film coatings should not display rub-off of the adhesive layer (hand made test) after immersion in water (7 min).
- Wet adhesion on glass (see Chapter 6): A release-free film coating is immersed in water (room temperature, distilled water) for a period of time of 7 min. Then the wet film coating is affixed on glass (bubble free and not floating). The adhesion of the wet adhesive on glass (the recovery of the adhesive properties over time) should be tested, trying to remove the

sample after different storage times (i.e., 1, 2, 5, 10, 15, 30, and 60 min). The shorter the time for adhesion recovery, the better the wet adhesion. Adequate values are less than 10 min; at least one test value at more than 30 min is required.
- Wet adhesion on polyethylene is important for plastic bottles. High speed label application should be used for plastic bottles based on polyethylene copolymers. Affixed labels should resist immersion in water or surface active agent solutions.
  - Dry samples should be affixed on squeeze bottles. After 24 hr the labeled bottles are immersed in a diluted solution of surface active agents (SAA) (0.5–1% common detergents, room temperature). Labels are peeled off after a wet storage for another 24 hr; peel forces are evaluated subjectively. No edge lifting is allowed.
  - Wet samples are tested the same way as dry samples, but wet labels (labels immersed for 7 min in water) should be used.

Summary of Adhesive Properties.

1. Peel adhesion:
   - The FINAT FTM 1 method could be simplified, but at least instantaneous and 24-hr peel values on a standard surface are needed.
   - At least one measurement from glass or stainless steel is needed.
   - Instantaneous polyethylene peel and 24-hr peel are the most important.
   - If values are similar, peel on solid polyethylene plates should be measured.
2. Tack:
   - For similar materials rolling ball tests should be repeated with the rolling cylinder test.
   - Loop tack on polyethylene is more important than rolling ball tack.
3. Shear:
   - If the measured shear values are too high, smaller size samples, samples with a higher coating weight, or hot shear measurements are to be used.
   - Films can be reinforced with paper in order to avoid elongation effects.

Tack/peel requirements for sheet/roll materials differ. Roll labels for high application speeds need a higher agressive tack and polyethylene peel. Film sheets need to display a better cuttability (see Figure 10.7). Figure 10.8 summarizes the relevant test methods used for the evaluation of the adhesive properties as a function of time, temperature and substrate.

Water Resistance

The water resistance is generally tested in order to characterize the water resistance/insolubility of a PSA laminate. Water resistance/insolubility is re-

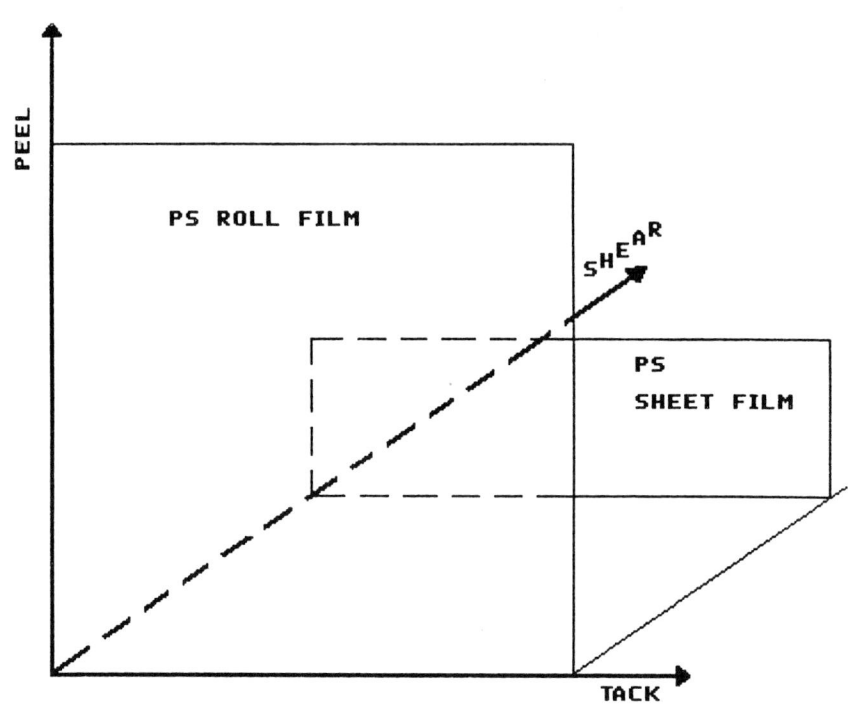

Figure 10.7 Requirements for PSA for roll/sheet labels.

quired when using labels under humid or wet conditions. For this application the appearance of the label and its adhesion/end-use properties under wet conditions are to be examined. With regard to water resistance paper and film labels should be included in the evaluation. Water solubility is required for wash-off labels, mainly based on paper face stock. Water resistance can be examined as the change of the adhesive properties under the influence of water (or water vapor), or as the change of the optical and dimensional characteristics of the label (tape) under the influence of water. The interdependence of the parameters characterizing the water resistance performance of a PSA or PSA laminate were discussed in Chapter 8. For certain applications a high water resistance (i.e., no influence of water on the PSA laminate) is required. For others a strong influence of water on the laminate is desired. One should differentiate the water resistance from the humidity resistance. Water-resistant products are special laminates; humidity resistance is generally required for most laminates. The evaluation of the water resistance, measured as the stability (or change) of the adhesive properties under the influence of

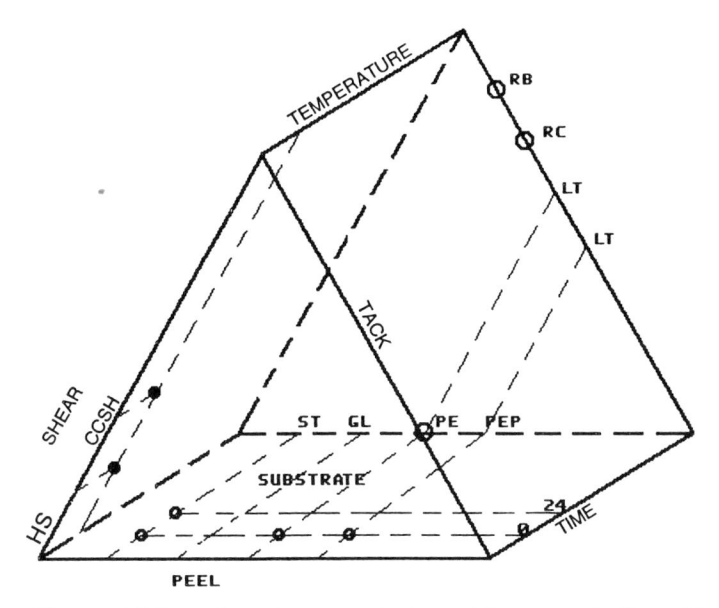

Figure 10.8 The main test methods used for the evaluation of the adhesive properties. RB, rolling ball; RC, rolling cylinder; LT, loop tack; ST, steel; GL, glass; PE, polyethylene film; PEP, polyethylene plate; HS, hot shear; CCSH, corrugated cardboard shear.

the humidity, is mainly based on the examination of the peel adhesion, tack, or shear after storage in a humid environment.

Water resistance can be evaluated by storing the laminates under water and removing them 24 hr prior to testing, to allow recovery of the adhesion. Water resistance was evaluated immediately after 24-hr water immersion or after a 7-day water immersion, followed by a 24-hr recovery period [95]. Another test measures water resistance after a 24 hr immersion, bonded to stainless steel for 30 min, than placed in room temperature tap water, followed by a 180° peel measurement 24 hr later [32]. In the Weyerhouser (shear) test a laminated Kraft paper strip (coated with PSAs) is immersed in water with a 350-g load attached; the time to failure is measured. Water-whitening tests measure the change of the optical appearance (see Section 2.2 of Chapter 6).

A clear Mylar label attached to stainless steel and immersed in water at 74°F for 48 hr shows neither whitening nor blushing [17]. There is an obvious objection to water whitening as a measure of water resistance. Highly water-sensitive materials like gelatin or soluble cellulosic films are not whitened at all. However, in water insoluble and slightly swollen latex films, the degree and speed of whitening are a good index of water sensitivity.

Humidity Resistance. As discussed in Section 1.2 of Chapter 6 and taking into account the strong influence of the environmental conditions on the adhesive properties of PSA laminates, test conditions are standardized and a standard value of 50% RH is proposed. Testing humidity resistance simulates real application environments, with increased humidity. The humidity resistance is tested for 7 days at 100°F, 85% relative humidity, and under condensing humidity conditions [95]. The humidity resistance was measured for PSAs coated on a 2 mil PET film, bonded to stainless steel for 30 min, then placed in a humidity cabinet at 100°F and 95% RH. After 7 days a 180° peel test is carried out [32].

Wash-Off Adhesives. Wash-off adhesives are designed to easily dissolve in water. The test method can be described as follows: the water removability is measured by applying a coated paper (8.2 × 3.7 cm) with 20 g/m^2 dry adhesive to a glass bottle, and then agitating the bottle gently in water at 60°C. The label should easily detach from the glass with the adhesive film still attached to the label, and being nontacky in the presence of water [57]. The water solubility should be tested for affixed labels as old as 3–6 months, stored at room or elevated (60–70°C) temperatues.

Washing Machine Resistance. Special labels of fabric (textile) used on clothing have to be tested for washing machine resistance. Such labels should resist more than 50 wash cycles and up to 10 chemical (dry) cleaning process cycles [134].

Deep Freeze Labels

A great deal of attention was paid to the performance of PSAs at low temperatures as adhesive failure at those temperatures may occur [135]. The surface temperature of the adherent is another factor. Most general purpose adhesives are formulated for tack when applied to surfaces with temperatures as low as 38–40°F. If the temperature of the adherent is lower, a higher degree of adhesive cold flow is required to provide proper wetout. Little concern needs to be given to products that will be labeled at room temperature and subjected to lower temperatures later. In general, low temperature properties are tested in function of the application field, but all of the adhesive properties have to be checked, in order to make a meaningful evaluation of the low temperature properties. For comparative purposes, static shear, peel adhesion, dynamic lap shear, and low temperature lamination were evaluated [102]. The measurement of the peel adhesion and of the tack at –5 and +5°C as well as at –20°C give a reasonable indication of the end-use performance, both for low temperature (i.e., chill) and deep freeze applications.

Static Shear Load Test at Low Temperatures. In the static shear load test at low temperatures a modified version of the PSTC test (1 in^2 contact surface on stainless steel, 1000-g weight) was suggested [102]. Slippage and delamination of the PSA-coated strips after five days storage in a constant temperature cold chamber were observed. For dynamic lap shear testing at low temperatures, 3 × 1-in. samples are coated with adhesive, equilibrated in an environmental chamber at the desired temperature and pulled apart in a 180° configuration at a speed of 1 in./min.

Low Temperature Adhesion (Peel). For the evaluation of cold peel adhesion, samples are applied when enclosed in an environmental chamber and allowed to equilibrate at the desired temperature for 10 min prior to being peeled in a 180° configuration at 12 in./min [102]. The temperature at which a PSA label is laminated to a substrate can be just as critical as the usage temperature. Therefore, in low temperature lamination tests, adhesive-coated strips and stainless steel panels are conditioned at −20°C for 30 min, then laminated together at this temperature. After warming the samples to 23° (room temperature), 180° peel adhesion values are measured [102]. According to common tests a 60-lb paper label and an adherent surface are put in a test chamber at 35°F for 30 min. Labels are applied to various surfaces, including a high density polyethylene (HDPE) panel, a polyester container, and a corrugated board; other test conditions also are suggested [59]. After a dwell time of 1 hr at 35°F, the labels are removed by hand from the surfaces at approximately a 90° angle and a rate of 10–12 in./min. Labels that tear upon removal are considered satisfactory [59].

Cold Tack. Cold tack is measured by attaching a label strip at the given temperature to a cardboard surface adjusted to the same temperature. The strip is applied with a 100-g roller and the laminate stored for 30 min at the required temperature in the deep freezer. Afterwards the label strip is removed slowly by hand under a defined angle and the percentage of cardboard fiber tear estimated. FINAT FTM 13 describes the standard method for low temperature performance.

Special Laminate Properties

Special laminate properties such as specific adhesion, environmental resistance, roll storage aging, and release force from liner may be tested for certain applications. Some of these performance characteristics and the corresponding methods used were described previously. Specific adhesion is the ability of the adhesive to develop a useful bond level on a particular surface to which it is applied. In most cases polyethylene adhesion is checked. Environmental resistance is the retention of bond strength and/or other properties upon exposure to water, high humidity, heat, cold, chemicals, sunlight, etc. Roll storage aging is the resistance to adhesive property changes upon long-term contact with the face material (and liner). Migration of plasticizers from PVC face materials, of slip agents, or the nature of surface agents, for example, can have a severe effect on some adhesives and relatively little on others.

The release force from the liner is the adhesive force to the liner determining possible predispensing of labels; it should not be so tight as to cause web break or liner tear.

REFERENCES

1. P. Caton, *European Adhesives and Sealants*, (12) 19 (1990).
2. National Starch Chem. Co., *Durotak*, Pressure Sensitive Adhesives; Technical bulletin, 4/86.
3. British Petrol, *Hyvis*, Technical bulletin, (1985).
4. BASF, *Prüfmethoden, Polymerdispersionen/Polymerlösungen*, Bestimmung der Auslaufzeit, PM-CDE 005d/July 1979, Ludwigshafen.
5. H. Hanke, *Coating*, (9) 261 (1986).
6. Allied Colloids, *Viscalex HV 30*, TPD, 6004 G.
7. W. R. Dougherty, 15th Munich Adhesive and Finishing Seminar, 1990, p. 70.
8. *ASTM D-2578-65; Standard Test Method for Wetting Tension of PE and PP films*.
9. K. Templer and F. Wultsch, *Wochenbl. f. Papierfabr.*, (11/12) 483 (1980).
10. E. Prinz, *Coating*, (10) 244 (1979).
11. A. Fau and A. Soldat, "Silicone addition cure emulsions for paper release coating", in *TECH 12, Advances in Pressure Sensitive Tape Technology*, Technical Seminar Proceedings, Itasca, IL, May 1989, p. 7.
12. G. Habenicht, M. Baumann, R. Penzl and K. Jalal, *Adhäsion*, (6) 17 (1988).
13. R. Heusch, *Fette, Seifen, Anstrichmittel*, (11) 123 (1969).
14. M. Osterhold, M. Breuchen and K. Armbruster, *Adhäsion*, (3) 23 (1992).
15. R. J. Roe, *J. Phys. Chem.*, (6) 2013 (1968).
16. *Coating* (2) 39 (1970).
17. *Surfynol Technical Bulletin*, Air Products and Chemicals Inc. (1991).
18. Byk-Mallincrodt, Technische Information, Entschäumer.
19. GAF, Technical bulletin, 7583-003 Rev. 1, p. 5.
20. Unibasic, *Emulsion Polymerization*, IV-VI, MD, USA, p. 18.
21. *Product and Properties Index*, Polysar, Arnhem, 02/1985.
22. *Pressure Sensitive Adhesives, Technical Bulletin*, Rohm & Haas, (1986).
23. R. P. Mudge (National Starch Chem. Co., USA), EPA 0225541/11.12.85.
24. P. K. Dahl, R. Murphy and G. N. Babu, *Org. Coatings Appl. Polymer Sci. Proceedings*, (48) 131 (1983).
25. *Adhäsion*, (10) 307 (1968).
26. J. Johnston, *Adhesives Age*, (12) 24 (1983).
27. H. Müller, J. Türk and W. Druschke, BASF, Ludwigshafen EP 0.118.726/11.02.83.
28. P. Tkaczuk, *Adhesives Age*, (8) 19 (1988).

29. A. Midgley, *Adhesives Age*, (9) 17 (1986).
30. *Coating*, (3) 360 (1985).
31. Ashland Oil Inc., Chemicals, *Bull.* No. 1496-1, 1982.
32. D. G. Pierson and J. J. Wilczynski, *Adhesives Age*, (8) 52 (1980).
33. T. H. Haddock (Johnson & Johnson, USA), EPA 0130080 B1/02.01.85.
34. P. A. Mancinelli, New Development in Acrylic HMPSA Technology, in *TECH 12, Advances in Pressure Sensitive Tape Technology*, Technical Seminar Proceedings, Itasca, IL, May 1989, p. 165.
35. C. P. Iovine, S. J. Jer and F. Paulp (National Starch Chem. Co., USA), EPA 0212358/04.03.87, p. 3.
36. *Tappi J.*, (9) 105 (1984).
37. *Technical Information*, LHM, Celanese Resins Systems, 991.
38. R. Wilken, *Finat News*, (1) 53 (1988).
39. R. Wilken, *Coating*, (8) 212 (1982).
40. D. Symietz, *Adhäsion*, (11) 28 (1987).
41. N.C. MacDonald, *Adhesives Age*, (2) 21 (1972).
42. H. Möhle, *Adhäsion*, (6) 260 (1966).
43. C. Watson, *European Adhesives and Sealants*, (9) 6 (1987).
44. M. A. Johnson, *Radiation Curing*, (8) 4 (1980).
45. *Coating*, (1) 29 (1968).
46. *Prüfung von Haftkleber, D-EDE/K*, BASF, Ludwigshafen, Juni/Juli, 1981.
47. *Coating*, (1) 16 (1984).
48. K. Taubert, *Adhäsion*, (10) 377 (1970).
49. *European Standard*, Entwurf, prEN, T-peel test, CEN, Brussels (1994).
50. *TLMI Manual*, 1994, p. 45.
51. J. R. Wilken, *apr*, (5) 122 (1986).
52. P. Dunckley, *Adhäsion*, (11) 19 (1989).
53. R. P. Muny, *Adhesives Age*, (12) 18 (1986).
54. *Adhäsion*, (6) 24 (1982).
55. W. Druschke, *Adhäsion*, (5) 30 (1987).
56. EP 0.244.997.
57. A. Kenneth, J. R. Stockwell and J. Walker (Allied Colloids), EP 0.147.067/ 29.11.83.
58. P. Hammond Jr., *ASTM Bulletin*, **360(5)**:123-133.
59. L. Krutzel, *Adhesives Age*, (9) 21 (1987).
60. H. Green, *Ind. Eng. Chem. Anal. Ed.*, (13) 632 (1941).
61. K. Kamagata, T. Saito and M. Toyama, *J. Adh. Soc. Jap.*, (6) 309 (1969).
62. F. Wetzel, *ASTM Bulletin*, (221) 64 (1957).
63. EP 3344863
64. *apr*, (14) 340 (1987).
65. W. Druschke, *Adhäsion und Tack von Haftklebstoffen*, AFERA, 1986, Edinburgh.
66. EP 0.100.046
67. A. Johnson, J. Morris, *Plast. Film Sheeting*, **4(1)**:50 (1988); in *CAS, Adhesives*, 25 (1988), 109:191649v.

68. J. P. Keally and R. E. Zenk (Minnesota Mining and Manufacturing Co., USA), Canadian Patent, 1224.678/19.07.82 (USP. 399350).
69. R. Jordan, *Coating*, (2) 37 (1986).
70. *Adhesives Age*, (7) 36 (1986).
71. C. Kemmenater and G. Bader, *Adhäsion*, (11) 487 (1968).
72. H. Liese, *Adhäsion*, (3) 110 (1966).
73. *Coating*, (6) 188 (1969).
74. *Adhäsion*, (3) 73 (1978).
75. C. W. Koch and A. N. Abbott, *Rubber Age*, (82) 471 (1957).
76. *Adhesives Age*, (10) 26 (1977).
77. U. Zorll, *Adhäsion*, (3) 69 (1976).
78. H. Wiest, *Adhäsion*, (4) 146 (1966).
79. J. Johnston, *Adhesives Age*, (11) 30 (1983).
80. G. R. Hamed and C. H. Hsieh, *Rubber Chem. Technol.*, **59**:883 (1986).
81. *Coating*, (1) 8 (1985).
82. M. Gerace, *Adhesives Age*, (8) 85 (1983).
83. *Coating*, (11) 316 (1984).
84. D. J. St. Clair and J. T. Harlan, *Adhesives Age*, (12) 18 (1975).
85. *Adhäsion*, (10) 399 (1965).
86. *Adhäsion*, (1/2) 27 (1987).
87. *Adhesives Age*, (5) 24 (1987).
88. P. Dunckley, *Adhäsion*, (11) 19 (1989).
89. *Adhesives Age*, (11) 28 (1983).
90. H. Möhle, *Adhäsion*, (6) 260 (1966).
91. O. Hahn, M. Schlimmer and D. Ruttert, *Adhäsion*, (12) 9 (1988).
92. D. J. James and H. C. Holyodke, *Adhesives Age*, (4) 23 (1984).
93. L. J. Smith, *Adhesives Age*, (4) 28 (1987).
94. C. M. Chum, M. C. Ling, R. R. Vargas (Avery Int. Co., USA), EPA 1225792/18.08.87.
95. R. Lombardi, *Paper, Film and Foil Conv.*, (3) 74 (1988).
96. *Glossary of Terms used in Pressure Sensitive Tapes Industry*, PSTC, Glenview, IL, 1974.
97. BASF, private communication.
98. A. Sustic and B. Fellow, *Adhesives Age*, (11) 17 (1991).
99. *Adhesives Age*, (12) 25 (1977).
100. *Wingtack, Technical Bulletin*, Goodrich, Akron.
101. R. Hinterwaldner, *Coating*, (3) 73 (1985).
102. L.A. Sobieski and T.J. Tangney, *Adhesives Age*, (12) 23 (1988).
103. Tackifier Resins Data Sheets, Akzo Eisele & Hoffmann, Mannheim.
104. WB PSA data sheets, Rhone-Poulenc.
105. Die Herstellung von Haftklebstoffen, Tl. 2.2; 15d, Nov. 1979, BASF, Ludwigshafen.
106. *Coating*, (7) 20 (1984).
107. E.G. Ewing and J.C. Erickson, *Tappi J.*, (6) 158 (1988).
108. *Adhesives Age*, (10) 24 (1977).

109. *Coating*, (7) 186 (1984).
110. *Technical Information, LAT 037/Aug 82; LAT 002/Aug 87; LAT 051/May 87; LAT 040/Aug. 82*, Doverstrand Ltd., Harrow (U.K.).
111. S. Price and J. B. Nathan Jr., *Adhesives Age*, (9) 37 (1974).
112. D. Becker and J. Braun, *Kunststoff Handbuch*, Bd. 1, K. Hanser Verlag, München-Wien, 1990, p. 931.
113. D. J. P. Harrison, J. F. Johnson, and J. F. Yates, *Polymer Eng. Sci.*, (14) 865 (1982).
114. R. Hinterwaldner, *Adhäsion*, (3) 14 (1985).
115. R. F. Grossmann, *Adhesives Age*, (12) 41 (1969).
116. *Adhesives Age*, (3) 36 (1986).
117. *Technical Service Report, 6110*, Firestone, August, 1986.
118. *Tappi J.*, 67 (9) 104 (1983).
119. S. Mitton and C. Mak, *Adhesives Age*, (1) 42 (1983).
120. *Resins*, Narez Technical Bulletin (Spain).
121. R. Mudge, Ethylene-Vinylacetate based, waterbased PSA, in *TECH 12, Advances in Pressure Sensitive Tape Technology*, Technical Seminar Proceedings, Itasca, IL, May 1989.
122. P. H. Gamlen and R. M. Lang (ICI), *Chemicals and Polymers, Research and Technology*, p. 8.
123. *Adhäsion*, (1) 11 (1974).
124. *Mobil Plastics Technical Bulletin*, MA 657/06/90.
125. C. W. Drechsler, *Coating*, (1) 10 (1985).
126. K. Goller, *Adhäsion*, (4) 101 (1974).
127. K. Goller, *Adhäsion*, (4) 126 (1973).
128. K. Goller, *Adhäsion*, (7) 266 (1973).
129. A. Dobman and J. Planje, *Papier u. Kunststoffverarb.*, (1) 37 (1986).
130. A. L. Bull, *Thermoplastic Rubbers, Technical Manual*, Shell Elastomers, TR 8.12, p. 13.
131. EP 0.213.860, p. 2.
132. I. J. Davis (National Starch, Chem. Co., USA), US Pat., 4.728.572/01.03.88.
133. R. Schuman and B. Josephs (Dennison Manuf. Co., USA), PCT/US86/02304/25.08.86.
134. *Coating*, (3) 65 (1974).
135. *Adhesives Age*, (11) 40 (1988).

Abbreviations and Acronyms

1 Compounds

A-CPE	amorphous copolyethylene
AC	acrylic
AN	acrylonitrile
APAO	amorphous polyalpha-olefin
APP	atactic polypropylene
APO	amorphous polyolefin
B	butadiene
BOPP	biaxially oriented polypropylene
BPO	benzoyl peroxide
BuAc	butyl acrylate
CMC	carboxy methylcellulose
CPP	cast polypropylene
CRA	control release agent
CSBR	carboxylated styrene-butadiene rubber
DEA	diethanol amine
DOP	dioctyl phthalate
EA	ethyl acrylate
EPC	epichlorohydrine
EHA	ethyl hexyl acrylate
EMA	ethyl methacrylate
EPDM	ethylene propylene diene monomer
EPR	ethylene propylene rubber
EPVC	emulsion PVC

EVAc	ethylene vinyl acetate
GR-S	a special grade of rubber
HDPE	high density polyethylene
HMA	hot-melt adhesive
HMPSA	hot-melt PSA
H-PVC	hard PVC
LDPE	low density polyethylene
M	maleinate
MAA	methacrylic acid
MEK	methyl ethyl ketone
MIBK	methyl isobutyl ketone
MMA	methyl methacrylate
NR	natural rubber
OPP	oriented polypropylene
PA	polyamide
PB	polybutylene
PBuAc	polybutyl acrylate
PC	polycarbonate
PE	polyethylene
PEB	poly(ethylene butylene)
PEHA	polyethyl hexyl acrylate
PET	polyethylene terephthalate
PIB	polyisobutylene
PP	polypropylene
Ppt	parts per thousand
PS	polystyrene
PSA	pressure-sensitive adhesive
Pts	parts
PTS	Papiertechnische Stiftung
PUR	polyurethane
PVA	polyvinyl alcohol
PVAc	polyvinyl acetate
PVC	polyvinyl chloride
PVE	polyvinyl ether
PVP	polyvinyl pyrrolidone
RR	rubber/resin
S	styrene
SAA	surface active agent
SBR	styrene-butadiene rubber
SBS	styrene-butadiene-styrene
SEBS	styrene-(ethylene/butene)-styrene

Abbreviations and Acronyms 571

SIS	styrene-isoprene-styrene
STS	stainless steel
TPE	thermoplastic elastomer
VAc	vinyl acetate
VAc/E	vinyl acetate/ethylene
VE	vinyl ether

2 TERMS

A	area or a constant
Ao	reference area
AFERA	Association des Fabricants Européens de Rubans Auto-Adhésifs
ASTM	American Society for the Testing of Materials
B	constant
BGA	Bundesgesundheitsamant
BWA	bonding wet adhesion
BWB	German military procurement office
BWB-TL	German military norm
C	constant or cuttability or cohesion
CC	carbon-carbon
CD	cross direction
CH	carbon-hydrogen
CI	cuttability index
CMD	cross-machine direction
DMA	dynamic mechanical analysis
DP	degree of polymerization
DSC	differential scanning calorimetry
DWA	debonding wet adhesion
EB	electron beam
EDP	electronic data processing
EN	European norms
EPSMA	European Pressure Sensitive Manufacturers Association
eV	electron volt
EV	viscosity test method
FINAT	Fédération Internationale des Fabricants Transformateurs d'Adhésif et Thermocallants sur Papiers et Autres Supports
FIPAGO	Fédération Internationale des Fabricants de Papiers Gommés
FTM	Finat Test Method
GID	gear-in-die
HC	hydrocarbon
HLB	hydrophilic-lipophilic balance

HM	hot-melt
HS	hot shear
HSG	hot shear gradient
IR	infrared
kp	kilopond
LVT	known Brookfield Viscosity method
MD	machine direction
MFI	melt flow index
MFR	melt flow rate
MFT	minimum film-forming temperature
MW	molecular weight
MWD	molecular weight distribution
MP	melting point
mPa·s	milliPascal · second
MV	viscosity test conditions
NC	viscosity test conditions
PN	Williams plasticity Number
PS	pressure-sensitive or polystyrene
PSTC	Pressure-Sensitive Tape Council
RB	rolling ball
RBT	rolling ball tack
RCT	rolling cylinder tack
RF	radiofrequency
RH	relative humidity
RT	room temperature
RVT	known Brookfield viscosity method
SAFT	shear adhesion failure temperature
SB	solvent-based
SF	solvent-free
SH	shear
SL	solventless
SP	softening point
SPS	slight panel staining
TD	transverse direction
TMA	thermo-mechanical analysis
TLMI	Tag and Label Manufacturers Institute
UV	ultraviolet
WB	water-based
WBA	wet bonding adhesion

Index

Acrylics
    adhesion/cohesion balance of, 128
        (*see also* Adhesion cohesion balance)
    adhesion towards polar sufaces of, 128
    advantages of, 128
    cohesion of, 237
    comparison of (*see also* Comparison)
        vs. CSBR, 127
        vs. EVAc, 127
    compatibility of, 334
    converting costs of, 247 (*see also* Converting properties)
    copolymers of, 131
    crosslinking of, 139, 212, 219 (*see also* Crosslinking)
    die-cuttability of, 128 (*see also* Cuttability)
    electron-beam cured, 150 (*see also* Electron-beam curing)
    on film, 247
    formulating of, 247 (*see also* Formulating)
    on paper, 247
    peel of (*see* Peel)

[Acrylics]
    plasticizer resistance of, 128 (*see also* Plasticizer resistance)
    solvent-based, (*see also* Solvent-based PSA)
        crosslinking of, 140 (*see also* Crosslinking)
    special features of, 246
    tackified (*see* Tackification)
    tackifying ability of, 33
    water-based, 150 (*see also* Water-based dispersions)
Additive, 327
    choice of, dependence on coating machine, 376
Adhesion
    build up, 188
    - cohesion balance, 11, 163, 225
        formulating of, 329 (*see also* Formulating)
    failure energy of, (*see* Separation energy)
    interfacial, 65
    on polyethylene, 127 (*see also* Peel)
    wet, 81, 389
Adhesive, 1

[Adhesive]
  anchorage of, 54, 63, 66, 242
    dependence on adhesion/cohesion, 197
    dependence on coating technology, 42
    improvement of with solvents, 426
    parameters of, 307
    for removable adhesives, 197
    wet, 389
  build up of, on the machine, 39, 43
  coating of, discontinous, 211 (*see also* Peel)
  coating machine of, 454 (*see also* Coating machine)
  converting properties of, 21
  degradation of, (*see* Aging)
  failure of, (*see* Mode of failure)
  flow out, 295 (*see also* Cuttability)
  hold out of, (*see* Coating weight)
  permanent, 23, 66
  plasticity of, 53
  properties of, 163
    balance of, 344
    dependence on coating technology, 376 (*see also* Coating technology)
    dependence on coating weight, 179 (*see also* Coating weight)
    influence on converting properties, 225 (*see also* Converting properties)
    influence on end use properties, 225
    influence on other properties, 225
    modification of, 244
    of PVC label, 388 (*see also* Label)
    test of (*see* Test method)
  removable, 17, 18, 23, 66 (*see also* Removable adhesive)
Aging
  resistance, 240
  test of, conditions of, 379

Agent
  antimicrobial, 327, 374, 432
  of controled release, 138 (*see also* Release coating)
  of crosslinking, 140, 410 (*see also* Crosslinking)
    in water-based PSA, 410 (*see also* Water-based dispersions)
  of detackifying, 374, 380, 383 (*see also* Peel, Cuttability)
  of formulating, 87 (*see also* Formulating)
  of hygroscopicity, 502
  of lowering the viscosity, 428
  protective, 410
  solubilizing, 432 (*see also* Water resistance)
  of stabilizing in water-based dispersions, 411 (*see also* Water-based dispersions)
  of wetting, (*see* Wetting agents)
Air
  brush, 458, 478, 484
  knife, 454, 456, 484 (*see also* Coating device)
Amorphous polyalphaolefins
  for migration resistance, 379 (*see also* Migration)
Antioxidants, 143, 361, 395 (*see also* Aging)
  for aquous dispersions, 395, 409
  for electron beam cured elastomers, 401
  for hotmelts, 395, 401
  level of, 395
  for natural rubber, 395
  for SBR latex, 126
  for thermoplastic elastomers, 401
Approval, 519
  BGA, 388, 429
  FDA, 388, 429
Arrheniuss low, 28

Ball tack test (*see* Rolling ball)

Index

BGA compliance (*see* Approval)
Blade coaters (*see* Coaters)
Bleeding, 60, 112, 155 (*see also* Migration)
   dependence of
      on coating weight, 109 (*see also* Coating weight)
      on application viscosity, 155
Bonding wet adhesion, 273, 389 (*see also* Adhesion)
Break down edge (*see* Stationary roll)
Butylene rubber, 122 (*see also* Polybutene)

Carboxylated styrene butadiene latex, 12, 123, 127
   advantages of, 127
   compatibility of, 334
   crosslinking of, 139 (*see also* Crosslinking)
   molecular weight of, 19, 81, 93
   shear values of, 225 (*see also* Shear strength)
   styrene content of, 83
   Tg values for, 83 (*see also* Glass transition temperature)
Cell depth, 183, (*see also* Gravure cylinder)
Cellulose, derivatives of, as face stock material, 264, 283
Chemical composition
   of formulated adhesive, (*see also* Formulating)
      influence on peel, 188 (*see also* Peel)
   of PSA, 119
      dependence on coating weight (*see* Coating weight)
      dependence on end use, 151 (*see also* End use properties)
      dependence on formulation, 147 (*see also* Formulation)
      dependence on physical status, 119, 147

[Chemical composition]
      dependence on solid state components, 156
      dependence on synthesis, 145
      factors of, 145
   of release coatings, 136 (*see also* Release coating)
Chloroprene latex (*see also* Neoprene)
   as shear modifier, 334
Coagulum
   build up of, (*see also* Adhesive build up)
      on machine, 257, 353
   level of, 257
   test of, 257 (*see also* Test method)
Coalescence, 89, 256
   dependence of
      on glass transition temperature, 90
      on viscosity, 90
Coatability, 21, 38, 68
   dependence of
      on foaming, 427 (*see also* Foam)
      on tackification, 354 (*see also* Tackification)
   of diluted systems, 254
   of HMPSA, 260
   of solvent-based PSA, 254
   of water-based PSA, 255
Coater
   of cartridge style, 456, 477 (*see also* Coating machine)
   choice of, 477
   die, 484 (*see also* Slot die)
   Duplex, 484 (*see also* Gear in die)
   kiss, 455
   knife, 457 (*see also* Knife)
   Park, 488 (*see also* Slot die)
Coating, 439
   defects of, 484
   device of, 455
      choice of, 455
      of laboratory samples, 524, 555
   Meyer bar (*see* Meyer bar)

[Coating]
  parameters influencing the usability of, 458
  requirements on, 476
  with stationary roll (*see* Stationary roll)
  for water-based PSA (*see also* Water-based dispersions)
  direct, 42, 155, 455
  of film (*see also* Film coating)
  metering device for, 62
  geometry of, choice of, 476
  in line, 450, 452 (*see also* Radiation curing, Siliconizing)
  machine of, 450, 453
    for HMPSA (*see* HMPSA)
    for labelstock, 45
    for silicones, 452 (*see also* Siliconizing)
    for solvent-based PSA (*see* Solvent-based PSA)
    for water-based PSA (*see* Water-based dispersions)
  on reverse roll, 42 (*see also* Reverse roll)
  rheology of, 35
    of HMPSA, 260
    of solvent-based PSA, 36
  speed of, 39, 453 (*see also* Running speed)
    dependence on tackification, 354
    of HMPSA, 260
    of solvent-based PSA, 255
  systems, 455 (*see also* Coating device)
  technology of, 450
  transfer, 39, 42, 60, 155
  versatility of, 254, 255, 257
    of HMPSA, 260
  weight of, 177
    adjustement of, 212
    of primer, 210
    dependence on adhesive quality, 185
    dependence on chemical composition, 179

[Coating]
  dependence on coating conditions, 179
  dependence on end use, 183
  dependence on face material, 185
  dependence on face stock porosity, 61
  dependence on substrate, 185
  dependence on surfactants, 181
  dependence on viscosity, 181
  on gravure cylinder, 183 (*see also* Gravure cylinder)
  of hot melts, 399 (*see also* Hot melts)
  influence on adhesive properties, 46
  influence on cold flow, 46 (*see also* Cuttability)
  influence on converting properties, 46
  influence on cuttability, 46
  influence on drying, 177, 221 (*see also* Drying)
  influence on peel, 178 (*see also* Peel)
  influence on tack, 178 (*see also* Tack)
  influence on shear, 178, 220, 221 (*see also* Shear)
  parameters of, 43, 48, 177, 18
  tolerances of, 43, 183, 329
  values of, 183, 185, 192, 210, 212
Cohesion, 112 (*see also* Shear strength)
  dependence of
    on additives, 220
    on crosslinking, 139 (*see also* Crosslinking)
    on molecular weight, 218
    on temperature, 3oo
Cohesive strength, 18 (*see also* Cohesion, Shear)
Cold flow, 23, 108
  dependence of
    on chemical affinity of face stock/liner, 307

Index

[Cold flow]
    on coating weight, 109, 306
    on creep modulus, 102
    on laminate, 306
    on shear rate, 110
    on smoothness of face stock/liner, 307
    on viscosity, 110
Colloids, 116 (*see also* Protective colloid)
Comonomer, in carboxylated latex, 126 (*see also* Monomers)
Comparison of PSA, 107
    on acrylics and other synthetic polymer-based elastomers, 227
    of adhesion on polar/nonpolar surfaces, 231
    on different chemical basis, 226
    in film application, 231
    of flow of, vs. plastics, 111
    formulation of, 245
    mechanical resistance of, vs. plastics, 114
    in permanent/removable paper label application, 231
    relaxation of, vs. plastics, 113
    rheology of
        vs. plastics, 113
        vs. other adhesives, 115
    on rubber-based vs. acrylic-based, 226
    on solvent-based acrylics vs. solvent-based rubber PSAs, 244
    on solvent-based acrylics vs. water-based acrylics, 438
    on solvent-based/waterbased/hot melt, 435, 438
    synthesis of, 245
    in water-based acrylics vs. other water-based PSAs, 244
    with thermoplastics and rubber, 108
Compatibility, 88
    of the resin, (*see also* Resin)
        dependence on melting point, 342
        dependence on molecular weight, 94, 342
        with PSA dispersions, 337

Composite structure, 256
    of PSA layer, 38, 352
    influence on adhesive properties, 70 (*see also* Adhesive properties)
Contact
    angle of, 40, 41, 256, 412 (*see also* Wetting out)
    measurement of (*see* Test methods)
    cement of (*see* Primer)
    hindrance of, 386
    surface of, reduction of, 211
Convertability, 68, 7 (*see also* Converting properties)
    dependence of
        on adhesive properties, 254, 260
        on adhesive physical state, 254
        on coating technology, 261
        on end use properties, 261
        on laminate, 22
        on solid state components of the laminate, 261
Converting properties, 245, 253
    dependence of
        on coating weight (*see* Coating weight)
        glass transition temperature, 88
        on laminate, 68
        on tackifier (*see* Tackification)
        on viscoelastic properties, 21
Corrugated board, shear test on, 214
Cratering, 64, 423 (*see also* Test methods)
Creep, 24, 211 (*see also* Cold flow)
    compliance, 14, 93, 211
    values of, 103
    dependence of
        on anchorage, 304 (*see also* Anchorage)
        on adhesive nature, 298, 303
Crosslinking
    chemical, of CSBR, 95, 139
    influence of
        on the peel (*see* Peel)
        on the shear, 136 (*see also* Shear)

[Crosslinking]
    of natural rubber, 95 (*see also* Natural rubber)
    physical, 95, 139, 219
    of rubber derivatives, 140
    of SIS, 84

Curing
    beam, 136, 140, 141 (*see also* Radiation curing, Silicone)
    generator of, 497
    of rubber-based PSA, 141, 497
    dual, 497
    electron beam, 34, 84, 85, 136, 219
    advantages/disadvantages of, 497
    curing time of, 496 (*see also* Siliconizing)
    influence on face stock, 496
    of silicones, 496, 497 (*see also* Release coating, Siliconizing)
    post, 137 (*see also* Silicones)
    radiation, 136 (*see also* Radiation curing, Silicone)
    of silicone, 136 (*see also* Silicone)
    thermal, (*see also* Silicones)
    ultraviolet, 34, 150, 212 (*see also* Silicones)

Curl, 286 (*see also* Lay flat)

Cuttability, 60, 293
    dependence of
        on adhesion/cohesion, 299
        on adhesive anchorage, 69, 304
        on adhesive properties, 300
        on bulk material, 56
        on coating weight, 47, 304
        on cold flow, 303, 304
        on composite structure, 308
        on creep, 304
        on cutting process conditions, 304
        environmental humidity, 59, 308
        on face stock smoothness, 68
        on formulation of tackifier dispersion, 359
        on fillers, 304
        on laminate stiffness, 305
        on laminate thickness, 313

[Cuttability]
    on peel, 303, 380
    on smearing, 295 (*see also* Smearing)
    on tack, 315
    on shear resistance, 214
    on time/temperature dependent rheology of PSA, 303
    dynamic, 294
    evaluation of, 295, 316
    of film paper/film laminate, 316
    of HMPSA (*see* HMPSA)
    improvement of, dependence on shear, 301
    parameters of, 60, 314
    static, 294
    of sheet materials, 294 (*see also* Sheet label)

Cutting
    angle of, 293 (*see also* Cuttability)
    on flying knife, 22
    on guillotine, 22, 294
    process of
    steps of, 295

Cylinder depth, 467 (*see also* Gravure depth)

Dahlgrens system, 479 (*see also* Coating machine)

Dahlquists criterion, 12 (*see also* Creep compliance)

Debonding wet adhesion, 389 (*see also* Water resistance)

Deep freeze
    adhesives
        migration of, 242
        test of, 397
    labels, 89

Defoamers, 220, 258, 428 (*see also* Foam)

Detachment
    angle, 54
    energy, 54 (*see also* Peel)

Index

Die coater, influence of, on coating quality, 356 (*see also* Slot die, Gear in die)
Die cuttability, (*see also* Cuttability)
  of acrylics, 246
  of films, 286
  dependence of
    on hygroscopicity, 354
    on shear, 214
    on tackification, 354
Diluent, reactive, 141, 150 (*see also* Radiation curing)
Diluting, 257, 423, 425, 426
  agents (*see* Formulation)
  response (*see* Water-based dispersions)
Dimensional stability, 291 (*see also* Srinkage)
  of films, 286
  parameters of, 292
Direct gravure, 456, 467
  advantages/disadvantages, 476
DMA, 10, 11, 92, 337
Doctor blade, 456, 457, 468
  oscillating, 476
  reverse angle, 474
Dryer
  electron beam, 492
  flying, 492
  ultraviolet, 492
  of water-based PSA, 258
Drying
  of coating, 489
  tunnel of, 489
  conditions of, 490.524
  dependence of
    on face stock, 374
    on carrier, 490
  energy requirements of, 490
  infrared, 491
  influence of, on shrinkage, 293
  methods of, 491
  radiofrequency, 492
  speed of
    dependence on coating weight, 177

[Drying]
  dependence on solids content, 375
  dependence on tackification (*see* Tackification)
  of solvent-based PSA, 37, 490
  temperature of, 259
  of polypropylene, 276
  of soft PVC, 269
  of water-based PSA, 259, 490 (*see also* Water-based dispersions)
DSC, 10 (*see also* Glass transition temperature)
Dwell time, 15 (*see also* Adhesive properties, Test method)
  values of, 201, 532, 533, 536, 538

Edge
  lifting of, 264
  flow (*see* Oozing)
Elastic modulus (*see* Modulus)
Elasticity/Plasticity Balance, influence of, on PSA properties, 33
Elastomers, 120, 134 (*see also* Raw materials)
  acrylic, 120, 127
  acryl-vinyl, 127
  crosslinked, 138
  ethylen vinyl acetate, 121
  natural, 121
  synthetic, 121, 122
    hydrocarbon-based, 122
    manufature of, 326
  termoplastic, (*see* Thermoplastic elastomers)
  uncrosslinked, 121
Emulsifiers, 411 (*see also* Surfactants)
  migration of, 242 (*see also* Migration)
Emulsion porperties
  improvement of, 411 (*see also* Formulation)
End use properties, 381
  dependence of
    on glass transition temperature, 88
    modulus, 75

[End use properties]
  on tackifier (*see* Tackification)
  on viscoelastic properties, 22
  influence on coating weight (*see* Coating weight)
Ethylene copolymers, 121, 131
  sequence distribution of, 84
Ethylene maleinate copolymers, 145 (*see also* Water-based dispersions)
  cohesion of, 152
Ethylene propylene diene copolymers, 123 (*see also* Thermoplastic elastomers)
Ethylene vinylacetate copolymers, 131, 132 (*see also* Water-based dispersions)
  compatibility of, 334
  glass transition temperature of, 133
  in raw materials, for hot melts, 96, 132, 134, 148 (*see also* Hot melts)
  solid content of, 133
  tack of (*see also* Tack)
  water-borne, 132, 134 (*see also* Water-based dispersions)
Eyrings model, 29
Face stock, 5, 49, 263 (*see also* Laminate)
  choice of, parameters of, 264
  cost effectiveness of, comparison of, 287
  of film, 268, 269
    advantages of, 268
    comparison of vs. paper, 284
    modulus of, 285
    profil tolerances of, 273
  damage of, through radiation, 150
  flexibility of
    influence on peel, 52, 66, 194 (*see also* Peel)
    influence on rheology, 51
    influence on tack, 168 (*see also* Tack)
  geometry of

[Face stock]
  influence on cuttability, 59
  influence on peel, 53
  influence on pressure-sensitive properties, 69
  influence of
    on adhesive choice, 62
    on adhesive properties, 50
    on anchorage, 65
    on coating technology, 62
    on converting properties, 55
    on cuttability, 55
    on migration, 60
    on peel, 50
    on removability, 66, 68
    on shear, 55
    on tack, 50
  in laboratory tests, 524
  of low surface energy, 276
  modulus of, influence on cuttability, 58
  of paper, 264 (*see also* Paper)
  plasticity of, influence on peel, 194 (*see also* Peel)
  stiffness of
    comparison of, 287
    influence on peel, 53, 194
  strengthening of, 211
  surface of
    influence on adhesive properties, 60
    influence on anchorage, 60
    influence on converting properties, 60
    inffluence on wetting out, 63 (*see also* Wetting out)
  thickness of, 69
    influence on peel, 69, 195
FDA compliance (*see* Approval)
Filler, 143, 220, 385
  in peel reduction, 211, 212, 385
  reactive, 385
    influence on shear, 223
Film
  coating of, metering device in, 62
  face stock, 264 (*see also* Face stock)
  plastic, 66 (*see also* Face stock)

Index

Filter, for water-based PSA, 433, 516 (*see also* Mechanical stability)
Fish eyes, 64, 428 (*see also* Wetting out)
Flagging, 226
   factors of, 272
   test of (*see* Test method)
Flexibilizers, 210 (*see* Removability)
Flow, test of, in plastics vs. PSA, 111
Fluorosilicones, 136
Foam, 39, 44, 257 (*see also* Water-based dispersions)
   dependence of
      on coating device, 428
      on running speed, 258
      surface tension, 428 (*see also* Surface tension)
      on tackification, 352 (*see also* Tackification)
   resistance, 366
   testing of (*see* Test method)
Formulation
   age of, influence on adhesive properties, 365
   ease and costs, of acrylics, 247
   of adhesive properties, 329
   on aging, 389 (*see also* Aging)
   for coating properties, 155, 374
   for converting properties, 377
   dependence of
      on adhesive technology, 397
      on coating technology, 375
      on face stock, 375
   for direct coating, 155
   for cuttability, 379 (*see also* Cuttability)
   of deep freeze PSA, 396 (*see also* Deep freeze label)
   for end use properties, 151, 381
   of film laminates, 247, 387
      permanent, 387
      removable, 388
   of hot melts, 377, 398, 442 (*see also* Hot melts)
      single component, 402
   for migration (*see* Migration)

[Formulation]
   opportunities of, 331
   of paper laminates, 247, 387
      of general purpose permanent, 387
      of removable, 387
   on peel, 329
   of permanent/removable labels, 383
   of PVC laminates, 388
   reasons of, 329
   for shear, 330
   for solid components of laminate, 156
   of solvent-based adhesives, 397, 435 (*see also* Solvent-based PSA)
   for special uses, 153, 389
   for tack, 329
   technological considerations of, 433
   for thermal resistance, 153, 396
   in transfer coating, 155
   of water-based adhesives, 407, 440
      dependence on coater configuration, 482
      special features of, 435
   of water resistance, 153, 396
Fox–Florys equation, 80, 88
Fungicide, 432 (*see also* Formulation)
Fracture energy, 189 (*see also* Peel)

Gear in die, 458, 466, 484 (*see also* Slot die)
   for HMPSA, advantages of, 488
Gel content, 245, 397
   of styrene butadiene copolymers, 126
   values of, 219
Glass transition temperature, 9, 75 (*see also* Viscoelastic properties)
   additivity of, 85, 87
   adjustment of, 85, 86
      in synthesis, 86
      in formulating, 87
      with plasticizers, 87 (*see also* Plasticizing)
      with tackifiers, 87 (*see also* Tackification)

[Glass transition temperature]
dependence of
on chemical composition/structure, 81
on crosslinking, 85
on flexible main chain, 81
on microstructure, 84
on molecular weight, 80
on monomers, 83 (*see also* Monomer)
on morphology, 84
on sequence distribution/length, 84
on side groups, 82
on tackifying resin, 333
factors of influencing, 80
of HMPSA, 88 (*see also* HMPSA)
influence of
on adhesive properties, 88
on converting properties, 89
on end use properties, 91
on peel, 89
on tack, 89
of acrylic HMPSA, 84
role of, in the characterization of PSA, 76
of silicone PSA, 89
values, 77, 79, 89, 91, 116, 119, 124, 126, 129, 131, 210, 329
Gordon–Taylors equation, 88
Gravity bleeding (*see* Migration)
Gravure
angle of, 469
depth of, choice of, 469
coater
adhesive viscosity on, 471
choice of cylinder for, 474
coating weight of, 453, 469, 471, 474
design of, 476
dry edge on, 464
direct (*see* Direct gravure)
indirect, 466 (*see also* Offset gravure)
induced structures (*see also* Striation)
offset (*see* Offset gravure)
reverse, 42 (*see also* Reverse gravure)
for paper-based laminates, 62

[Gravure]
roll coating
on film, 60
on paper, 60
on rotation (*see* Rotogravure)
Grit
build up of, (*see also* Coagulum)
test of (*see* Coagulum, test)

Hold strength, twenty degree (*see* Shear strength)
Holding power, 112, 218, 358 (*see also* Shear, test)
dependence of
gel content, 219
on melting point of the resin, 218
values of, 219
Hookes law, 6, 19, 215, 221
Hot melt compounding, 402 (*see also* Hot melt PSA)
requirements for, 403
Hot melt PSA, 1, 10, 34, 60, 64
acrylic, 130, 154, 406
viscosity of, 28
advantages/disadvantages of, 405
coating of, 486
energy consumption for, 406
machine of, 454
speed of, 406 (*see also* Running speed)
high temperature performance of, 86
radiation cured, 406 (*see also* Radiation curing)
recipe for, 399 (*see also* Formulation)
removable, 401 (*see also* Test of removability of HMPSA)
SBS, 100
screening of, 520
stabilizers of, 401
tackification of (*see* Tackification)
viscosity of, 400, 401
water soluble, 154 (*see also* Water resistance)
weather resistant, 154

Index

Hot shear, 299 (*see also* Shear strength, Test method)
  gradient of, 300
    dependence of on formulation, 301
  influence of, on cuttability, 369
Humidity, rest of
  influence on cuttability, 369
  influence on lay flat, 270 (*see also* Lay flat)
  influence on peel, 193
  influence on speed, 256
  tolerances of, 270
Hydrocarbon-based tackifier, 365 (*see also* Resin, Tackifier)
  advantages/disadvantages of, 370
  applications for, 370
  dispersion of
    compounding of, 369
    foam resistance of, 366 (*see also* Foam)
    mechanical stability of, 366
    wetting agent level for, 366
  influence of
    on the adhesive properties of PSA, 365
    on aging of PSA, 366
    on converting properties of PSA, 366
  level of, 366
  price of, 369, 370
  raw materials for, 370
  replacement of rosin with, 372
Hydrophilic-lipophilic balance, 413 (*see also* Surfactants)

Interfacial energy, 197 (*see also* Peel)
Isobutylene rubber, 122 (*see also* Polyisobutylene)
Isocyanate, (*see also* Polyisocyanate)
  crosslinking agent, 385

Kiss coating (*see* Coating device)
Knife
  of air (*see* Air knife, Coating device)
  floating, 459

[Knife]
  holding of, 460
  over roll, 455, 457, 458
    coating weight on, 459
    parameters of, 459
    running speed of, 459
    types of, 459, 460
    viscositiy on, 458
  roller, 459
  rotating roll knife, 455
  rubber blanketed, 459
Label, 1
  application technology of, 22 (*see also* Labeling)
  coating technology for, 42 (*see also* Coating machine)
  deep freeze, 13 (*see also* Deep freeze)
  dispensing of, 286 (*see also* Labeling)
  end use properties of, 389 (*see also* End use)
  film, 266 (*see also* Film coating)
  manufacturing of (*see* Manufacture)
  multilayer, 263
  reel, 263
  removable (*see* Removable label)
  see through, 151 (*see also* Film label)
  sheet, 263
    requirements for, 559
  stiffness of, 314 (*see also* Face stock)
  surface quality of, 293
  thickness of, influence on the adhesive properties, 69 (*see also* Adhesive properties)
  wash of, 153, 560 (*see also,* Test method, Water resistance)
Labeling, 22, 225, 262, 449
  guns, 22
Laminate, 262
  build up of, 263
  components of, 263
  of film, 199
  flexural resistance of, 69 (*see also* Stiffness)

[Laminate]
  paper, general purpose,
    manufacturing of, 62
  properties of, 241
    build up of for roll material, 22
    build up for sheet material, 22
  PVC, 388
  removable (*see also* Removable label)
  roll, 263
  sheet, 262
Laminating, 449
  properties, 89
Latex
  of natural rubber (*see* Natural rubber)
  synthetic, 126
Lay flat, 290 (*see also* Test method)
  dependence of, on humidity, 270, 291
  of printed PVC, 270
Layer number
  influence of
    on the stiffness (*see* Stiffness)
    on adhesive properties, 70
Legginess, 205, 225, 438 (*see also* Removability)
Lifting, test of (*see* Test method)
Line screen, 183, (*see also* Gravure)
  influence of, on coating weight, 469
Lodges theory, 8
Loop tack, 20, 164, 174 (*see also* Tack)
  test of, 164 (*see also* Test method)
Loss modulus, 17, 92, 94, 337 (*see also* Viscoelastic properties)
  dependence of, on molecular weight distribution, 94
  frequency and temperature, 167
  master curve of, 12
  peak temperature of, 11
Loss tangent delta peak, 9, 10, 93, 94 (*see also* Viscoelastic Properties)

Manufacture of pressure sensitive label, 449
Manufacture of PSA, 323
  hot-melt, 402 (*see also* Hotmelt PSA)

[Manufacture of PSA]
  raw materials, 323 (*see also* Raw materials)
  natural, 323
  synthetic, 325
  simultaneous, of PSA and laminate, 494, 496
Manufacture of release liner, 498 (*see also* Release liner)
Mastication, 33, 94, 119, 324, 398 (*see also* Natural rubber)
  for hot melts, 403
Mass transfer to the web (*see* Roll coating)
Maxwells model, 186
Mechanical stability
  of water-based PSA, 257
  of water-based resin dispersions, 257, 366
Melt flow
  index of, 112
  rate of, 111
Memory effect, 15, 114
Metering, (*see also* Coating machine)
  bar of, 42
    choice of, 462
  device of, 37
  rod of, 459, 462
  roll of (*see* Roll coating)
Meyer bar, 456 (*see also* Coating device)
  in film coating, 60
Migration, 22, 377 (*see also* Aging, Bleeding, Penetration, Test method)
  of acrylics, 242
  dependence of
    on adhesive characteristics, 378
    on coating technology, 42
    on face stock porosity, 60, 379
    on thickener molecular weight, 378, 379
  of CSBR, 242
  of EVAc, 242
  parameters of, 60, 61

Index

[Migration]
   resistance to, 241
   on soft PVC, 242
   test of, 293
   of water-based PSA (see Water-based dispersions)
Milling, 19, 122, 324 (see also Mastication)
   influence of, on viscosity, 325
Minimum film forming temperature, 25, 77, 86, 89 (see also Coalescence)
   dependence of
      on glass transition temperature, 90
      on experimental conditions, 91
Mirror test (see Test of removability)
Mixed systems, 474 (see also Crosslinking)
Mode of failure, (see also Test method)
   for peel, 197, 205, 206, 536, 537
   for tack, 528
   for shear, 540
Models
   viscoelastic, 166 (see also Tack)
Modulus
   adjustement of, 99
      in formulation, 99
      in synthesis, 99
      in crosslinking, 100
   creep, 102
   influence on peel, 189
   dependence on
      adhesive thickness, 98
      chemical composition, 94
      filler, 97, 98
      glass transition temperature, 97
      material characteristics, 93
      molecular weight, 27, 93, 94
      other parameters, 98
      sequence distribution, 95
      side chain length, 95
      stress rate, 33
      tackifier resin, 95, 98, 333
   dynamic, 102
   factors influencing it, 92

[Modulus]
   flexural, influence on peel, 194
   influence of, on stiffness (see Stiffness)
   loss (see Loss modulus)
   role of, of water-based acrylics, 99
      in PSA characterization, 91
   storage (see Storage modulus)
   values of, 101, 102
Molecular weight, 76, 93
   adjustment of, 119
   of base elastomers, 80
   distribution of, 76, 93
      in solution polymers, 149
   of tackifier resin, 81
   values of, 81
Money viscosity, 19, 122, 324
Monomers, (see also Chemical composition)
   for acrylics, 81, 245
      and vinyl copolymers, 128, 129
   choice of, 81, 83, 85
   in crosslinking, 385
   emulsifier, 386
   for ethylene vinylacetate copolymers, 131, 245
      level of, 131
   hard, 129, 330
   influence of, on adhesion build up, 188 (see also Adhesion build up)
   in plasma treatment, 279 (see also Treatment)
   in radiation curing, 406 (see also Radiation curing)
   toxicity of, 497
   ratio of, 82
   for shear, 217
   soft, 130, 145
   in styrene butadiene rubber, 245
   in thickeners, 425
Mullins effect (see Stress softening)
Multiroll coating, 455, 457

Natural rubber, 119, 121, 122
   adhesives

[Natural rubber]
  latex, 87, 93, 122
    plateau modulus of, 93 (*see also* Plateau modulus)
    for removability, 153 (*see also* Removability)
    compatibility with resins, 88, 121 (*see also* Resin)
Neoprene latex
  carboxylated, 127
  tackification of, 218, 351, 411
Newtonian systems, 6, 19, 31, 221
Non-Newtonian systems, 20, 28, 29, 31, 36, 352

Offset gravure, 457, 467 (*see also* Indirect gravure)
  advantages of, 468
  angular, 467
  coating weight in, 468
  differential, 468
  normal, 468
  vertical, 467
Oil, (*see also* Hot melts)
  processing, 377, 384
  rheology modifier, 338
    compatibility of, 399
    influence on modulus, 400
Oozing, 22, 60, 108 (*see also* Bleeding)
  dependence of, on viscosity, 110
  test method of (*see* Test method)
Orchards Equation, 41 (*see also* Wetting out)

Paper
  cuttability of (*see* Cuttability)
  face stock of, 266 (*see also* Face stock)
    comparison of vs. film, 284
  labels of (*see* Labels)
  modulus of
    dependence on humidity, 98, 267
    parameters of, 98
  printing of, 266, 267 (*see also* Printability)

[Paper]
  for release liner, 288, 499
  sensitivity of, towards humidity, 266
  stiffness of, 268, 309 (*see also* Stiffness)
Particle size, (*see also* Test method)
  for CSBR, 126
  for water-based PSA, 149
Peel
  adhesion, 2, 46, 186 (*see also* Test method)
  adjustment of, 334
  angle of, 14, 532, 533
    dependence on face stock flexibility, 51
    normalized, 52, 532
    influence on peel, 195
  of CSBR, 237
  cyclical, 204
  dependence of
    on adherend, 202, 203
    on adhesive geometry, 187, 189
    on adhesive molecular weight, 189
    on adhesive nature, 187
    on adhesive state, 193
    on application pressure, 204
    on chemical composition of base elastomer, 187
    on coating technology, 193
    on coating weight, 52, 67, 190
    on cohesion, 202, 204
    on contact surface (*see* Contact surface)
    on crosslinking, 188, 330
    on dwell time, 15, 187, 201
    on elastic modulus, 14, 52, 53, 195
    on experimental conditions, 201
    on face stock, 52, 53, 69, 193
    on filler, 188
    on glass transition remperature, 17 (*see also* Glass transition temperature)
    on laminate construction, 198
    on layer number, 198, 199
    on layer structure, 193

Index

[Peel]
  on peeling angle, 202
  on peeling rate, 16, 187
  on primer, 210 (*see also* Primer)
  on release liner, 197, 202
  on shape of the adhesive layer, 192
  on strain rate, 15, 535
  on substrate, 54, 187, 536
  on tackification, 188
  on temperature, 18
  on viscoelastic properties, 14
  on viscosity, 14
  of EVAc, 237
  factors of, 187
  improvement of, 204
    by tackifier, 213
  influence of, on other characteristics, 226
  low temperature (*see* Deep freeze)
  measurement of, 187
    test rate for, 202
  modifiers of, for water-based PSA, 409
  for permanent/removable PSA, 204
  on PET, 67 (*see also* Substrate)
  rate of, 7, 532
  reduction of
    by coating weight, 212 (*see also* Coating weight)
    by fillers, 212, 385 (*see also* Fillers)
    by flexibilizers (*see* Flexibilizers)
    by primers (*see* Primer)
    by stress resistant polymers (*see* Stress resistant polymers)
  on release liner, 17
    dependence on peel angle, 51
  of removable PSA, values of, 385
  of tackified acrylic PSA, 231
  test method of, (*see also* Test method)
    at high speed (*see* Test method)
    at 90°, 52, 192, 201
    in T peel, 532
  values of, 201, 205, 210, 212, 213
    for UV cured PSA, 150

Peelability, (*see also* Removability)
  dependence of, on tackifier level, 188
  parameters of, 383
Peel force from release liner, 202 (*see also* Test method)
  dependence of
    on peeling angle, 202
    on peeling rate, 202
    on test conditions, 202
  values of, 202
pH, 364
  adjustment of, 429
  influence of, on tackification of water vased PSA, 408
Phase separation temperature, 97 (*see also* Tackification)
Penetration, 60 (*see also* Migration)
  dependence of, on modulus, 92
Pipelines, for water-based PSA, 433
Plasticizer, 24, 25, 26, 87, 120, 142, 143
  aromatic, 373
  choice of, 373
  compatibility of, 25, 373
  for electron beam cured PSA, 373
  for hot melts, 373 (*see also* Hot melts)
  influence of
    on cohesion, 372
    on creep, 211 (*see also* Creep)
    on glass transition temperature, 372
    on tack, 372
    on viscosity, 374
  level of, 156, 211, 374 (*see also* Formulation)
  macromolecular, 331 (*see also* Polybutene)
  migration of, 242 (*see also* Migration)
  oils (*see* Oil)
  sensitivity towards, 372
  volatile, 86
  for water-based PSA, 409
Plasticizing, 87, 372
  effect, of comonomers, 82

Plateau modulus, 93, 101 (*see also* Viscoelastic properties)
dependence of
on formulating additives, 142 (*see also* Formulating)
on tackifier resin, 98, 372
influence of, on tackifying, 96 (*see also* Tackification)
of SBR, 95
of SBS, 96
Polishing bar (*see* Coating machine)
Polyacrylate rubber, 128, 129, (*see also* Acrylics, Raw Materials)
temperature resistance of, 129
Polyalphaolefin, amorphous, 123
Polyamide, face stock, 269, 284
Polyaziridine, 140, 141, 210, 411 (*see also* Crosslinking)
Polybutadiene, 150
Polybuthene (*see also* Butylene rubber)
aging of, 384
as tackifier, 171, 351, 373, 379
Polycarbonate, face stock (see Face stock)
Polyester
as face stock, 264, 276, 283 (*see also* Polyethylenterephtalate)
as raw material for HMPSA, 148
Polyethylene
additives (*see* Formulation)
face stock of, 263, 269, 274
treatment of, 277, 281 (*see also* Treatment)
release liner of, 274, 499
Poly(ethylenebutylene), crosslinking of, 95 (*see also* Thermoplastic elastomers)
Polyethyleneterephtalate
as face stock, 269
as release liner, 499
as standard surface face stock, 67, 196, 524, 527, 528
Polyisobuthylene, 122, 372
aging of, 384
for removability, 384

Polyisocyanate, 386 (*see also* Crosslinking)
Polyisoprene, molecular weight of, 81, 123
Polymer synthesis, 86
comparison of, for PSA raw materials, 245
of styrene block copolymers, 125
Polyken Tack, 13, 173 (*see also* Test method)
Polyolefin
amorphous, 124
face stock of, 207, 264, 274
Polypropylene
atactic, 123
face stock of, 66, 269, 275
comparison vs. polyethylene, 275
treatment of, 278
release liner of, 499
surface treatment of, 276
Polystyrene
compatibility of, 88
face stock, 264, 283
Poly(styrene-butadiene-styrene), 124, 125, 148
in hot melts (*see* Hot melts)
sequence distribution of, 96
influence on cuttability, 96
polystyrene content of, 96
Poly(styrene-ethylene-butylene-styrene), 123, 125
aging resistance of, 125
compatibility of, 399
in hot melts, 399 (*see also* Hot melts)
Poly(styrene-isoprene-styrene), 124, 125, 148
aging of, 218
crosslinkable on EB, 84, 125, 188
in hot melts, 399 (*see also* Hot melts)
molecular structure of, 84
polystyrene content of, 148
sequence distribution of, 84, 96
tack of, 171 (*see also* Tack)
Polytack (*see* Test method)
Polyurethane, 131, 135
crosslinking agent, 386

Index

Polyvinylacetate, 25, 116, 131
   dispersions, 82 (*see also* Water-based dispersions)
Polyvinylethers, 131, 134, 372
   solubility of, 134, 153
   solvents for, 135
   in tackifier, 135, 344, 351
   resistance of
      to aging, 135
      to plasticizer, 135
Polyvinylchloride
   face stock of, 87, 120, 193, 207, 264, 269
   advantages/disadvantages of, 270
   comparison of vs. polyolefins, 276
   flagging of, 270 (*see also* Flagging)
   plasticized, 67, 242, 269
   properties of, 273
   shrinkage of, 270 (*see also* Shrinkage)
   surface tension of, 270 (*see also* Surface tension)
   versatility of, 283
Polyvinylpyrrolidone, thickener, 379, 425
Polytack, 397 (*see also* Tack)
Postadditives, 413 (*see also* Formulation)
Premetering, 478 (*see also* Coating device)
Pressure-sensitive adhesive, 1 (*see also* Adhesives)
   acrylic (*see* Acrylic adhesives)
   chemical composition of, 119 (*see also* Chemical composition)
   definition of, 1
   formulating of, 327 (*see also* Formulating)
   history of, 1, 120, 124, 126, 130, 132, 399, 461, 496, 501, 529, 532
   laminate, definition and construction of, 262
   manufacturing of (*see* Manufacture of PSA)
   permanent, 130, 2o4
   physical basis of for the viscoelastic behavior, 75 (*see also* Viscoelastic properties)

[Pressure-sensitive adhesive]
   removable, 33, 204 (*see also* Removable adhesive)
   silicone-based, 135
      adhesive characteristics of, 135
      single component, 123
   viscoelastic properties of, 2 (*see also* Viscoelastic properties)
Pretreatment (*see* Treatment)
Primer, 24, 207, 265
   coating of, 450, 475, 500
   coating weight of, 210
   influence of
      on peel, 207
      on removability, 207, 386
      on PVC, 207
   recipes of, 207
Printability, 49, 289
   dependence of
      on antioxidants, 409
      defoamers, 429
      on surfactants, 418
   of paper (*see* Paper)
   of polyolefins, 207, 274, 289
   of PVC, 269, 273
Probe tack (*see also* Polyken tack, Test method)
   tester of, 529
Protective colloids, 412, 425
Protective films, 257, 324, 500
   release coating for, 498
Pumps, for water-based PSA, 433

Quick stick, 20, 164 (*see also* Tack)
   test of (*see also* Test method)
   values of, 124

Radiation curing, 138
   advantages of, 496
   of PSA, 494
   of silicone, 494, 495
      electron beam, 496 (*see also* Siliconizing)
      ultraviolet, 497 (*see also* Siliconizing)

Raw materials, (*see also* Manufacture of PSA)
   of film labels, 152
   of HMPSA, 147
   of permanent/removable labels, 152
   of PSA, 120
   of reel/sheet label stock, 151
   of resins, 370
   of solvent-based PSA, 35, 148
   of water-based PSA, 149
Readhering adhesives, 210, 388
Reel stability (*see* Converting properties)
Relaxation, 24, 113
Release
   coating
      aquous, 136
      radiation cured, 142
      silicone-based, 136
      silicone free, 498
      solvent-based, 136
   controlled, 142, 499
      agents of, 500
   force
      adjustement of, 382
      aging test of (*see* Test method)
      dependence on adhesive nature, 307, 383
      dependence on paper thickness, 496
   liner, 287, 288
      conditioning of (*see* Test method)
      film-based, 288, 499
      humidity of, 291
      manufacturing of (*see* Manufacture)
   nature of, 498
   wettability of (*see* Test method)
Remoisturization, 455, 500, 502
Removable PSA
   elastomers in, 384
   migration of, 213
   pell values of, 205
   resins in, 384
   shear values of, 212

Removability, 21, 23, 24, 201 (*see also* Test of removability)
   criteria of, 23, 205
   dependence of
      on adhesive anchorage, 206
      on energy dissipation, 205
      on face stock flexibility, 206
      on viscosity/modulus, 24
   of water-based PSA (*see* Water-based dispersions)
   problems of, 212
Repositioning, 389
Residence time, 405 (*see also* Hot melt compounding)
Resin, (*see also,* Chemical basis, Tackifier, Tackification)
   acid, 352
   acidity of, influence on anchorage, 364
   aging stability of, 327
   aliphatic, 326, 337, 342, 356
   aromatic, 326, 337
   compatibility of, 88, 100
   for beam curing, 141 (*see also* Beam curing)
   choice of, 337
   colofonium derivatives, 142, 344, 399 (*see also* Rosin)
   compatibility of, 342
   concentration of, (*see also* Tackifier level)
   influence on modulus, 95, 98
   coumarone-indene, 337, 399
   cyclo-aliphatic, compatibility of, 88
   hydrocarbon-based, 143, 326, 342, 399 (*see also* Hydrocarbon resin-based tackifier)
   aging of (*see* Aging, Test method)
   compatibility of, 342, 369
   converting properties of, 366
   hydrogenated, 342, 356, 382
   liquid, 338, 357, 382
   manufacture of, 325
   mixtures of, 350
   natural
      manufacture of, 325

Index

[Resin]
    tackifying formulations of
    nature of, influence on adhesive
        properties, 344
    level of (see Tackifier level)
    molecular weight of, influence on
        modulus, 98
    molten, 351, 397
    phenolics, 337, 356
    polyterpene, 355, 357
    reactive, 356, 411
    rosin (see Rosin resin)
    rosin acid (see Rosin)
    rosin ester, 342 (see also Rosin)
        for CSBR tackification, 126
        softening point of, 340
    softening point, influence on adhesive properties of PSA, 344
    solutions, 351
    synthetic, manufacture of, 326
    tall oil, 338
    terpene phenol, 357
Resistance
    light, 152
    mechanical, 114
    migration (see Migration, Test method)
    thermal
        of acrylics, 219
        formulating for, 382
        of natural rubber adhesives, 219
    to plasticizer, 152 (see also Test method)
    to rewetting, 419
Reverse adhesion test, 196 (see also Peel)
Reverse gravure coating, (see also Coating device)
    with closed chamber, 474
    coating weight in, 454, 481, 482
    components of, 479
    with doctor blade (see also Doctor blade)
        coating weight in, 474
        viscosity in, 474
    running speed of, 479, 481, 482
    for water-based dispersions, 479

Reverse roll coating, 455, 458, 479, (see also Dahlgren system)
    rotogravure, (see also Rotogravure)
    coating weight in, 481
    for HMPSA, 486
    viscosity values of, 42, 481
    running speed in, 480
    viscosity in, 480
Rheology of pressure sensitive adhesive, 5
    bonded, 44
    dispersions, 34, 38
    solutions, 34, 35
        dependence on adherent, 37
        dependence on polymer, 36
        dependence on solvent, 36
        dependence on technology, 37
        special features of, 36
    uncoated, 5
Rheology of pressure sensitive laminate, 44
    dependence of
        on liquid components of the laminate, 46
        on solid components of the laminate, 48
Ribbing (see Striation)
Ring and ball
    softening point, 343, 379
    test method of, 384 (see also Test of softening point)
Rod coating, 461 (see also Meyer bar)
Roll
    coating, 465 (see also Rotogravure)
    methods of, 466
    coating weight regulation of, 469
    coating weight values of, 469
    with Doctor blade, 474
    of HMPSA, 486
    of solvent-based PSA, 477
Rolling ball, (see also Test method)
    dependence of, on softening point of the resin, 352
    tack, 175 (see also Tack)
Rolling cylinder, (see also Test method)
    comparison of, vs. rolling ball, 531

Rosin resin tackifier, 356, 357
  acid, 358
  in acrylics, 338
  dispersion of, (see also Tackifier
      dispersions)
    wetting agent level in, 366
  ester, 364
    compounding of, 364
    cuttability of, 355
    in natural rubber, 345
  price level of, 365
Rotating drum (see Tack test method)
Rotogravure, 423
  direct, 457
  for silicones, 496
  speed of, 423
Rubber, natural (see also Elastomer)
  -based PSA (see Rubber/resin
      adhesives)
  manufacture of, 324
  mastication of (see Mastication)
  mechano-chemical destruction of
      (see Mastication)
Rubber/resin adhesives, 36, 88, 93, 120,
      323 (see also Solvent-based PSA)
  solvents for, 37 (see also Solvents)
Running speed, parameters of, 258, 259

SAFT (see also Shear strength, Test method)
  dependence of
    on face stock surface, 222
    on softening point of the resin, 218
    on tackifing, 350
Sandwich structure, 262 (see also
    Laminate)
Self reinforcement, 95 (see also Natural
    rubber)
Separation energy, 7, 11, 12, 15, 23, 31
    (see also Peel adhesion)
  dependence of, on surface tension, 68
  as tack, 164
Separation rate, 202 (see also Peel
    adhesion)
Shear, on the coating machine, 37, 256,
      462, 481, 484

Shear modulus, 37
  calculation of, 100
  dependence of, on crosslinking, 94
      (see also Crosslinking)
  dynamic, 20
Shear resistance, 2, 46, 213
  of acrylics, tackified, 240
  of automotive PSA (see Test method)
  dependence of
    on adhesive/cohesive properties, 216
    on adhesive nature, 217
    on coating weight, 220
    on composite status, 22o
    on crosslinking, 100, 172, 218
    on face stock, 222
    on modulus, 19
    on molecular weight, 218
    on sequence distribution, 219
    on softening point of the resin, 350
    on strain rate, 20
    on substrate, 222
    on tackifying (see Tackification)
    on temperature, 21
    on tensile strength, 215
    on time, 20
    on viscoelastic properties, 18
    on viscosity, 19
  factors of, 217
  improving of, 223
  influence of
    on cuttability (see Cuttability)
    measurement of, 214
    as Williams plasticity, 215 (see
        also Williams plasticity)
  of solvent-based PSA, 20
  test method of (see also Test method)
    dead load hot strength, 214
    dynamic, 214
    hot, 214, 216
    room temperature, 214
    SAFT, 214 (see also SAFT)
    static, 214
    20°hold strength, 214
  values of, 124, 214, 216, 223, 237, 523

Index

Shear thinning, 21 (*see also* Shear on the coating machine)
Shrinkage (*see also* Plasticizer reistance, Test method)
   of acrylics, 242
   dependence of, on printing, 272
   of face stock materials, 292
   parameters of, 292
   of PVC, 271
      dependence on plasticizer migration, 272
      dependence on relaxation, 271
      dependence on technology, 271
   values of, 242, 550
Sieve residue, test of (*see* Test method)
Silicone
   acrylics, 495
   fluids, 288
   defoamers, 428 (*see also* Formulating)
   pressure sensitive adhesives, 120
      tack of, 172
      peel of, 192
   release, 136
      addition cured, 136
      catalyst system of, 136, 499
      coating technology of, 138 (*see also* Siliconizing)
      coating weight of, 288, 496, 497, 498, 500, 501
      crosslinked, 136
      emulsions, 138
      moisture cured, 136
      postcure of, 136, 496
      radiation cured, 136, 137
      solvent-based, 138
      solventless, 39, 138
      thermal cured, 136
      UV cured (*see* Radiation curing)
      water-based, 501
Siliconizing
   coating machines of, 500
      for solventless systems, 500
      speed of, 500, 501
   radiation (*see also* Radiation curing)
      coating speed of, 496, 497

[Siliconizing]
   electron beam, 496
   ultraviolet, 497
   viscosity in, 496
   technology of, 39, 497
      energy consumption of, 497
      for solvent-based systems, 500
      for solventless systems, 501
Slip agents
   in polyolefins, 281
   migration of, 283
Slot die, 455, 458, 484
   for HMPSA, 488
      advantages/disadvantages of, 488
Smearing
   dependence of
      influence on cuttability, 295 (*see also* Cuttability)
      on relaxation, 302
      on tack (*see* Tack)
   parameters of, 56
Smoothing bar, 455, 461, 480 (*see also* Coating machine)
Softening point
   of the adhesive, test of (*see* Test method)
   of the resin, test of (*see* Test method)
Solids content
   of solvent-based PSA, 37, 324, 438
   of water-based PSA, 149, 263
      improvement of, 351
Solubility
   diagramm of, 342
   parameter of, 65, 66
Solvent, 258
   hydrocarbon-based, 37
   influence of
      on anchorage (*see* Anchorage)
      on water resistance, 429
   recovery of, 398, 437, 493
   in solvent-based PSA, 398
   in WBPSA, 263, 351, 431
Solvent-based PSA
   acrylics

[Solvent-based PSA]
    advantages/disadvantages, 436, 437 (*see also* Acrylics)
    molecular weight of, 81
    special features of, 244
    anchorage of, 65
    coating machine of, 454 (*see also* Coating machine)
    equipment for, 437
    molecular weight of, 36, 81
    rool coating of
        running speed of, 477
        solid content of, 477
        viscosity of, 477
    solid content of, 81, 397
    surface tension of, 36
    viscosity of, values of, 36
Stability
    dimensional (*see* Shrinkage)
    thermal, 241
    test of, 123
Stabilizers
    in HMPSA (*see* Formulation)
    in PVC, 156
    for ultraviolet, 394, 395
Staining, 341, 356, 378, 540 (*see also* Migration, Penetration)
Stationary roll, 464, 479 (*see also* Coating device)
Stiffening
    effect of, 7
    of face stock, 550
Stiffness
    of laminate, 69, 314
        parameters of, 309
    of liner, 313
Storage
    modulus, 8, 10, 14, 92, 94, 337, 338 (*see also* Viscoelastic properties)
    tanks (*see* Tanks)
    influence of, on peel, 435
    of water-based PSA (*see* Water-based dispersions)
Stress
    softening, 113

[Stress]
    restricting polymers, 206 (*see also* Removability)
    resistant polymers, 210
Striation, 352, 463
Strike through (*see* Penetration)
Stringing (*see* Legginess)
Strip coating, 455
Stripe structure, 256 (*see also* Structure)
Structure, textured, 257
Styrene butadiene rubber, 122 (*see also* Raw materials)
    carboxylated, (*see* Carboxylated styrene butadiene rubber)
    compatibility of, 88
    latex of, 122, 123
        glass temperature of, 126
        shear of, 126
        solid content of, 126
        styrene content of, 126
    tack of, 95, 338
    tackified, peel of, 134
Styrene copolymers, 124 (*see also* Thermoplastic elastomers)
    molecular weight of, 124
    styrene content of, 124
    tackifier for, 342 (*see also* Hot melts, Tackifying)
Substrate, 5
    elasticity/plasticity balance of influence on peel, 54
    for laboratory tests, 524
    surface of, influence on peel, 68
    for test, 202
    of peel, 202, 534, 536, 537
    of tack, 527, 528
Sulfosuccinates, 414 (*see also* Surfactants, Wetting out)
    level of, 412
    solubulity of, 414
    thickeners, 364
Supplier, selection of (*see* Tackifier dispersion)
Surface
    energy of, 64

Index

[Surface]
  standard, for loop tack, 164
  textured, 37 (*see also* Structure)
Surface tension, 35, 64
  dependence of
    on viscosity, 41 (*see also*
      Orchards Equation)
  dynamic, 41, 413
  of fluorsilicones, 136
  influence of
    on foam (*see* Foam)
    on shear, 222
    on wetting, 40, 256
  of PVC, 270
  static, 41
  values of, 245, 270, 413
Surfactants, 126, 220, 412, 416, 416 (*see also* Emulsifier, Wetting agent)
  anionic, 412
  choice of, 413
  fluorinated, 416, 420
  influence of
    on drying, 419 (*see also*
      Composite structure)
    on the properties of PSA
      laminate, 418, 419
  interaction of, with other layers of the laminate, 418
  level of, 419, 420, 434
    parameters of, 420, 421
  migration of, 418 (*see also* Migration)
  plasticizing effect of, 418
  as thickeners, 425
Tack, 2, 11, 14, 46, 163 (*see also* Test method)
  in acrylic block coplymers, 342
  of acrylics, 229
  deadeners (*see* Detackifying agents)
  dependence of
    on adhesive nature, 171
    on coating weight, 170, 172, 173
    on creep compliance, 91, 103
    on crosslinking, 172 (*see also* Crosslinking)

[Tack]
  on experimental parameters, 13
  on face stock, 174
  on fillers, 172
  on gel content, 126, 171
  on glass transition temperature, 164
  on modulus of elasticity, 12, 164
  on molecular weight (*see* Chemical basis)
  on plasticizer, 171 (*see also* Plasticizing)
  on rheology, 167
  on strain rate, 13 (*see also* Viscoelastic properties)
  on substrate, 528
  on surface active agents, 172
  on surface tension, 175
  on tackifier, 171 (*see also* Tackification)
  on tackifier melting point, 171 (*see also* Resin)
  on test method and conditions, 174 (*see also* Test method)
  on time/temperature, 13, 173, 174 (*see also* Rheology)
  on viscoelastic properties, 11
  on viscosity, 11
  on wetting out, 175
  energy, 173
  factors of, 170
  improvement of, 176, 334 (*see also* Tackification)
  index of, 169
  influence on cuttability (*see* Cuttability)
  level of, 170, 337
    of acrylics, 170, 229
    comparison of, 170
    of EVAc, 170, 231
    of SBR, 170
    of uncoated PSA, 172
  measurement of, 164 (*see also* Test method)
    as coefficient of friction, 164

596 Index

[Tack]
    as cohesion, 168
    as fracture energy, 169
    as peel, 167, 169
    as plasticity, 169
    on polyethylene, 175
    relative, 168
    of rubber resin adhesives, 338
    test method of
        as loop tack, 168 (*see also* Loop tack)
        as rolling ball, 164 (*see also* Rolling ball)
        as rolling cylinder, 164 (*see also* Rolling cylinder, Test method)
    values of, 175
    Werle tester of (*see* Test method)
    wet, 164, 397
Tackification, 88, 99, 100, 331
    of acrylics
        ease of, 229 (*see also* Tackifying)
        water-based (*see* Water-based adhesives)
    for cost reduction, 333
    of CSBR, 95
    dependence of, on the physical state of the tackifier, 351
    of hot melt PSA, 400
    with resins, 331
        liquid, 435
        molten, 435
        solution, 434
    for shear, 340
    with solvent-based tackifier, 434
    special features of, 356
    of water-based PSA, 351, 408
    with water-based resin dispersion, 434
Tackifier, 120, 142 (*see also* Plasticizer, Resin)
    in acrylic block coplymers, 342
    choice of, 341, 343
    concentration/level of (*see* Tackifier level)
    in CSBR, 343, 350
    degradation of, 156

[Tackifier]
    dispersion of, 150
    requirements for, 407
    in electron beam crossslinkable SIS, 343
    in ethylene vinylacetate copolymers, 344 (*see also* Formulation)
    in hot melts, 342
    hybrid, 371
    hydrocarbon resin-based (*see* Resin)
    influence of
        on converting (*see* Converting properties)
        on end use properties (*see* End use properties)
        on shear (*see* Shear)
    in natural rubber, 342, 343, 350
    in polyvinylacetate, 344
    in styrene block copolymers
    rosin resin-based (*see* Rosin)
Tackifier level, 329
    in acrylics, 229, 334
    dependence of, on elastomer nature, 338
    in electron beam cured PSA, 338
    in ethylene/maleinate copolymers, 338
    in natural rubber, 338, 345
    in polybutadiene, 338
    in SIS rubber, 338
    influence of
        on migration, 435 (*see also* Migration)
        on peel, 334, 340
        on shear, 33o, 340
        on tack, 338
    for screening, 338
Tackifying
    ability of, 333
    ease of, 333
    response, 329
        of acrylics, 237
        of CSBR, 237
        of EVAc, 237
        of water-based PSA (*see* Water-based dispersions)

Index

Tanks, for water-based PSA, 433
Tapes, 1, 10, 20, 66, 155
  adhesive strength of, 126, 212
  aging of, test of, 547
  block copolymers for, 96, 126
  coating technology of, 257, 356, 454, 479, 494
  coating weight of, 62
  double sided, test of, 527
  health care, 154
  masking, 523
  of polyvinylchloride, 207
  primer for, 210
  wet adhesive, test of, 532
Test methods, 509
  of accelerated weathering, 549
  of adhesive nature, 513
  of adhesive properties, 523, 557, 559
    combined, 531
    evaluation of, 543
    reference material of, 523
    standard methods for, 523
    specimen preparation for, 523
  of adhesive residue 553 (*see also* Test of removability)
  of aging properties, 542, 546, 549
    evaluation of adhesive degradation, 548
    evaluation of migration, 549 (*see also* Migration)
    of PVC face stock, 550
    of release, 549 (*see also* Release liner)
    standard conditions in, 546
  of coagulum (*see* Coagulum)
  coating conditions for, 525
  of coating weight, 518, 546
  of cold flow, 521
  of compatibility, 337
  of compliance, 522 (*see also* Approval)
  conditions of, 525
  of contact angle, 512
  of deep freeze labels, 562 (*see also* Deep freeze)

[Test methods]
  cold tack of, 563
  low temperature adhesion of, 563
  of static shear load test at low temperatures of, 563
  of dimensional stability, 550
  drying conditions for, 517, 525
  of cuttability, 340, 552
  of edge ooze, 552
  of film coating, 555
    adhesive properties of, 557
    adhesive screening in, 555
    adhesive versatility in, 555
    loss of transparency of, 558
    water whitening of, 558
    wet adhesion of, 558
    wet anchorage of, 558
  of foam, 513
    weight of, 515
  of grit content, 516
  of HMPSA, 510
  of humidity resistance, 562
  of laminate properties, 522
    general, 522
    special, 552, 563
  of lifting, 554
  of liquid adhesive, 509
  of migration, 551
  of peel strength, 67, 202, 212, 531, 557
    on cardboard, 537
    with constant peel force, 525
    high speed, 535
    90° peel adhesion, 534
    on plasticized PVC film, 537
    on polyethylene plate, 537
    from release liner, 202, 535
    standard, 533
    standard substrate for, 202
    T-peel, 534
    test conditions, 202
  of plasticizer resistance, 550
  of removable labels, 552
  of removability, 552
    of HMPSA, 553

[Test methods]
  of repositionability, 554
  of shear strength, 112, 537
    automotive, 543
    dynamic, 541
    dynamic lap shear, 541, 562
    experimental conditions for, 543
    hot shear, 540, 558
    shear adhesion failure temperature, 540
    standard, 538, 539
    statical/room temperature, 538
    twenty degree hold, 542
    Weyerhousers (*see* Test of water resistance)
  of shrinkage, 551
  of sieve residue, 516
  of softening point, 519
  of solid adhesive, 518
  of solvent-based PSA, 51o
  standard, 509
  of surface tension, 512
    dynamic, 513
  of tack, 526 (*see also* Tack)
    Bulls tack, 530
    Hammonds probe tack, 529
    Kendalls probe tack, 529
    loop tack, 527, 529 (*see also* Loop tack)
    Polyken tack, 529
    quick stick, 526 (*see also* Quick stick)
    rolling ball, 530 (*see also* Rolling ball)
    rolling cylinder, 557 (*see also* Rolling cylinder)
    toothed wheel, 530
    Werles, 532
    Wetzels probe tack, 529
    Zosels probe tack, 529
  of tensile strength, 522
  of viscosity
    for hot melts, 510
    for solvent-based PSA, 510

[Test methods]
  for water-based PSA, 511, 555 (*see also* Thixotropy index)
  of washing machine resistance, 562
  of wash off adhesives, 562
  of water-based PSA, 511
    adhesive properties of, 557
    coating properties for, 511
    coating weight of, 518
    density of, 518
    drying ability of, 516
    foaming of (*see* Test of foam)
    mechanical stability of, 516
    particle size of, 517
    screening of, 517
    solids content of, 518, 555
    wetting characteristics of, 512, 555
  of water resistance, 559, 561
  of water solubility, 560, 562
  of water whitening, 561 (*see also* Water whitening)
  of wettability, of release liner, 513
  of wetting angle, 512, 513
    for adhesive extract, 513
  of wetting out, 512
    direct, 512
    indirect, 512
    vinyl wetting test of, 512
  of Williams plasticity, 520
Thermoplastic rubbers, 100, 123, 124
  compatibility of, 100
  in raw materials for hotmelts, 96 (*see also* Hotmelts)
Thickeners, 254, 421
Thickening, dependence of, on PH, 421
Thixotropy, 257, 421
  index of, 511
Time-Temperature Superposition Principle, 20, 21, 97
Transparency, loss of, test of (*see* Test method)
Treatment, of surface
  chemical, 277
  Corona, 66, 277, 278, 500

Index

[Treatment, of surface]
   flame, 277, 280
   fluor, 281
   influence on shear, 222
   plasma, 222, 280
   shelf life of, 278, 281
Two roll gravure coater (see Reverse gravure)
Two sided coating (see Offset gravure)

Ultraviolet
   drying (see Drying)
   stability to, 240
      of rubber/resin adhesives, 242
   stabilizers (see Formulation)

Versatility, of solvent-based PSA (see Solvent-based PSA)
Viscoelastic models, 166 (see also Tack)
Viscoelastic properties, 5
   dependence on
      chemical composition/structure, 27
      experimental and environmental conditions, 28
      material characteristics, 26
      time, 30
   factors of, 25, 299
   influence on
      applied label, 25
      adhesive properties, 11
   physical basis of, 75
Viscosity, 6
   adjusting of, 421
   dependence of
      on molecular weight, 26
      on shear rate, 31
      on storage time, 181 (see Water-based dispersions)
      on tackifier resin, 28
      on temperature, 28
   of base elastomers, 27
   of HMPSA, 155
   influence of
      on adhesive coating weight (see Coating weight)

[Viscosity]
      on coating versatility, 42
      on coating weight, 43
      on wetting, 42
   of polymer solutions, 26, 36
      time/temperature dependence of, 37
   of PSA dispersions, 39
      factors of, 44
   stability of, 257
   values of, 257
      of acrylic HMPSA, 154
      of HMPSA, 155
      on knife over roll, 155
      on Meyer rod, 155
      in rotogravure, 155
Viscous components, for PSA, 142
Voigts model, 186 (see also Viscoelastic properties)

Water-based dispersions, 407
   acrylics, (see also Acrylics)
      modulus of, 99
      molecular weight of, 81
      special features of, 244
   coating machine of, 455, 477
      coating weight of, 477
      coating weight tolerance of, 477
      requirements for, 477
   coating device of, 478
   drying of, 258
   formulation of (see Formulation)
   improvement of the adhesive properties, 407
   ready to use, 38
   shelf life of, 38
   special additives in, 431
   viscosity of, values of, 257, 258
Water resistance, 154, 389
Water whitening, 389 (see also Test method)
Weathering performance, 356
Web
   control of, 455
   tension of, 463
Wet elongation, 291 (see also Lay flat)

Wettability
  of tackifier dispersions, 353
  of release liner (*see* Test methods)
Wetting agents, 412, 413 (*see also* Surfactants)
  choice of, 414
  level of, 413
Wetting out, 11, 12, 35, 256
  angle of, test of (*see* Test method)
  dependence of
    on face stock, 63
    on viscosity, 421
    on pH, 420
  dynamic, 38, 40
  parameters of, 256
  of solvent-based PSA, 254
  static/dynamic, 40, 256
  theoretical basis of, 39
  of water-based dispersions, 38, 39, 254

Window of performance, 8, 10, 331, 337 (*see also* Viscoelastic properties)
Williams–Landel–Ferrys equation, 21, 29, 30, 167
Williams plasticity, 112, 169 (*see also* Test method)
Wing up, (*see* Flagging)
Wire wound rod, 461 (*see also* Meyer bar, Rod coating)
  coating weight on, 462
  running speed of, 463
  viscosity of, 462
Work of
  adhesion, 7, 65, 67
  detachment, 53
Wrap angle, 463
Youngs modulus, 6